Unlocking the
Stratigraphical Record

Unlocking the Stratigraphical Record

Advances in Modern Stratigraphy

Edited by

PETER DOYLE and MATTHEW R. BENNETT

School of Earth & Environmental Sciences,
University of Greenwich, UK

JOHN WILEY & SONS

Chichester · New York · Weinheim · Brisbane · Singapore · Toronto

Copyright © 1998 by John Wiley & Sons Ltd,
Baffins Lane, Chichester,
West Sussex PO19 1UD, England

National   01243 779777
International   (+44) 1243 779777
e-mail (for orders and customer service enquiries): cs-books@wiley.co.uk
Visit our Home Page on http://www.wiley.co.uk
or http://www.wiley.com

Other Wiley Editorial Offices

John Wiley & Sons, Inc., 605 Third Avenue,
New York, NY 10158-0012, USA

WILEY-VCH Verlag GmbH, Pappelallee 3,
D-69469 Weinheim, Germany

Jacaranda Wiley Ltd, 33 Park Road, Milton,
Queensland 4064, Australia

John Wiley & Sons (Asia) Pte Ltd, 2 Clementi Loop #02-01,
Jin Xing Distripark, Singapore 129809

John Wiley & Sons (Canada) Ltd, 22 Worcester Road,
Rexdale, Ontario M9W 1L1, Canada

Library of Congress Cataloging-in-Publication Data

Unlocking the stratigraphical record: advances in modern stratigraphy /
edited by Peter Doyle & Matthew R. Bennett.
    p.   cm.
Includes bibliographical references and index.
ISBN 0 471 97766 7 (cloth). — ISBN 0 471 97463 3 (pbk. : acid-free paper)
    1. Geology, Stratigraphic.   I. Doyle, Peter.   II. Bennett, Matthew (Matthew R.)
QE651.U55   1998   98–41823
55.7—dc21   CIP

British Library Cataloguing in Publication Data

A catalogue record for this book is available from the British Library

ISBN 0-471-97463-3
ISBN 0-471-97766-7

Typeset in 10/12pt Palatino from the author's disks by Vision Typesetting, Manchester
Printed and bound in Great Britain by Bookcraft (Bath) Ltd

This book is printed on acid-free paper responsibly manufactured from sustainable forestation, for
which at least two trees are planted for each one used for paper production.

Contents

Contributing Authors

Matthew R. Bennett, School of Earth & Environmental Sciences, University of Greenwich, Medway Campus, Pembroke, Chatham Maritime, Kent ME4 4AW, UK

M. Andrew Bussell, School of Earth & Environmental Sciences, University of Greenwich, Medway Campus, Pembroke, Chatham Maritime, Kent ME4 4AW, UK

Paula J. Carey, School of Earth & Environmental Sciences, University of Greenwich, Medway Campus, Pembroke, Chatham Maritime, Kent ME4 4AW, UK

Beris M. Cox, British Geological Survey, Keyworth, Nottingham NG12 5GG, UK

Peter Doyle, School of Earth & Environmental Sciences, University of Greenwich, Medway Campus, Pembroke, Chatham Maritime, Kent ME4 4AW, UK

Gerhard Einsele, Geologisch-Paläontologisches Institut, Universität Tübingen, Sigwartstrasse 10, 72076 Tübingen, Germany

Jane E. Francis, Department of Earth Sciences, University of Leeds, Leeds LS2 9JT, UK

Andrew S. Gale, School of Earth & Environmental Sciences, University of Greenwich, Medway Campus, Pembroke, Chatham Maritime, Kent ME4 4AW, UK

Peter Gutteridge, Cambridge Carbonates Ltd, 11 Newcastle Drive, The Park, Nottingham NG7 1AA, UK

Anthony Hallam, School of Earth Sciences, University of Birmingham, Edgbaston, Birmingham B15 2TT, UK

David A. T. Harper, Department of Geology, University College Galway, Galway, Ireland

Malcolm J. Hole, Department of Geology & Petroleum Geology, University of Aberdeen, Meston Building, King's College, Aberdeen AB9 2UE, UK

Charles H. Holland, Department of Geology, Trinity College Dublin, Dublin 2, Ireland

David I. M. Macdonald, Cambridge Arctic Shelf Programme, West Building, Gravel Hill, Huntingdon Road, Cambridge CB3 0DJ, UK

John M. McArthur, Department of Geological Sciences, University College London, Gower Street, London WC1E 6BT, UK

Paul N. Pearson, Department of Geology, University of Bristol, Queens Road, Bristol BS8 1RJ, UK

Duncan Pirrie, Camborne School of Mines, University of Exeter, Redruth, Cornwall TR15 3SE, UK

Michael G. Sumbler, British Geological Survey, Keyworth, Nottingham NG12 5GG, UK

Stephen J. Vincent, Cambridge Arctic Shelf Programme, West Building, Gravel Hill, Huntingdon Road, Cambridge CB3 0DJ, UK

Alf Whittaker, British Geological Survey, Keyworth, Nottingham NG12 5GG, UK

Preface

The story of our planet is perhaps one of the most exciting and fundamental components of modern geology. Stratigraphy is the key to understanding the earth, its materials, structure and past life; it provides the means by which to unlock the secrets of our geological past. Our personal enthusiasm for this story led us to write our previous stratigraphy book *The Key to Earth History* (Wiley, 1994) as an introduction to the subject and to redress the traditional student view of stratigraphy as simply a dull recital of names and dates. Here we explore the advanced tools with which to order and interpret the stratigraphical record. Each of the chapters has been written by specialists in the field, providing both examples of good practice and discussion of recent debates. Collectively and individually they provide a platform from which the reader can reach the tools, techniques and literature of modern stratigraphy. We are indebted to the contributors who have made this book possible and we would like to express our thanks for their co-operation and support.

Peter Doyle and Matthew R. Bennett
Chatham Maritime, 1997

1
Introduction: Unlocking the Stratigraphical Record

Peter Doyle and Matthew R. Bennett

Stratigraphy is the key to understanding the earth, its materials, structure and past life. It encompasses everything that has happened in the history of the planet. Simply defined, stratigraphy is the study of rock units and the interpretation of rock successions as a series of events in the history of the earth. The role of the stratigrapher is to describe, order and interpret rock units in terms of events and processes, and to correlate this information in time in order to build up this record of earth history. Fundamentally, stratigraphy provides the perspective of time; a perspective which makes geology almost unique within the physical sciences.

All geologists are in some way stratigraphers, since almost all pure geological research is an attempt to unlock the secrets of the earth. Stratigraphy provides the frame of reference which underpins this research. Applied studies also rely on stratigraphy, as, for example, it is crucial in the exploration for oil and gas reserves, and it provides a sound basis for the understanding of the properties and extent of geological units in civil engineering. The aim of this book is to provide reviews of some of the most important principles and practices in modern stratigraphy. All of the chapters have been written by specialists in their field, providing both examples of good practice, and discussion of recent debates. Collectively and individually they provide a platform from which the reader can reach, both in practical and intellectual terms, the tools, techniques and literature of modern stratigraphy. This first chapter serves as an introduction to these contributions and provides the framework for the volume as a whole.

Unlocking the Stratigraphical Record: Advances in Modern Stratigraphy. Edited by P. Doyle and M.R. Bennett.
© 1998 John Wiley & Sons Ltd.

1.1 STRATIGRAPHY AS A MODERN DISCIPLINE

Geology has a long history dating back to the Renaissance, but, arguably, formal acceptance of the guiding principles of the subject – uniformitarianism, superposition and relative chronology – followed from the publication of Charles Lyell's (1797–1875) classic book *Principles of Geology* in 1830. Its publication heralded the 'golden age' of geology, riding on the crest of the great wave of public enthusiasm for the natural sciences (Bowler 1992). Text and classbooks on geology proliferated, and geologists such as Lyell and his contemporaries were public figures. Examining one of these texts today, such as Lyell's own *Students' Elements of Geology* (1878) or Page's *Advanced Textbook of Geology* (1861), one is struck by a simple fact: that the study of geology was synonymous with stratigraphy. Each text dispenses quickly with the mundanity of petrological description and gets straight to the perceived heart of the matter: the examination and reconstruction of past environments and life through time. Clearly, stratigraphy was the intellectual driving force of geology, and the major advances and debates were almost all stratigraphical in nature (e.g. Rudwick 1985; Secord 1986).

The most important contribution from this time was the development of the Chronostratigraphical Scale, a truly international standard which provides a framework of relative time within which all rock units may be compared. The systems that make up this scale were mostly established within the first 50 years of the nineteenth century. For example, the Cambrian, Silurian, Devonian and Permian systems were all erected by Roderick Murchison (1792–1871) and Adam Sedgwick (1785–1873) within a period of just six years, between 1835 and 1841 (Berry 1986). The rapid development of the Chronostratigraphical Scale was made possible by refinement of the basic principles of biostratigraphy, first laid down by William Smith (1767–1839) at the turn of the nineteenth century, and was driven by the need for tools to establish and interpret the sequence of rock units on a global scale. At the same time that the Chronostratigraphical Scale was emerging, the application of the uniformitarian principles championed by Charles Lyell and James Hutton (1726–1797), and the development of the concept of the facies by Armanz Gressley (1814–1865) provided the mechanism for the interpretation of rock units as the products of ancient processes.

The nineteenth century was clearly a period of rapid advance not only in terms of the development of the tools of stratigraphy, but also in terms of our understanding of earth history (Bowler 1992). By the mid-twentieth century, however, stratigraphy had lost its place as the intellectual heart of geology and the subject slowly degenerated into a simple catalogue of units and names. Some of the worst and least imaginative aspects of the subject date from this period. The source of this decline is complex, but it primarily reflects a failure of the geological community to provide a convincing explanation for global tectonic activity. From the days of James Hutton, stratigraphers had recognized the great importance of angular unconformities as evidence of denudation of fold mountains followed by subsequent marine flooding episodes, but what created the mountains? James Hall (1811–1898) and James Dwight Dana (1813–1895) in America evolved their geosynclinal theory in the mid to late nineteenth century, through the examination of the linear fold belts welded to the ancient cratonic interiors of most of the continents. Challenged only by the then bizarre notion of continental drift by Alfred Wegener (1880–1930) in the early part of the twentieth century, there was no dynamic force in stratigraphy. There was an absence of interpretative models capable of making sense of the complex, often conflicting picture, of earth history that was emerging (Hallam 1990). Geologists retreated to their introspective studies and were content to catalogue their local successions.

The discovery of sea-floor spreading through the work of scientists such as Fred Vine, Drummond Matthews and Harry Hess in the 1960s did much to revolution-ize geology and to breathe new life into stratigraphy. Indeed, new techniques in the subject, particularly in the study of the record of the earth's magnetic reversals (magnetostratigraphy), led to the recognition of the alternate magnetic 'strips' of the spreading ridge basalts, and the widespread acceptance of plate tectonics. The advent of the 'new global tectonics', as it was then called, reinvigorated the subject of stratigraphy. Stratigraphical studies provided the means by which plate tectonic theory learned from the present day could be applied to the geological past. Perhaps the most seminal work in this respect was the stratigraphical and palaeon-tological studies of the North American geologist J. Tuzo Wilson, who confirmed the existence of an early Palaeozoic, proto-Atlantic ocean; the Iapetus Ocean of later authors (Wilson 1966). Placed in a plate tectonic context, this led to the concept of the Wilson Cycle, the opening and closing of ocean basins, which provides the interpretative framework so long demanded by the stratigraphical fraternity. Gone was the need to invoke elaborate 'land-bridges' in order to explain away difficult palaeogeographical and palaeobiogeographical conundrums thrown up by the stratigraphical record (e.g. Oakley & Muir-Wood 1949; Wills 1951). Stratigraphy was once again placed at the heart of geology, as it was needed to provide the framework within which to apply plate tectonics in the geological past; complex tectonic belts could be teased apart and placed into a new global tectonic setting. Furthermore, precision in relative and absolute dating of events, and more detailed interpretational models were needed to meet this new chal-lenge.

As the models of our global environment – plate tectonic, climatic and biological – have become ever more sophisticated, the demand for high-resolution stratig-raphical records has accelerated. In recent years there has been a rapid advance in the sophistication and resolution of stratigraphical tools (e.g. Hailwood & Kidd 1993). Cyclostratigraphy, chemostratigraphy, seismic stratigraphy and sequence stratigraphy have all been developed over the past two decades, allowing the more detailed subdivision and accurate correlation of rock successions across the globe. International co-operation is redefining the Chronostratigraphical Scale with greater precision, and developing globally accepted protocols for litho- and bio-stratigraphy (e.g. Hedberg 1976; Whittaker *et al.* 1991). Stratigraphy has new challenges, and needs new and refined techniques to meet these challenges. Yet the basic principles and aims of stratigraphy remain the same as they did in the early days of the subject: those of establishing the sequence of rock units and then interpreting them as events within earth history.

1.2 THE STRATIGRAPHICAL TOOL KIT

As a consequence of recent advances in stratigraphy driven by increasingly more specific questions posed by the integrative models of earth history, modern stratig-raphers are faced with a bewildering array of tools and techniques at their dis-posal. This 'stratigraphical tool kit' (Doyle *et al.* 1994) becomes ever more complex as these ideas advance continually. Generally, however, it is possible to recognize two types of stratigraphical tool: (1) those tools primarily involved with establish-ing the sequence of rock units; and (2) those primarily involved in the interpreta-tion of rock units as events in the development of the earth. The principal subjects, techniques or tools within the 'stratigraphical tool kit' are shown in Figure 1.1, and are discussed in outline below.

1.2.1 Establishing the Sequence

Broadly, the establishment of stratigraphical order may be carried out using two main types of stratigraphical tool: (1) those which establish the relative age order of rock units, that is *sequencing tools*; and (2) those which can be used to correlate rock sequences and establish their age, that is *time tools* .

Sequencing tools

The first duty of any geologist working in an unknown area is to establish the geological sequence in that area. In order to achieve this, a set of 'sequencing tools' is used (Figure 1.1). This involves the observation, description and distinction of the main lithological units, through the process of lithostratigraphy. Lithostratig-raphy recognizes basic chronology, using the principle of superposition and way-up criteria. Other methods of lithostratigraphical analysis and correlation can be achieved through electrical and magnetic properties of rocks, which may be derived from subsurface boreholes and wireline logging techniques.

In complex terrains, lithostratigraphy may not be clear, and here the recognition of a relative chronology of rock units is only possible with those sequencing tools that will allow the tectonic history to be deciphered. In all cases, recognition of lithostratigraphical units is usually carried out by all field geologists and usually leads to the development of a geological map, but today, the lithostratigraphical exploration of large tracts of impenetrable or inhospitable terrain may be success-fully carried out using remote sensing techniques which serve as valuable se-quencing tools within the stratigraphical repertoire.

Time tools

The initial recognition of lithostratigraphical units and the determination of their relative chronology is insufficient for correlation with either other units in adjacent regions, or with the Global Chronostratigraphical Scale. Time tools are not

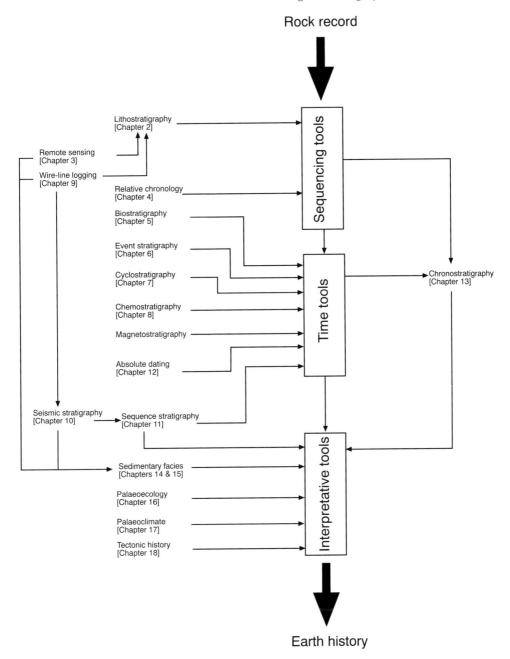

Figure 1.1   *Schematic illustration of the principal tools within the 'stratigraphical tool kit' and of the relevant chapters within this volume*

normally subject to diachroneity (the formation of the same lithology at different intervals of geological time) and must be sufficiently widespread to be of use in both regional and global correlation. Traditionally, biostratigraphy has provided the basis for most correlations, and is still the most important method of correlation with the Global Standard Stratigraphy (Figure 1.1). Recently, however, a raft of new techniques has helped refine correlations where recourse to fossils is difficult because of preservational or environmental conditions. These include the recognition of depositional event horizons, rhythmic Milankovitch cycles, and the use of stable isotopes from carbonate and other sedimentary rock sequences (Figure 1.1). Seismic and sequence stratigraphies are concerned with the correlation of geometrical packages of sediments, initially from seismic sections, but now commonly using onshore rock successions. These correlation tools also provide an extremely important mechanism for the interpretation of the development of sedimentary basins, and in the analysis of basin fill, and provide a cross-over into interpretational tools (Figure 1.1).

Finally, unstable isotopes provide not only a continuously refinable method of absolute dating of the rock record through comparison of the ratio of parent to daughter nuclides, but also an efficient method in some cases of relative chronology in multiple igneous bodies and metamorphic events.

1.2.2 Interpreting the Record

The interpretation of rock units as the products of events in earth history is an extremely important part of stratigraphy, and it forms the basis for our understanding of the global environment through time. As such, it is reliant upon the recognition of chronology, because it is important to be able to observe geological units in a series of equivalent time-slices. Interpretation of the ancient environments in each time-slice, and through each time-slice in a succession, is reliant upon facies analysis; that is, the interpretation of the sum total of the characteristics of a rock body in order to determine its environment of deposition. Geometry, rock type, sedimentary structure and fossils are just some of the environmental indicators which help decipher past environments. Facies analysis is therefore the most important interpretational tool (Figure 1.1). The recognition of relative changes in sea-level through time is possible using basic facies analysis, and through the interpretation of the sedimentary package geometries provided by seismic and sequence stratigraphy techniques.

Fossils provide an important interpretative tool, as, using uniformitarian principles, it is possible to reconstruct ancient ecologies and to extrapolate the ecological tolerances of living to fossil organisms. In many cases simple tasks, such as distinguishing between marine and non-marine sediments, are only actually possible through the presence/absence of marine fossils, for example (Figure 1.1). Fossils are also extremely important in our interpretation of the evolution of orogenic belts, particularly through the mapping out of the distribution patterns of ecologically restricted taxa. For example, this tool was particularly important in the discovery of the former existence of the Iapetus Ocean, first identified through the recognition of separate trilobite 'provinces' for northern and southern Britain and Ireland by J. Tuzo Wilson in 1966 (Wilson 1966).

1.3 SUMMARY

Stratigraphy is an essential and dynamic discipline in modern geology. As our understanding of global processes accelerates, there is a clear and increasing demand for high-resolution stratigraphical data. In recent years this has led to the development of a bewildering array of specialist, and increasingly sophisticated, analytical tools. Despite this, the fundamental aims of stratigraphy remain unchanged: those of establishing the sequence of rock units and then interpreting them as events in earth history. This volume provides an introduction to, and foundation in, not only the fundamental principles, but also the ever more specialized array of stratigraphical tools; its chapters provide the key with which to help unlock the stratigraphical record.

REFERENCES

Berry, W.B.N. 1986. *Growth of the prehistoric time scale based on organic evolution.* Blackwell Scientific Publications, Oxford.

Bowler, P.J. 1992. *The Fontana history of the environmental sciences.* Fontana Press, London.

Doyle, P., Bennett, M.R. & Baxter, A.N. 1994. *The key to earth history. An introduction to stratigraphy.* John Wiley, Chichester.

Hailwood, E.A. & Kidd, R.B. (eds) 1993. *High resolution stratigraphy.* Geological Society Special Publication No. 70.

Hallam, A. 1990. *Great geological controversies.* Oxford University Press, Oxford.

Hedberg, H. D. 1976. *A guide to stratigraphic classification, terminology and procedure.* John Wiley & Sons, New York.

Lyell, C. 1830. *Principles of geology.* John Murray, London.

Lyell, C. 1878. *Student's elements of geology,* 3rd edition. John Murray, London.

Oakley, K.P. & Muir-Wood. H.M. 1949. *The succession of life through geological time.* British Museum (Natural History), London.

Page, D. 1861. *Advanced textbook of geology,* 3rd edition. William Blackwood, Edinburgh and London.

Rudwick, M.J.S. 1985. *The great Devonian controversy.* University of Chicago Press, Chicago.

Secord, J.A. 1986. *Controversy in Victorian geology: the Cambrian–Silurian dispute.* Princeton University Press, Princeton.

Whittaker, A., Cope, J.C.W., Cowie, J.W., Gibbons, W., Hailwood, E.A., House, M.R., Jenkins, D.G., Rawson, P.F., Rushton, A.W.A., Smith, D.B., Thomas, A.T. & Wimbledon, W.A. 1991. A guide to stratigraphical procedure. *Journal of the Geological Society of London,* **148**, 813–824.

Wills, L.J. 1951. *Palaeogeographical atlas of the British Isles and adjacent parts of Europe.* Blackie, London.

Wilson, J.T. 1966. Did the Atlantic close and then re-open? *Nature,* **211**, 676–681.

Part I
ESTABLISHING THE SEQUENCE

2
Lithostratigraphy: Principles and Practice

Beris M. Cox and Michael G. Sumbler

Lithostratigraphy is the branch of stratigraphy that is concerned with subdivision of the rock succession into units on the basis of gross lithology or rock type. It involves the differentiation, delineation, classification and formal definition of such units, which are known as lithostratigraphical units. It has its roots in field observation, and its traditional manifestation is in geological maps and cross-sections. Indeed, mappability – be it tracing a unit at outcrop or at depth – is one of the fundamental tests for the validity and utility of a lithostratigraphical classification.

In any given area, lithostratigraphy provides the primary and most objective stratigraphical classification and the framework within which other more interpretative stratigraphical disciplines are developed. When investigating a geologically unknown area, a sensible first approach is to devise a lithostratigraphical classification of the rocks seen at outcrop. This allows the spatial relationships of rock units to be expressed in maps and sections which may delineate economically important mineral resources or aquifers. After this first step, an assessment of the geological structure and geological history of the area may then be made.

Although there are maps dating from before 1800 which show the surface occurrence of certain mineral deposits (e.g. papyrus maps dating from the thirteenth century BC showing gold fields in Egypt), these did not take account of the rock succession, and were only concerned with a two-dimensional representation of a mineral body. Not until the work of William Smith (1769–1839) and his contemporaries in continental Europe did maps become documents that recorded

Unlocking the Stratigraphical Record: Advances in Modern Stratigraphy. Edited by P. Doyle and M.R. Bennett.
© 1998 John Wiley & Sons Ltd.

the rock succession and therefore, indirectly, an area's geological history. The earliest geological maps of England and Wales by William Smith (Shepherd 1917; Cox 1942) showed rock units that had been recognized at outcrop, in coastal cliffs and other natural exposures, as well as in artificial excavations such as road and canal cuttings. They were given names, such as Chalk, Green Sand, Blue Marl or Oaktree Soil, Forest Marble and Clay, Red Marl, Millstone, Magnesian Limestone, 'Coal Measures' and Derbyshire Limestone, and were based on gross or distinctive features of the lithology or characteristics of the outcrop. Each named unit was distinguished by overall lithology, and their outcrops could be traced by their topographical expression. Although these were some of the earliest lithostrati-graphical units to be delineated, the fact that many form the basis of modern, formally defined, units shows the basic soundness and objectivity of Smith's criteria for subdivision. At a similar time (1811), Leopold ('Georges') Cuvier and Alexandre Brongniart produced a geological map of the Paris Basin, and during the 1820s through to the early 1840s, small-scale geological maps were compiled for many European countries (Dudich 1984). The first geological maps of America were produced by William Maclure in 1809 (Dudich 1984; Robson 1986).

At the time of this early mapping, three basic principles of stratigraphy were already established, mainly as a result of the work of Niels Stensen (otherwise known as Nicolaus Steno, 1631–1687). The first and most important of these states that in any sequence of stratified rocks, the lower layers must be older than the upper layers. This has been called the law or principle of superposition, and has been attributed to both James Hutton (1726–1797) and William Smith as well as Steno some one hundred years earlier. Hutton and Smith certainly played a key role in establishing it as a fundamental of stratigraphy; Smith through observation and application, and Hutton through combining it with other known facts of his day into a coherent 'Theory of the Earth'. The determination of sequence in stratified rocks is an essential first step in unravelling the geological history of a sedimentary succession. The second principle, the so-called principle of original horizontality, states that stratified rocks were originally deposited approximately horizontally; depositional surfaces, the interface between the accumulating sediment and the overlying water or air, were essentially horizontal because of gravity. Whilst one must take account of cross-bedding and the significant depositional dips that may occur in some environments, if a sedimentary rock succession no longer displays near-horizontal bedding, then it can be assumed that the strata have been disturbed by post-depositional events. The third principle is that of lateral continuity which states that, irrespective of the present limits of the outcrop of a body of stratified rock, when it originally formed it extended laterally until either it terminated against the edge of the depositional basin or it thinned to zero thickness, or it changed character by merging into another deposit.

These basic principles of stratigraphy, first outlined several centuries ago, are still pertinent today, and the law of superposition in particular is fundamental in lithostratigraphy. Since these early times, many new techniques have evolved, particularly for mapping at depth where, apart from borehole cores or rock chippings, there may be little direct evidence of lithology. Downhole wireline geophysical logging is now a powerful tool in this respect (see Chapter 9), and seismic data can be processed to produce cross-sections through the succession, beneath the land and under the sea (see Chapter 10). However, the availability of these and other modern techniques should not be allowed to lead to over-sophisticated definitions of lithostratigraphical units based on obscure criteria; a lithostratigraphical scheme should merely formalize what is essentially the natural categorization of the rock succession.

Lithostratigraphy is potentially applicable to all stratified rocks because all such rocks can be described and classified in terms of their lithology. Many other types of stratigraphical classification are dependent on the presence of a particular secondary character which may or may not be available; for example, biostratigraphy (see Chapter 5) depends on the presence of fossils, and magnetostratigraphy on the presence of measurable magnetic characteristics. In addition, event stratigraphy (see Chapter 6) and sequence stratigraphy (see Chapter 11) have their special applications.

In this chapter, we first consider the basis of lithostratigraphical classification and its associated nomenclature, and then discuss the role of fossils in lithostratigraphy, and the practice and problems of lithostratigraphical correlation using examples from Britain.

2.1 LITHOSTRATIGRAPHICAL CLASSIFICATION AND NOMENCLATURE

In the eighteenth and early nineteenth centuries, many geologists, largely influenced by the German Abraham Gottlob Werner, took the principle of lateral continuity to an extreme, believing that geological strata, having been precipitated from a primeval ocean which once covered the globe, extended indefinitely around the world as 'universal formations'. This 'Neptunist' doctrine was eventually seen to be ill-founded, and after Werner's death in 1817 declined in favour as it was realized that individual rock units are, in fact, of relatively local extent (Geikie 1905; Robson 1986), and most are confined to a particular sedimentary basin or part thereof. The definition of lithostratigraphical units is therefore seldom the subject of international debate or dispute, in contrast to chronostratigraphical units (see Chapter 13). Nevertheless, a common and uniform approach to lithostratigraphical classification and nomenclature is desirable in order to foster mutual understanding amongst the international geological community. Under the auspices of the International Subcommission of Stratigraphic Classification (ISSC) and its predecessor the International Subcommission on Stratigraphic Terminology (ISST), the first international guide on such matters was produced by Hedberg (1976). Subsequently, derivatives of this guide have been produced nationally; for example, in the UK, the Geological Society of London has produced two editions of a guide to stratigraphical procedure (Holland *et al.* 1978; Whittaker *et al.* 1991), and in the USA, the American Association of Petroleum Geologists issued a code on behalf of the North American Commission on Stratigraphic Nomenclature (1983). More recently, a second edition of Hedberg's (1976) guide has been published (Salvador 1994). Although these guides vary in their scope and detail, there is an internationally agreed hierarchy of formal lithostratigraphical units, and recommendations concerning the naming and definition of such units. These should be followed as closely as possible, but it should be noted that these publications provide guidelines for best practice, not mandatory rules. Consequently, where the need arises, there is still some degree of flexibility and scope for discretion and personal interpretation in nomenclatural practice.

2.1.1 Lithostratigraphical Standard Units

The standard units in lithostratigraphy are the bed, the member, the formation, the group and the supergroup. Of these, the formation is fundamental, and is the primary building block of lithostratigraphical classification. Complete division of the rock succession into formations should be regarded as a mandatory requirement in the practice of lithostratigraphy. Once the formations are established, it may be appropriate to define additional subdivisions (members and beds) or groupings (groups and supergroups).

The formation is a body of rock that is distinct from adjacent strata by merit of its lithology or lithofacies. The term lithofacies is used to indicate the sum total of the rock's lithological and gross faunal/floral characteristics that together are the product of the particular environment in which it formed (see Chapter 14). Division of the rock succession into formations is therefore dependent on the natural characteristics of the rocks themselves and, because of this, a formation might be of any thickness from less than one metre to many hundreds or even thousands of metres. It has commonly been defined as the smallest unit that is mappable at the surface or traceable in the subsurface but, more realistically, it is a unit readily mappable at a scale appropriate for the region in which it is to be used. Traditionally, mappability in this context meant field mapping by a geologist walking over the ground and marking the outcrop on a base map using criteria such as soil type, topographical expression, and exposures (if any) where the rocks can be seen *in situ*. However, mappability defined in this way may be restricted by factors not connected with the properties of the rock succession such as the erosional history of the region, the extent of superficial deposits, the amount of vegetation or urban development, and even the degree of competence of the geologist concerned. It is therefore quite wrong to define formations or classify borehole cores strictly on the basis of what can be readily mapped at the surface. A better concept of mappability is that a formation should be potentially traceable in the rock succession both at the surface and at depth using any or all of the means at the disposal of the modern geologist, including geophysical techniques, with the proviso that a formation *must* be distinguishable from adjacent formations by significant differences in overall lithology. As a consequence, different formations, and the boundaries between them, should be readily recognizable in exposures and borehole cores with the minimum of specialist knowledge. Within its area of occurrence, a formation should be laterally continuous or have once been so, and its boundaries may be sharp, or fixed arbitrarily in a gradational sequence. Where the boundaries of a formation are not directly mappable because of a lack of exposure or topographical expression, but where the formation can be recognized in cored boreholes, they should be mapped out tentatively by extrapolation.

A member is a subdivision of a formation, and must always be part of a formation. It has lithological characteristics that distinguish it from the rest of the formation but the differences are insufficient to warrant its classification as a formation in its own right, and it need not fulfil the mappability criterion demanded of formations. Its lateral extent may be less than that of its parent formation or it may conceivably be larger, and extend laterally from one formation into another. It may be of any thickness, and is commonly lenticular or wedge-shaped. Formations need not be divided into members, or they may be partially or completely divided into them. The constituent members of a particular formation need not be the same in all areas.

A bed is the smallest unit within the scheme of formal lithostratigraphical classification. All sedimentary rock successions can be divided into sedimentological beds (i.e. sedimentary layers, from a centimetre to several metres in thickness, separated by bedding planes), but a formally defined lithostratigraphical bed may comprise any number of sedimentological beds. It is generally only those units which are sufficiently distinctive to be useful for correlation that are given formal status as lithostratigraphical beds. Examples are coal seams in the Coal Measures, and 'marker' beds of marl or flint in a thick chalk succession. A lithostratigraphical bed is a subdivision of either a member or a formation, and may pass laterally from one member or formation to another, without changing its name, but not vertically. Strata which, in the past, have been grouped together and loosely termed 'Beds' – such as the Baggy Beds (Devonian), Kellaways Beds (Jurassic) and Bracklesham Beds (Eocene) – can virtually all now be given the formal status of formation or member, and the word 'Beds' replaced accordingly. Some authors, whilst acknowledging the formational (or otherwise) status of their 'Beds', prefer to retain, rather than replace, that term. However, clumsy and tautological expressions such as 'Kellaways Beds Formation' are unnecessary; Kellaways Formation is the correct form.

The unit above the rank of formation is the group, which comprises two or more adjacent formations with significant unifying lithological and/or genetic features. The boundaries of a group should be marked by a major change in lithology and/or inferred depositional environment, or by a substantial non-sequence or break in sedimentation. For example, the Mercia Mudstone Group (Triassic) comprises a succession of terrestrial and fluvio-lacustrine, predominantly mudstone, 'red-bed' formations formed during a major episode of basin-filling. This was terminated by a marine incursion and deposition of entirely different facies, which marks the upper boundary of the group. Its lower boundary with the Sherwood Sandstone Group is marked by a major but gradational lithological change. The constituent formations of a group need not be the same in all areas. A formation does not have to be part of a group and, in certain cases, may pass laterally from one group into another. In an area where the stratigraphy has yet to be fully determined, a group may contain some rocks that have not been assigned to a formally defined formation, but the group must at least partially comprise named formations, otherwise it should itself be treated merely as a formation.

Groups have sometimes been divided into subgroups although most of the published codes recommend that these should be avoided, or used only as informal subdivisions. However, they have proved to be a helpful addition to the lithostratigraphical hierarchy in some instances, for example within the Dalradian Supergroup (Precambrian–Cambrian).

The highest rank in the hierarchy of formal lithostratigraphical units is the supergroup. This is used only in those cases when it is necessary, for pragmatic reasons, to combine two or more adjacent groups or an assemblage of groups and formations that have some unifying characteristics; for example, the Windermere Supergroup (Ordovician–Silurian) which combines a natural grouping of folded and cleaved, predominantly marine, sedimentary rocks comprising three groups and a number of additional formations (Kneller *et al.* 1994). It is often the case that a logically defined 'group' has to become a supergroup because the particular package of strata already includes units defined as groups, the latter perhaps having been prematurely given an unnecessarily high rank in the lithostratigraphical hierarchy; these are cases where personal interpretation and opinion have led to differences of approach. An example is provided by the Wealden (Lower Cretaceous), variously accorded the status of group or supergroup.

Where the lithologies that characterize a formation pass laterally into other lithologies, it will eventually become necessary to designate another formation or group. Whilst, in reality, the change may involve lateral interdigitation of lithofacies, the formations and groups must conform to the law of superposition and follow an ordered succession. Boundary criteria must be chosen accordingly. The examples shown in Figures 2.1 and 2.2 illustrate this more fully. The various units are defined by their lithofacies, the boundaries of which are of various types: erosional disconformities, sharp conformities, vertical gradations, lateral gradations and lateral interdigitations. However, the boundaries between the corresponding lithostratigraphical units are necessarily defined, in any one section, at one specific point. For example, in Figure 2.1, *lithofacies* characteristic of the Hampen Formation are gradually replaced by those characteristic of the Rutland Formation over an outcrop distance of some 40 or 50 km, but the boundary between the *lithostratigraphical units* of the Hampen and Rutland formations is defined 'precisely' by reference to some arbitrary geographical locality or, in practice, at the junction of two adjoining geological maps. Similarly, *lithofacies* characteristic of the Hampen Formation interdigitate with those of the Taynton Limestone Formation, but the Hampen Formation and Taynton Limestone as *lithostratigraphical units*, do not. Note also, in Figure 2.1, how some formations are replaced laterally by members of a different formation, or several formations may combine into another. Note too how the arbitrary lateral boundary between the Sharp's Hill Formation and Rutland Formation corresponds with the change in status of the Taynton Limestone Formation to Wellingborough Member (of the Rutland Formation): generally a wedge of one formation within another is to be discouraged because the law of superposition ceases to apply, and it would be confusing on a map depicting the formations' outcrops.

Special problems may be presented in successions of cyclical sedimentation (see Chapter 7), such as the British Silesian where lithologically indistinguishable units

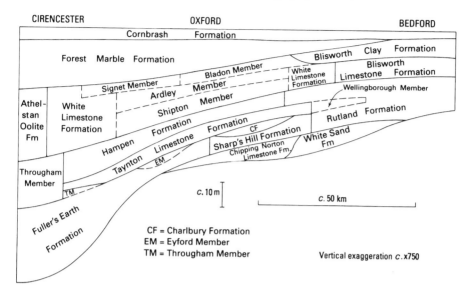

Figure 2.1 *Schematic section illustrating lithostratigraphical relationships within the Great Oolite Group (Middle Jurassic) between Cirencester and Bedford, southern England. The nomenclature used is a mixture of formalized traditional terms (e.g. Cornbrash, Forest Marble, White Limestone) and newer names chosen to accord with modern recommendations (e.g. Rutland Formation)*

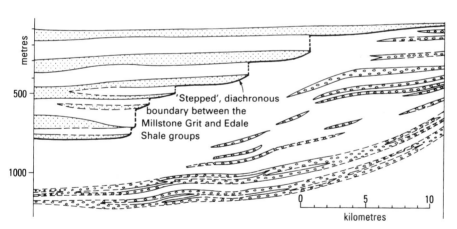

Figure 2.2 *Schematic section showing the lateral boundary between the Millstone Grit and Edale Shale in the Carboniferous of the southern Pennines, UK. These are both traditional names which have been redefined and formalized as groups. Whilst both groups are dominated by mudstone (unshaded), feldspathic sandstones (fine stipple) characterize the Millstone Grit Group, and quartzitic sandstones (coarse stipple) characterize the Edale Shale Group. The boundary between the two groups had to be defined in such a way that it could be drawn objectively and consistently in any one section. In this case, the base of the Millstone Grit Group has been defined at the first appearance of feldspathic sandstone within the mudstone succession.*
[Modified from: Evans et al. (1968); new classification by Rees and Wilson (1998)]

(for example, deltaic sandstones) occur in successive cycles, and in structurally complex areas, such as parts of the Lower Palaeozoic in the Southern Uplands of Scotland, where tectonostratigraphical units are more easily mapped out (see Chapter 4). Volcanic successions also present difficult classification problems. The products of a brief eruptive episode, such as the Llewelyn Volcanic Group (Ordovician), may be enveloped in sedimentary rocks that, in the absence of the volcanic succession, are also classified as a group. Lateral continuity of deposits may be difficult to demonstrate because of syn-volcanic faulting, contemporaneous erosion and redeposition. Potentially good marker beds, like pyroclastic flows, change in character as they pass from subaerial to submarine palaeoenvironments, and the outcrop is broken by both syn-volcanic and later faulting. Lateral thickness changes are abrupt; the intracaldera facies of the Pitts Head Tuff Formation (Ordovician) is 700 m thick and passes sharply into an extensive outflow subaerial and submarine facies of less than 100 m thickness (Howells *et al.* 1991). Original horizontality cannot be assumed in subaerial successions: airfall tuffs, vitally important in correlation, will be draped over the original topography which may show considerable and rugged relief. Finally, when a lava flow locally intrudes older lake sediments, the law of superposition may apparently cease to apply.

2.1.2 Lithostratigraphical Nomenclature

Ideally, the name of a lithostratigraphical unit should have three components, made up of a place name, a lithology and a standard hierarchical term. Sherwood Sandstone Group (Triassic) is a good example. Inclusion of a single lithological term is recommended but not mandatory because the use of a single rock type might be inappropriate. Where it is included, the singular of the noun is preferred, as in Mercia Mudstone (not Mudstones) Group (Triassic) and Sandringham Sand (not Sands) Formation (Jurassic–Cretaceous). In practice, most lithostratigraphical units are referred to in speech by two-part names, for example Kimmeridge Clay rather than Kimmeridge Clay Formation (Jurassic), but, in written text, it is best practice always to use the full name. The use of a lithological adjective in place of a noun, for example Snowdon Volcanic Group (Ordovician) and hence Snowdon Volcanics, is not generally to be encouraged. In offshore or other remote areas, suitable geographical names may not be available, in which case some other proper name is used. In the North Sea, for example, the names of bathymetric features and oil fields have been used, and also the names of explorers (e.g. the Amundsen and Cook formations) and Scottish lochs (e.g. the Etive and Rannoch formations). The latter examples reflect bad nomenclatural practice as they are legitimate geographical names of no direct relevance to the units concerned. The same geographical or proper names should not be used for more than one unit. Whether names are three- or two-part, they bear initial capitals. Qualifying terms such as lower, middle and upper are for informal use only, and are therefore used without initial capitals. However, long-standing subdivisions, particularly of formations, usually take an initial capital where no advantage is to be gained by overturning well-established practice. Examples include the Lower, Middle and

Upper Coal Measures (Carboniferous), and the Lower and Upper Kimmeridge Clay (Jurassic), although these units have no formal status in the lithostratigraphical hierarchy.

Applying these rules of nomenclature to new areas of geological investigation is relatively straightforward, but many stratal units were named before the conventions of modern lithostratigraphy were expounded. These names have great historical standing and familiarity of usage but do not necessarily conform with the recommendations; for example, Blue Lias (Jurassic) and Millstone Grit (Carboniferous), both of which include old and poorly defined lithological terms and no place name. A common-sense approach must be adopted when dealing with such long-established names. Wholesale abandonment of well-established stratal names merely for the sake of procedural dogma is not desirable and it may be better to retain the traditional name (if necessary, in a formalized variant) so as to foster nomenclatural stability, widespread understanding and clarity of concept. Although such names may have been in use for generations, few have been properly defined, and it should be the task of any stratigrapher who studies these formations in detail, to provide definitions in line with modern good practice.

With increased understanding and knowledge of the geology of an area, the names of some lithostratigraphical units will prove to be synonyms of others. Common sense and regard for practical utility should determine which name is to have priority. The junior name(s) need not be abandoned in favour of the more senior and, conversely, long-established names like Blue Lias need not be rejected just because of their old-fashioned appearance.

2.1.3 Definitions and Specifications

When a study of a rock succession necessitates a new or revised lithostratigraphical scheme, it is important that the details are made known to other workers, ideally by publication in appropriate geological literature. For a formation or other formal lithostratigraphical unit, the eight categories listed below constitute its definition and provide the essential diagnostic information. A group does not necessarily require this degree of definition if it is completely divided into formations, but it is helpful to give an indication of its type area (where a full set of its component formations is developed) and its gross lithology, as well as a general statement listing the constituent formations and the rationale behind the make-up of the group. Of course, where a group is not completely divided into formations, a definition of that group's lower boundary may be required. Similarly, supergroups may be defined in this more abbreviated way. The eight categories are as follows:

1. *Name.* Whether it be a new name or an existing one, the derivation of the unit's name should be indicated and bibliographical references should be given indicating the nomenclatural history. In the case of entirely new units, the place name should be a recognized geographical locality or area where the unit is typically developed, taking account of historical precedent. The same place

name should not be used for more than one unit even if that unit is of a different rank.

2. *Lithology*. The characteristic and distinguishing gross lithology(ies) need to be stated. Subordinate lithologies should be included, as well as brief comments on features such as colour, sedimentary structures, bed-thickness and cyclicity. Distinctive wireline log responses might also be usefully included; indeed, in offshore areas, where the lithological characterization of lithostratigraphical units depends largely on well data in the form of rock cuttings or chippings, side-wall cores and continuous cores (although rare), the wireline log signatures (see Chapter 9) may be a crucial part of the definition of a unit.

3. *Type and reference sections*. Provision of type and reference sections allows geologists to observe the characteristics of lithostratigraphical units at first hand. All formal lithostratigraphical units of formation or lower rank must have a designated type section. This should be the best representative section, which shows most typical lithological characters of a unit and, ideally, also its basal boundary. This section is designated at a specific geographical locality, preferably but not necessarily that which gives its name to the unit. Cored boreholes – preferably ones for which there is a full suite of curated specimens, sliced core and geophysical logs – may be used where there is no suitable exposure. It is often found that no one section exposes the whole of a particular lithostratigraphical unit (including lower and upper boundaries). In such cases, a type area can be indicated instead; however, the lower and upper boundaries must always be explicitly identified in specified sections within that area. A type section is often referred to as a stratotype which, in the context of lithostratigraphy, means a specific package of strata in a specific section or area; a unit-stratotype covers the whole unit whereas a boundary-stratotype includes only the top or base of the unit.

   Reference sections, which supplement the type section or type area, may be proposed when the latter are, for example, not fully representative of the range of lithologies. Again, sections showing unit boundaries are particularly important. Reference sections are essential in cases where the original or 'historical' type section is no longer accessible. For example, access to the type section of William Smith's Blue Lias at Saltford railway cutting, between Bath and Bristol (Torrens & Getty 1980) is impractical because the line is used by high-speed trains. However, the Blue Lias is well exposed on the coasts of Dorset, Somerset and Glamorgan, and the cliffs there can be used as reference sections; in particular, the fine coastal sections near Lyme Regis in Dorset, which show both lower and upper boundaries, can be designated as a primary reference section. This example makes the point that the type section cannot be changed; it can only be supplemented by reference sections.

4. *Boundaries*. The boundaries of lithostratigraphical units are placed at positions of lithological change. They may be marked by a sharp lithological contrast or positioned at some level within a gradational sequence. In either case, enough detail should be given so that there is no doubt or ambiguity about where the boundary is taken in the stratotype, and where it should be drawn elsewhere, whether at outcrop or in borehole cores. The names of the underlying and

overlying lithostratigraphical units should be included in the description. The name alone may suffice to define the upper boundary which is, of course, defined by the basal boundary of the overlying unit. For example, the boundary specifications for the Redcar Mudstone Formation (Lower Jurassic) can be given as follows (Powell 1984):

Lower boundary: where grey, silty mudstone with thin beds of limestone and calcareous sandstone rest with sharp and irregular contact on grey-green soft mudstone with thin siltstone laminae (Cotham Member, Lilstock Formation, Penarth Group); in the BGS Felixkirk Borehole, this is at depth 288.87 m.
Upper boundary: base of the Staithes Sandstone Formation.

Boundaries of lithostratigraphical units are not necessarily isochronous surfaces (see Section 2.3) but increasingly there is a trend towards drawing lithostratigraphical boundaries at an 'event' or sequence stratigraphical boundary (see respectively Chapters 6 and 11). This may be appropriate where a lithological change coincides with such an event boundary but, in general, it is bad practice as it negates the principles of lithostratigraphy, the purpose of which is to classify the rock succession on the basis of lithology. As a consequence, lithostratigraphical boundaries should only be drawn at disconformities or unconformities if there is an associated lithological change. However, successive units of similar lithology should preferably not be combined if they are known to be separated by a regional unconformity or major hiatus.

5. *Subdivisions*. An indication of named formal or informal subdivisions should be given with suitable bibliographical references if available.
6. *Thickness*. Thickness details should include the typical range and extremes, and specify the areas in which those extremes occur. Often the form 'up to *x* metres' is used.
7. *Distribution*. Area(s) of occurrence should be indicated in terms of present geography and, ideally, also in terms of contemporaneous sedimentary basins. For example, the Burnham Formation (Upper Cretaceous) is developed in Lincolnshire and Yorkshire, within the East Midlands Shelf and southern margin of the Cleveland Basin.
8. *Age*. The age of the unit should be indicated by reference to the standard scheme of periods/epochs/ages (Harland *et al.* 1990) prefixed, as appropriate, by Early, Mid and Late.

Other useful details which should be given are the name of the next higher unit in the lithostratigraphical hierarchy (e.g. the group to which a formation belongs), a general statement about fossil content and useful fossil markers, and an indication of lateral boundaries and correlative units.

2.2 FOSSILS IN LITHOSTRATIGRAPHY

Generally, fossils should play no part in the formal definition of a lithostratigraphical unit except at a very general level; only when they are so abundant that they may be treated as component clasts and are a characteristic part of the gross lithology should they be included in the definition. However, it is helpful to give an indication of the character of the faunal/floral assemblage of a lithostratigraphical unit, and to specify fossil marker beds, in order to aid recognition. In practice, often the quickest way to find out where one is placed in an established lithostratigraphical succession is to pick up a fossil; for example, in the British Mesozoic, a fossil ammonite could readily indicate whether a particular unit of mudstone belonged to the Lower Jurassic Lias Group, the Middle/Upper Jurassic Oxford Clay Formation, the Upper Jurassic Kimmeridge Clay Formation, or the Lower Cretaceous Gault Formation. When dealing with successions in offshore areas, it is common practice to indicate biostratigraphical markers (notably first downhole occurrences and first downhole acme occurrences of microfossils) that assist with placement in the lithostratigraphical succession.

Some older stratal names include the name of a particularly abundant or characteristic fossil. When these stratal names are retained for formal lithostratigraphical units, the fossil name should not be italicized and should not change its name even if subsequent palaeontological research shows that the name of the relevant taxon has been superseded. Where a binomial taxon has been used, the generic name should be abandoned, and the specific name used with initial capital. For example, Supracorallina Bed is correct, and *Astarte supracorallina* Bed is not, even though the bivalve originally called '*Astarte supracorallina*' is now identified as *Nicaniella extensa* (Phillips). Other examples where fossils give their name to lithostratigraphical units include Clypeus Grit Member (named after a genus of echinoid) and Clavellata Formation (named after the bivalve *Myophorella clavellata*).

In practice, it is much easier to operate a 'fossil-free' lithostratigraphy in sandstone and limestone successions than in mudstones, where differentiation of adjacent formations without the guidance of fossils is more difficult. For example, at the surface, all Mesozoic mudstone formations tend to weather down to a brown clay, and they may present a problem to the field mapper unless there are plenty of borehole or pit sections available. However, fossils such as oysters, which often survive in the soil, can be used to differentiate individual formations. For example, in the Late Jurassic mudstones of the UK, the oysters *Gryphaea dilobotes*, *G. lituola*, *G. dilatata* and *Deltoideum delta* can be used to aid recognition and differentiation, at outcrop, of the Oxford Clay, West Walton and Ampthill Clay formations (Horton *et al.* 1995).

Marine bands in the Upper Carboniferous of the UK and elsewhere are treated as lithostratigraphical units, normally of bed status (although this hierarchical term is not used). A marine band which is characterized by a fossil that distinguishes it from all other marine bands is named after that fossil; for example, the Subcrenatum Marine Band is a bed in the Coal Measures named after the ammonoid *Gastrioceras subcrenatum*. These fossil-specific marine bands are particularly valuable for correlation as they are unique event horizons (see Chapter 6) within a repetitive succession of sedimentary cycles (see Chapter 7). A marine band that lacks any uniquely diagnostic fossil is named after a locality where it is well developed. For example, the Honley Marine Band is named after an exposure near

Huddersfield, Yorkshire, and the Burton Joyce Marine Band after an occurrence in a borehole at Burton Joyce, Nottinghamshire.

2.3 LITHOSTRATIGRAPHICAL CORRELATION

Lithostratigraphical correlation involves the tracing and identification of a lithostratigraphical unit away from its type section and outside its type area on the basis of its lithological character. This is most simply done by tracing the outcrop across country, perhaps by utilizing some associated secondary feature such as its topographical expression, characteristic soil or vegetation patterns. In offshore areas, seismic data can provide a crucial tool for correlation between boreholes and for mapping the three-dimensional form of a sediment body (see Chapter 10), and geophysical methods which involve the electrical and other properties of a rock succession can also be used to establish lithological/lithostratigraphical correlation (see Chapter 9).

Although a lithostratigraphical unit will originally have formed a continuous body of sediment, it is generally now more or less fragmented as a result of subsequent geological and geomorphological processes such as faulting and erosional dissection, and exposure may be poor or interrupted. Other disciplines, particularly biostratigraphy (see Chapter 5), have a major part to play when dealing with the correlation of discontinuous, isolated or structurally complicated deposits and, in these cases, their stratigraphical position relative to other lithostratigraphical units becomes an increasingly important aid to identification. Such problems are notable in fluvial, glacial and other terrestrial deposits, and are particularly acute in deposits of Quaternary age ('Drift'), for it is rare to find them in an ordered stratified succession. Typically, they rest unconformably on bedrock and their top is marked by the modern ground surface; their stratigraphical context is not therefore closely constrained. Lateral continuity of units is also difficult to establish and fossils cannot be used as a correlation tool in the way that they are in older parts of the geological column compared with which, given the short time-span of the Quaternary (*c.* 2 Ma), the succession is divided in great detail.

Unlike most other types of stratal units, lithostratigraphical units may be diachronous; lines of lithostratigraphical correlation delimit lithological equivalence, and are not necessarily 'time-lines'. A classic example of diachronism of a lithostratigraphical unit comes from the British Jurassic, and was first demonstrated by Buckman (1889) on the basis of ammonites. A thick unit of sandstone, representing a southward-prograding sand bar, is of Early to Late Toarcian age in the Cotswolds but, traced southwards to the Dorset coast (Figure 2.3), it is of Late Toarcian to Aalenian age. In different areas, it has been given different names – Cotteswold Sands (Lycett 1857), Midford Sands (Phillips 1871), Yeovil Sands (Hudleston *in* Buckman 1879) and Bridport Sands (Woodward 1888) – but, as one formerly continuous sand body, it is appropriate that it should have a single lithostratigraphical name (Figure 2.4). Although Cotteswold Sands has priority by some 30 years, Bridport Sands is preferred over all the other names because of the magnificence of its coastal type section and the widespread usage of the term.

Figure 2.3 *Coastal exposure of Bridport Sand Formation, West Bay, Dorset, UK – the youngest part of a diachronous body of Jurassic marine sands (see Figure 2.4). The prominent bands are carbonate-cemented beds, and pick out sedimentary rhythms. [© NERC 1972; source British Geological Survey A12038]*

SSW DORSET COAST	MENDIP HILLS	COTSWOLD HILLS	Sub-zone	Zone	Age
			Aalensis	Opalinum	AALENIAN
	absent on		Moorei		
Bridport Sands Yeovil Sands	Mendips Shelf		Levesquei	Levesquei	
			Dispansum		LATE TOARCIAN
	Midford		Fallaciosum		
	Sands	?	Striatulum	Thouarsense	
	Cotteswold			Variabilis	
	Sands		Crassum		EARLY TOARCIAN
			Fibulatum	Bifrons	
			Commune		

Figure 2.4 *Diachronism of the Bridport Sand Formation in south-west England, over a distance of some 150 km. The zones and subzones are based on ammonite faunas. [Modified from: Torrens (1969) and Callomon & Cope (1995)]*

2.4 LITHODEMIC UNITS

In recent years, certain igneous intrusive and metamorphic rock bodies have been excluded from the remit of lithostratigraphy because they are not stratified or do not conform with the law of superposition. Instead, these rock bodies are referred to as lithodemic units rather than lithostratigraphical units. This distinction, which was first introduced in the code of the North American Commission on Stratigraphic Nomenclature (1983), has not been agreed by everyone (although it is widely used in the USA and Australia) but is becoming increasingly accepted. Like lithostratigraphical units, lithodemic units are defined, distinguished and delineated on the basis of lithological character. The basic unit (equivalent to the formation of lithostratigraphy) is the *lithodeme*, which comprises a body of predominantly intrusive, highly deformed and/or metamorphosed rock (from the Greek *lithos* meaning 'stone' and *demos* meaning 'body'). The next higher rank is the *suite* (equivalent to the group) and then the *supersuite* (equivalent to the supergroup). There are no named subdivisions of a lithodeme; if it is to be divided, the divisions become new lithodemes, and the original lithodeme then becomes a suite. As with lithostratigraphical units, names of lithodemic units generally comprise a place name followed by a lithological term. The term *complex* is used for an assemblage of rocks which generally includes two or more genetic (i.e. sedimentary, igneous or metamorphic) types; it commonly has a highly complicated structure. The term is not included in the lithodemic hierarchy, and may sometimes be replaced by the term terrane. A complex that contains a mixed assemblage of extrusive volcanic rocks with related intrusions and their weathering products is termed a volcanic complex, and a largely or wholly igneous body that contains different rock types (e.g. granite, diorite, granodiorite) is termed an igneous complex. The term structural complex is used where tectonic processes have produced a complicated mass of different rock types (Figure 2.5).

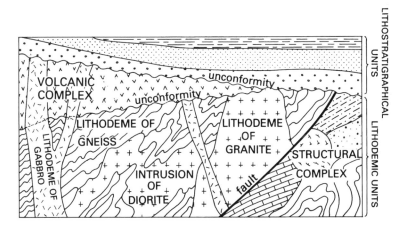

Figure 2.5 *Diagrammatic cross-section showing various lithodemic units. The lithodeme of gneiss and intrusion of diorite were deformed together, and may be regarded as a complex. [Modified from: North American Commission on Stratigraphic Nomenclature (1983)]*

2.5 CONCLUSIONS

Lithostratigraphy is the primary and most objective stratigraphical classification of the rock succession, and the framework within which other more interpretative stratigraphical disciplines can be developed. The fundamental unit of lithostratigraphical classification is the formation, which should be mappable at the surface or traceable in the subsurface. Its boundaries should be defined solely on the basis of lithology or lithofacies, and should be readily recognizable in exposures and borehole cores with the minimum of specialist knowledge. It is only appropriate to draw a lithostratigraphical boundary at an 'event' or sequence stratigraphical boundary when the latter coincide with a lithological change. The internationally agreed hierarchy of formal lithostratigraphical units and recommendations concerning the naming and definition of such units should be adhered to as closely as practical in order to foster mutual understanding amongst the geological community, and to maintain a high standard of best practice in lithostratigraphy.

ACKNOWLEDGEMENTS

We thank our colleagues at the British Geological Survey, in particular Peter M. Allen, Allan Brandon, J. Ian Chisholm and Robert W. O'B. Knox, for helpful discussion. Dr Allen kindly supplied the text identifying some of the special problems in the lithostratigraphical classification of volcanic successions. The chapter is published with the approval of the Director, British Geological Survey (NERC) but we take full responsibility for the views expressed herein.

REFERENCES

Buckman, J. 1879. On the so-called Midford Sands. *Quarterly Journal of the Geological Society of London*, **35**, 736–743.

Buckman, S.S. 1889. On the Cotteswold, Midford, and Yeovil Sands, and the division between Lias and Oolite. *Quarterly Journal of the Geological Society of London*, **45**, 440–474.

Callomon, J.H. & Cope, J.C.W. 1995. The Jurassic geology of Dorset. In Taylor, P.D. (ed.) *Field geology of the British Jurassic*. The Geological Society, Bath, 51–103.

Cox, L.R. 1942. New light on William Smith and his work. *Proceedings of the Yorkshire Geological Society*, **25**, 1–99.

Dudich, E. (ed.) 1984. *Contributions to the history of geological mapping*. Proceedings of the 10th INHIGEO Symposium Budapest 1982. Akadédmiai Kiadó, Budapest.

Evans, W.B., Wilson, A.A., Taylor, B.J. & Price, D. 1968. Geology of the country around Macclesfield, Congleton, Crewe and Middlewich. *Memoir of the Geological Survey of Great Britain*, Sheet 110 (England and Wales).

Geikie, A. 1905. *The founders of geology*. Macmillan, London.

Harland, W.B., Armstrong, R.L., Cox, A.V., Craig, L.E., Smith, A.G. & Smith, D.G. 1990. *A geologic time scale*. Cambridge University Press, Cambridge.

Hedberg, H.D. 1976. *International stratigraphic guide: a guide to stratigraphic classification, terminology, and procedure*. John Wiley & Sons, New York.

Holland, C.H., Audley-Charles, M.G., Bassett, M.G., Cowie, J.W., Curry, D., Fitch, F.J., Hancock, J.M., House, M.R., Ingham, J.K., Kent, P.E., Morton, N., Ramsbottom, W.H.C.,

Rawson, P.F., Smith, D.B., Stubblefield, C.J., Torrens, H.S., Wallace, P. & Woodland, A.W. 1978. *A guide to stratigraphical procedure*. Special Report of the Geological Society of London, 10.

Horton, A., Sumbler, M.G., Cox, B.M. & Ambrose, K.A. 1995. Geology of the country around Thame. *Memoir of the British Geological Survey*, Sheet 237 (England and Wales).

Howells, M.F., Reedman, A.J. & Campbell, S.D.G. 1991. *Ordovician (Caradoc) marginal basin volcanism in Snowdonia (north-west Wales)*. HMSO for the British Geological Survey, London.

Kneller, B.C., Scott, R.W., Soper, N.J., Johnson, E.W. & Allen, P.M. 1994. Lithostratigraphy of the Windermere Supergroup, northern England. *Geological Journal*, **29**, 219–240.

Lycett, J. 1857. On the sands intermediate the Inferior Oolite and Lias of the Cotteswold Hills, compared with a similar deposit upon the coast of Yorkshire. *Annals and Magazine of Natural History*, **20**(2), 170–177. [Also in *Proceedings of the Cotteswold Naturalists' Field Club*, **2**, 142–149 (1860)].

North American Commission on Stratigraphic Nomenclature 1983. North American stratigraphic code. *Bulletin of the American Association of Petroleum Geologists*, **67**, 841–875.

Phillips, J. 1871. *Geology of Oxford and the valley of the Thames*. Clarendon Press, Oxford.

Powell, J.H. 1984. Lithostratigraphical nomenclature of the Lias Group in the Yorkshire Basin. *Proceedings of the Yorkshire Geological Society*, **45**, 51–57.

Rees, J.G. & Wilson, A.A. 1998. Geology of the country around Stoke on Trent. *Memoir of the British Geological Survey*, Sheet 123 (England and Wales).

Robson, D.A. 1986. *Pioneers of geology*. Natural History Society of Northumbria, Newcastle upon Tyne.

Salvador, A. 1994. *International stratigraphic guide: a guide to stratigraphic classification, terminology, and procedure*. The International Union of Geological Sciences, Trondheim and The Geological Society of America, Inc., Boulder.

Shepherd, T. 1917. William Smith: his maps and memoirs. *Proceedings of the Yorkshire Geological Society*, **19**, 75–253.

Torrens, H.S. (ed.) 1969. *International field symposium on the British Jurassic. Excursion No. 1. Guide for Dorset and south Somerset*. Geology Department, Keele University, Keele.

Torrens, H.S. & Getty, T.A. 1980. The base of the Jurassic System. *Special Report of the Geological Society of London*, **14**, 17–22.

Whittaker, A., Cope, J.C.W., Cowie, J.W., Gibbons, W., Hailwood, E.A., House, M.R., Jenkins, D.G., Rawson, P.F., Rushton, A.W.A., Smith, D.B., Thomas, A.T. & Wimbledon, W.A. 1991. A guide to stratigraphical procedure. *Journal of the Geological Society of London*, **148**, 813–824.

Woodward, H.B. 1888. Further note on the Midford Sands. *Geological Magazine*, **5** (Decade 3), 470.

3
Remote Sensing and Lithostratigraphy

M. Andrew Bussell

Remote sensing refers to the determination of terrain characteristics without direct physical contact. This is achieved through the detection, enhancement, imaging and interpretation of reflected and emitted electromagnetic radiation. Cameras and multispectral sensors may be carried on airborne or satellite platforms, and aerial photographs or satellite images are typical products. At its simplest, and most traditional, remote sensing is used whenever a field geologist studies a distant mountainside and, using experience of rock weathering, geomorphology, texture and outcrop geometry, attempts to make geological sense of what is seen. Modern methods of remote sensing can simulate this process in the laboratory, providing a quick and effective reconnaissance technique covering large areas quickly and cheaply. Tens of square kilometres may be covered by a single aerial photograph costing a few pounds, a current Landsat Thematic Mapper (TM) scene covers in excess of 24 000 km^2 and may be purchased for less than US$4400, while archive data are even cheaper. Imagery may be available at considerable discounts for academic users, as is the case for Landsat and SPOT data for the UK. At a more local scale, the geologist can provide field photographs or field spectral data to use as a basis for recording and interpreting stratigraphical successions.

Whatever the scale of image used, the scope for extraction of stratigraphically relevant information is enormous. Under favourable circumstances the succession of photostratigraphical units can be determined. Unconformities, pinchouts, on-lap, offlap, facies transitions and large-scale structures, such as channels and reefs, can be recognized. Lithologies may sometimes be identified and good estimates of

Unlocking the Stratigraphical Record: Advances in Modern Stratigraphy. Edited by P. Doyle and M.R. Bennett.
© 1998 John Wiley & Sons Ltd.

thickness obtained. Similarly, in basement or plutonic terrains, sequences of deformations or intrusions among lithodemic units (see Chapter 2) may be determined by interpretation of the remotely sensed images. All this can be achieved in advance of fieldwork, so that a provisional stratigraphical interpretation can be taken into the field for testing and refinement.

This chapter aims to demonstrate how remote sensing is used as a basic tool for reconnaissance stratigraphical investigation covering the major areas of data characteristics, photogrammetry, image enhancement and image interpretation. It begins by describing the characteristics of the raw material: the aerial photographs and digital images. The techniques of photogrammetry and image processing are explained and their value in maximizing information available for photostratigraphical interpretation is emphasized. Photogrammetry provides methods by which strike azimuth and dip data can be obtained, and allows the display of powerful three-dimensional perspective models as aids to visualization and interpretation. Image processing is important because digital images are often virtually useless in their raw state; image enhancement is essential for the discrimination of photostratigraphical units and emphasis is laid on the experimental and interactive nature of image processing. Principles for photostratigraphical interpretation of remotely sensed images are then developed to show, for example, how the photostratigraphical column may be determined, facies changes identified and unconformities recognized. Finally, it should always be remembered that the results from remote sensing methods are only as good as the quality of the accompanying field verification permits.

3.1 THE RAW MATERIAL

All remotely sensed images are generated by the detection of radiation that has been reflected or emitted from the earth's surface. Solar radiation that is incident on solid materials may be either reflected or absorbed: the higher the ratio of reflection to absorption, the higher the albedo of the material is said to be. The wavelength of reflected solar radiation is predominantly shorter than 3 μm while that due to the earth's thermal emission is of longer wavelength (Figure 3.1A). Incident radiation may be differentially absorbed according to wavelength, depending on the reflecting material, thereby imparting a 'spectral signature' to the reflected radiation (Figure 3.2). The geometry of reflections varies from mirror-like (specular) to diffuse (Lambertian) but in practice neither of these ideals is met and an intermediate condition is achieved (Figure 3.3). The precise nature of this non-Lambertian behaviour will depend on the surface material, its morphology and the wavelength of incident radiation under consideration.

The images used to interpret stratigraphical relationships are of two fundamental types: analogue and digital. Analogue imagery is produced by conventional photographic methods while digital data are obtained by radiation-sensitive sensors which scan the ground and record variations in radiation numerically. A computer is then used to process and print the images. Aerial photographs can be converted to digital format using scanning devices and they may then be enhanced

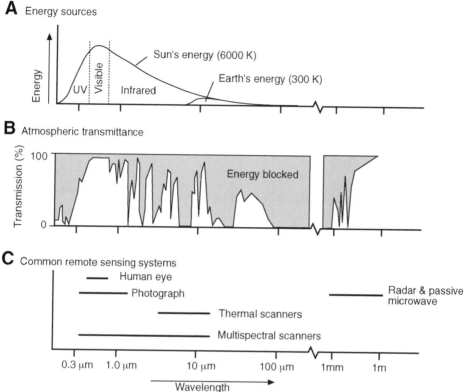

Figure 3.1   *The electromagnetic spectrum: sources, atmospheric transmittance and remote sensing systems [Modified from: Lillesand & Kiefer (1994)]*

using many of the techniques applied to primary digital data. Detection of radiation is a common theme and Figure 3.1B and C shows atmospheric transmittance and the portions of the electromagnetic spectrum detected by various sensors.

In the rest of this section consideration is given to film type, aerial survey format, the geometry of aerial photographs, the acquisition of digital multispectral data and radar data. Aerial photographic surveys are conducted using a variety of film types and the significance of tonal variation for photogeological interpretation will depend on the film used. Similarly the geometric characteristics of the terrain and geological boundaries, viewed on a single aerial photograph or a stereopair, vary according to the method of aerial survey. The spatial and spectral characteristics of digital imagery depend on the sensing system that acquired the data and the image processing techniques used to enhance the imagery before printing.

3.1.1 Photographic Film

The commonest photographs in aerial survey archives are panchromatic, with film emulsion sensitive to radiation across the visible part of the electromagnetic

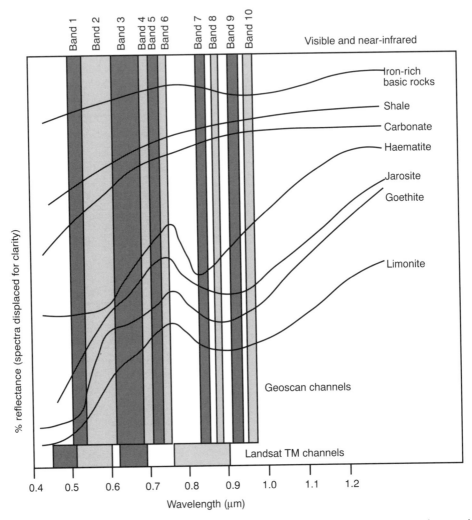

Figure 3.2 *Spectral reflectance curves (spectral signatures) for a variety of rocks and minerals between 0.4 and 1.2 μm compared with the spectral resolution of the Geoscan and Landsat TM sensors. [Modified from: Agar et al. (1994)]*

spectrum (Figure 3.4A). Clarity of the resulting black and white prints is often enhanced by the use of a filter to reduce the haze which results from atmospheric scattering of short wavelength radiation. The principal drawback of panchromatic prints is their lack of spectral discrimination: pale grey and white tones record strong reflectances from surfaces that may be any colour in the visible spectral range. An alternative form of grey scale imagery is produced when film sensitive to infrared radiation is used (Figure 3.4A). Such photographs are often sharper than panchromatic prints because of the reduction of atmospheric scatter effects. Chlorophyll in vegetation reflects near-infrared radiation very strongly and with wide variation according to vegetation type. Consequently these images are more

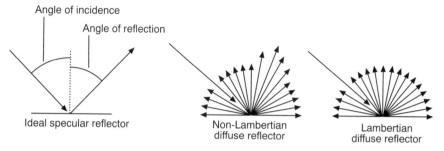

Figure 3.3 *Illustration of the variety of reflection geometries, ranging from ideal specular to Lambertian. [Modified from: Lillesand & Kiefer (1994)]*

useful than panchromatic, particularly for the discrimination of contrasting vegetation types which may relate to substrate.

Colour film contains much more information and provides a more powerful basis for rock discrimination than most grey tone imagery. It may be either true colour or false colour. True colour is produced by three-layer film in which each layer is sensitive to radiation from different spectral ranges within the visible part of the spectrum – typically blue, green and red (Figure 3.4B). The development process ultimately leads to the reproduction of true colours on the print. False colour film is produced by three-layer film sensitive to the green, red and infrared parts of the spectrum. Intensities of these three layers are developed as intensities of blue, green and red respectively in the final prints, hence the term false colour (Figure 3.4C). False colour photographs are often better than true colour prints for discriminating vegetation patterns related to substrate because of the sensitivity of infrared radiation to vegetation.

3.1.2 Survey Format and Geometry of Photographs

Oblique aerial photographs, with an inclined line of sight, are often used for panoramic shots and perspective views. Systematic survey photography to be used for photogrammetry and photogeological interpretation is invariably in vertical format, i.e. with a vertical camera axis. Important orientation information is recorded for each photograph; the fiducial marks enable the principal point to be found for each print and the marginal titling strip often contains photo counter, altimeter, clock, focal length and spirit level information (Figure 3.5). Additionally, the survey agency may be identified. If the photography is to be used for digital photogrammetry then data from the camera calibration certificate are essential.

Negatives are usually in 9″ × 9″ (22.5 cm × 22.5 cm) format and the survey is built up as a series of overlapping strips of photography with sidelap between the strips commonly of the order of 30%. Each strip is composed of photographs taken from a constant altitude along a flight line, with approximately 60% overlap between adjacent photographs (Figure 3.6). Flying height is usually adjusted before the start of each flight line to ensure a constant altitude above the mean ground height during the acquisition of each strip. This is done to achieve similar

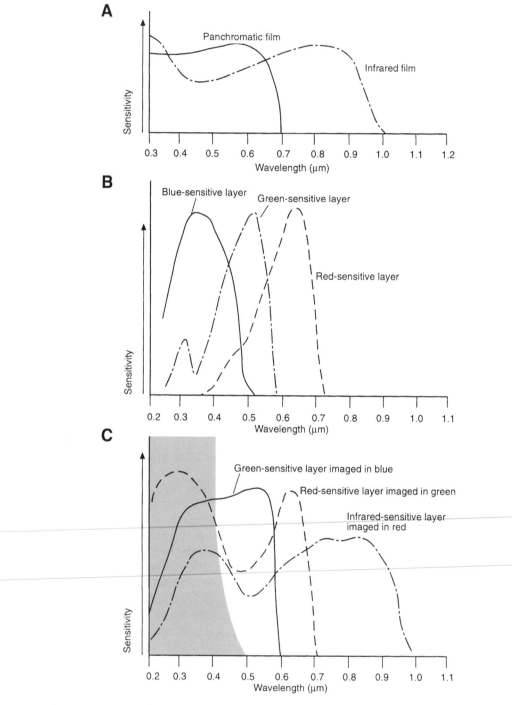

Figure 3.4 *Spectral sensitivity of **A**: panchromatic and infrared (IR) film; **B**: colour film; **C**: false colour infrared film. [Modified from: Wolf (1983)]*

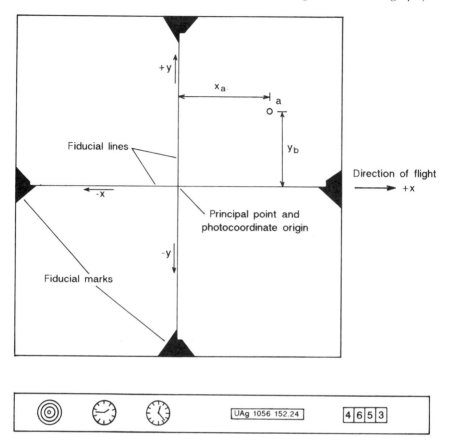

Figure 3.5   *Frame of vertical aerial photograph showing fiducial marks and how they are used to construct x–y photocoordinates. The titling strip shows a spirit level, an altimeter, a clock, a camera identifier, the focal length of the camera and a counter. [Modified from: Wolf (1983)]*

mean scales for each strip. Defects found in vertical surveys include tilt, drift and crab (Figure 3.6) and can result in complete loss of ground cover and/or loss of stereoscopic viewing capability. Tilt is recognized using the titling strip spirit level. This defect can make stereoviewing difficult and the photos cannot be used for photogrammetric purposes such as control plot construction (see Section 3.2.3) without rectification to compensate for the tilt.

The scale of each vertical aerial photograph is related to flying height above sea level (H = altimeter reading), camera focal length (f) and mean ground height (h_g) in the following way:

$$\frac{f}{H - h_g} = \text{Representative Fraction}$$

For flat ground, scale becomes larger as $H - h_g$ decreases, but this simple relation-

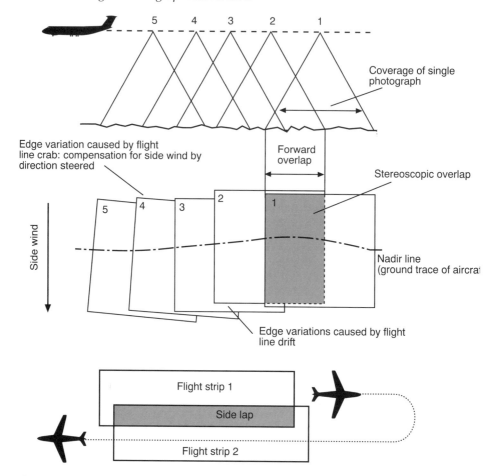

Figure 3.6 *Characteristics of vertical aerial survey. [Modified from: Lillesand & Kiefer (1994)]*

ship becomes less relevant once slope gradient and aspect are taken into account; for example, ground which slopes towards a principal point (PP) will be imaged at a larger scale than ground which slopes away (Figure 3.7). This is a consequence of the phenomenon of radial relief displacement: the apparent radial displacement away from the principal point of high ground points relative to low ground points. For these reasons the scale of even a single aerial photograph varies between wide limits. A very important consequence of radial relief displacement for the geologist is illustrated in Figure 3.8. Here the geometric distortion of outcrop pattern in hilly terrain gives a misleading impression of dip direction on a single print (Raasveldt 1959).

As an alternative to aerial photographs, the geologist may acquire the imagery by field photography. Simple multiple overlapping Polaroid photographs of stratigraphical sections can provide a very effective basis for recording complex or inaccessible sections quickly and effectively. Moseley (1981) has drawn attention

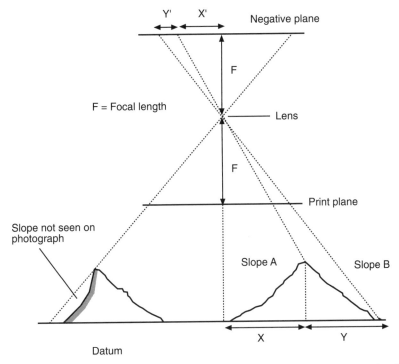

Figure 3.7 *Distortion of scale due to slope aspect on a single aerial photograph. Slopes A and B have identical dimensions on a map but their average scales on the negative are X'/X and Y'/Y respectively. Very steep slopes directed away from the principal point may not be imaged at all*

to these methods which gain added power from the capacity of modern digital scanners and image processing software to mosaic, enlarge and enhance such images. If a photogrammetric survey camera is available then accurate ground-based stereoscopic imagery may be acquired which may be photogeologically interpreted with a stereoscope, enhanced by computer for display, or modelled using digital photogrammetry (Stirling 1990).

3.1.3 Digital Data Acquisition

Satellite and airborne sensing systems all rely on an array of radiation-sensitive detectors which scan the ground beneath the aircraft or satellite track to build up a raster array of data. This raster array can be considered as a series of rows and columns of ground elements, each representing an instantaneous field of view (IFOV), for which a value related to ground brightness has been collected by the detector. This value is commonly, but not always, an eight-bit number giving a possible recording range of 0 to 255. These values are used to control the brightness of individual pixels on a computer screen, to create a black and white (grey scale) image of the scanned terrain.

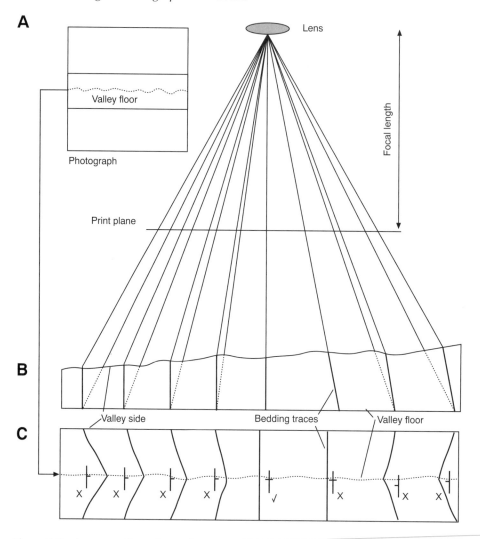

Figure 3.8 *Apparent dip and true dip on a single vertical aerial photograph.* **A.** *A single print transected by a valley.* **B.** *The geometrical relationship between valley-side bedding traces and their optical path to the camera during photo acquisition.* **C.** *The appearance of the bedding traces on the photograph together with symbols indicating correct and incorrect apparent dip directions*

The spatial and spectral resolution, and the spectral range, of such images are key characteristics for the geologist. Spatial resolution refers to the size of the ground cell represented by each data value. Ideally these need to be as small as possible. Spectral range and resolution refers to the number, and the wavelength width, of spectral bands for which data are collected. As the number of spectral bands increase then so does the number of values per pixel so that the raster matrix becomes multi-layered, similar to a multi-page spreadsheet. Improved spectral resolution permits better discrimination of rock materials because narrower spectral absorption and reflection features can be identified (Figure 3.2). Unfortunately, other factors being constant, improved spectral resolution can reduce the energy flux below the sensitivity threshold of the detector, resulting in the need to increase the size of the IFOV. Such a degrading of spatial resolution increases the likelihood that the area of the IFOV is a 'mixel', with spectral characteristics related to the several different materials that occur within it. Consequently there is a trade-off in most sensing systems between spatial and spectral resolution, and acquisition strategies have focused on attempting to get the best of both worlds.

Figure 3.9 provides information on some data types of value for geologists. At one extreme, Advanced Very High Resolution Radiometer (AVHRR) data are of interest for synthesis of large regions at low cost, but offer limited spatial discrimination. Landsat Multispectral Scanner (MSS) provides limited spectral resolution but offers the capacity for synthesis of smaller regions at better spatial resolution. The Airborne Thematic Mapper (ATM), GEOSCAN and AVIRIS instruments combine progressively more powerful spectral resolution with a spatial resolution determined only by the flying height of the aircraft. The costs of these data can only be borne by commercial organizations or large research grants. In the middle ground, Landsat Thematic Mapper (TM) and SPOT panchromatic data probably represent the currently preferred option for the geological community. TM gives good spatial and spectral discrimination, and data from old archives can cost only a few hundred pounds. The spatial discrimination of these data can be further enhanced by using merging techniques in conjunction

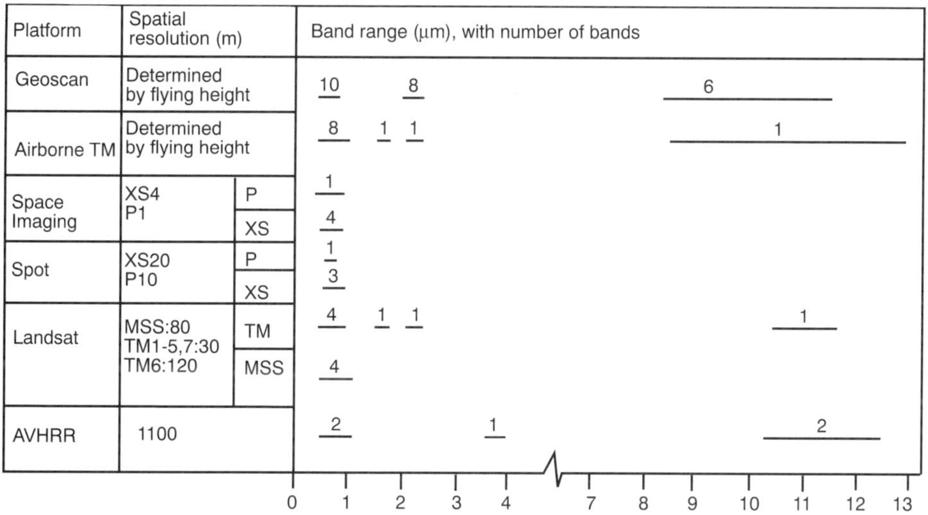

Platform	Spatial resolution (m)		Band range (μm), with number of bands
Geoscan	Determined by flying height		10, 8, 6
Airborne TM	Determined by flying height		8, 1, 1, 1
Space Imaging	XS4 P1	P	1
		XS	4
Spot	XS20 P10	P	1
		XS	3
Landsat	MSS:80 TM1-5,7:30 TM6:120	TM	4, 1, 1, 1
		MSS	4
AVHRR	1100		2, 1, 2

(Band range axis: 0 1 2 3 4 7 8 9 10 11 12 13)

Figure 3.9 *Comparison of spatial and spectral resolutions of orbiting and airborne sensors: Advanced Very High Resolution Radiometer (AVHRR), Landsat Multispectral Scanner (MSS) and Thematic Mapper (TM), SPOT multispectral (XS) and panchromatic (P), Space Imaging multispectral (XS) and panchromatic (P), Airborne Thematic Mapper (ATM) and GEOSCAN sensors*

with SPOT panchromatic data. SPOT images have the further advantage of stereoscopic capability (Barton *et al.* 1994) and the data can be used to generate digital elevation models (DEMs).

By the end of the millennium we will see improved spectral and spatial resolution from spaceborne sensors. Existing sensors will be improved: Landsat TM will acquire a 10 m resolution panchromatic band and spatial resolution of the thermal band will be improved to 60 m. Post-millennium, SPOT 5 is expected to provide 2–3 m panchromatic resolution and 10 m in multispectral mode. Many new sensors are also planned; for example, the CSIRO consortium's ARIES scanner will provide up to 64 channels of spectral data in the visible–near infrared–short-wave infrared, opening up the possibility of hyperspectral data from space. In an exciting development, December 1997 will see the launch of a new commercial satellite sensor by Space Imaging that will achieve 1 m resolution in panchromatic mode and 4 m resolution in multispectral mode with bands equivalent to TM1-4 (Dykstra 1996; Furniss 1996). As with SPOT, stereoscopic viewing and the generation of digital elevation models add to the power of the data. Competing sensors will be Quickbird and Orbimage 3, both providing broadly similar types of data (Plumb *et al.* 1997).

As an alternative to airborne or satellite-mounted sensors, the geologist can acquire digital data in the field by using a portable spectrometer. Field applications of gamma logging techniques offer clear potential (see Chapter 9), while, for the wavelength range of satellite sensors, Lamb *et al.* (1996) have shown how the PIMA instrument can be used as a rapid field identification tool when mapping mineral alteration associated with ore deposits. Lang *et al.* (1990) used similar methods to characterize various siliciclastic and carbonate strata in the southern Bighorn Basin area of the United States, and there seems little doubt that this technique can provide an important additional dataset for the examination of stratigraphical sequences. In conjunction with traditional field logging and ground-based digital photogrammetry, field spectrometry offers the prospect of total stratigraphical documentation for onshore successions.

3.1.4 Radar Data

Radar contrasts with the visible, near infrared and thermal infrared radiation captured by Landsat TM in its longer wavelength (Figure 3.1), its reliance on an artificial radiation source and its capability to penetrate cloud. Regular pulses of radiation illuminate a narrow strip of terrain to one side of the platform and perpendicular to the flight path. These are reflected back to a detector. Distance of a ground reflection location along the illuminated strip is determined by the return time of the reflected pulse, and reflection strength is governed by the orientation of the surface relative to the sensor and its texture. As the platform moves forward, the image is built up as a series of strips.

Radar imagery alone is unlikely to be of great value for stratigraphical interpretation but it can provide a valuable data layer for combining with other types of imagery, particularly multispectral satellite imagery. Geomorphology related

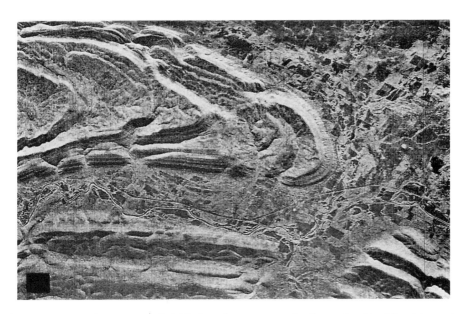

Figure 3.10 *Radar image of folded sedimentary rocks from the Ouchita Mountains, Oklahoma. Note the clear banding on the scarp faces which results from alternations of sandstone and shale. [Reproduced with permission from John Wiley & Sons, from: Lillesand & Kiefer (1994)]*

to geological structure is commonly well-defined on radar images and can be helpful in providing evidence of structural succession complementary to the material-discriminating properties of multispectral data (Figure 3.10). Surface texture can also sometimes be an aid to lithological discrimination (He & Wang 1990).

3.2 TOOLS FOR IMAGE INTERPRETATION

The main aims of stratigraphically oriented remote sensing are to identify photo-stratigraphical units, to establish their strike and dip and thereby to develop a photostratigraphical succession. Tools to achieve these aims fall into four broad categories. First, there are the principles of qualitative interpretation of photo-lithology used traditionally by the photogeologist and based on years of experi-ence in integrating fieldwork with aerial photographic interpretation (Ray 1960; Allum 1966). Secondly, there are the modern techniques of digital image enhance-ment and display that can be used to discriminate lithology (Drury 1993). Thirdly, remotely sensed data can be analysed photogrammetrically to give quantitative information on strike, dip and formation thickness (Foster & Beaumont 1992). Finally, the capacity to derive digital elevation models from aerial photography and drape the image data over perspective digital elevation models (DEMs) leads to powerful down-plunge, along-strike visualization techniques for qualitative and quantitative photogeological interpretation.

3.2.1 Lithological Interpretation from Aerial Photographs

Many of the aerial photographic characteristics of a stratigraphical succession can be linked to lithological characteristics and sometimes directly to rock type through the experience of the observer and appreciation of basic principles. Most importantly, criteria for lithological identification are area-specific because photo-geological appearance of a rock type is the result of the interaction of a complex set of local variables over time. These include climatic characteristics, seasonality, erosion, weathering and soil-forming processes, vegetation types, geomorphology and Quaternary land surface history. This is particularly important in mountain-ous regions where similar stratigraphical units may occur in a variety of climate zones, or zones with contrasting neotectonic and geomorphological histories. Criteria for rock identification must be developed locally and are not for export. Key characteristics for identification of photolithological units consist of tone, geomorphology and surface texture.

Tone on panchromatic images, or colour on true or false colour images, often relates directly to lithology and illumination. The latter is controlled by slope inclination and aspect relative to solar azimuth and zenith. Determination of the solar illumination direction for a set of aerial photographs is therefore an essential precursor to the assessment of tonal variation. Since slope aspect controls the intensity of illumination, the same lithology will vary in its tone in any area of strong topographic relief (Figure 3.11). Furthermore, because of the non-Lamber-tian nature of reflections, the same slope may have a different tone on two adjacent aerial photographs. Sometimes lithological control of colour or image tone is simple, as in the examples of highly reflective chalk, moderately reflective sand-stone, poorly reflective black shales (Figure 3.12) and red-coloured sandstones. Stratigraphical variation in colour or tone within a unit can give rise to tonal form lines which may or may not have geomorphological expression too (Figure 3.11A). These give a direct indication of bedding thickness and dip direction. More commonly, the development of weathered surfaces or residual soils, lichen or more fully developed vegetation cover result in photographic tones or colours which are characteristic of the presence of the substrate but bear no direct relation-ship to the bare rock.

Important geomorphological characteristics include drainage density and pat-tern, within formation relief and relief contrast with adjacent photolithological units. Drainage density and pattern are a function of permeability and structure. Permeability, in conjunction with rainfall, controls the density of surface drainage channels (Figure 3.12) while the joint system, bedding dip and nature of the succession within the photolithological unit influences the drainage pattern (Fig-ure 3.13). Local relief within the outcrop of the unit will also be related to the nature of the succession and erosion history. Significant relief contrast with adjac-ent units is commonly related to a contrast in their susceptibility to erosion, and the identification of strike-parallel scarp features and dip slopes is often crucial to interpretation (Figures 3.10, 3.12 and 3.13). An important characteristic is the development of form line features through the erosion of beds with contrasting resistance to erosion (Sgavetti 1992). The size of these scarplet features gives a clue

Figure 3.11 *Horizontal Lower Miocene outer ramp carbonates near Ragusa, southeast Sicily (H.M. Pedley pers. comm. 1997) showing contouring form lines. Note the variations of tone due to contrasting vegetation densities (A and B) and variable sun angle (B and C). [Print 4108, titling strip: 3NA/98/3PG/21.5.43/1530/F24/30,000 S. Supplied by Air Photo Library, Department of Geography, University of Keele. Reproduced by permission from: Ministry of Defence]*

to bedding frequency and offers the chance to determine dip accurately. Sgavetti shows how different types of photofacies relief contrast can be recognized in aerial photographs to help indicate the nature of facies transitions and to build up the weathering profile of the stratigraphy (Figure 3.14).

Surface texture, following petrography, can be defined as the size, shape, orientation and mutual relationships of objects on the ground surface. Texture may vary with the season and the scale of photography. It is, therefore, a mixed bag of characteristics with geological significance that may be direct, indirect, very subtle or completely lacking. The most important element of texture is the tendency for a lithology to form a smooth weathered surface or a rough rocky surface. As a rule of thumb, the larger the blocks, the coarser the surface texture (Figure 3.14). This texture may have a shape element controlled by joint pattern and frequency.

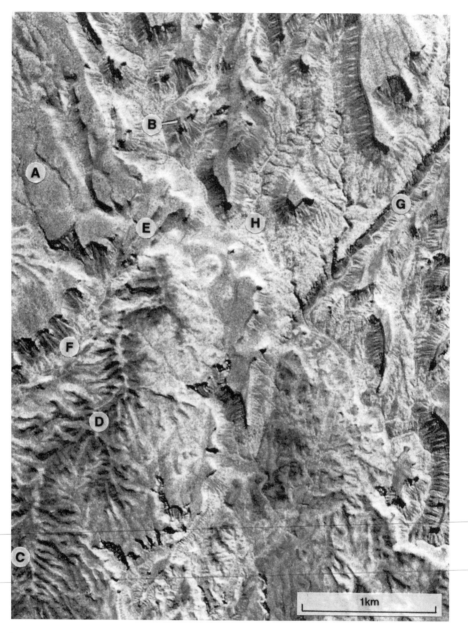

Figure 3.12   *Faulted and gently dipping shales and sandstones, Utah. Scarp faces and low drainage densities on dip slopes are developed in limestone (A) which overlies fine-grained sandstones and shales at B. Shales have a darker tone and a more closely spaced drainage pattern (C and D) and overlie paler sandstone (H). Geomorphological evidence of faulting is seen at E, F and G. [Reproduced with permission from US Geological Survey, from: Ray (1960)]*

Figure 3.13 *Photogeological units distinguished by drainage densities, joint patterns and tone in a gently dipping sedimentary succession, Utah. Massive sandstones have pale tone and are well-jointed (B). Overlying mudstones and sandstones have a darker tone and form a scarp face and dip slopes (A). Poorly resistant limy sandstone, mudstone and limestone are easily eroded and have fine drainage texture (C). [Reproduced with permission from US Geological Survey, from: Ray (1960)]*

Texture may be strongly influenced by vegetation type and density, elements which often have geological significance because of their control by water availability. For example, in the carbonate sequences of the Costa Blanca in Spain, water retentivity is significantly enhanced by marl components in the succession, and Figure 3.15 shows a subtle contrast in texture across the Amadorio Fault line. The contrast is in the size and frequency of small trees and bushes, and the control is provided by their greater density on marl compared to well-drained limestone. Classic African examples show substrate control of surface vegetation distribution which helped pick out a fold structure in a region of deep surface weathering and poor exposure (Figure 3.16).

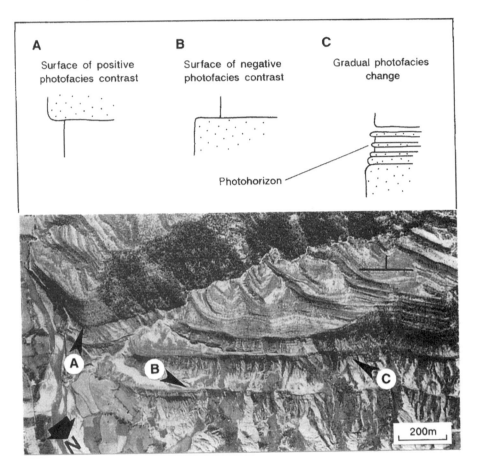

Figure 3.14 *Photostratigraphical characteristics of erosion profiles in Upper Cretaceous tur-bidite sequence from the south-central Pyrenees. [Modified from: Sgavetti (1992)]*

3.2.2 Image Processing for Lithological Discrimination

Before any photolithological discrimination can be achieved using digital imagery, the effects of electronic noise, atmospheric contributions to the imagery and geometric errors must be removed by pre-processing. Contrast enhancement of grey scale and colour images may then lead to the identification of photogeological units. Inspection of interband correlation coefficient matrices can be used to optimize the information content of colour composites for lithological discrimination, and the incorporation of thermal infrared data in colour displays may be explored. Knowledge of spectral signatures of target materials can be used to develop colour composite and band ratio images for their identification. Principal component methods provide a means of discriminating photolithologies and a procedure for the contrast enhancement of colour composites. Finally, image

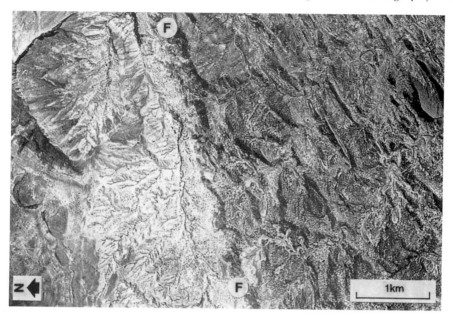

Figure 3.15 *Limestones north of the Amadorio Fault line (F–F) are better drained and support fewer bushy trees than the marls and limestones to the south. Inward-facing scarp features of the Amadorio Dome can be seen in the north. [Print 10449 (VV AST6 Roll118 1370PMG 4JULY56 54AM78). Reproduced with permission from: Servicio Geografico del Ejercito, Madrid]*

classification offers a route towards the generation of a geological map from a digital image. Different scenes require different strategies and in practice various combinations of techniques may be required.

Pre-processing

Radiometric correction of imagery is performed in order to remove electronic noise and atmospheric effects from the raw data. Examples of noise include (1) speckle, that is the random scatter of anomalously high or low data values; (2) line drop outs, that is scan lines of anomalously high or low values; and (3) striping, that is recurrent patterns of anomalously high and low scan lines due to variation in sensitivity among an array of scanning detectors. A variety of cosmetic image processing techniques are available for correcting these defects. More significantly, the data will need correcting for atmospheric contribution which typically increases with decreasing wavelength. By sampling areas of deep shade in the image, which would give zero data values for a planet with no atmosphere, the atmospheric contribution can be assessed for each band and subtracted from the raw data. Since the atmospheric contribution varies with wavelength, this is an important correction to make prior to the use of band/band ratios for rock discrimination (Crippen 1988).

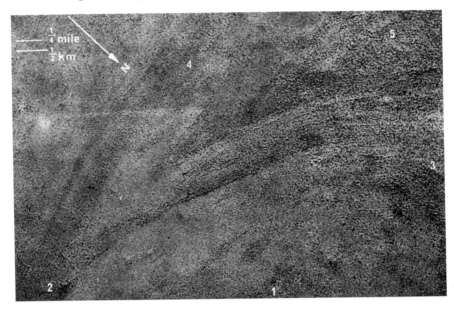

Figure 3.16 *Soil tones and variation in vegetation density help to pick out structure and variations in the photostratigraphical succession on a deeply weathered erosion surface, Botswana. [Reproduced with permission from J.A.E. Allum & Pergamon Press, from: Allum (1966)]*

Geometric correction is performed in order to correct for geometric distortions introduced during data acquisition. The correction aims to place image points in their correct mutual positions within a cartographic grid system. This is achieved by matching image ground control points (GCPs) to their co-ordinates on a map. Polynomial equations relate the GCP image co-ordinates to the map co-ordinates and are used to calculate new grid co-ordinates for the rectified image. This can be envisaged as the rotation and warping of the old image pixel array to fit beneath a new map grid. The new pixel locations in the new map grid are then assigned data values either by proximity, using the data value of the nearest pixel from the warped image (nearest neighbour), or by some form of interpolation (Richards 1993). Interpolation procedures help to avoid the development of geometric artefacts in the output image but nearest-neighbour methods should be used if the data numbers are to be used later to characterize lithologies, for example by ratio or classification methods.

Lithological discrimination by single band display

Contrast enhancement is invariably necessary for sensible image display because the image data histogram commonly shows positive skew and a limited range. For example, Landsat TM data have a digital resolution from 0 to 255, but a skewed data range of, for example, 5 to 115 will result in an image with poor display characteristics because it uses only 110 display grey levels out of a possible 256.

Manipulation of the relationship between input data numbers and output display values results in contrast enhancement. Such enhancements can be automatic, as with histogram equalization, by which equal numbers of pixels are assigned to all 256 available output tones; alternatively the enhancement can be controlled inter- actively by the user. Trial and error will show, for example, which of the seven grey scale images that can be produced using Landsat TM data give information that is stratigraphically significant.

Lithological discrimination using band ratios

The same two principal factors that govern data number distribution in a digital image also control tone variation in an aerial photograph: surface material and slope aspect. Laboratory measurements of reflectance spectra for different rock types show that there are strong contrasts in the shape of spectral reflectance curves so that they may be regarded as 'spectral signatures' (Figure 3.2). There are a variety of obstacles in the way of using these to help analyse satellite data (Cervelle 1991; Price 1994) and coarser 'spectral signatures' may be constructed from data values over known rock outcrops. These signatures will provide more realistic criteria for lithological discrimination because they come from real rock surfaces in the area of study, with a range of weathering veneers and lichen covers. Such spectral signatures collected from an image will show a range of variation from one location to another due to variation in slope aspect; well-illuminated slopes will have elevated values for all visible–NIR bands and vice versa for shady slopes.

As a partial solution to this aspect effect, band ratios may be calculated, which effectively 'cancel out' the effect of differential illumination intensity. The solution is only partial because of the variable non-Lambertian nature of reflection from most natural materials, atmospheric scatter and illumination of terrain by diffuse reflections from adjacent surfaces. The bands chosen for ratio calculation will be those most effective at discriminating between the lithologies present; a classic example is the use of TM 3/1 ratios to discriminate iron minerals (Figure 3.17). The appearance of bright pixels on a 3/1 image may well relate to iron minerals in the stratigraphical succession, although there is also commonly a strong 'background' pattern due to geomorphology and human activity which influence variable vegetation cover, gullying and landslides, all of which may expose red soil. Additionally, although a band ratio may be carefully designed to discriminate a specific rock type, there is no guarantee that other spectrally similar materials may not also be present: they may be different rocks with the same characteristic ratio, but a different albedo, or they may be materials of no geological significance at all. The use of data with improved spectral resolution provides the opportunity for designing ratios that will more sensitively and unambiguously discriminate ma- terials by their spectral signatures (Figure 3.2).

As an extension of the ratio concept, three-component triangular plots can be used to characterize the relative values of three satellite data bands. This approach was used by Damanti (1990) to analyse sediment sources and identify sediment mixing in a series of Quaternary fans in Argentina. Alternatively, a hue value can

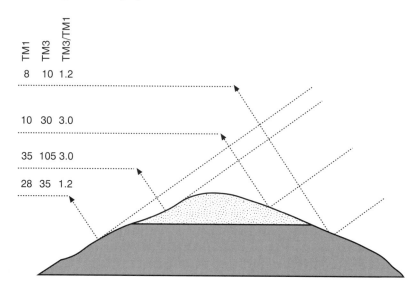

Figure 3.17   *Cross-section showing red sandstone overlying shale, and the effect of slope aspect relative to illumination direction. The ranges of TM1 and TM3 data overlap for the two rock types but the ratios discriminate independent of illumination*

be generated from three input bands using an intensity–hue–saturation transform (Liu & Moore 1990). Three input bands are coded to red, green and blue, and a hue is calculated for each pixel using the relevant band data values. The resultant hue, in the continuum red–yellow–green–cyan–blue–magenta, can be regarded as a significant measure of the ratios of the three input bands. Saturation and intensity are related to slope aspect and illumination effects, and are discarded.

All these approaches use a single value for each pixel to control display tone. However, the greater the range of data that contributes to an image, the more useful it is likely to be for geological interpretation. This can be achieved by colour display.

Lithological discrimination using colour images

Colour display increases the information contained by an image because the intensity of the three primary display colours can be controlled by either three raw data bands, three simple ratios or three different hue values.

The simplest colour images are generated by choosing three spectral data bands to control the intensities of red (R), green (G) and blue (B) in the display. The colour of each pixel depends on the blending of these three primary colours, controlled by the three different data values. As with single band imaging, the contrast of each colour in a three-band image requires enhancement of individual bands to produce an image with maximum contrast. A common display combination uses data from Landsat TM bands 3, 2 and 1 (spectral red, green and blue, abbreviated to RGB) to control the intensities of red, green and blue in the display (321:RGB), thereby simulating real colour. There is no a priori reason for choosing this, or any other, band combination, and band selection to optimize discrimination becomes an important issue. Experimentation with the relatively small numbers of combinations possible using TM data can give very useful results for lithological discrimination (Crosta & Moore 1989a). In the case of hyperspectral data the researcher is confronted by large numbers of bands and so experimentation, at least in these cases, is not a viable option. Two approaches can be taken: (1) the three display bands can be chosen to maximize their variance; or (2) band choice may be directed to known spectral characteristics of rock materials.

Data bands are often highly correlated, and the display of three highly correlated bands will result in an image which is pseudo-grey scale with little colour variation. A method is required which overcomes this problem and maximizes the information content of the three bands selected for display. Sheffield (1985) proposed the use of the band combination with the greatest data volume in the resultant scatter plot ellipsoid and concluded that bands 1, 4 and 5 most commonly provided optimum information content. Similarly, Drury (1993) proposed that the band correlation matrix for, for example, the six visible–short wave infrared TM bands may be inspected, and the three bands with the lowest mutual correlation chosen as inputs to the RGB image. This is most likely to be successful where geological materials dominate the scene. Another argument commonly made is that a TM 147 colour composite optimizes information because of the wide spectral separation of the bands chosen (Ferrari 1992): spectrally separated bands, being less likely to be correlated, will provide complementary rather than repetitious information. For TM imagery another useful approach is to take the two most poorly correlated visible–near infrared (NIR) bands, and to use these in combination with band 6, the thermal infrared (TIR) band. Since thermal characteristics are often poorly correlated with visible–NIR reflectance (Drury 1993), this approach will maximize the variability of the image. The inclusion of band 6 degrades the sharpness of the image because of its 120 m pixel size but this may be a worthwhile sacrifice to achieve good rock discrimination on a regional scale.

Another way of making band selections is to compare the 'spectral signatures' of the photolithological units within the scene and to choose band : colour allocations that optimize the separation of the rock types. The effects of differential illumination will still be present in such images but this may be mitigated by the use of three band ratios, each discriminating a lithological characteristic, used as inputs to a colour image. Choice of optimum ratios can be made on a purely statistical basis, after masking vegetation areas (Chavez *et al.* 1982). Alternatively, the ratios can be chosen for their relationship to known reflection and absorption features of geological materials. For example, Drury (1993) suggests for TM imagery that 3/1 and 5/4 ratios are both expected to be high for ferric and ferrous iron minerals respectively, while 5/7 will be high for clays. This certainly seems to be a useful technique, at least for sequences with a significant volcanic component (Davis & Berlin 1989; Beratan *et al.* 1990). Unfortunately, geologically significant geomorphological information is lost in the ratioing process. Crippen *et al.* (1988) show how topographic shading can be retained in such colour ratio images.

As an alternative to colour control by three ratios, three hues can be generated from the same dataset using three different input colour : band combinations.

These encapsulate a large amount of the spectral variation in the scene and the three resultant hues can then be used to control an RGB colour display, generating a hue:RGB image (Liu & Moore 1990).

Lithological discrimination by principal components

All the difficulties encountered in the construction of colour images for material discrimination relate to two separate fundamentals: the problem of data redundancy in multispectral imagery and the need to remove topographically related variations in illumination from the data values. Principal component (PC) methods offer an alternative strategy for approaching these problems.

To show how PC methods work we can consider a simplified example. A scene contains a wide variety of materials among which are materials O and X. These occur on sunny slopes and shady slopes and have been scanned by an imaginary sensor with two spectral bands I and J. The materials have strongly contrasting spectral signatures (Figure 3.18A) and plot in different parts of a band I/band J scatter plot (Figure 3.18B). Principal component analysis transforms the axes of the data to be parallel to the two principal axes of the data ellipse. That parallel to the major ellipse axis is called the first principal component axis (PC1) and that parallel to the minor axis is the second principal component axis (PC2). The PC1 axis measures the main variance in the data, while PC2 provides a measure of scatter perpendicular to the PC1 axis. All pixels now have a PC1 and a PC2 value in the transformed co-ordinate system. By considering the positions of the 'sunny' and 'shady' pixels for materials O and X, it is apparent that PC1 values are similar

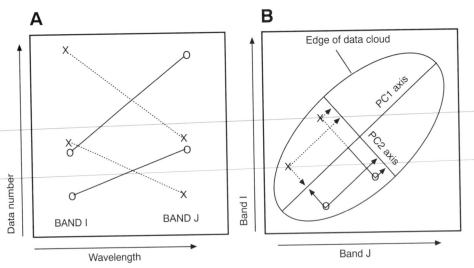

Figure 3.18 **A.** *Spectral signatures of materials O and X under conditions of good and poor illumination as detected in spectral bands I and J.* **B.** *Band I/J scatter plot showing data for poorly and well-illuminated materials O and X, and their projected positions on the PC1 and PC2 axes*

for similarly illuminated materials irrespective of their spectral characteristics. By contrast, PC2 functions as an effective discriminator of materials O and X irrespective of illumination. It should be noted that while, in this example, PC1 is an index of illumination, it may also be a measure of albedo. High-albedo materials will have high values for bands I and J and consequently high PC1 values; low-albedo materials will have low PC1 values. The principal component values in this synthetic example could be used to control image grey tones, generating a principal component image. A PC1 image will be an albedo and illumination image, while a PC2 image will be an effective discriminator of materials O and X.

For a three-band sensor, like SPOT XS, similar arguments can be made with a three-dimensional scatter plot. A triaxial data ellipsoid provides the axes from which principal components 1, 2 and 3 are calculated. If we consider Landsat Thematic Mapper (TM) data, six principal components can be calculated for the six wavebands from the visible–short wave infrared spectral range, but graphical illustration becomes impossible once more than a three-band dataset is considered. Nevertheless the principal component values can be of real help in discriminating materials. Values for principal components 1–6 measure progressively greater departures from correlation and encompass progressively smaller amounts of the variance. Following the arguments developed above, all of principal components 2–6 may be of assistance in discriminating materials. As the higher order principal components contain progressively less correlated data then information related to noise, speckle, striping and line drop outs tends to dominate. Consideration of eigenvector loadings, measures of the contribution of each of the original data bands to each principal component, can give greater insights into the interpretation of principal component images (Crosta & Moore 1989a,b; Loughlin 1991).

To illustrate the usefulness of PC analysis for stratigraphical reconnaissance, Figure 3.19 shows the southern margin of the Miocene Sorbas Basin, southern Spain (the approximate location is given in Figure 3.32). PC1 shows topography but is poor for rock discrimination, while PC2 helps locate the basal Messinian unconformity and the outcrop of marls of the Abad Member which pinches out southwards towards the basin margin. Pale tones within the dark-toned Abad Member discriminate an outlier of Yesares gypsum and previously unmapped inliers of the Azagador Member and Tortonian sediments (McNulty 1995; J. Wood pers. comm. 1996). PC6 is dominated by noise but still discriminates lithology better than PC1.

The discrimination power of principal component images is optimized if they are used as inputs to colour composites. For example, with Landsat TM data, principal components 2, 3 and 4 are commonly good material discriminants, and their combination as a colour composite can provide a powerful means of optimizing colour variation of an image (Settle 1984; Quari 1992). The resultant colour variations in the imagery can, however, be difficult to interpret in terms of the spectral characteristics of the materials, although field checking will often reveal the geological significance of the colour distributions. Alternatively, principal components can provide a means of optimizing contrast enhancement of standard three-band colour images for material identification. Normally, because of inter-band data correlation, vast amounts of data display space within the triaxial RGB

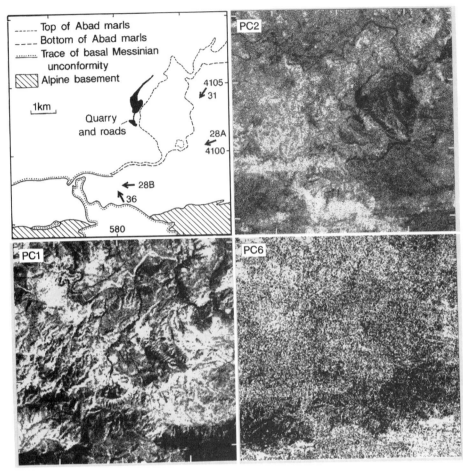

Figure 3.19   *Illustration of the discrimination of stratigraphical relationships near the southern margin of the Sorbas Basin, southern Spain, using principal component analysis. The map shows selected geological boundaries within the upward Messinian succession from the Azagador (limestone), through Abad (marl) to the Yesares (gypsum) members, and the location of gypsum quarries and roads (modified from: Instituto Geologico y Minero de Espana 1974). PC1 shows relief and the overstep of basal Messinian strata onto basement rocks but fails to discriminate the Abad marls. PC2 shows the overstep of Tortonian and Alpine basement rocks (dark and mid-grey tones) by lowest Messinian strata (brightest). The outcrop and southerly pinchout of marls of the Abad Member is also clearly seen (darkest tones). Pale tones within this dark-toned region relate to an outlier of gypsum and unmapped inliers of Azagador Member and Tortonian sediments. PC6 picks out the unconformity trace but is otherwise poor and dominated by noise. The approximate location of the area is shown in Figure 3.32. Arrows show directions of perspective views in Figures 3.28, 3.31 and 3.36. The images were produced using ER Mapper software*

C Common remote sensing systems

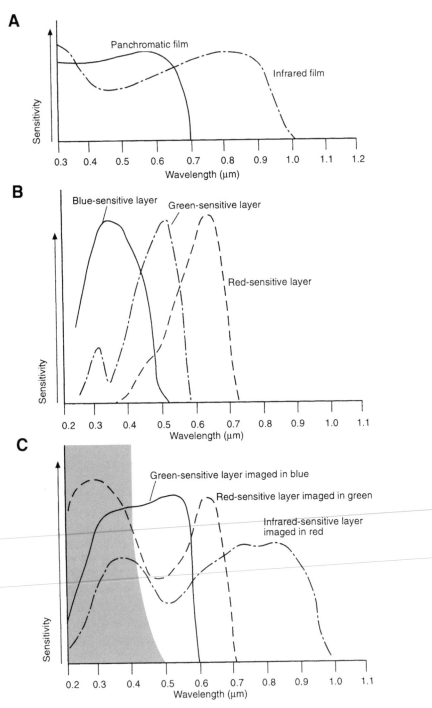

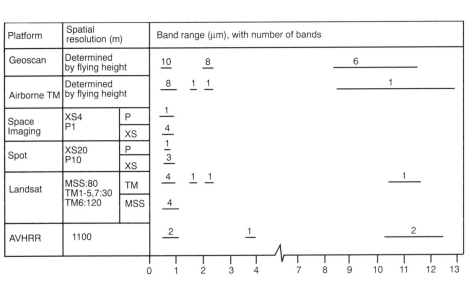

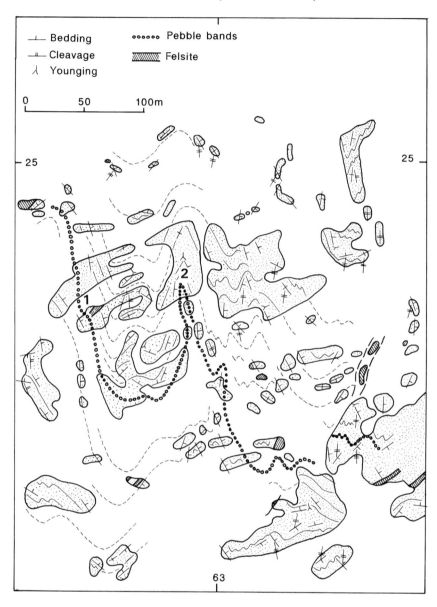

Figure 4.4   *Geological map of Cairn Poullachie, Strathnairn, showing a conglomeratic horizon (the Gairbeinn Psammite) within the Grampian Group of the Scottish Central Highlands deformed by two major generations of folds. Pebbles on the limbs of the folds are deformed to give axial ratios in the order of 19:1 and bedding is strongly attenuated such that any cross-bedding is completely obliterated (Locality 1). In the cores of the later set of folds, cross-bedding with recognizable cut-offs is preserved (Locality 2 in Figure 4.3)*

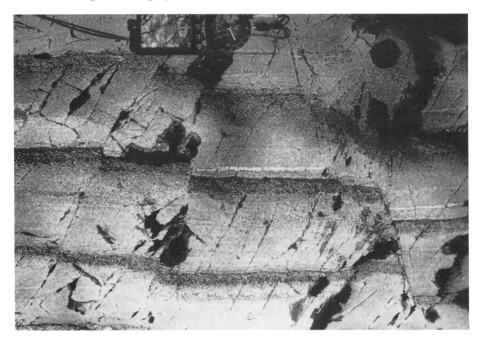

Figure 4.5   *Repeated graded units or rhythmites from the Grampian Group, Central Highlands of Scotland. [Photograph: P. J. Carey]*

able to demonstrate that the layering is in fact bedding. This is best done by demonstrating the presence of other sedimentary structures associated with the layering.

Distinguishing between sedimentary and tectonic structures is never easy. A structure resembling cross-bedding can result from tight folding, and mullion structures can resemble sedimentary flame structures or load structures. Conversely, folds can occur as the result of slumping or de-watering in sedimentary rocks. Hobbs *et al.* (1976) list a number of criteria for distinguishing between soft sediment and tectonic structures, of which three are considered to be the most reliable:

1. the presence of undeformed burrows within the deformed layer;
2. the presence of a metamorphic foliation or mineral grains deformed around the hinge of the folds; and
3. the presence of deformed fossils showing variations in strain that are systematic around the fold.

Other criteria pointing to a sedimentary origin for structures include:

1. folds closing in opposite directions in close proximity;
2. truncation of folds by overlying undeformed beds;
3. structures with orientations unrelated to the regional tectonic stress systems;

4. the absence of related joints and veins;
5. a lack of axial planar fabrics;
6. the occurrence of folds restricted to single beds; and
7. the chaotic nature of the structures.

Unfortunately, under certain circumstances tectonic folds can show many of these features and these criteria cannot be used singly to indicate a sedimentary origin of the layering.

Thicknesses of individual beds and lithologies may have been substantially altered by deformation. Many beds will have suffered layer-parallel shortening prior to folding, and during folding the limbs of the fold may suffer homogeneous flattening and thence thinning (Figure 4.6). Shearing will also result in thinning of the layering, even if shearing is oblique to the layering. The extent of these changes is difficult to estimate without the presence of strain markers. These can vary from deformed pebbles to reduction spots or deformed fossils. There are several techniques, beyond the scope of this chapter, that can be used to establish the amount of strain and its orientation (Ramsay & Huber 1983).

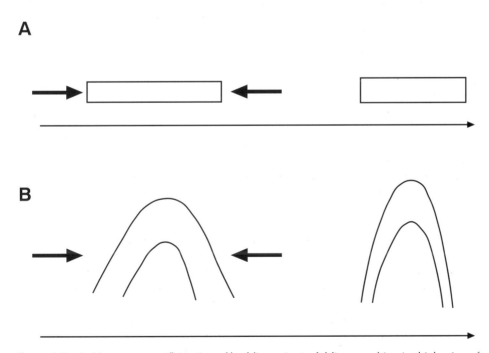

Figure 4.6 **A.** *Homogeneous flattening of bedding prior to folding, resulting in thickening of beds.* **B.** *Homogeneous flattening during folding, resulting in thickening at the hinge of the fold and thinning of the limbs*

4.1.3 Chemostratigraphy

Geochemistry is an important tool in distinguishing between different strati-graphical horizons, particularly if the chemistry can be related to variations in the nature of the original sediment. Its use as an aid to stratigraphical correlation in the highlands of Scotland, on both a regional and a local scale, has been discussed in detail by Lambert *et al.* (1981, 1982), Winchester & Max (1996), Hickman & Wright (1983) and Haselock (1984). Geochemistry proved very useful in resolving the controversy over the presence or absence of Lewisian Inliers within the Moine (Winchester 1971).

It is generally considered that low-grade, greenschist or lower amphibolite facies, metamorphism does not appreciably alter trace element concentrations within sediments (Krauskopf 1979). From whole rock, trace and major element analysis of samples from a type area of an individual stratigraphical unit, it is possible to establish a characteristic chemical composition of a unit and determine probable stratigraphical positions of unknown rock units (Hickman & Wright 1983; see Chapter 7).

4.1.4 Magnetostratigraphy

Magnetostratigraphy is generally considered to be the use of reversals of the earth's magnetic field to determine the stratigraphical age of rocks or the use of the orientation of remnant magnetism in a rock to determine polar wandering curves and thence palaeolatitude. However, variations in the magnetite concentration in rocks can be used as an aid to lithostratigraphical correlation via ground magnetic surveys of total magnetic field strength (Haselock & Leslie 1992). Magnetite occurs in sediments as detrital grains often concentrated in heavy mineral seams. It is therefore a reflection of the provenance of the clastic portion of the lithology and a good lithostratigraphical criterion (Figure 4.7). The presence of magnetic minerals such as magnetite increases the magnetic susceptibility of a particular lithology and increases the total magnetic field strength measured in the field with a proton precession magnetometer. The susceptibility is largely unaltered in low- to medium-grade metamorphic rocks.

4.1.5 Recognition of Unconformities within Polydeformed Terranes

There have been a number of cases in which the description of unconformities within polydeformed terranes has led to some controversy. The problems arise as the result of trying to correlate fold phases from one area to another and one group of rocks to another. Essentially, the presence or absence of an unconformity within a complex terrane rests on the recognition of more phases of folding within the 'basement' rocks than in the 'cover' sequence, or the recognition of structures in the basement have been reworked by structures found in the cover. With major crustal shortening a situation could have arisen where rocks of the same age were

91

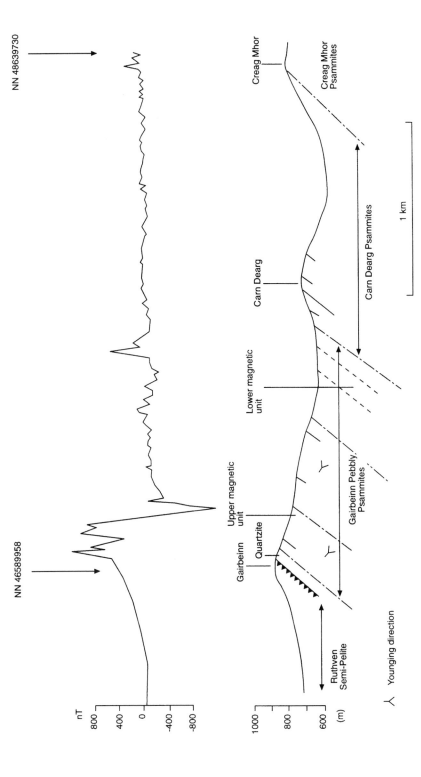

Figure 4.7 Total magnetic field strength varying with stratigraphy. [Modified from: Haselock & Leslie (1992)]

deposited a large distance from each other and only one set of rocks were affected by orogenic events. Later events could then have juxtaposed the two groups, producing an apparent angular unconformity. Alternatively, during a single orogenic event the deformation may have produced recognizable fabrics only in the less competent lithologies. Such an explanation was put forward to explain differences in fabrics found in the Skiddaw Slates and Borrowdale Volcanics of the English Lake District (Soper & Moseley 1978). Here the existence of an unconformity had been proposed because penetrative fabrics were recognized in the Skiddaw Slates but were absent from the Borrowdale Volcanics.

Unconformities within polymetamorphic terranes are typically marked by thrusts or slide zones as they act as a focus for high strain and mylonite formation. It is therefore difficult to pinpoint the precise contact. Examples include the Moine/Lewisian boundary in the Northern Highlands of Scotland, and the Hjemsoy unconformity in Finnmark (Ramsay *et al.* 1985); in both these cases the primary unconformity is preserved only locally.

4.2 STRUCTURAL ANALYSIS

The structural history of an area can be determined from a study of the distribution and geometry of minor structures or small-scale structures. It is important therefore to collect as many data in the field as possible and to recognize the limitations in the distribution of those data. For this reason it is important to show the approximate limit of outcrop, in complex terranes, so that the extent of the factual data is clear. Form surface mapping is also an important technique, where the trace of any penetrative surface is shown on the map (Figure 4.4). The form line on the map does not necessarily represent a single surface but the overall structure. As many as possible of the orientation data should be accurately plotted on the map at the locality at which they are recorded and preferably at the time they are measured. Data should also be plotted on an equal-area stereographic projection, again preferably as they are recorded in the field notebook. Later stereographic analysis may mean dividing the data into subareas, but plotting data in the field can often allow preliminary interpretations to be made which will assist in the collection of further data.

4.2.1 Vergence of Minor Structures

One of the fundamental assumptions of the study of minor structures is defined as Pumpelly's Rule, which states that the amount and direction of plunge of a major structure is indicated by the orientation of the axes of the minor folds associated with it. Minor folds in profile have an S, M or Z geometry depending on which part of the major structure they occur. This geometry can be expressed in terms of fold vergence (Figure 4.8; Bell 1981). Careful mapping of the location and geometry of minor folds can lead to an interpretation of the geometry of the major structures. Folds must be viewed in a consistent direction, down the plunge of the fold axis. In

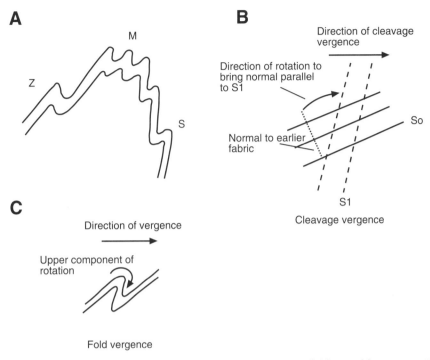

A

M

Z

S

B

Direction of cleavage
vergence

Direction of rotation to
bring normal parallel
to S1

Normal to earlier
fabric

So

S1

Cleavage vergence

C

Direction of vergence

Upper component of
rotation

Fold vergence

Figure 4.8   **A.** *Geometry of minor structures around a major fold.* **B.** *Fold vergence.* **C.** *Cleavage vergence [Modified from: Bell (1981)]*

complex areas it is important to be able to demonstrate that Pumpelly's Rule does actually apply or at least establish the relationship between the minor and the major structures.

When only one phase of folding is involved, the relationship between cleavage and bedding can be used to establish the geometry or vergence of the major folds and hence establish the presence of anticlines or synclines. Overturned limbs of reclined or recumbent folds can also be readily identified as the cleavage related to the fold dips more shallowly than the overturned bedding (Figure 4.9). The fabrics must be observed in the fold profile, i.e. in a plane perpendicular to the fold axis or cleavage/bedding intersection.

Fabrics in deformed rocks can be defined by discrete fractures in the rock, preferred orientation of elongate or platy grains, a crystallographic preferred orientation of grains, a preferred orientation of grain aggregates, or compositional variations. The fabrics can be planar (S), defining foliations, linear (L) or any combination of linear and planar (L/S) (Flinn 1965). Foliations fall into three broad types: gneissosity, schistosity and cleavage. Gneissosity is the result of compositional layering of metamorphic and/or deformational origin and involves the large-scale recrystallization of the rock that occurs at relatively high grades of metamorphism. Schistosity occurs at middle grades of metamorphism and is the result of a parallel alignment of tabular or elongate minerals such as micas or

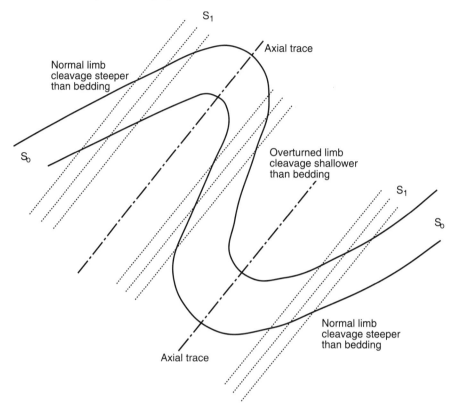

Figure 4.9   *Relationship between axial planar cleavage S_1 and bedding S_0 on the limbs of an overturned fold*

amphiboles. The term cleavage covers a variety of structures which may be penetrative, affecting all parts of the rock, or non-penetrative and spaced. This classification depends, however, on the scale of observations, as penetrative structures in hand specimen may be spaced in thin section. The same fabric may be penetrative in one lithology but spaced in another because of contrasts in competency associated with composition. The main type of penetrative fabric is slaty cleavage, while spaced fabrics include crenulation cleavage produced by microfolding, fracture cleavage produced by closely spaced fractures or joints, and pressure solution cleavage produced by differential solution of, for example, quartz and mica.

Cleavages and schistosity are generally parallel to the XY plane of the finite strain ellipsoid and this plane coincides with the axial surface of a fold (Price & Cosgrove 1990) However, the foliations are commonly not exactly parallel to the axial surface and may form convergent or divergent fans. These fans can vary from lithology to lithology as a result of competency contrasts and cleavage refraction (Figure 4.10). Cleavages can also form as conjugate sets symmetric about the axial plane of a fold (Price & Cosgrove 1990).

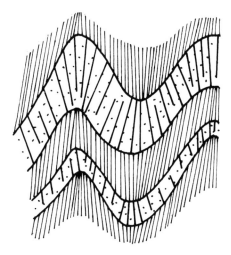

Figure 4.10   *Cleavage refraction between beds of contrasting lithology (alternating psam-mites and semipelites)*

Wilson (1982) describes the use of polishing or grooving and displacement of early veins to indicate movement between bedding planes during flexural slip folding. In a flexural slip fold the upper beds must slip over the lower beds and therefore it is possible to tell if the beds have been inverted (Figure 4.11). In areas of more than one phase of deformation where sedimentary structures are not preserved it is only possible to tell whether beds have been overturned as a result of the folding to which the minor structures relate, and does not necessarily imply that the beds are stratigraphically inverted. Way-up or younging evidence is required before any stratigraphical interpretation can be made.

4.2.2 Facing Directions

A fold faces in a direction normal to its axis, along the axial plane, and towards the younger beds. This coincides with the direction towards which the beds face at the hinge. A normal, upright fold faces upwards, while an anticline closing downwards, faces downwards. If a cleavage is parallel to the axial plane of the fold, the direction towards which the beds young within the cleavage plane will show the direction in which the fold itself will face (Shackleton 1958; Figure 4.12). A fold will face downwards only if it affects beds that have already been inverted by an earlier fold event or if an earlier fold closure has been re-articulated by a later deformation. The facing of a fold will change as one passes across the axial trace of an earlier or a later fold. Therefore recording the cleavage facing within an area can pick out the axial traces of the major structures. Reversals and duplication of the stratigraphy can thence be identified. Shackleton (1958) used the concept of facing to elucidate the structure and stratigraphy of the highland border area of the South-West Highlands of Scotland (Figure 4.13). The structure is dominated by the

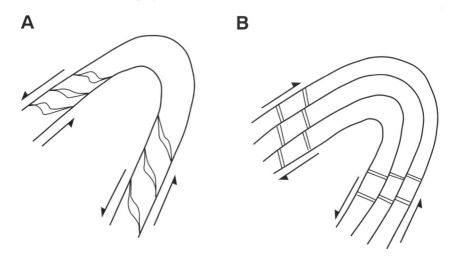

Figure 4.11   *Indicators of overturned beds in flexural slip folds. **A.** Development of tension gashes. **B.** Displacement of early veins. [Modified from: Wilson (1982)]*

inverted limb of the D_1/D_2 Tay Nappe, consistently facing subhorizontally to the SSE across the gentle D_2 Cowal Antiform. The Tay Nappe or Aberfoyle anticline faces steeply down to the SSE immediately north of the Highland Boundary Fault (Holdsworth & Roberts 1984).

4.2.3 Fold Generations and Interference Structures

Mapped folds and related structures should be grouped together as a generation if they are thought to have formed during the same phase or stage of deformation. The phase of deformation may be part of a continuous and progressive deformation or successive phases may be separated by millions of years. The grouping should be done on the basis of style, orientation and overprinting relationships. A description of the fold style includes all the geometric features of the fold such as its amplitude to wavelength ratio, layer-parallel thickness variation, interlimb angle, variation in curvature, relationship between successive layers and the presence of related foliations and lineations. Style varies considerably with rock type and with amount of strain, so the concept must be treated with some caution. However, a combination of style and orientation may define a generation of structures.

There are many types of lineations that may occur in deformed rocks and it is important to recognize those which are important to the structural history and those which are not: (e.g. those formed by wind blasting or glacial striations). Park (1989) suggests that lineations can be divided into five groups:

1. slickenside striations which give an indication of the direction of movement on faults or on the limbs of flexural slip folds;
2. axes of parallel crenulations or small-scale folds;

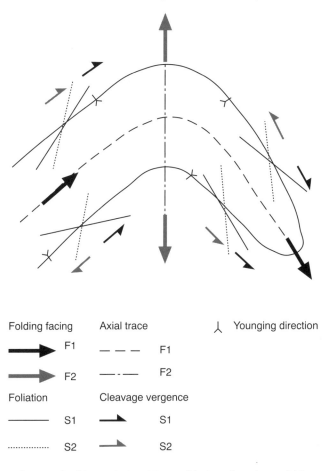

Folding facing   **Axial trace**   人 Younging direction

➡ F1   — — — F1

➡ F2   —·— F2

Foliation   **Cleavage vergence**

———— S1   ➤ S1

·············· S2   ➤ S2

Figure 4.12 *Cleavage/bedding relationships and facing directions within a refolded fold*

3. elongation of deformed objects such as pebbles or ooids;
4. mineral lineations resulting from the parallel orientation of elongate minerals; and
5. intersection lineations resulting from the intersection of two sets of planes.

   Overprinting may result in the destruction of an earlier structure particularly in the case of fabrics but also, depending on the relative orientation of successive folds, may result in interference patterns. Interference patterns resulting from the superposition of two sets of folds have been divided into three groups (Ramsay 1967), depending on the relationship between the axial planes and axes of the two phases and with complete gradation between them (Figure 4.14). The three inter-ference groups are as follows:

Type 1: results from two phases of folding in which their axial planes and axes are
    perpendicular to each other and the axes of each generation are perpendicular to

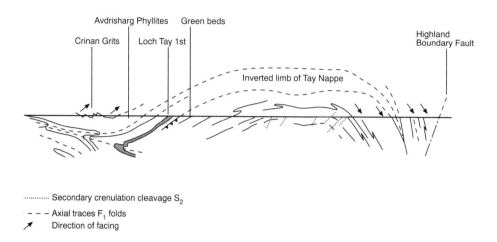

Figure 4.13 *Diagrammatic section across the Cowal Antiform from Loch Awe to Loch Lomond, south-west Highlands of Scotland. [Modified from: Shackleton (1958)]*

the axial planes of the other. The resulting pattern can be recognized as a series of domes and basins.

Type 2: formed from folds which have perpendicular axes but the axial planes are parallel. The resulting patterns have been termed 'angel wings' or 'mushrooms'.

Type 3: resulting from folds which are coaxial and which produce hook-like interference patterns.

As it is easier to fold a folded sequence about an axis parallel to the existing axis, which may be considered by analogy with an attempt to fold corrugated iron, Type 3 interference structures are probably the most common (Figure 4.15) but they are also the easiest to recognize in the field as both generations of fold can be seen in the same profile plane. The other two types of structure require good three-dimensional exposure.

Interference patterns may be the result of non-cylindrical folding and not the result of multiple phases of folding. Non-cylindrical folds are the result of intense shearing during the formation of a single fold phase (Figure 4.16; Cobbold & Quinquis 1980) and these folds, commonly known as sheath folds, can be recognized by strong lineations developed on all limbs of the fold parallel to the axis of the sheath (Holdsworth & Roberts 1984).

4.2.4 Recognition and Interpretation of Fabrics

In many complexly deformed areas, fold closures or hinges of major or minor folds and interference structures are difficult to find (logically the percentage area or outcrop of hinges must be much smaller than the area of the limbs). The fabrics

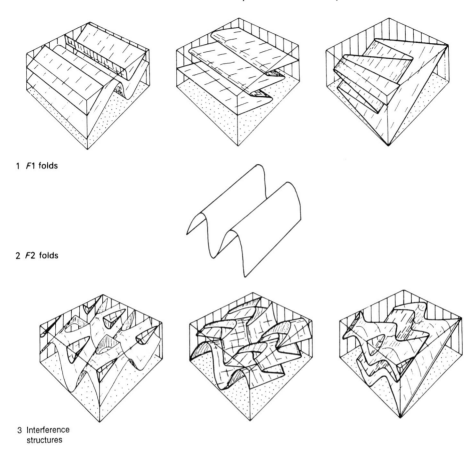

1 *F*1 folds

2 *F*2 folds

3 Interference
   structures

Figure 4.14 *Interference patterns generated by the superimposition of upright flow folds on previous folds of varying attitude. [Reproduced with permission from Blackie & Sons, from: Park (1989)]*

that are developed in association with the folding are, therefore, crucial to the interpretation of the fold history.

The objective in the interpretation of fabrics is to establish a chronological sequence. If the fabric is pervasive or penetrative it affects the whole rock and earlier fabrics would not be recognized. Non-penetrative fabrics can preserve evidence for earlier deformation in, for example, the hinges of the microfolds in crenulation cleavage.

Most workers in polydeformed terranes identify a bedding-parallel fabric. Bedding is classified as S_0 and the fabric as S_1, suggesting that it is related to a phase of folding which must, of necessity, be intepreted as isoclinal recumbent folds. The problem with this hypothesis is that in many areas neither minor nor major folds associated with such fabrics have been identified. This has led to the suggestion that the fabrics are associated with diagenesis or are primary structures due to

Figure 4.15 *Type 3 (Hook) interference pattern in psammitic rocks from the Moine near Loch Erribol, Scotland. [Photograph: P. J. Carey]*

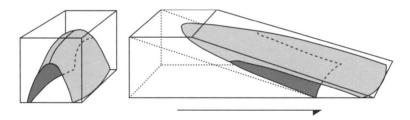

Figure 4.16 *Development of shear folds through simple shear. [Modified from: Ramsay (1980)]*

initial compaction of the sediments and alignment of the phyllosilicates in the sediment (Maltman 1981). Crenulation cleavages affecting these bedding-parallel fabrics may well be first fabrics rather than second, and the fold chronology for many areas may be simpler than originally thought.

Metamorphic mineral growth in relation to fabrics is also important in distinguishing between phases of folding. Pre-tectonic minerals are wrapped by the fabric (Figure 4.17), and have strain shadows producing eye structures or augen. Syn-tectonic fabrics contain spiral inclusion trails indicating rotation during growth, and post-tectonic minerals grow across the fabric.

A   B   C

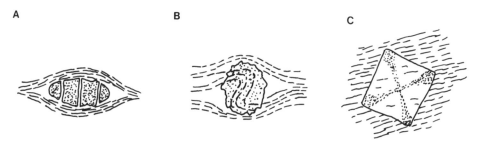

Figure 4.17   *Metamorphic mineral growth in relation to deformation.* **A.** *Pre-tectonic garnet.* **B.** *Syn-tectonic garnet.* **C.** *Post-tectonic chiastolite*

4.2.5 Stereographic Analysis of Orientation Data

One of the most useful tools in structural analysis is the stereographic or equal-area projection. The equal-area projection is most commonly used as it allows measurements of the density of the spatial data. For details of projection techniques, readers are referred to Phillips (1960), Ragan (1985) and Rowland & Duebendorfer (1994), amongst a number of other standard structural geology texts.

Different generations of structures can commonly be distinguished on the basis of their orientation. Fold and fabric orientations must be very carefully measured in the field, then plotted on maps and on stereographic projections. Refolded relationships can then become more obvious. Poles to bedding, measured around a cylindrical fold, will plot on a great circle, the pole to which defines the fold axis. The distribution of poles should give an indication of the style of the fold, particularly its interlimb angle, whether it is straight limbed or whether it has a gently curving hinge. If the fold is curvilinear the poles will spread from the simple great circle depending on the degree of curvilinearity. This curvilinearity may be an original feature of the fold or may be the result of refolding. At this stage the analysis is best accomplished by subdividing the whole area into small homogeneous domains. Each domain should have a uniform or simple outcrop pattern on the map and the data for each fabric element (fold axes, different generations of foliation, lineation, etc.) should be plotted separately (Turner & Weiss 1963; Ramsay 1967). The subdivision into relatively homogeneous areas is not always easy but is best done by inspection of the map data. Changes from one area to the next can be determined by comparing plots and producing synoptic diagrams. A simple example of this technique and the resulting stereographic projections are shown in Figure 4.18.

Lineations or linear fabric elements such as fold axes will be deformed by successive fold events, and their resulting orientation will be determined by the mechanism of deformation. Very simple flexural slip of buckle folds will result in a small circle distribution of lineations but shear folds will result in great circle distributions (Figure 4.19). These two mechanisms are, of course, theoretical end members and most distributions will be more complex. Detailed discussion of these distributions is given by Ramsay (1967).

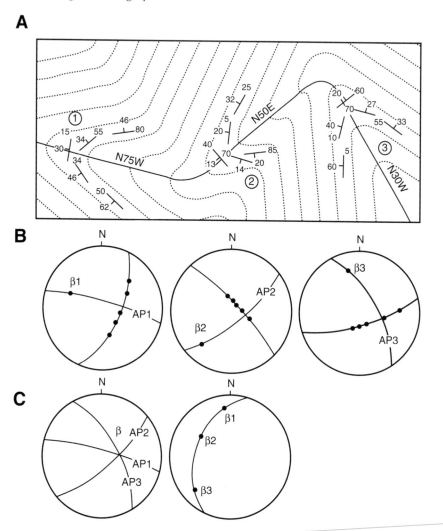

Figure 4.18   A. *Idealized map of superposed folds. Subareas 1, 2 and 3 are recognizable by the apparent traces of the hinge surface which are rectilinear.* **B.** *Stereograms of the data from subareas 1, 2 and 3.* **C.** *Synoptic diagrams showing the axis of the second folds defined by the axial planes from the three subareas and the first fold axes lying on a great circle suggesting similar folding (Figure 4.19). [Modified from: Ragan (1985)]*

4.2.6 Cross-cutting Relationships

Cross-cutting features must be later than the rocks which they cut. Cross-cutting intrusions such as pegmatites, granites or dykes are particularly useful as they can generally be readily dated. However, where rocks have been deformed subsequent to the intrusion, it is often difficult to establish exactly when the intrusion was emplaced within the structural sequence. This problem has recently led to a number of reinterpretations of the age of sequences and structures.

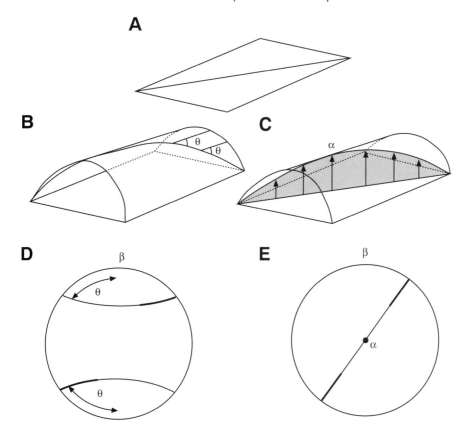

Figure 4.19 *Effect of different fold mechanisms on lineations:* **A.** *Original lineation.* **B.** *Flexural slip folding.* **C.** *Similar folding.* **D.** *Stereographic projection of B.* **E.** *Stereographic projection of C.*

Post-tectonic intrusions are the easiest to recognize and slot into their true stratigraphical position. They can generally be reliably dated and indicate the youngest age for the cessation of tectonic history. Syn-tectonic and pre-tectonic intrusions are much more difficult to fit precisely into the structural history (Figure 4.20; see Paterson & Tobisch 1988). In both cases, cross-cutting relationships, if they were ever present, are commonly obliterated by subsequent deformation, much as the angle of cross-bedding is reduced. Igneous bodies are generally more homogeneous than their host sediments and therefore respond differently to the imposed stress, perhaps acting as passive markers and possibly preserving evidence of their original igneous texture even after intense deformation. However, other bodies, particularly the smaller ones, will take up all the fabrics and structures seen in the host rocks and appear to be concordant. In some cases their igneous nature can only be determined through chemical analysis.

There are a number of problems associated with the study of fabrics in and around major intrusive bodies, and the establishment of a chronological sequence (Berger & Pitcher 1970; Hutton 1988). Fabrics may occur within intrusive bodies as

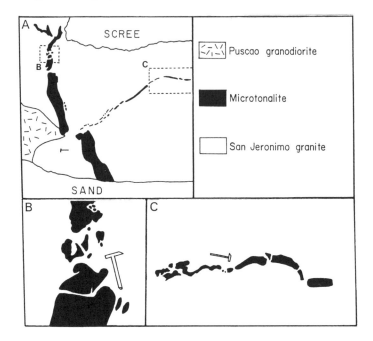

Figure 4.20 **A.** *Field sketch of an outcrop illustrating synplutonic dykes in the Rio Huara ring complex, Peru. Exposures in the southern wall of Quebrada Huamilache, near Sayan. **B** and **C** show further detail of selected parts indicated in A. Reprinted from Bussell (1991) in* Enclaves and granite petrology, *Didier & Barbarin (eds), 155–156 with kind permission of Elsevier Science – NL, Sara Burgerhart Straat 25, 1055 KV Amsterdam, The Netherlands*

a result of emplacement mechanisms, processes within the pluton, emplacement into a regional stress regime (syn-tectonic), and as the result of subsequent tectonic activity. For example, Bussell (1991) describes the deformation of microdiorite magma intruded into a granitic body which is still consolidating (Figure 4.21). In many plutons foliations are seen to cross-cut the margins of the pluton and the simplest view of this relationship is that the fabric post-dates intrusion. However, it has been pointed out that these discordances can originate as a result of the stresses exerted on the early emplaced margins of the pluton by further intrusion of magma into the central portion of the body. If the ductility contrast between the pluton and the country rocks is low then a cross-cutting fabric will result. If the ductility contrast is high then the fabrics will be parallel or at a low angle to the pluton walls. If the pluton is intruded during regional compression and the ductility of the pluton is similar to that of the country rocks, the fabrics within the pluton would be sub-parallel to those in the country rock. If there was a high ductility contrast then the fabrics in the pluton may well be discordant and the pluton would be more deformed than the host (Berger & Pitcher 1970).

Plutons can be emplaced as permitted intrusions where space is created for the intrusion through a variety of mechanisms (Hutton 1988). Alternatively, plutons can be forceful and make space for themselves by pushing aside the country rock. Forceful emplacement results in folding and fracturing of the surrounding country

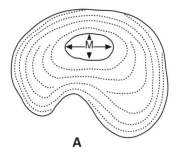
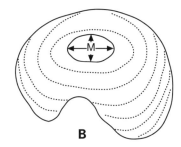

A

B

Figure 4.21 *Schematic representation of the origin of the differing relations of internal fabric to the margins of plutons. **A.** Ductility of the pluton is higher than that of the envelope. **B.** Low ductility contrast. M represents mobile material still being added to the plutons. [Modified from: Berger & Pitcher (1970)]*

rock. An excellent example of a forceful pluton is the Arran granite pluton in south-west Scotland (Read & Watson 1962). An arcuate fold following the northern margin of the pluton and a number of arcuate faults are superimposed on the regional structures of the Dalradian country rocks.

The following examples illustrate the problems of structural correlation of early igneous bodies. Both the West Highland Granitic Gneiss and the Ben Vuirich Granite have previously been considered to be syn-tectonic, but the evidence is equivocal, and in the case of the Ben Vuirich Granite at least, recent re-interpretations suggest a pre-tectonic emplacement.

The West Highland Granitic Gneiss

Several sheet-like bodies of granitic gneiss occur within the Moine metasediments of the Northern Highlands of Scotland (Figures 4.22 and 4.23). The largest of these bodies is the Ardgour Gneiss, which has been dated at 1028 ± 43 Ma (Brewer *et al.* 1979). The importance of this Grenvillian age for the regional Moine geology has been controversial because the granitic gneisses have been variously interpreted as metosomatic, magmatic or as tectonically emplaced slices of basement (see Brown 1991). In other words, the age may give a minimum Grenvillian age for deposition of the Moine if the Gneiss is pre-tectonic or, if the gneiss was tectonically emplaced or metasomatic, a Grenvillian age for the early orogenesis of the sequence, making the Moine deposition much older. These two possibilities have important repercussions in terms of correlation with orogenic episodes either south of the Great Glen Fault or with North America and Greenland. Barr *et al.* (1985) have concluded that the Ardgour Gneiss formed from a granite emplaced magmatically either before or during the first phase of deformation of the Moine. The gneiss shows two foliations S_1 and S_2. The strong S_1 foliation also occurs as the first fabric within metasedimentary xenoliths within the gneiss. Local melting and pegmatite segregations are coeval with the S_2 foliation (Figure 4.23). The granitic gneiss sheets have concordant margins on outcrop-scale as a result of the subsequent deformation but are cross-cutting on a regional scale.

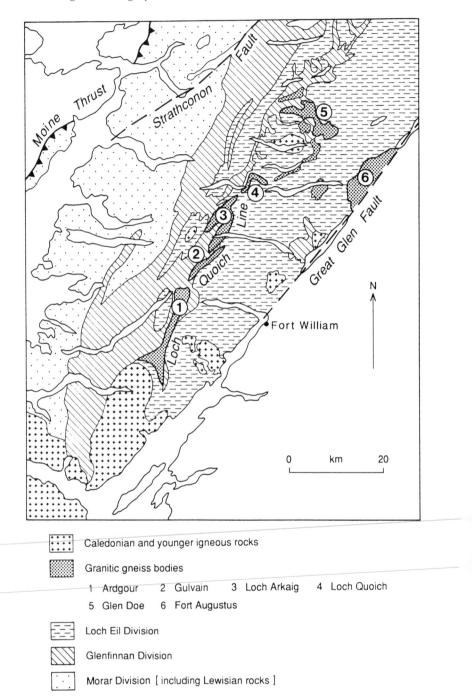

Figure 4.22   *Distribution of granitic gneiss bodies in the Northern Highlands of Scotland. [Reproduced with permission from the Geological Society, from: Brown (1991)]*

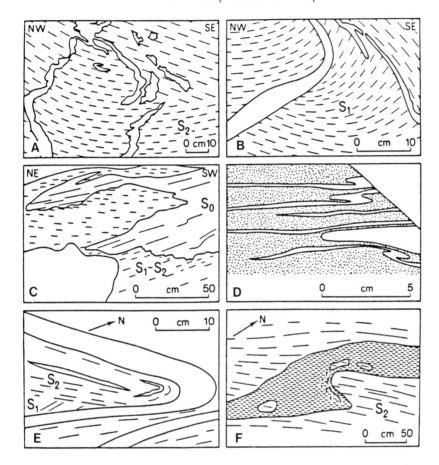

Figure 4.23 *Structures in the West Highland granitic gneiss.* **A.** *Early pegmatitic lits cut by the second foliation.* **B.** *Early pegmatitic lit cutting first foliation.* **C.** *Psammitic xenolith: banding in psammite (S_0) is truncated by granitic gneiss. Composite foliation ($S_1 - S_2$) terminates at boundary with xenolith.* **D.** *Intrafolial F_2 folds in pegmatitic lits.* **E.** *F_2 fold of pegmatitic lits: S_2 is the axial plane foliation which has largely obliterated S_1.* **F.** *Amphibolite sheet with xenoliths of granitic gneiss. [Modified from: Barr et al. (1985)]*

The Ben Vuirich Granite

Within the Dalradian Sequence of Perthshire, Scotland, the Ben Vuirich Granite has been used as an important time-marker for the Grampian and Caledonian Orogenies. The granite was thought to intrude already deformed Dalradian metasediments, cutting across early tectonic structures but itself deformed by the later structures. It was therefore interpreted as syn-tectonic (Bradbury *et al.* 1976). Radiometric dating, particularly by U/Pb zircon and Rb/Sr whole rock methods, suggested an emplacement age of 514 +6/−7 Ma and hence a minimum age for the Grampian D1 event (Pankhurst & Pidgeon 1976). The granite was interpreted as post-D2 and pre-D3 in the structural history of the Dalradian as it cross-cut the syn-D2 Killiecrankie slide but carried an LS fabric which was continuous with an S3 fabric in the country rocks. In 1989, Rogers and his co-authors published a revised age for the Ben Vuirich Granite, based on abraded zircons, suggesting an emplacement age of 595 ± 2 Ma (Rogers *et al.* 1989). This Precambrian age for the granite, and therefore the deformation associated with it, threw many of the Dalradian workers into confusion, as the Leny Limestone, considered part of the Upper Dalradian, had yielded Cambrian trilobites and the undoubtedly Dalradian Tayvallich Limestone and Shales had yielded supposed Cambrian acritarchs. The Dalradian sequence was therefore being deformed before it had been deposited! However, Harris (*in* Rogers *et al.* 1989) re-examined the evidence for a Cambrian age for the top of the Dalradian and concluded that it was highly equivocal, so that Dalradian sedimentation could be restored to the Precambrian.

   Ben Vuirich serves as a very good illustration of the difficulties associated with interpreting isotopic data of such syn-tectonic intrusions and placing the intrusion within its correct structural position. Tanner & Leslie (1994) suggest that the granite was intruded earlier in the structural history. As a result of careful re-mapping of the granite and surrounding area, these authors concluded that the fabric that cuts Ben Vuirich is the regional D2 fabric, not D3 as proposed by Bradbury *et al.* (1976). An S1 bedding-parallel fabric defined by phyllosilicates is cut by the granite at locality A (Figure 4.24). S1 is deformed by the D2 folds and overprinted by the main regional S2 fabric. S2 is defined by the preferred orientation of biotite and elongate aggregates of quartz grains accompanied by garnet growth. At locality A, the S2 fabric crosses into the granite and continues as a coplanar penetrative fabric. There is also clear evidence that cordierite–andalusite contact metamorphism associated with the granite preceded the D2 metamorphism and is pre-dated by a fabric that may be of regional D1 age (Locality B in Figure 4.24). Tanner & Leslie (1994) conclude that the Ben Vuirich Granite was probably intruded during the later part of, or following, the D1 tectonic event. Tanner (1996) has developed the story further through re-examination of fabrics within the hornfels at Locality B. He discusses three alternative interpretations of the fabrics:

1. the early fabric is pre-tectonic, a bedding parallel compaction fabric or an extensional fabric related to Dalradian basin formation. In this way the granite could be entirely pre-tectonic intruded into a previously undeformed passive margin sequence;
2. the fabric is tectonic and the regional D1 fabric. The granite was emplaced between two orogenic episodes; and
3. the fabric was formed largely as a result of the mechanical emplacement of the granite body. Evidence for alternatives 1 or 2 would have been obscured by subsequent deformation and metamorphism.

Tanner (1996) concludes that the fabric evidence, in combination with the regional tectonic picture of development of the Dalradian sedimentary basin and emplacement of the Tay Nappe, supports the hypothesis that the Ben Vuirich Granite was emplaced before any orogenic event had affected the Dalradian rocks. It cannot be

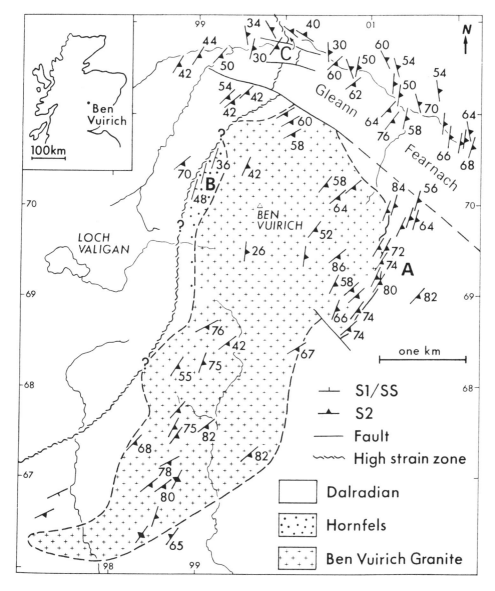

Figure 4.24 *Geological map of the Ben Vuirich Granite showing the positions of localities A and B referred to in the text. [Reproduced with permission from the Geological Society, from: Tanner & Leslie (1994)]*

used as evidence for a regional Precambrian structural event affecting the Appin to Southern Highland groups of the Dalradian Supergroup.

4.3 TECTONIC STRATIGRAPHY

Even though a sequence of lithologies may preserve excellent sedimentary structures, with a consistent younging direction indicating no repetition by folds, the sequence may still represent a tectonic stratigraphy rather than the true stratigraphy. Units may have been repeated by low-angle normal faults or lags, thrusts or shear zones. It is therefore important to establish criteria for recognizing the presence of such structures within the stratigraphical column.

First, it is necessary to have an understanding of the geometry of thrust zones (Boyer & Elliott 1982; Butler 1982). Thrusts generally propagate along a staircase path through the stratigraphy made up of ramps and flats (Figure 4.25). The thrust sheet slides along a relatively weak bedding plane and then cuts up through the stratigraphy, at an angle of approximately 30°. In this way older strata are placed upon younger, and the sequence is duplicated as many as ten times (Hobbs *et al.* 1976). Thrusts may develop in sequence either forwards or backwards from the first thrust. Where later thrusts develop in the footwall (forward propagation), earlier thrusts are carried along piggyback. This is considered to be the normal mode of propagation, but many thrust belts contain out-of-sequence thrusts that originate at the lowest or sole thrust and cut up through the thrust stack, often using pre-existing structures. This may mean that the normal rule of stratigraphical repetition is broken and locally stratigraphy may be missing. It may be very difficult to recognize the presence of a thrust if it is contained within a flat as the zone of deformation may be very thin and confined to a single weak layer. The presence of thrusting is much more obvious where the angular discordance associated with a ramp is recognized.

One of the principal techniques in studying thrust zones is to produce balanced sections of the zone (Dahlstrom 1969). This method depends on having a good understanding of the stratigraphy of the area so that an undeformed template can be constructed, but the technique is iterative so that the stratigraphical and structural interpretation of the area can remain internally consistent. Such balanced sections are useful for evaluating amounts of shortening across thrusts belts for tectonic reconstructions and also for examining the relationship between thrust development and the active sedimentation in the downwarped basin in front of the propagating thrust sheet.

In major fold and/or thrust belts, distinct, well-defined stratigraphical successions can commonly be recognized within large-scale thrust or fold sheets or nappes. Many nappes are compressional in origin with thrusts or shear zones at their base, but others have been attributed to gravitational sliding along a low-angle normal fault or lag. Stratigraphical sequences can usually be correlated between the nappes and the nappes themselves set into a regional sequence (e.g. Ramsay *et al.* 1985). The nappes are generally named after the structure that marks the base of the sheet, for example the Moine and Sgurr Beag nappes of the

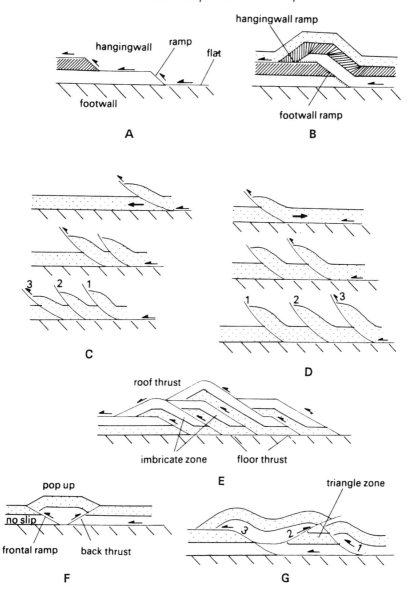

Figure 4.25 *Structures and terminology of thrust zones.* **A.** *Shape of thrust surface: ramps and flats.* **B.** *Hangingwall geometry: a fold in the hangingwall must result from a ramp.* **C.** *Piggyback thrust sequence (new thrusts develop in the footwall).* **D.** *Overstep thrust sequence (new thrusts develop in the hangingwall).* **E.** *Structure of duplex-imbricate thrust slices are contained between a floor thrust and a roof thrust.* **F.** *'Pop-up' structure formed by backthrusting.* **G.** *Triangle zone formed by backthrusting.* [Reproduced with permission from Blackie & Sons, from: Park (1989)]

Northern Highlands of Scotland (Roberts *et al.* 1987). Individual nappes may have undergone considerable horizontal displacements and the rocks within them are considered allochthonous as opposed to autochthonous.

A thrust nappe consists of more or less undisturbed material carried along on a major basal thrust plane. Within the nappe the stratigraphy may be duplicated as many as ten times with no significant inversion (Hobbs *et al.* 1976). Fold nappes consist of major recumbent folds in which the lower overturned limb may or may not be sheared out. The latter case may result in extensive areas of inverted beds; such nappes are generally formed at high grades of metamorphism deep in an orogenic belt and are rooted in zones of vertical beds. Rootless nappes form at higher levels in the crust and are generally less metamorphosed than the rooted structures. The fold may be generated at a break of slope down which the material is sliding and then develop into a rootless recumbent fold.

As nappes represent allochthonous sequences for stratigraphical reconstruction and correlation, it is necessary to determine the direction of transport of the nappe. There are a variety of minor structures that can be used to determine the sense of movement or shear sense of major thrusts or shear zones. These must be viewed perpendicular to the foliation developed during the shearing and parallel to any structural lineation. The shear sense is usually reported as dextral (right-lateral or top to the right) or sinistral (left-lateral or top to the left).

Many rocks in shear zones possess composite planar fabrics defined by two foliations at a low angle to each other produced simultaneously as a result of the shearing. One fabric is generally penetrative at all scales and is termed an S surface and the second fabric (C) is a grain-shape fabric consisting of spaced shear surfaces marked by zones of grain-size reduction. The asymmetry of the S foliation relative to the C surface gives the sense of shear (Figure 4.26). Porphyroblasts are rotated within the shear zone and can develop asymmetric tails which show the sense of rotation or shearing.

4.4 STRATIGRAPHICAL CORRELATION IN COMPLEX TERRANE: THE EXAMPLE OF THE NORTHERN GRAMPIAN HIGHLANDS

The Grampian Highlands are part of the Caledonian orogenic belt of Scotland, which lies between the Great Glen Fault and the Highland Boundary Fault. The rocks of the area consist of lithologically diverse metasediments and mafic meta-volcanic rocks of late Precambrian age making up the Dalradian Supergroup (Figure 4.27). The major structure consists of large recumbent folds passing north-westwards into a steep belt or root zone. Further north-west the structure is dominated by large-scale reclined folds associated with lags or low-angle normal faults (Thomas 1979; Soper & Anderton 1984). The structure and stratigraphy of the south-western part of the area is fairly well understood but at deeper crustal levels to the north-east both structure and stratigraphy become less certain, with the possibility that there are outcrops of a pre-Dalradian basement. The geology of the northern part of the Grampian Highlands (the Central Highlands) illustrates

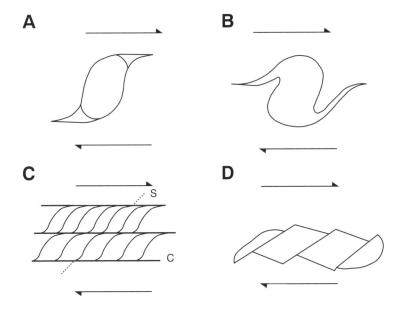

Figure 4.26 *Sense of shear indicators.* **A.** *σ-structure.* **B.** *δ-structure.* **C.** *S–C band structure.*
D. *Bookshelf sliding on antithetic fractures*

some of the problems associated with the recognition of unconformities within polydeformed terranes. The issues revolve around the stratigraphical significance of contrasts in structural style and response to metamorphism between contrasting lithological units, and whether these contrasts are sufficient to warrant recognition of an angular unconformity.

Using a combination of field and isotopic data, Piasecki & Van Breeman (1979) and Piasecki (1980) proposed that the rocks of this area should be divided into a basement Central Highland Division, consisting of coarse pelitic gneisses, and a cover Grampian Division, consisting of largely non-migmatized psammitic meta-sediments. They recognized two phases of folding in the basement that were not present in the cover and concluded that the Central Highland Division had suffered Grenvillian orogenesis prior to the deposition of the Grampian Division. High strain zones or slides affect both basement and cover, at or close to an unconformity and have obliterated the original depositional surface in all but very small pockets (Piasecki & Temperley 1988). The high strain zones are intruded by pegmatites that have yielded 740 Ma ages and metagabbros that have yielded ages greater than 750 Ma (Highton 1992), and it has been suggested that these zones are associated with deformation related to the Morarian orogeny (Van Breeman *et al.* 1974; Piasecki *et al.* 1981).

Unfortunately, within the Central Highlands the simple basement/cover theory has a number of difficulties. The top of the Grampian Division appears to young without break into rocks of the Appin Group of the Dalradian (Treagus 1974; Hickman 1975) and structures within the Grampian Division have been directly correlated within structures within the Dalradian which are of Grampian Age. For

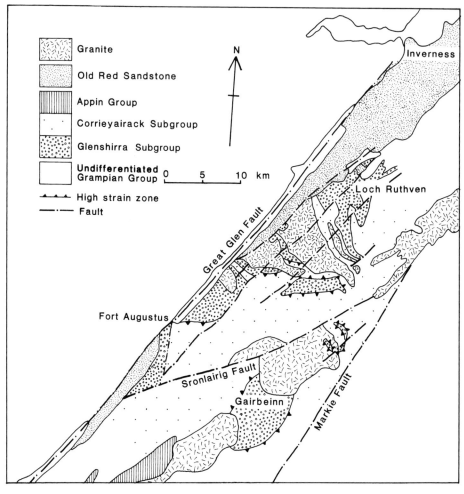

Figure 4.27   *Geological map of the Central Highlands of Scotland*

these reasons, Harris *et al.* (1978) suggested that the Grampian Division should be included within the Dalradian and renamed the Grampian Group. Deposition of the Upper Dalradian may be as late as 595 ± 2 Ma (Halliday *et al.* 1989), whereas Piasecki (1980) suggested that the Grampian Division was deposited in the 1000–750 Ma interval, giving at least 150 million years of sedimentation without a recognizable break. Harris *et al.* (1981) suggests that Piasecki's basement/cover model could be extended into the Strathnairn area of the Central Highlands.

More recent detailed mapping in the Strathnairn and the Corrieyairack areas, although resulting in the recognition of two distinct lithological successions – the Glenshirra Subgroup and the Corrieyairack Subgroup – locally separated by a zone of high strain, has concluded that the whole sequence should be included within the Grampian Group (Haselock *et al.* 1982; Figure 4.27). All the metasediments in this area have been subjected to three episodes of deformation and metamorphism at low to middle amphibolite facies, but show large contrasts in apparent structural and metamorphic complexity as the result of contrasts in composition and ductility. A number of stratigraphical tools have been used to come to this conclusion.

The Glenshirra Subgroup

In the type area south of the Corrieyairack, the Glenshirra Subgroup consists of three formations, separated on a lithostratigraphical basis: the Creag Mhor Psammite, the Carn Dhearg Psammite and the Gairbeinn Pebbly Psammite. The base of the subgroup has not been recognized as the lowest formation occurs in the core of an upward-facing antiform. The Gairbeinn Pebbly Psammite Formation is a very distinctive unit, comprising a well-bedded granular arkosic sandstone with local conglomerates. It contains abundant well-preserved sedimentary structures indicative of fluvial sedimentation in a braided stream (Figures 4.1A,B and 4.3). The formation can be traced across extensive unexposed tracts because abundant, magnetite-rich heavy mineral seams lend it a distinctive magnetic signature (Figure 4.7; Haselock & Leslie 1992). Over most of its outcrop the top of the formation is marked by a zone of high strain, the Gairbeinn Slide. However, north-east of the Foyers Granite on Garbhal Mhor the upper boundary of the pebbly psammite has been interpreted as a little-modified stratigraphical transition via interbedded quartzites and semipelites into the overlying Corrieyairack Subgroup.

Corrieyairack Subgroup

The Corrieyairack Subgroup consists of five formations dominated by the Glen Doe Psammite, which is a sequence of grey well-bedded micaceous psammites typical of shallow marine turbiditic sandstones in which way-up evidence consists of small-scale graded units or rhythmites (Figure 4.5) and very rare cross-laminations. The Ruthven Semipelite is the lowest formation recognized within the Subgroup and it has a transitional stratigraphical contact with the underlying Gairbeinn Formation well exposed on Garbhal Mhor. The semipelites are coarsely crystalline and contain quartzo-feldspathic segregations concordant with the dominant phyllosilicate foliation (Figure 4.28). Locally, calc-silicate bands, psammitic ribs, abrupt variations in grain size, relative abundance of quartzo-feldspathic segregations and concentrations of muscovite pseudomorphs after kyanite may record original lithological variation. Garnetiferous amphibolite bodies are common, particularly within the northern outcrops, and are concentrated in layers, which possibly represent disrupted sills. These gneisses are lithologically similar to and can be traced laterally into rocks which have been assigned to the Central Highland Division by Piasecki (1980) and Harris *et al.* (1981).

Detailed studies of the tectonic fabrics have led to the recognition of a consistent history of deformation throughout the Glenshirra and Corrieyairack Subgroups, and indeed southwards to Glen Garry and Strathspey (Lindsay *et al.*

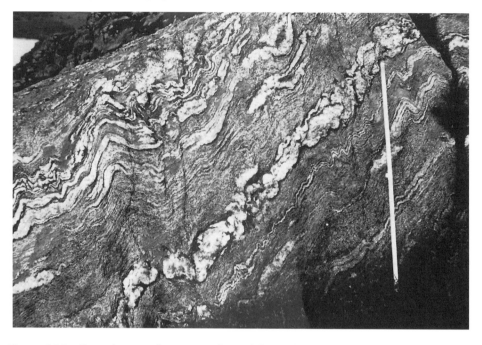

Figure 4.28   *Coarsely crystalline semipelites of the Ruthven Formation. [Photograph: P. J. Carey]*

1989). The structural interpretation was greatly assisted by the results of a ground magnetic survey covering approximately 150 km² of very poorly exposed ground around the post-tectonic Foyers Granitic Complex (Haselock & Leslie 1992). These results were particularly important in elucidating the interference pattern between D1 and D2 folds south of the Sronlairig Fault, and also delineating the extent of the Foyers Complex and its relationship to the country rocks.

The earliest structures recognized are associated with the formation of the gneissic foliation and a bedding-parallel schistosity, more strongly developed in the north of the area. Minor folds carrying this foliation as an axial planar fabric are present within the Ruthven Semipelite and Gairbeinn Pebbly Psammite in the north of the area and also in the area south of the Sronlairig Fault. A major D1 closure in the latter area consists of an anticline which is refolded by a D2 structure so that it is downward facing in the north and upward facing in the south.

In the northern area, E–W trending folds fold the gneissic foliation and in the more pelitic lithologies develop a migmatitic, spaced, axial planar fabric. Farther south the D2 folds are characterized by an axial planar crenulation cleavage widespread in the pelitic lithologies but represented by a grain shape fabric in the psammitic lithologies.

The outcrop pattern of the lithostratigraphical units in the area is largely controlled by major, upright, tight folds which swing from NE–SW in the south to

N–S in the north. These folds have a well-developed crenulation or crenulation cleavage associated with them, which is commonly difficult to separate from the earlier S2 fabric. Locally, a conjugate set of crenulations is developed. The cleavage shows marked refraction between psammitic and pelitic lithologies on a local and regional scale.

A high strain zone is locally developed between the two subgroups, but high strain also occurs on the limbs of D2 and D3, folds well illustrated by the pebble bands on Cairn Poullachie in the north of the area where pebbles show axial ratios up to 2:1:0.14 on the limbs of the D3 fold. All signs of sedimentary structures are destroyed and the bedding has a finely laminated tramline appearance. In the hinge of the fold, however, pebbles are crenulated and the sedimentary structures, although deformed, are still useful as way-up indicators (Figures 4.3 and 4.4).

The structural history is common to all the metasedimentary rocks of the area regardless of their metamorphic/migmatitic state and in none of the migmatitic rocks do the D1–D3 structures described overprint earlier deformation or metamorphic fabrics. This absence of overprinting could be explained in terms of transposition of the earlier structures and complete recrystallization if it were not for the stratigraphical position of the migmatized rocks overlying as they do psammites which preserve relatively undeformed sedimentary structures and a clear D1–D3 structural history. There is therefore no evidence to support the presence of a basement complex in this area as previously suggested, and the study represents a cautionary tale in the structural interpretation of contrasting lithologies producing over-complex stratigraphical stories.

There are still large areas of migmatized rocks to the east of Strathnairn which have proved difficult to correlate or subdivide. These have been termed the Central Highland Migmatite Complex (Harris *et al.* 1994) and there is a growing body of isotopic data which supports some sort of deformational event at *c.* 750 Ma which affected Grampian Group rocks. The base of the stratigraphical pile seems therefore to have been deformed prior to the deposition of the top, possibly on a series of low-angle normal faults representing the extension of the sedimentary basin (Soper & Anderton 1984).

4.5 CONCLUSIONS

Unravelling the structure and stratigraphy of complex terranes requires a very careful analysis of field data. A good field geologist should be able to record structural data, and compile careful descriptions of structural features and lithologies. The production of detailed maps and cross-sections is essential in order to appreciate the spatial relationships involved. Even with very careful observations, the answers are rarely unequivocal and a solution that is internally consistent between the structure and stratigraphy may be just one of a number of solutions. However, such solutions are preferable to those which invoke a complex structural history to allow a simple stratigraphical story, or a complex stratigraphy to allow a simple structural story (Goodman *et al.* 1997).

ACKNOWLEDGEMENTS

The diagrams were drawn for the most part by Matthew R. Bennett.

REFERENCES

Bailey, E.B. 1922. The structure of the South-West Highlands of Scotland. *Quarterly Journal of the Geological Society of London*, **78**, 82–127.

Bailey, E.B. 1930. New light on sedimentation and tectonics. *Geological Magazine*, **117**, 77–92.

Barr, D., Roberts, A.M., Highton, A.J., Parson, L.M. & Harris, A.L. 1985. Structural setting and geochronological significance of the West Highland Granitic Gneiss, a deformed early granite within the Proterzoic Moine rocks of NW Scotland. *Journal of the Geological Society of London*, **142**, 663–675.

Bell, A.M. 1981. Vergence: an evaluation. *Journal of Structural Geology*, **3**, 197–202.

Berger, A.R. & Pitcher, W.S. 1970. Structures in granitic rocks: a commentary and a critique on granite tectonics. *Proceedings of the Geologists' Association*, **81**, 441–461.

Boyer, S.E. & Elliott, D. 1982. Thrust systems. *Bulletin of the American Association of Petroleum Geologists*, **66**, 1196–1230.

Bradbury, H.J., Smith, R.A. & Harris, A.L. 1976. 'Older' granites as time-markers in Dalradian evolution. *Journal of the Geological Society of London*, **132**, 67–68.

Brewer, M.S., Brook, M.P. & Powell, D. 1979. Dating of the tectonometamorphic history of the SE Moine, Scotland. In Harris, A.L., Holland, C.H. & Leake, B.E. (eds) *The Caledonides of the British Isles reviewed*. Geological Society of London, Special Publication No. 8, 129–137.

Brown, P.E. 1991. Caledonian and earlier magmatism. In Craig, G.Y. (ed.) *The geology of Scotland*. Geological Society, London, 229–281.

Bussell, M.A. 1991. Enclaves in the Mesozoic and Cenozoic granitoids of the Peruvian Coastal Batholith. In Didier, J. & Barbarin, B. (eds) *Enclaves and granite petrology*. Elsevier, Amsterdam, 155–166.

Butler, R.W.H. 1982. The terminology of structures in thrust belts. *Journal of Structural Geology*, **4**, 239–245.

Cloos, E. 1947. Oolite deformation in the South Mountain Fold, Maryland. *Bulletin of the Geological Society of America*, **58**, 843–918.

Cobbold, P.R. & Quinquis, H. 1980. Development of sheath folds in shear regimes. *Journal of Structural Geology*, **2**, 119–126.

Dahlstrom, C.D.A. 1969. Balanced cross-sections. *Canadian Journal of Earth Science*, **6**, 743–747.

Evans, R.H.S. & Tanner, P.W.G. 1996. A late Vendian age for the Kinlochlaggan Boulder Bed (Dalradian)? *Journal of the Geological Society of London*, **153**, 823–826.

Flinn, D. 1965. On the symmetry principle and the deformation ellipsoid. *Geological Magazine*, **102**, 36–45.

Goodman, S., Crane, A., Krabbendam, M., Leslie, A.G. & Ruffell, A. 1997. Correlation of depositional sequences in a structurally complex area: the Dalradian of Gleann Fernach to Glen Shee, Scotland. *Transactions of the Royal Society of Edinburgh*, **88**, in press.

Halliday, A.N., Graham, C.M., Aftalion, M. & Dymoke, P. 1989. The depositional age of the Dalradian Supergroup: U–Pb and Sm–Nd isotopic studies of the Tayvallich Volcanics, Scotland. *Journal of the Geological Society of London*, **146**, 3–6.

Hambrey, M.J. 1983. Correlation of Late Proterozoic tillites in the North Atlantic Region and Europe. *Geological Magazine*, **120**, 209–320.

Harris, A.L., Baldwin, C.T., Bradbury, H.J., Johnson, H.D. & Smith, R.A. 1978. Ensialic basin sedimentation: the Dalradian Supergroup. In Bowes, D.R. & Leake, B.E. (eds) *Crustal evolution in Northwest Britain and adjacent regions. Geological Journal Special Issue*, **10**, 115–138.

Harris, A.L., Parson, L.M., Highton, A.J. & Smith, D.I. 1981. New/Old Moine relationships between Fort Augustus and Inverness (abstr.) *Journal of Structural Geology*, **3**, 187–188.

Harris, A.L., Haselock, P.J., Kennedy, M.J. & Mendum, J.R. 1994. The Dalradian Supergroup in Scotland, Shetland and Ireland. In Gibbons, W. & Harris, A.L. (eds) *A revised correlation of Precambrian rocks in the British Isles*. Geological Society, London, Special Report No. 22, 33–53.

Haselock, P.J. 1982. The geology of the Corrieyairack Pass area, Inverness-shire. Unpublished PhD thesis, University of Keele.

Haselock, P.J. 1984. The systematic geochemical variation between two tectonically separate successions in the southern Monadhliaths, Inverness-shire. *Scottish Journal of Geology*, **20**, 191–205.

Haselock, P.J. & Leslie, A.G. 1992. Polyphase deformation in Grampian Group rocks of the Monadhliath defined by a ground magnetic survey. *Scottish Journal of Geology*, **28**, 81–87.

Haselock, P.J., Winchester, J.A. & Whittles, K.H. 1982. The stratigraphy and structure of the Southern Moriadhliath Mountains between Loch Killin and Glen Roy. *Scottish Journal of Geology*, **18**, 275–290.

Hickman, A.H. 1975. The stratigraphy of the late precambrian meta-sediments between Glen Roy and Lismore. *Scottish Journal of Geology*, **11**, 117–142.

Hickman, A.H. & Wright, A.E. 1983. Geochemistry and chemostratigraphical correlation of slates, marbles and quartzites of the Appin group, Argyll, Scotland. *Transactions of the Royal Society of Edinburgh, Earth Sciences*, **73**, 251–278.

Highton, A.J. 1992. The tectonostratigraphical significance of pre-750 Ma metagabbros within the northern Central Highlands, Inverness-shire. *Scottish Journal of Geology*, **28**, 71–76.

Hobbs, B.E., Means, W.D. & Williams, P.F. 1976. *An outline of structural geology*. John Wiley. Chichester.

Holdsworth, R.E. & Roberts, A.M. 1984. A study of early curvilinear fold structures and strain in the Moine of the Glen Garry region, Inverness-shire. *Journal of the Geological Society of London*, **141**, 327–338.

Hutton, D.H.W. 1988. Granite emplacement mechanisms and tectonic controls. *Transactions of the Royal Society of Edinburgh, Earth Sciences*, **79**, 245–255.

Krauskopf, K.B. 1979. *Introduction to geochemistry*, 2nd edition. McGraw-Hill, New York.

Lambert, R.St.J., Winchester, J.A. & Holland, J.G. 1981. Comparative geochemistry of pelites from the Moinian and Appin group (Dalradian) of Scotland. *Geological Magazine*, **118**, 477–490.

Lambert, R.St.J., Winchester, J.A. & Holland, J.G. 1982. A geochemical comparison of the Dalradian Leven Schists and the Grampian Division Monadhliath Schists of Scotland. *Journal of the Geological Society of London*, **139**, 71–84.

Lindsay, N.G., Haselock, P.J. & Harris, A.L. 1989. The extent of Grampian orogenic activity in the Scottish Highlands. *Journal of the Geological Society of London*, **146**, 733–735.

Maltman, A.J. 1981. Primary bedding-parallel fabrics in structural geology. *Journal of the Geological Society of London*, **138**, 475–483.

Pankhurst, R.J. & Pidgeon, R.T. 1976. Inherited isotope systems and the source region prehistory of the early Caledonian granites in the Dalradian Series of Scotland. *Earth and Planetary Science Letters*, **31**, 55–68.

Park, R.G. 1989. *Foundations of structural geology*, 2nd edition. Blackie & Son, Edinburgh.

Paterson, S.R. & Tobisch, O.T. 1988. Using plutons to date regional deformations: problems with common criteria. *Geology*, **16**, 1108–1111.

Phillips, F.C. 1960. *The use of stereographic projection in structural geology*, 2nd edition. Edward Arnold, London.

Piasecki, M.A.J. 1980. New light on the Moine Rocks of the Central Highlands of Scotland. *Journal of the Geological Society of London*, **137**, 41–59.

Piasecki, M.A.J. & Van Breeman, O. 1979. The 'Central Highland Granulites': cover-basement tectonics in the Moine. In Harris, A.L., Holland, C.H. & Leake, B.E. *The Caledonides* of the British Isles Reviewed. Geological Society of London, Special Publication No. 8, 139–144.

Piasecki, M.A.J., Van Breeman, O. & Wright, A.E. 1981. The late Precambrian geology of Scotland, England and Wales. In Kerr, J.W. & Ferguson, A.J. (eds) *Geology of the North Atlantic Borderlands*. Canadian Society of Petroleum Geology Memoir, **7**, 57–94.

Piasecki, M.A.J. & Temperley, S. 1988. The northern sector of the Central Highlands. In Allison, I., May, F. & Strachan, R.A. (eds) *An excursion guide to the Moine geology of the Scottish Highlands*. Scottish Academic Press, Edinburgh.

Price, N.J. & Cosgrove, J.W. 1990. *Analysis of geological structures*. Cambridge University Press, Cambridge.

Ragan, D.M. 1985. *Structural geology. An introduction to geometrical techniques*. John Wiley, Chichester.

Ramsay, D.M. 1980. Shear zone geometry: a review. *Journal of Structural Geology*, **2**, 83–99.

Ramsay, D.M., Stuart, B.A., Jansen, Ø., Andersen, T.B. & Sinha-Roy, S. 1985. The tectono-stratigraphy of western Porangerhalvoya, Finnmark, north Norway. In Gee, D.G. & Sturt, B.A. *The Caledonian Orogen – Scandinavia and related areas*. John Wiley, Chichester, 611–619.

Ramsay, J.G. 1967. *Folding and fracturing of rocks*. McGraw-Hill, New York.

Ramsay, J.G. & Huber, M.I. 1983. *The techniques of modern structural geology. Volume 1: strain analysis*. Academic Press, London.

Read, H.H. 1922. Discussion of Bailey, E.B. The structure of the S. West Highlands of Scotland. *Quarterly Journal of the Geological Society of London*, **78**, 129–130.

Read, H.H. & Watson, J. 1962. *Introduction to geology*. Macmillan, London.

Roberts, A.M., Strachan, R.A., Harris, A.L., Barr, D. & Holdsworth, R.E. 1987. The Sgurr Beag nappe: a reassessment of the stratigraphy and structure of the Northern Highland Moine. *Bulletin of the Geological Society of America*, **98**, 497–506.

Rogers, G., Dempster, T.J., Bluck, B.J. & Tanner, P.W.G. 1989. A high precision U/Pb age for the Ben Vuirich granite: implications for the evolution of the Scottish Dalradian. *Journal of the Geological Society of London*, **146**, 789–798.

Rowland, S.M. & Duebendorfer, E.M. 1994. *Structural analysis and synthesis. A laboratory course in structural geology*, 2nd edition. Blackwell Scientific, Oxford.

Shackleton, R.M. 1958. Downward-facing structures of the Highland border. *Quarterly Journal of the Geological Society of London*, **113**, 361–392.

Soper, J. & Moseley, F. 1978. Structure. In Moseley, F. (ed.) *The geology of the Lake District*. Yorkshire Geological Society Occasional Publication No. 3, 45–67.

Soper, N.J. & Anderton, R. 1984. Did the Dalradian slides originate as extensional faults? *Nature*, **307**, 205–211.

Tanner, P.W.G. 1996. Significance of the early fabric in the contact metamorphic aureole of the 590 Ma Ben Vuirich Granite, Perthshire, Scotland. *Geological Magazine*, **133**, 683–695.

Tanner, P.W.G. & Leslie, A.G. 1994. A pre-D2 age for the 590 Ma Ben Vuirich Granite in the Dalradian of Scotland. *Journal of the Geological Society of London*, **151**, 209–212.

Tanton, T.L. 1930. Determination of the age relations in folded rocks. *Geological Magazine*, **117**, 73–76.

Thomas, P.R. 1979. New evidence for a Central Highland root zone. In Harris, A.L., Holland, C.H. & Leake, B.E. (eds) *The Caledonides of the British Isles reviewed*. Special Publication of the Geological Society of London No. 8, 205–211.

Treagus, J.E. 1969. The Kinlochlaggan Boulder Bed. *Proceedings of the Geological Society of London*. **1654**, 55–60.

Treagus, J.E. 1974. A structural cross-section of the Moine and Dalradian rocks of the Kinlochleven area, Scotland. *Journal of the Geological Society of London*, **130**, 525–544.

Turner, F.J. & Weiss, L.E. 1963. *Structural analysis of metamorphic tectonites*. McGraw-Hill, New York.

Van Breemen, O., Pidgeon, R. & Johnson, M.R.W. 1974. Pre-Cambrian and Palaeozoic pegmatites in the Moines of northern Scotland. *Journal of the Geological Society of London*, **130**, 493–507.

Vogt, T. 1930. On the Chronological order of deposition of the Highland Schists. *Geological Magazine*, **117**, 68–72.

Wilson, G. 1982. *Introduction to small-scale geological structures.* George Allen & Unwin, London.

Winchester, J.A. 1971. Some geochemical distinctions between Moinian and Lewisian rocks, and their use in establishing the identity of supposed inliers in the Moinian. *Scottish Journal of Geology*, **7**, 327–344.

Winchester, J.A. & Max, M.D. 1996. Chemostratigraphic correlation, structure and sedimentary environments in the Dalradian of the NW Co Mayo inlier, N.W. Ireland. *Journal of the Geological Society of London*, **153**, 779–802.

5
Evolutionary Concepts in Biostratigraphy

Paul N. Pearson

Biostratigraphy is the discipline by which the relative stratigraphical positions of sedimentary rocks are determined with reference to their fossil content. Different types of fossil are characteristic of different sedimentary environments and, because of organic evolution, they also characterize different periods of time. As a consequence of the latter, it is possible to assess the relative ages of sedimentary rocks using palaentological evidence. The age-determination of rocks using fossils is properly known as biochronology. This chapter explores the relationship between evolutionary processes and the nature of biostratigraphical and bio-chronological correlations.

It is worth remembering that the first successful biostratigraphies were developed before the fact of evolution was widely accepted. It remains true that for a rough and ready result it is unnecessary to consider the processes behind the pattern, just as lithostratigraphical subdivisions can be made on rock descriptions alone. However, as biostratigraphical studies have become increasingly sophisticated in response to the demand for ever greater resolution, so it has become fruitful to consider them in the light of evolutionary theory. Phenomena of particular significance are speciation processes, controls on biogeography, patterns of phyletic evolution, and extinction processes.

The most important link in the relationship between evolution and biostratigraphy is Linnaean taxonomy, the firmly established set of procedures that is used for identifying and classifying organisms and their remains, including fossils. Taxonomy is an integral part of the biostratigraphical method, because its procedures

Unlocking the Stratigraphical Record: Advances in Modern Stratigraphy. Edited by P. Doyle and M.R. Bennett.
© 1998 John Wiley & Sons Ltd.

underpin the formal definition of biostratigraphical units. Evolutionists working with some fossil groups have long acknowledged a degree of arbitrariness and ambiguity in the taxonomic method, especially in relation to gradually evolving populations. The implications of this ambiguity for biostratigraphical correlations have often been unacknowledged. To redress this balance, the links between evolution, taxonomy and biostratigraphy are the central theme of this chapter, which culminates in a new classification of biohorizons based on evolutionary processes. It will be necessary, however, to begin by reviewing the fundamental principles of the subject.

5.1 BIOHORIZONS, BIOZONES, BIOCHRONS

Most of the terms and ideas of biostratigraphy have a long and tortuous history, and have often been used in different senses at different times (Hancock 1977; Schoch 1989). Several modern texts aim to provide a practical guide to the correct usage of biostratigraphical nomenclature, of which the most authoritative are the *International Stratigraphic Guide* (Hedberg 1976) and the *North American Stratigraphic Code* (North American Commission on Stratigraphy 1983). Unfortunately, these guides disagree on some points of detail. In this chapter, the *International Stratigraphic Guide* is followed.

The development of a biostratigraphy is achieved by a method known as biostratigraphical zonation ('biozonation' for short), whereby strata are organized into units which depend on their fossil content. A hypothetical example is given in Figure 5.1. The first step in establishing a biozonation involves the sampling of stratigraphical sequences and the identification of the fossils found within them. The vertical distribution through the rocks of a particular kind of fossil is known as its 'stratigraphical range'. Diagrams that document the stratigraphical distributions of some or all the fossil taxa present at an individual site are called 'stratigraphical range-charts'. From a range-chart, biostratigraphical 'events' can be determined, such as the first and last occurrences of species, subspecies or higher taxa. Other types of event are also sometimes used in biostratigraphy, such as the peak abundance of a species or the appearance of a particular assemblage of species.

The level of any biostratigraphical event can be thought of as linked from site to site by an actual surface running through the rock. This surface is known as a biostratigraphical horizon (or 'biohorizon'). In Figure 5.1, two events have been recognised at neighbouring sites, corresponding to two biohorizons. A biostratigraphical zone ('biozone') is the stratigraphical interval between two specified biohorizons. Since a biohorizon has a physical reality as a surface within the rock, a biozone should be thought of as an actual volume of rock, often having a complex three-dimensional shape. Being an actual object, a biozone is not a unit of time, so it is incorrect to talk of sediment being formed 'before', 'during' or 'after' a particular biozone; one should refer to strata being 'below', 'within' or 'above' it. Nevertheless it is possible to conceive of a unit of time which corresponds to the total period in which the biozone was formed, which is known as a biochron.

Biozones may be of different types depending on the events used to define their

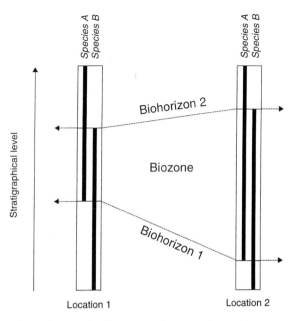

Figure 5.1 *A hypothetical example illustrating the procedure of biozonation. Vertical black bars represent the stratigraphical ranges of two species at different sites. In this case, the first appearance of Species A corresponds to Biohorizon 1, and the last occurrence of Species B corresponds to Biohorizon 2. Biohorizons are shown as dashed lines because their exact shape and position can be determined only by studying intervening sites. A biozone (in this case, a concurrent range zone) is the stratigraphical and geographical interval between two biohorizons, and extends beyond these locations for as far as the overlap in the species can be observed*

lower and upper boundaries. Restricting the discussion to those delimited by either first appearances or last occurrences (i.e. excluding abundance changes and so on), there are five logical possibilities, as shown in Figure 5.2. The most obvious type is the *taxon range biozone* (sometimes known as the 'total range biozone'), which is defined as 'the body of strata representing the total range of occurrence (horizontal and vertical) of specimens of particular taxon' (Hedberg 1976: 53). Note that it is not only defined in terms of its stratigraphical ('vertical') thickness, but also its geographical ('horizontal') extent. It is easy to determine whether a sample belongs within a given taxon range biozone, simply by determining whether the named taxon is present.

A second sort of biozone is the *concurrent range biozone* which is the interval of overlap (stratigraphical and geographical) between two taxon range biozones. This is defined as 'the concurrent or coincident parts of the range-zones of two or more specified taxons' (Hedberg 1976: 55). The remaining possibilities shown in Figure 5.2 are various types of *interval biozone*. An interval biozone is the 'interval between two distinctive biostratigraphical horizons' but 'not itself the range zone of any taxon or concurrence of taxons' (Hedberg 1976: 60). Once again, concurrent range biozones and interval biozones should be thought of as having both a stratigraphical and geographical extent.

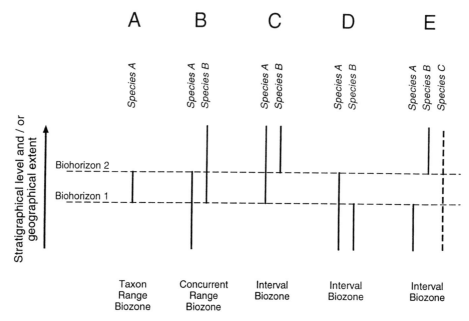

Figure 5.2 *Nomenclature of biostratigraphical zones following the* International Stratigraphic Guide *(Hedberg 1976). Note that all five examples are regarded as 'interval zones' by the North American Commission on Stratigraphy (1983) which uses the term in a more general sense to denote 'the body of strata between two specified, documented lowest and/or highest occurrences of single taxa' (North American Commission on Stratigraphy 1983). The interval biozone illustrated in case E is sometimes referred to as a 'partial range zone' (e.g. Berggren et al. 1995), but this term has been used inconsistently*

When standing in a quarry or at a cliff face, everyone can agree that a distinctive lithological boundary, such as a prominent bedding plane, has a physical meaning as a 'partition' within the rock sequence. What is not so obvious is that a biohorizon is also a real physical surface. In this respect, biohorizons are similar to metamorphic isograds (surfaces linking rocks of the same metamorphic grade). Isograds are identified by petrologists on the presence or absence of particular minerals. Similarly, careful study by a palaeontologist can enable the positions of biohorizons to be precisely delimited in the rock by the presence or absence of particular fossils. Part of the power of biostratigraphy lies in the ability to develop a detailed subdivision of a given sedimentary succession in a sequence which otherwise appears homogenous to the eye.

Since biozones have a geographical as well as stratigraphical extent, they can be mapped on a regional scale. A biozone might be folded, broken by faults, interrupted by unconformities, or partially or wholly obliterated by metamorphism. Geological maps and cross-sections showing biozones may be just as useful as lithological maps, for example, in elucidating the structural features of an oil field. In many instances, biostratigraphy has been crucial to the understanding of tectonically complex areas. A celebrated early example of this was Lapworth's (1878) unravelling of the geological structure of the Southern Uplands of Scotland using graptolite distributions. As an important part of the 'stratigraphical tool kit' (Doyle *et al.* 1994), biostratigraphy will always be of great academic and commercial importance.

Various practical difficulties are commonly encountered in applying a given biozonation to a new area. Firstly, sampling and preservational biases must be considered. Sampling density and sample sizes may vary considerably, and are sometimes beyond the control of the biostratigrapher. Fossils may be very abundant in some beds, but in other beds they may be scarce or difficult to extract. The mode of preservation may vary from place to place. For example, external moulds may be found in one place and internal moulds at another; a particular problem for some molluscan groups such as gastropods. Alternatively, some of the key fossils may have been dissolved away or abraded. Interval biozones are particularly susceptible to such problems because they are defined at least partly on negative evidence; that is, it is necessary to confirm the absence of a particular fossil or fossils to locate a given sample within an interval biozone.

Another common difficulty with some biozonations is that ancient erosion and re-sedimentation may have reworked fossils, making last-occurrence biohorizons difficult to identify with confidence. Similarly, organisms may have been living in crevices within an earlier formed sediment, or perhaps, in the case of microfossils, introduced by infiltrating fluids, which would cast doubt on first-appearance biohorizons. The procedures of sample recovery, especially in drilling and coring operations, may have caused some disturbance and mixing of samples. Laboratory contamination is also sometimes a serious concern, particularly for micropalaeontologists, and special measures must be adopted to minimize the problem.

A routine example of the practical application of biostratigraphy to a lithologically monotonous sequence is given in Figure 5.3. This is one of several sites drilled during Ocean Drilling Program Leg 144 on seamounts in the Pacific Ocean. At this site (ODP Hole 872C), 140 m of white pelagic ooze was recovered from the top of a Cretaceous volcanic seamount. Despite careful study by sedimentologists, only two lithological units can be identified in the cores, based on minor variations in the percentages of the microfossil types included (Figure 5.3A: 1, nannofossil–foraminifer ooze; 2, foraminifer ooze). However, many planktonic foraminifer species are present, and their stratigraphical ranges were determined. This allows the sequence to be divided into 20 biozones (Figure 5.3B; labelled P22 through to N22). Wavy lines in the figure are disconformities inferred from 'gaps' in the biostratigraphy but otherwise invisible in the core. Additional geologically important features that can be identified from the presence of 'out of sequence' fossils in the cores are down-hole contamination caused in the drilling process (inverted arrows), and sea-floor reworking (plus signs) (Figure 5.3B). Once a biostratigraphical zonation has been produced, the position of geological stage boundaries can be approximated (Figure 5.3C). Finally, it is possible to produce a biochronological age–depth plot (Figure 5.3D). This relies on the fact that biostratigraphical events have previously been assigned absolute ages by calibrating them at other sites to the geomagnetic polarity reversal time scale. In this case, the age–depth plot indicates temporal changes in sedimentation rate on top of the seamount, expressed as changes in the gradient of the curve.

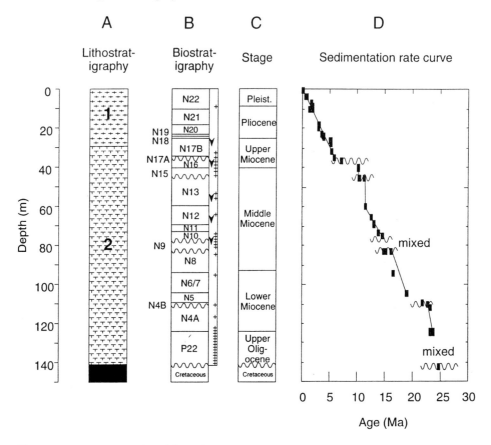

Figure 5.3 *Biostratigraphy of the pelagic sediments on Lo-En Guyot in the Marshall Islands group of seamounts (Ocean Drilling Program Hole 872C). Lithostratigraphy is from Premoli Silva, Haggerty, Rack et al. (1993). Biostratigraphical data are from Pearson (1995)*

5.2 DIACHRONEITY

The word 'biohorizon' seems to imply that the surface is in some way level, as does the less formal term 'biostratigraphical datum' which is sometimes used in its place. However, following the *International Stratigraphic Guide*, it cannot be assumed that a biohorizon is 'level' in time. Similarly, the term 'biostratigraphical event' seems to imply that whatever happened occurred simultaneously everywhere, but there is no reason why this must necessarily have been the case. Therefore, it is important to stress that the word 'event' is used in biostratigraphy more loosely than by sedimentologists, for whom an 'event-bed' is a lithological unit laid down in a fixed and relatively short period of time (e.g. see Chapter 6 and Ager 1973). A useful and underused word for rocks that occupy the same biostratigraphical position without implying that they were formed at the same time is 'homotaxis' (Huxley 1862).

A biohorizon, then, is effectively a surface connecting rocks with homotaxial assemblages. It may be nearly synchronous, i.e. following an imaginary 'time-line' in the rock; or it may be diachronous, cutting across 'time-lines'. A major objective of biochronology is to determine the degree of diachroneity on individual biohorizons, because this has a direct influence on the confidence with which an age can be assigned to a particular fossil assemblage.

One method of assessing the diachroneity of biohorizons is to compare the order of 'events' at more than one site. Obviously, if two events (e.g. two extinctions) occur in a different order at neighbouring sites, one or both biohorizons must be diachronous. When more than two events are considered, it becomes possible to determine, with caution, which is the more likely to have been diachronous, and to what extent. As further events and sites are compared, so greater confidence can be attached to selected biohorizons. This is the essence of graphical correlation (Shaw 1964; Martin *et al.* 1993).

Another method is to compare the levels of biohorizons with respect to sedimentological event-beds, such as volcanic ash bands or turbidites, which are believed to trace out approximate 'time-lines' in the rock. This approach is well exemplified by the recent work of Goldman *et al.* (1994) on the middle Ordovician of eastern North America. In this region, geologists have long been puzzled by a substantial mismatch between the major sedimentary formations and the graptolite biohorizons. It had been previously concluded that the biohorizons are highly diachronous and a poor guide to biochronological correlation. However, Goldman and colleagues were able to trace a group of bentonites (ash bands) across the region, each corresponding to an individual volcanic eruption. The graptolite biohorizons were found to parallel these bentonites, proving that they are, in fact, virtually synchronous over the entire area, whereas the sedimentary formations are markedly diachronous.

Given such evidence, some workers are inclined to take the view that certain biohorizons actually correspond to time-lines, as do lithological event-beds. This may be a useful practical step for unravelling certain stratigraphical complexities (e.g. Mitchell & Maletz 1995), but unfortunately it is often impossible to specify an instantaneous cause for a biostratigraphical 'event' in the same way as it is possible, for example, to envisage a discrete volcanic eruption as having given rise to an ash band. Without a detailed understanding of the mechanisms involved, it might always be argued that closer study of a biostratigraphical 'event' would reveal some degree of diachroneity. An exception to this could occur if a particular biohorizon, such as an extinction, could be confidently linked to a sudden cause such as a bolide impact. On the whole, however, palaeontologists remain ignorant of the causes of most of the extinctions used in biostratigraphy, which often occur without any apparent disruption to other species in the fauna. Similarly, with regard to evolutionary first appearances, it is difficult to envisage a mechanism for the instantaneous dispersal of a new species, which must have occurred if a given first-appearance biohorizon is to be regarded as an synchronous 'event'.

5.3 EVOLUTION AND BIOCHRONOLOGY

With the above considerations in mind, Figure 5.4 investigates the anatomy of a hypothetical biozone (for simplicity, a taxon range biozone) with regard to the time and space in which it was laid down (remembering, of course, that a biozone cannot itself be said to have a duration). The species defining the biozone is envisaged as originating at one particular time and place, and subsequently dispersing to its maximum geographical range. Its final extinction also occurs at a specific time and place, corresponding to the last living specimen. In Figure 5.4, the extinction is preceded by a period of contraction of the geographical range, which is interrupted by a phase of re-dispersal. The biochronological ranges of the taxon at different sites are represented by vertical bars in Figure 5.4. A real biozone might correspond to almost any shape on such a plot, depending on the rapidity of dispersal and extinction. The 'ideal' shape is of course a rectangle, in which dispersal and extinction are instantaneous over the entire area in which the biozone is found, but this cannot be assumed.

The situation illustrated in Figure 5.4 in effect corresponds to 'allopatric speciation', wherein a species arises in a small, geographically restricted and isolated population. This is thought to be the most common mechanism of speciation in many groups, including most of those used in biostratigraphy. However, in other

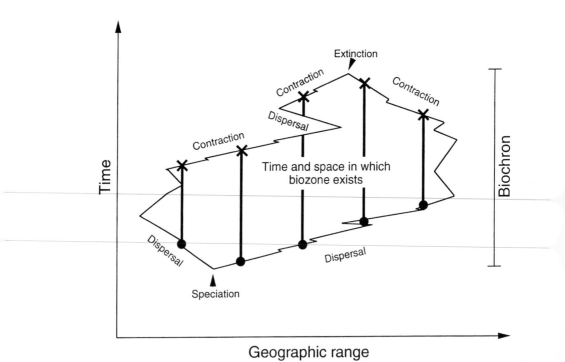

Figure 5.4   *Anatomy of a taxon range biozone illustrating the interaction of evolution and biogeography*

groups, speciation may take place within very large populations, involving much or all of their geographical range without clear isolation ('sympatric speciation'). For example, planktonic protistans such as radiolaria and foraminifera often live in huge concentrations that are constantly being mixed by ocean currents. Geographical barriers, such as those between water masses, are unlikely to be sufficiently persistent or unbridgeable to isolate a population long enough for speciation to occur (Lazarus 1986; Lazarus *et al.* 1995; Pearson *et al.* 1997).

In the allopatric model, the process responsible for a first-appearance biohorizon is dispersal (unless one is lucky enough to sample the exact place of origin of the new taxon). An examination of modern biogeographical distributions reveals countless examples of regional populations that might, at some time, expand or contract their range, producing diachronous biohorizons. Dispersal may occur very quickly if the organisms are mobile in some part of their life cycle and circumstances for the new species are favourable. Alternatively, re-invasion of the territory of the ancestral species may occur only after divergent evolution in the two forms allows them to coexist by exploiting different ecological niches. The progress of the re-invading species may be retarded by a hybridization zone, or a subtle environmental barrier such as temperature tolerance (Scheltema 1977; Valentine 1977; Kirby & Saul 1995), or it may be abrupt (e.g. Feist & Petersen 1995). Obviously such processes have great significance for the usefulness of a biohorizon that may be subsequently defined on the event.

In the sympatric case, dispersal is not an important factor. Rather, a first-appearance biohorizon is due to a combination of morphological change and reproductive isolation in place throughout the population, such that a new fossil taxon can be recognized. At present, despite much effort, palaeontologists are still a long way from a satisfactory understanding of the processes involved in sympatric speciation and the rates at which they occur. Sympatric speciation is signposted as a key area for future research.

Eventually, all species become extinct. Like the mechanisms of speciation, the causes of extinction have been much discussed. An optimum biohorizon is provided by a species which becomes extinct simultaneously everywhere. A devastating and rapidly transmitted virus or an extraterrestrial impact may conceivably make a species extinct rapidly with respect to the geological time scale. With respect to shelf faunas, it is believed that pulses of sea water anoxia may produce striking almost-synchronous extinction horizons that stand out from the 'background' evolutionary pattern. After such an extinction, invasion of new forms from deeper water sparks off renewed evolution on the shelf, producing a new group of species known as a 'biomere' (e.g. Palmer 1984). Alternatively, species may become finally extinct after the gradual diminution of their geographical range. For example, Jenkins (1992) has proposed that many planktonic foraminifera became progressively restricted to the tropical belt before their eventual extinction.

Another factor that must be considered when assessing a biohorizon, whether it is a first or last appearance, is the abundance pattern of the species in question. If a species is very rare after its initial evolution, and then only gradually increases in abundance, it may be difficult to correlate precisely. Similarly, if the decline in numbers before an extinction was gradual, the biostratigrapher would have to search very hard for the last remaining individual at every site, and correlations may be very sensitive to even small amounts of reworking (Haq & Worsley 1982). As a consequence of these problems, and if the fossil record is good, it is often preferable to quantitatively identify the abundance fluctuations of a population rather than search for its absolute appearance or disappearance level. The quantitative approach is commonly used for identifying nannofossil biohorizons, in which a species has to exceed an arbitrary abundance level (e.g. Raffi *et al.* 1993). A further refinement of this method is the use of 'cross-over events' in the identification of biohorizons in which the abundance of one species exceeds the abundance of another. For example, Backman & Shackleton (1983) argued that the cross-over in abundance of the Miocene–Pliocene nannofossil *Ceratolithus rugosus* and its precursor *C. acutus* is a much more reliable 'event' for regional correlation than the appearance or disappearance of either form. Recent work has extended this approach to many other nannofossil events, yielding significant improvements in biostratigraphy and biochronology.

5.4 TAXONOMY, SPECIES, AND THE DEFINITION OF BIOSTRATIGRAPHICAL UNITS

It is now time to consider in more detail how biostratigraphical units are defined and recognized. Some stratigraphical units, such as geological formations, are defined by the selection of a 'stratotype', which is a physical locality where the unit is declared to exist and to which all other places where that unit is found must be correlated (see Chapter 2). Although biostratigraphers do recognize 'reference sections' where a particular series of biozones is well displayed, the formal stratotype approach cannot be applied. Consider the hypothetical example in Figure 5.4, imagining for this purpose that the vertical dimension is stratigraphical thickness rather than time. In this diagram, it is impossible to find the complete biozone at any one place. Local first appearances and last occurrences do not necessarily correspond even closely to the global appearance and extinction levels. Therefore, as is made clear in the *International Stratigraphic Guide*, 'the concept of the biozone depends entirely on the concept of the taxa which are used to delimit its upper and lower boundaries' (Hedberg 1976). It is because of this that the procedure of fossil classification (taxonomy) is an integral part of the biostratigraphical method.

Biostratigraphy can be applied at any level in the taxonomic hierarchy. Even the most general kind of fossil information, such as the presence of an unidentified graptolite, may be of use for investigating problematic rock units, or in reconnaissance mapping. However, most practical biostratigraphy is applied at the species level and therefore requires specialist knowledge. Vitally important biohorizons often rely on fine morphological distinction between species, such as the number of lobes on an ammonoid suture or the density of thecae on a graptolite stipe. In this way, biostratigraphy is brought head-to-head with the 'species problem' in palaeontology. How do palaeontologists distinguish species in practice? Do these units correspond to species as generally understood in the biological sense? What processes are responsible for the appearance and disappearance of these units in the fossil record, and therefore the biohorizons that are defined upon them?

Few controversies in biology have been as contentious or long-running as that over definitions and concepts of 'species' (Mayr 1982). Early workers such as John Ray and Carl Linnaeus believed that each individual organism was an imperfect reflection of an ideal form or 'essence' that characterized the species to which it belonged (the 'essentialist' species concept). When naming a new species, a single 'type' specimen was selected which was believed to exhibit the most important features of the species (leading to a 'typological' classification). Nowadays, of course, biologists are agreed that there is no ideal form for a species. Rather, species are simply populations of individuals which may show great variability owing to their genetic differences, and may evolve through time. Mayr's definition is one of the most commonly quoted: 'species are groups of actually or potentially interbreeding natural populations which are reproductively isolated from other groups' (Mayr 1942: 120).

Individuals within a species are never identical in their morphology. This is partly because of the unique environmental effects that have influenced their growth and development ('ecophenotypic variation'), and partly because almost all species have substantial genetic variability. Natural selection may act on this variation so that populations at different times are significantly different from one another genetically and morphologically, but nevertheless belong to the same line of descent. A single line of descent is known as an 'evolutionary lineage', and can be thought of as an extension of the biological species through time. The concept of the lineage goes back at least to Huxley (1870), and was formalized by Simpson (1961: 153) who defined the evolutionary species or lineage as 'an ancestral–descendant sequence of populations evolving separately from others with its own unitary evolutionary role and tendencies'. Unfortunately, some modern workers use the term 'lineage' very loosely, but it is desirable to retain the strict meaning of Huxley and Simpson, in which populations must be connected by lineal descent. It is a key concept in palaeontology for which there is no more appropriate word.

Morphological evolution within a lineage is commonly known as 'anagenesis' (i.e. single line origin), to distinguish it from the process of branching, or 'cladogenesis' (i.e. branching origin). Anagenesis must have occurred to give rise to all the morphological differences we see between species, because cladogenesis, on its own, would simply produce a batch of cryptic species. In some groups, most anagenesis apparently happens rapidly around the time of cladogenesis ('punctuated equilibrium'), followed by long periods of stability ('stasis'), whereas in others it occurs over longer periods of time and is not necessarily associated with cladogenesis ('phyletic gradualism'). The relative frequency of these two patterns has been much debated (Gould & Eldredge 1993). It seems likely that gradualism may be common in those organisms which belong to huge well-mixed populations such as the marine plankton, for whom speciation is difficult to achieve by geographical isolation (Pearson *et al.* 1997). Another proposal is that gradualism may be dominant in those populations which occupy more constant environments, whereas stasis and punctuation occur in more widely fluctuating conditions. This hypothesis has been dubbed the *'plus ça change'* model by Sheldon (1996) after the saying *'plus ça change, plus c'est la même chose'* (the more that something changes, the more that it is the same thing). Whatever the relative frequency of punctuation and gradualism, it is clear that both patterns must have been to some extent responsible for producing biostratigraphical events.

5.5 EXPLORING MORPHOSPACE

Having considered the processes of evolution, it is now necessary to discuss the way in which the complexities of an evolutionary history are perceived by palaeontologists engaged in the practical task of fossil classification. Taxonomy is a collective endeavour, achieved by the scientific community through continuous improvement. Very often, initial discoveries of a fossil group are haphazard chance finds, and early investigators may be unaware of each other's contributions. Once the potential value of a group is recognized, this phase is followed by specialist treatment, resulting in monographs in which the whole group is reviewed and coherent classifications are presented. Further discoveries may be integrated into such schemes or may force a re-evaluation. With increased understanding, so the anatomy of the fossil group in question becomes better understood, and taxonomists learn which features are most useful in making reliable species discriminations. Standards of description and illustration inevitably improve, and while early work may have been conducted on just a few specimens, later studies often involve detailed analysis of large populations in well-ordered stratigraphical series with an emphasis on well-preserved material. At any time, the current body of taxonomic knowledge amounts to a 'descriptive filter' through which fossil information is processed (Pearson 1993). This descriptive filter imparts a subtle, sometimes even unconscious, influence on the basic data of biostratigraphy.

According to Mayr, the replacement of the essentialist species concept of Linnaeus and others by 'population thinking' was one of the most important foundations for modern biology (Mayr 1982). While this may be true in theory, it is a curious anomaly that the typological approach is still firmly imbedded in standard taxonomic procedures. When a new species is named, a type specimen (the 'holotype') must be selected which is considered to be representative of the species. According to the rules (i.e. the *International Code of Zoological Nomenclature*), if a type specimen is not selected, the species is not valid. The typological method remains fundamental to taxonomy although it no longer reflects an essentialist philosophy. It has survived because it provides a powerful and flexible system for assembling taxonomic information. New species can be named and described on very fragmentary evidence, such that further discoveries can later add to our knowledge of the species. Species can be named for asexual organisms such as bacteria, and even for parts of organisms, such as the various conodont elements which make up the animal's feeding apparatus, or for the roots and fruits of a particular plant.

A useful idea to consider in this context is 'morphospace', a conceptual volume in which all the shapes of a group of fossils (e.g. the individuals within a genus) are distributed. Morphospace may have many dimensions, one for each character that varies in the group in question. Morphospace is unlikely to be occupied continuously and homogeneously, because of the evolutionary processes described above. Instead, individuals will tend to occur in clusters which reflect their genetic and ecotypic similarities. Putting aside for the moment considerations of sibling species, hybridism and so on, Figure 5.5 illustrates a simplified case, in which members of biospecies intergrade with one another morphologically, but are separated from the individuals of other biospecies by empty morphospace.

Taxonomists are effectively explorers of morphospace. If a morphology is discovered that is sufficiently different from any previously described form, a new species is named. This involves the selection of a type specimen (the holotype) which, like any individual, has a unique morphology and can be thought of as occupying a single point in morphospace. Naming a new species is, therefore, like planting a flag in morphospace, just as explorers might claim a new territory. In future studies, all subsequently discovered morphologies which lie close to this 'flag' (i.e. the point in morphospace defined by the holotype) can be referred to that species.

Ideally, as knowledge of a fossil group increases, taxonomists can ensure that a set of 'flags' is fairly evenly spaced through morphospace, such that any individual can be easily classified to a species by reference to the nearest holotype. The effect of this is to overlay a 'taxonomic grid' on the range of variation that exists in nature (Figure 5.5A). When viewed in this way, typological species can be thought of as *sectors of morphospace* centred on the morphology of their holotype and delimited by proximity to the holotype morphologies of other formally described species (Pearson 1993).

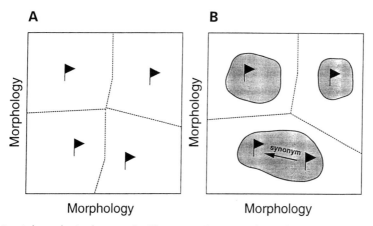

Figure 5.5 A hypothetical example illustrating how typological classification works in a single time plane. Morphospace is multidimensional, but is shown here in two dimensions. *A.* Flags represent the positions of holotype specimens of a newly discovered genus. Other individuals are assigned to the species with the most similar (nearest) holotype. Dotted lines indicate the 'taxonomic grid' overlying morphospace that results from this typological approach. *B.* Shaded areas represent morphospace actually occupied, as subsequently discovered by studying population variability. In this case, two nominal species are confirmed as synonyms, causing a reorganization of the taxonomic grid

Clearly, it is desirable that just one typological species is named for every biological species; that is, in this simplified case, every cluster of shapes that exists in morphospace. In the hypothetical example given in Figure 5.5, detailed study following the initial descriptive phase reveals that two of the typological 'species' are in fact fully connected by intermediate forms. Since they belong to a single cluster, it is reasonable to propose that they are merely variants of a single biospecies, so one of the nominal species is superfluous. Such species are known as 'subjective synonyms', and the rule is that the earliest named species has priority. Much of the craft of taxonomy involves the naming of new species, and the recognition of synonyms, such that a useful classification scheme results which is consistent with the diversity known to have occurred in real populations.

So far, the typological method can be made to work in a satisfactory way, such that typological species have a one-to-one relationship with biological species. Of course, there are many potential difficulties which may be encountered, such as the existence of markedly different growth stages in some species (consider caterpillars and butterflies), sexual dimorphs, cryptic species and so on. It is even possible to envisage members of a biological species which, by virtue of the great variability of that species, are actually closer in morphology to the holotype of another species than they are to the holotype of their own. Such occasions are rare, however, and can be dealt with individually. Despite the philosophical clash between typological taxonomy and population thinking, on the whole the approach can be made to fit with the biological species concept, provided only a single time plane is considered.

A useful tool sanctioned by the rules of nomenclature is that a worker can formally select a series of specimens (in addition to the holotype) that are believed to belong to the same species, in order to illustrate the range of variation encountered. Such specimens are known as 'paratypes', and, in effect, their selection allows a taxonomist to more carefully delimit the sector of morphospace occupied by the species in question. However, a subsequent investigator may find that some or all paratypes belong to different species than the holotype – something that happens all too often. In that case, the holotype alone remains the true arbiter of the species concept.

5.6 CHRONOCLINES AND PSEUDOSPECIES

The real difficulty in squaring the typological approach with the biological species concept comes when the time dimension is considered. The root of this is that there is no reason why morphology should be bounded in the time dimension. Consider a lineage which evolves though time such that the morphologies produced eventually lie outside the range of variation that was initially encountered. Many such lineages (known as 'chronoclines') have been described by palaeontologists, especially in those groups that are useful in biostratigraphy. How can successive populations from such a chronocline be classified? If no divisions are made, then wholly different fossils will have to be classified together, and a great deal of information will be lost to biostratigraphy. If the taxonomist breaks the chronocline down into 'species', they will inevitably have artificial boundaries that cut through the morphological continuum in one direction or another. Such artificial species are termed 'pseudospecies' and there are two main types (Figures 5.6A and B).

The approach to breaking down a chronocline that is easiest to reconcile with the biological species concept is to divide the chronocline using the time (or stratigraphical) dimension, i.e. according to the range of variation encountered in successive populations (Figure 5.6A). This kind of procedure can be conducted using various statistical procedures, and is of great potential use for the detailed biostratigraphy of well-known lineages. This type of pseudospecies can be called a 'chronospecies'. Examples of this approach are given in the studies of Baarli (1986) and Wei (1987). However, a problem with chronospecies is that individuals having identical morphologies will be classified in different species simply because they come from different times, or more precisely because they are drawn from populations that have different morphological distributions. A large amount of population information is therefore required to make such divisions workable, and this may not be available in practice. Often the biostratigrapher has to make do with a small number of fossils, or wants to know the stratigraphical significance of even a single specimen. On a more practical level, he or she may not have enough time to indulge in detailed morphometric studies every time a biostratigraphical age-determination is required.

An alternative solution is to subdivide the chronocline vertically, using divisions in morphospace (Figure 5.6B). As discussed previously, this is in fact the inevitable result of the typological method. As a consequence, a number of pseudospecies belonging to a single lineage are distinguished by their morphology using the established formal procedures. This type of pseudospecies is known as a 'morphospecies', and each is defined by proximity to a type specimen. For the biostratigrapher, it is desirable to subdivide chronoclines as finely as possible,

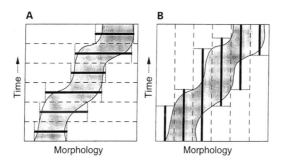

Figure 5.6 *Options for the subdivision of a chronocline into pseudospecies. In these diagrams, morphospace has been reduced to just one dimension. The shaded area corresponds to a lineage evolving through time. **A.** Populations can be distinguished according to the range of variation encountered at different stratigraphical levels (chronospecies). **B.** Individuals are distinguished according to their morphological similarity (morphospecies). The latter type of subdivision is the inevitable result of the typological approach. The vertical dotted lines represent extensions through time of the 'taxonomic grid' of Figure 5.5*

such that the greatest amount of stratigraphical resolution is achieved. Often this results in multiple overlapping ranges of morphospecies, which all intergrade, and in fact belong to the same evolving lineage. Inevitably, these morphospecies are artificial units (Simpson 1937; Bown & Rose 1987; Blatt *et al.* 1991).

The adoption of morphospecies may offend the sensibilities of biologically minded taxonomists, but the concept is straightforward to apply. It should also be remembered that the use of the term 'species' for typologically defined units of classification has priority over its use in the modern 'biological species concept'. The rationale of the morphospecies approach has been very ably explained by Fordham (1986, 1995) who argues that typologically defined fossil 'species' (he calls them 'phena') are conceptually distinct from evolutionary lineages, and should be classified separately to avoid confusion. Provided that a clear distinction is made, and no attempt is made to pass morphospecies off as biological species, then the approach is sound. Likewise, it is acceptable to name morphospecies based on parts of organisms, or distinct growth stages, or even trace fossils, and such units can be used to great effect in biostratigraphy.

It has been claimed that because biostratigraphy proceeds by the recognition of the stratigraphical ranges of species, it necessarily provides support for the punctuated equilibrium model of evolution. For example, Eldredge & Gould (1977: 29) stated that 'by the mere recognition of *any* nontrivial stratigraphical range of *any* morphologically defined taxon at or near the species rank, we are necessarily implying a stability or stasis in species-specific differentia'. However, it can be seen from Figure 5.6 that just because a morphospecies can be given a stratigraphical range, it does *not* imply stasis in the population as a whole, as Eldredge and Gould claimed. Instead, if the descriptive filter of taxonomy is not appreciated for what it is, the existence of stratigraphical ranges might give the impression of punctuation and stasis in cases where the pattern does not exist (Sheldon 1987).

As typologically defined species can be thought of as sectors of morphospace centred on their holotypes, so the stratigraphical ranges of those species can be thought of simply as the stratigraphical thicknesses over which sectors of morphospace are occupied (Pearson 1993). This is a neutral way of viewing them, in which their biological significance in terms of anagenesis and cladogenesis can be deferred for a dedicated study of lineages. Indeed, disagreements about the particular way in which morphospecies should be grouped together into lineages may in themselves have no direct biostratigraphical implications. This point was demonstrated recently by Neraudeau *et al.* (1995), in their study of fossil water voles. In principle, the evolutionary appearance of a morphospecies in the record is always due to anagenesis, which may or may not correspond to biological speciation (cladogenesis). If it does not, the event is known as a 'pseudospeciation'. Similarly, the disappearance of morphospecies may be due to 'pseudoextinction', or it may indicate a real extinction. In either case, a workable biostratigraphy can be developed, but it is important to recognize that the resulting biohorizons will have a totally different character. Figure 5.7 explores some of the relationships between the stratigraphical ranges of morphospecies and a hypothetical sequence of biological speciations and extinctions, following Fordham's (1986) line of argument.

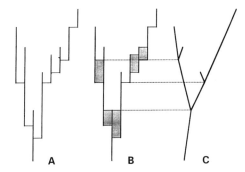

Figure 5.7 *Possible relationships between typological species and evolutionary lineages.* **A.** *The dendrogram records the stratigraphical ranges of morphospecies which are linked to show their supposed evolutionary relationships.* **B.** *Further study shows that some of the morphospecies are linked by intermediates at various times (shown by grey shading).* **C.** *A lineage phylogeny is revealed by this approach, which documents the actual times of cladogenesis and extinction (as opposed to pseudospeciation and pseudoextinction). Note that cladogeneses may correspond to pseudospeciations, pseudoextinctions, or even may occur midway in the stratigraphical ranges of two morphospecies. Pseudoextinctions may or may not correspond to extinctions, and anagenetic transitions may be gradual or sudden.* [Modified from: Pearson (1993)]

5.7 A NEW CLASSIFICATION OF BIOHORIZONS

The *International Stratigraphic Guide* attempts the formal recognition of biostratigraphical units defined on gradual evolutionary transitions by including the concept of a *lineage biozone*. This is defined as 'a type of range [bio-]zone consisting of the body of strata containing specimens representing a segment of an evolutionary or developmental line or trend, defined above and below by changes in features of the line and trend' (Hedberg 1976: 58–59). Unfortunately, the definition as given has several weaknesses. One problem is that the concept is restricted to range biozones, but it is equally possible to imagine types of interval biozone that are based on gradual transitions. Indeed, two of the hypothetical examples given in the guide *are* interval biozones. Also, to be a lineage biozone by this definition, it has to be defined above and below by gradual transitions, but it is possible to imagine biozones that are defined at one end by a gradual transition and at the other by a more objective event such as a genuine extinction. Another hypothetical 'lineage zone' given by the guide is of this type.

Instead of attempting to classify biozones based on the mode of evolution which defines their boundaries, it would seem a better approach to retain the biozone nomenclature presented in Figure 5.2, but classify individual biohorizons on evolutionary principles. This helps to focus discussion on the causes of individual biostratigraphical events. The following classification is restricted to biohorizons defined on the end-points of the stratigraphical ranges of morphospecies. Of the four types proposed, the first two are 'objective' in the sense that all workers would be expected to place them at the same level in each stratigraphical section given the same observations; and the second two are 'subjective', relying on an arbitrary taxonomic subdivision of a morphological continuum.

1. *Dispersal biohorizon.* If a species evolves in one place and then spreads geographically it may provide a useful and unambiguous biostratigraphical event. At any individual site, a dispersal biohorizon is marked by the first appearance of a species with no evidence of gradual transition from a pre-existing species. It may be more or less diachronous, depending on factors such as the mobility of the organisms in question, their environmental tolerance, and the presence or absence of competing species. This type of biohorizon is anticipated in groups that evolve by allopatric speciation, but is expected to be rare in groups that tend to evolve sympatrically. The reappearance of a species after its disappearance from a particular area is a special case of a dispersal biohorizon that may be useful for local correlations.

2. *Extinction biohorizon.* If a species disappears altogether from a particular area without leaving any directly descendent species, it will provide an unambiguous event for biostratigraphical correlation. An extinction biohorizon is marked by the last occurrence of a species that leaves no direct descendants. It may be more or less diachronous, depending on the causes of the extinction. In some cases, a species may locally disappear only to reappear again, which is a special case of an extinction biohorizon. Note that the term 'extinction' in this context refers to the local disappearance of a taxon. The final genetic extinction of a lineage cannot be determined with confidence from any individual area.

3. *Pseudospeciation biohorizon.* A pseudospeciation is deemed to have occurred whenever a morphospecies appears in the stratigraphical record by continuous evolutionary transition in place from a pre-existing form. Its stratigraphical placement depends on a subjective judgement as to where the correct boundary between the ancestral and descendent taxa should be placed. The degree of diachroneity on a pseudoextinction biohorizon depends on complex morphological and genetic factors. If a population shows little geographical variability, and evolves as a coherent unit, the biohorizon may be fairly synchronous. However, if substantial geographical variability occurs, diachroneity will result. In particular, the movement of geographical clines as a result of environmental change will inevitably result in stratigraphical clines, which will produce diachronous pseudospeciation biohorizons.

4. *Pseudoextinction biohorizon.* Just as morphospecies may appear in the stratigraphical record at a particular site by anagenesis, so they may disappear in the same way. A pseudoextinction occurs when a morphospecies disappears from the record but fully intergrades with a descendent form in an unbroken lineage. Diachroneity of a pseudoextinction biohorizon may be caused by the same factors as cause the diachroneity of a pseudospeciation biohorizon.

As previously mentioned, the above list is not intended as a classification of all biohorizons, but only those that are determined by the first appearances or last occurrences of typologically defined taxa (morphospecies). Nevertheless, biohorizons defined on size changes within a lineage can be thought of as being either pseudospeciation or pseudoextinction biohorizons (each size class being, in effect, a morphospecies), likewise those defined statistically on the appearances or disappearances of chronospecies. However, excluded from the classification are various other types of biohorizon, such as those defined on abundance changes, cross-over events, assemblage changes and so on. Also additional to this classification are biohorizons that may potentially be defined on cladogenetic events, as perceived by palaeontologists as the breakdown of a pre-existing morphological continuum (see Fordham 1995).

It is useful to recognize at a formal level the difference between, for example, a dispersal biohorizon and a pseudoextinction biohorizon. Traditionally, both would be viewed simply as 'first appearances', but the processes involved are quite different. It is possible to imagine, for example, a species which evolved by pseudospeciation in one area and subsequently dispersed. This would result in two distinct biohorizons of entirely different character, which it is useful to distinguish. Similarly, it is important not to confuse a pseudoextinction biohorizon with an extinction biohorizon.

In general, pseudospeciation and pseudoextinction biohorizons are likely to be much more difficult to identify with confidence at a high resolution, because they are based on gradual transitions. For example, the Palaeogene/Neogene working group of the International Commission on Stratigraphy recently recommended defining the Oligocene–Miocene boundary at the level of first appearance of the planktonic foraminifer *Globorotalia kugleri* in the Rigorosa Formation of northern Italy (Steininger 1994). Unfortunately, the selected event is a pseudospeciation (*G. kugleri* having evolved by continuous transition from *G. pseudokugleri*) and is therefore very difficult to identify consistently at a temporal resolution less than several tens of thousands of years, even in well-preserved material.

Since different investigators inevitably draw their distinctions between morphospecies at subtly different places, this must result in biohorizons being identified, unwittingly, at different stratigraphical levels. Progress in improving the correlation of transitional biohorizons like the first appearance of *G. kugleri* may be best achieved by establishing a better consensus as to the permitted bounds of particular morphospecies. It is necessary for each worker to carefully describe the criteria that are used to separate the defining taxa, because this is an integral part of the definition of the biohorizon. It is worth bearing in mind that although they may be 'arbitrary' events, many gradual transitions can, in principle, be defined unambiguously if common agreement can be reached.

Finally, it may be asked what proportion of real biohorizons are defined on 'arbitrary' events like pseudospeciations and pseudoextinctions. This is a difficult question to answer unless the fossil record of the group in question is known in great detail. Probably the best known of all fossil groups is the planktonic foraminifera, on which much morphometric work has been conducted. For these it is possible to provide an estimate: of the 58 planktonic foraminifer biohorizons used by Berggren *et al.* (1995) for Cenozoic correlations, as many as 31 are probably gradual transitions. The proportion is bound to vary from group to group depending on the balance between allopatric and sympatric speciation, and the frequency of phyletic gradualism. However, even if gradual transitions in other groups are much rarer than in planktonic foraminifera, the proposed nomenclature will still be of value in highlighting anomalous cases.

5.8 CONCLUSION

As Romer (1958) remarked, fossils would have been as useful to the earliest biostratigraphers had they been distinctive assortments of nuts and bolts rather then the remains of once-living organisms. Modern biostratigraphy, however, relies intimately on a biological understanding of fossils. Only in an evolutionary context is it possible to intelligently assess the reliability of correlations based on those 'nuts and bolts'.

Although quantitative studies are becoming increasingly common, most biostratigraphy still relies simply on presence/absence data for particular fossil taxa. Whatever the approach, the most fundamental question a biostratigrapher can ask is what processes affect the likelihood that a given fossil taxon will be found in a particular bed? Part of the answer to this question involves answering a second, related question: what criteria are used to include a fossil in that particular taxon, as opposed to other, similar, taxa? If all life evolved by an extreme form of punctuated equilibrium, in which species appeared fully formed and easily distinguished from one another, and remained constant until their extinction, a wholly objective typological taxonomy would be a realistic aim. However, we know that gradual evolution can and does occur, many examples having been described from fossil groups commonly used in biostratigraphy. Gradualism is by no means the enemy of biostratigraphy; in fact it provides substantially more information on which to base correlations than arises simply from cladogenesis and extinction. However, if the traditional methods of end-point biostratigraphy and typological taxonomy are to be retained, as they certainly can be, it is necessary to formally distinguish those biohorizons defined on phyletic transitions from those resulting from genuine first appearances and extinctions. This will help focus attention on the evolutionary causes of individual biohorizons.

ACKNOWLEDGEMENTS

I thank Stuart McKerrow for stimulating an interest in these subjects, and Ian Billing and Andrew Shortland for the ensuing discussion. This chapter was written while in receipt of a Royal Society University Research Fellowship at the University of Bristol.

REFERENCES

Ager, D.V. 1973. *The nature of the stratigraphical record*. Macmillan, London.
Baarli, B.G. 1986. A biometric reevaluation of the Silurian brachiopod lineage *Stricklandia lens/S. laevis. Palaeontology*, **29**, 187–205.

Backman, J. & Shackleton, N.J. 1983. Quantitative Biochronology of Pliocene and early Pleistocene calcareous nannofossils from the Atlantic, Indian and Pacific Oceans. *Marine Micropalaeontology*, **8**, 141–170.

Berggren, W.A., Kent, D.V., Swisher, C.C. & Aubry, M. 1995. A revised Cenozoic geochronology and chronostratigraphy. In Berggren, W.A., Kent, D.V., Aubry, M. & Hardenbol, J. (eds) *Geochronology, time scales and global stratigraphic correlation*. Society of Economic Paleontologists and Mineralogists, Special Publication No. 54, 129–212.

Blatt, H., Berry, W.B.N. & Brande, S. 1991. *Principles of stratigraphic analysis*. Blackwell, Boston.

Bown, T.M. & Rose, K.D. 1987. *Pattern of dental evolution in early Eocene Anaptomorphine primates (Omomyidae) from the Bighorn Basin, Wyoming*. The Paleontological Society, Memoir No. 23.

Doyle, P., Bennett, M.R. & Baxter, A.N. 1994. *The key to earth history*. John Wiley & Sons, Chichester.

Eldredge, N. & Gould, S.J. 1977. Evolutionary models and biostratigraphic strategies. In Kauffman, E.G. & Hazel, J.E. (eds) *Concepts and methods of biostratigraphy*. Dowden, Hutchinson & Ross, Stroudsberg, 25–40.

Feist, R. & Petersen, M.S. 1995. Origin and spread of *Pudoproetus*, a survivor of the late Devonian trilobite crisis. *Journal of Paleontology*, **69**, 99–109.

Fordham, B.G. 1986. Miocene–Pleistocene planktic foraminifers from DSDP Sites 208 and 77, and phylogeny and classification of Cenozoic species. *Evolutionary Monographs*, **6**, 1–200.

Fordham, B.G. 1995. Advantages of Comprehensive Taxonomy for routine systematic documentation in conodont biostratigraphy. *Boletin de la Academia Nacional de Ciences*, **60**, 469–482.

Goldman, D., Mitchell, C.E., Bergstrom, S.M., Delano, J.W. & Tice, S. 1994. K-bentonites and graptolite biostratigraphy in the middle Ordovician of New York state and Quebec: a new chronostratigraphic model. *Palaios*, **9**, 124–143.

Gould, S.J. & Eldredge, N. 1993. Punctuated equilibrium comes of age. *Nature*, **366**, 223–227.

Hancock, J.M. 1977. The historic development of concepts in biostratigraphic correlation. In Kauffman, E.G. & Hazel, J.E. (eds) *Concepts and methods of biostratigraphy*. Dowden, Hutchinson and Ross, Sroudsberg, 3–22.

Haq, B.U. & Worsley, T.T. 1982. Biochronology – biological events in time resolution, their potential and limitations. In Odin, E.S. (ed.) *Numerical dating in stratigraphy*. John Wiley & Sons, Chichester.

Hedberg, H.D. (ed.) 1976. *International stratigraphic guide*. John Wiley, New York.

Huxley, T.H. 1862. Geological contemporaneity and persistent types of life. In Huxley, T.H. 1908. *Discourses, biological and geological*. Macmillan, London, 340–388

Huxley, T.H., 1870. Palaeontology and the doctrine of evolution. In Huxley, T.H. 1908. *Discourses, biological and geological*. Macmillan, London, 340–388

Jenkins, D.G. 1992. Predicting extinction of some extant planktonic foraminifera. *Marine Micropalaeontology*, **19**, 239–243.

Kirby, M.X. & Saul, L.R. 1995. The Tethyan Bivalve *Roudairia* from the upper Cretaceous of California. *Palaeontology*, **38**, 23–38.

Lapworth, C. 1878. The Moffat series. *Quarterly Journal of the Geological Society of London*, **34**, 240–346.

Lazarus, D.B. 1986. Tempo and mode of morphologic evolution near the origin of the radiolarian lineage *Pterocanium prismatum*. *Paleobiology*, **12**, 175–189.

Lazarus, D.B., Hilbrecht. H., Spencer-Cervato, C. & Thierstein, H. 1995. Sympatric speciation and phyletic change in *Globorotalia truncatulinoides*. *Paleobiology*, **21**, 28–51.

Martin, R.E., Neff, E.D., Johnson, G.W. & Krantz, D.E. 1993. Biostratigraphic expression of Pleistocene sequence boundaries, Gulf of Mexico. *Palaios*, **8**, 155–171.

Mayr, E. 1942. *Systematics and the origin of species*. Columbia University Press, New York.

Mayr, E. 1982. *The growth of biological thought*. Belknap Press, Cambridge, MA.

Mitchell, C.E. & Maletz, J. 1995. Proposal for adoption of the base of the *Undulograptus austodentatus* Biozone as a global Ordovician stage and series boundary level. *Lethaia*, **28**, 317–332.

Neraudeau, D., Viriot, L., Chaline, J., Laurin, B. & van Kolfschoten, T. 1995. Discontinuity in the Plio-Pleistocene Eurasian water vole lineage. *Palaeontology*, **38**, 77–85.

North American Commission on Stratigraphy 1983. North American stratigraphic code. *Bulletin of the American Association of Petroleum Geologists*, **67**, 84–87.

Palmer, A.R. 1984. The Biomere problem: evolution of an idea. *Journal of Paleontology*, **58**, 599–611.

Pearson, P.N. 1993. A lineage phylogeny for the Paleogene planktonic foraminifera. *Micropaleontology*, **39**, 193–232.

Pearson, P.N. 1995. Planktonic foraminifer biostratigraphy and the development of pelagic caps on guyots in the Marshall Islands Group. *Proceedings of the Ocean Drilling Program, Scientific Results*, **144**, 21–59.

Pearson, P.N., Shackleton, N.J. & Hall, M.A. 1997. Stable isotope evidence for the sympatric divergence of *Globigerinoides trilolous* and Orbulina universa (planktonic foramihifera). *Journal of the Geological Society of London*, **154**, 295–302.

Premoli Silva, I., Haggerty, J., Rack, F. *et al.* 1993. *Proceedings of the Ocean Drilling Program, Initial Reports*, **144**.

Raffi, I., Backman, J., Rio, D. & Shackleton, N.J. 1993. Plio-Pleistocene nannofossil biostratigraphy and calibration to oxygen isotope stratigraphies from Deep Sea Drilling Project Site 607 and Ocean Drilling Program Site 677. *Paleoceanography*, **8**, 387–408.

Romer, A.S. 1958. Darwin and the fossil record. In Barnett, S.A. (ed.) *A Century of Darwin*. Heinemann, London, 130–152.

Scheltema, R.S. 1977. Dispersal of marine invertebrate organisms: paleobiogeographic and biostratigraphic implications. In Kauffman, E.G. & Hazel, J.E. (eds) *Concepts and methods of biostratigraphy*. Dowden, Hutchinson and Ross, Sroudsberg, 73–107.

Schoch, R.M. 1989. *Stratigraphy: principles and methods*. Van Nostrand Reinhold, New York.

Shaw, A.B. 1964. *Time in stratigraphy*. McGraw-Hill, New York.

Sheldon, P.R. 1987. Parallel gradualistic evolution in Ordovician trilobites. *Nature*, **330**, 561–563.

Sheldon, P.R., 1996. Plus ça change – a model for stasis and evolution in different environments. *Palaeogeography, Palaeoclimatology, Palaeoecology*, **127**, 209–227.

Simpson, G.G. 1937. Patterns of phyletic evolution. *Bulletin of the Geological Society of America*, **48**, 303–314.

Simpson, G.G. 1961. *Principles of animal taxonomy*. Columbia University Press, New York.

Steininger, F.F. 1994. *Proposal for the Global Stratotype Section and Point (GSSP) for the base of the Neogene (the Paleogene/Neogene boundary)*. International Commission on Stratigraphy, Subcommission on Neogene Stratigraphy: Working Group on the Paleoegene/Neogene boundary, Vienna.

Valentine, J.W. 1977. Biogeography and biostratigraphy. In Kauffman, E.G. & Hazel, J.E. (eds) *Concepts and methods of biostratigraphy*. Dowden, Hutchinson and Ross, Stroudsberg, 73–107.

Wei, Kou-Yen 1987. Multivariate morphometrical differentiation of chronospecies in the late Neogene planktonic foraminiferal lineage *Globoconella*. *Marine Micropaleontology*, **12**, 182–202.

6
Event Stratigraphy: Recognition and Interpretation of Sedimentary Event Horizons

Gerhard Einsele

The correlation of sedimentary sequences via marker beds or event horizons is the subject of event stratigraphy. It uses the idea that isochronous events are recorded in the stratigraphical record and can be used to correlate both within and sometimes between depositional basins. In this chapter a range of events are explored and the sedimentological criteria by which they can be recognized are reviewed.

In geology and palaeontology the term 'event' is often used in different senses. For example, events have been postulated in the evolution and extinction of life, and as a result of volcanic activity, climate change and palaeogeography. Some of these events may last as long as several millions of years and as a consequence cannot be used as isochronous events in stratigraphy. This chapter, however, defines events as shorter, mainly physically controlled, incidents which vary in both character and duration. The most relevant of these to stratigraphy are depositional, non-depositional and erosional events. However, there are a number of episodes in which the faunal response to physical processes is important, and therefore many physical event beds can be defined by both sedimentological and faunal criteria. In this chapter, four following main types of physical event horizons are distinguished and discussed.

1. *Depositional events.* Depositional events occur practically instantaneously, that is within hours and days. They usually represent rare intervals of rapid deposition

Unlocking the Stratigraphical Record: Advances in Modern Stratigraphy. Edited by P. Doyle and M.R. Bennett.
© 1998 John Wiley & Sons Ltd.

within a system of relatively slow background sedimentation. Other terms used for similar phenomena are 'depositional episodes' (Dott 1983) or, for particularly catastrophic cases, 'convulsive geological events' (Clifton 1988). Volcanic ash layers and other tephra deposits belong to this group.

2. *Non-depositional and erosional events.* Erosional events may be either single episodes of short duration, such as intense fluvial erosion on land or storm wave and current erosion in shallow seas; or they occur repeatedly, separated by intervals of variable duration. Repeated erosional events lead to multiple winnowing and reworking of the pre-existing sediment surface. Non-deposition, omission, condensation and sediment starvation are commonly described events in marine sediments. They occur over longer time periods and are frequently associated with sea-level changes, variations in the oceanic current systems, or fluctuations in the terrestrial sediment input and biogenic sediment production. These types of events are often not strictly isochronous and are therefore of limited value in stratigraphical correlation.

3. *Other rare physical events.* These include earthquakes producing specific sedimentary structures (seismites) and meteorite impacts generating characteristic sediment layers or leaving behind traces of specific elements.

4. *Biological events.* A typical example of a biological event is the sudden appearance of new taxa or faunal elements that were not formerly present in the area studied (e.g. Ernst *et al.* 1983; Kauffman 1988; Kauffman *et al.* 1991). These phenomena usually reflect drastic changes in environmental conditions, such as the opening of oceanic gateways, the modification of current systems, and variation in nutrient supply or in redox conditions. Frequently, the resulting event beds are controlled by both physical and biological processes and are therefore referred to as composite events (Kauffman *et al.* 1991). Some biological events, such as in the evolution of life (e.g. Walliser 1995), usually span longer time periods than most other events.

This chapter mainly deals with physical events, that is event types 1 and 2 in the above list. These are manifest as event beds or event horizons in almost all sections of continental and marine sediments. Glacial events recorded by dropstone horizons in lacustrine and marine sediments are not considered. Type 3 events are only briefly discussed, and those of type 4 are beyond the scope of this chapter. Key definitions of the important terms used are collected in Table 6.1.

6.1 GENERAL CHARACTERISTICS OF STRATIGRAPHICAL EVENTS

Significant events, whether physical, biological or composite, are recorded by discrete thin stratal units or discontinuities in sedimentary sections. Depositional events usually form sedimentary layers of centimetre to sub-metre thickness, although they may also form beds which are tens of metres thick. In all cases these layers are referred to as *event beds*. Event beds normally deviate in composition, texture, structure and fossil content from their host sediments (Einsele *et al.* 1991).

Table 6.1 *Definitions of key terms*

Amalgamation   Upper part or entire pre-existing event bed is reworked and redeposited by subsequent depositional event (mixing of older and younger material).

Calcareous tempestites   Skeletal carbonate from carbonate ramps and platforms is redeposited as graded beds at the site of erosion and in deeper water.

Condensation   Thinning of a sedimentary deposit without interruption of sediment accumulation.

Condensed section   Relatively thin stratigraphical succession which is uninterrupted, especially at water depths below the storm wave base, and comprises a considerable time span.

Debris flows   Flow mass typically contains a large proportion of clasts supported by sandy and muddy matrix. Calcareous types resulting from semi-lithified carbonate buildups may be poor in fine-grained matrix. Their deposits are often referred to as sedimentary 'mega-breccias'.

Deep-sea fan association   Sediments and facies architecture of deep-sea fans, including channel fills, fan lobes and various turbidites.

Event bed   Relatively thin stratal unit with specific characteristics not present in the host sediment.

Event horizon   Discontinuity surface in 'normal' sediment buildup, or layer of unusual fossil content without distinct change of physical sedimentary characteristics. If used in a wide sense, this term may also include event beds.

Lag sediment (lithoclasts, non-skeletal)   The fine-grained portion of a primarily mixed sediment layer is sorted out and removed by current or wave action leaving behind the coarsest and most resistant particles.

Marker bed, key bed, marker horizon   Event bed or event horizon with characteristics markedly different from the host sediment. Key beds or marker beds and marker horizons can be easily traced over long distances.

Mega-turbidites   Exceptionally thick and widely extended turbidites as well as compound mass flow/turbidite deposits.

Mud flows   Flow mass consists predominantly of muddy matrix and a limited number of floating clasts. Pebbly mudstone is a specific type of a lithified mud flow deposit.

Olistoliths   Extremely large clasts of olistostromes.

Olistostromes   Large-scale debris flow deposits with intra-clasts and clasts from sources outside the basin.

Omission   Interrupted sediment accumulation. This term is also used when part of a section is missing as a result of tectonic movements.

Preservation potential   Defines the chances of freshly deposited sediment of being preserved after burial under younger sediment.

Proximal and distal tempestites   Storm deposits close to or farther away from their sediment source, respectively.

Sediment gravity flows   Gravity forces on slopes overcome internal strength (cohesion and internal friction) of sediment mass which moves downslope as laminar to turbulent mass flow, resulting in disturbance and mixing of the original strata.

Skeletal lags (shell beds)   Concentrations of relatively large biogenic skeletal remains caused by single depositional episodes, winnowing of the pre-existing fine-grained matrix, or multiphase winnowing and reworking.

Starvation   A basin is sediment-starved if the rate of subsidence of the basin floor is higher than the sedimentation rate. This term may also be used for short time intervals and is then similar to condensation.

Storm muds (mud tempestites)   Same origin as storm sands, but commonly thinner and often overlooked, may extend farther basinward.

Storm sands (sandy tempestites)   Siliciclastic sands and pebbles, derived from the beach and foreshore zone, are redeposited as graded sand layers in the foreshore zone and deeper water.

Tsunami deposits (tsunamiites)   Sediment layers of mixed origin (from land, coastal and subtidal zones), grain size, and composition, produced by one or several pulses of huge catastrophic waves affecting coastal areas and shallow seas.

Turbidites   Mostly thin (2–50 cm) beds of graded sand and silt, originating from shallower environments, transported and redeposited by turbulent bottom currents. Modifications: calcareous turbidites, muddy turbidites.

However, non-depositional events produce a distinct break in the sedimentary record, for example a surface settled by benthic organisms, or an irregular erosional surface. These features are here referred to as stratal discontinuities or *event horizons*, because they do not form a bed of sediment in a strict sense.

In a deep-marine sequence (Figure 6.1), the sedimentary section may contain numerous minor and major depositional events. All these beds contribute considerably to the buildup of the sequence. Minor events create 'depositional noise', deposited in addition to a more or less permanent pelagic rain, and form beds which can barely be distinguished and/or traced over long distances. It is normally a very rare, unusually thick, coarse-grained or otherwise distinct marker bed or discontinuity which can serve as a *marker horizon* in event stratigraphy. In addition, specific successions of beds, distinct volcanic ash layers, and relatively short periods of black shale deposition or redox events (Figure 6.1) can frequently be used as marker horizons (see Chapters 7 and 15). The event deposition concept also implies that the sedimentation rates of thin units in a vertical section may vary

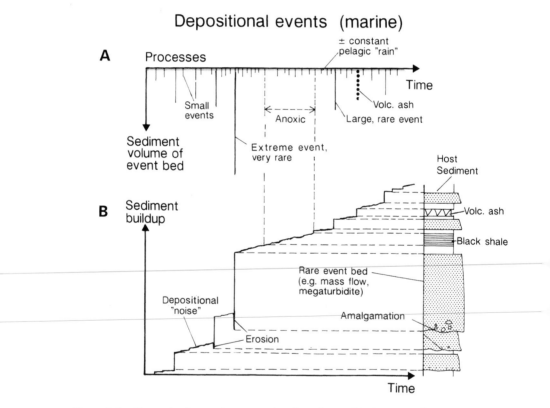

Figure 6.1   *General scheme demonstrating the relationship between depositional events and their sedimentary record. **A.** Succession of minor and major events versus time. **B.** Sediment buildup versus time. Only the large and very rare depositional events are useful as stratigraphic marker beds. The 'normal' depositional processes may be superimposed by other types of events, such as anoxic events or tephra events*

greatly from one unit to the next, particularly if they contain differing numbers of event beds of changing thicknesses.

Event beds and marker horizons are mostly restricted to individual sedimentary basins and can be traced only in specific cases from one basin to another. Exceptions to this rule are extensive volcanic ash layers, erosional unconformities and other features associated with global sea-level changes, as well as rare meteorite-induced layers.

Depositional events commonly return after irregular, stochastic intervals, and are referred to as non-periodical or discyclic events. However, some other events appear to occur periodically with a certain, more or less constant recurrence interval and these are referred to as cyclic events (see Chapter 7). Distinguishing the two is of importance, but may be difficult in many cases. The presence of closely spaced event beds and event horizons of varied nature allows a considerable refinement of stratigraphical correlation in a given sedimentary basin. This so-called 'high-resolution stratigraphy' (e.g. Kauffman *et al.* 1991) is thought to be at least equivalent to the finest biostratigraphy.

To utilize event beds in stratigraphy within a basin or, if possible, from basin to basin, they should meet two basic requirements: (1) they should have a wide lateral extent; and (2) they should possess specific characteristics (textural, mineralogical, geochemical, faunal, etc.) which qualify them as marker beds or marker horizons.

6.2 TYPES OF DEPOSITIONAL EVENTS AND THEIR SIGNIFICANCE

Depositional events occur in both continental and marine environments and have been described in a number of textbooks and special volumes (e.g. Einsele *et al.* 1991; Shiki *et al.* 1996). Common types of these event beds are listed in Table 6.2. The contribution of tempestites, turbidites, sediment gravity flow deposits, and

Table 6.2 *Types of event deposits. [Modified from: Einsele* et al. *(1996)]*

Marine	Marine and continental	Continental and lacustrine
Tsunami deposits (submarine and coastal)	*In situ* earthquake structures (seismites)	Flood deposits
Storm deposits, supratidal	Volcanic ash fall deposits	
Tempestites, subtidal (siliciclastic and biogenic sand and mud)	Deposits caused by meteorite impacts	
Turbidites (siliciclastic and biogenic sands, silts and muds, with organic matter; volcanic ash)	Rockfall, slide and slump deposits	Lacustrine turbidites
	Sediment gravity flow deposits (mud, grain, debris, pyroclastic flow deposits, olistostromes)	

flood deposits to the filling of various basins is often very significant (Einsele *et al.* 1996). In particular, basins with a high input of terrigenous and volcaniclastic material, such as deep-sea fans, deep-sea trenches, fore-arc and back-arc basins, foreland basins, and some adjacent basins, tend to be dominated by event deposits. Alluvial fans and fan deltas as well as many lake basins are also characterized by a great number of event beds. If muddy tempestites and mud turbidites which are commonly thin and frequently obscured by bioturbation are included, then the contribution of event deposits to basin filling becomes even greater.

In the following discussion, some general characteristics of depositional events in various, mostly marine environments are briefly discussed before evaluating the significance of selected beds or bed successions as marker horizons in stratigraphy.

6.2.1 Shallow-marine Storm Sands (Tempestites) and Tsunami Deposits

The effects of storm waves on rocky coasts and sandy beaches can be directly observed and have therefore been known for a long time. Often they produce supratidal storm deposits of limited extent and poor preservation potential. However, storm waves also have a strong impact on the subtidal sediments where they stir up sand and pebbles, seaweed, various shells and skeletal debris, and finer-grained material. After the storm has waned the suspended material is re-deposited either directly at the site of wave erosion, or in deeper water as 'storm sands', 'storm deposits', or 'tempestites' (from the Latin *tempestus* meaning 'storm').

Catastrophic tsunami waves, generated by submarine earthquakes or large volcanic eruptions, produce tsunami deposits or 'tsunamiites' which resemble tempestites to some degree. However, tsunamiites appear to be less common in the geological record than tempestites.

Storms and storm action

Heavy storms develop under special climatic and geographic conditions. Tropical storms (hurricanes) are initiated in low-latitude regions within the trade wind belt. They travel westward and are deflected toward the poles by Coriolis forces. Extratropical storms are common in zones of mid-latitudinal atmospheric circulation along polar fronts. They migrate eastward, are more consistent in their direction, and can reach the same magnitudes as tropical storms. A third mechanism producing landward-directed storms are the monsoons during the summer season.

Apart from the generation of large waves with a deep wave base, onshore winds can also produce surface currents and drive a net mass flux of water toward the coast. The set-up of water along a wide section of the coast is enhanced by low barometric pressure, abundant rainfall, converging shorelines, and broad, shallow shelves (Figure 6.2A; Nummedal 1991; Seilacher & Aigner 1991). The excess water volume can flow back through rip currents (carrying some sediment in suspen-

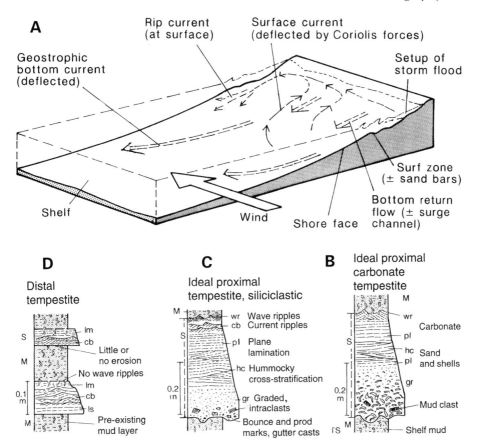

A. Rip current (at surface)
Surface current (deflected by Coriolis forces)
Geostrophic bottom current (deflected)
Setup of storm flood
Surf zone (± sand bars)
Bottom return flow (± surge channel)
Shelf
Wind
Shore face

D Distal tempestite

S
M
0.1 m
M

lm
cb
Little or no erosion
No wave ripples
lm
cb
ls
Pre-existing mud layer

C Ideal proximal tempestite, siliciclastic

M
S
T
0.2 in

wr Wave ripples
cb Current ripples
pl Plane lamination
hc Hummocky cross-stratification
gr Graded, intraclasts
Bounce and prod marks, gutter casts

B Ideal proximal carbonate tempestite

M
S
T
0.2 m
M
S

wr
pl
Carbonate
hc Sand
pl and shells
gr
Mud clast
Shelf mud

Figure 6.2 **A.** *Wind-driven storms producing coastal set up of storm flood, foreshore bottom return flow, and geostrophic bottom currents affecting the inner and outer shelf. As a result of Coriolis forces, all currents are deflected to the right in the Northern Hemisphere.* **B** *and* **C.** *Sedimentary structures of individual proximal carbonate and siliciclastic storm deposits (tempestites).* **D.** *Distal tempestite. S, sand; M, mud; gr, graded and massive; lm, laminated mud.* [Modified from: Einsele (1992)]

sion) along the surface, as well as by bottom currents which may carry sediment in higher concentrations. Bottom return flows in the surf and shoreface zones tend to be channelized and may produce surf channels. Unconfined, non-channelized bottom currents (geostrophic currents) on the shelf, generated by downwelling water masses, are also deflected by Coriolis forces and therefore flow obliquely away from the coastline. In regions with high tides, geostrophic flows can be augmented by tidal ebb currents.

Storm waves and storm-induced geostrophic currents operate simultaneously and cause a combined flow system (Snedden *et al.* 1988; Myrow & Southard 1996). Back-and-forth oscillation of the ground wave is superimposed on the quasi-steady bottom current. The net shear stress imparted on the sea floor may erode and move sediment during one-half of the wave stroke, but will be insufficient to do so during the other half. The shoreface and parts of the inner shelf are dominated by wave-induced oscillatory shear, while the deeper environments are controlled mainly by steady, obliquely offshore or almost shore-parallel currents.

Storm-generated bed forms and facies patterns

As a result of the changing flow regime, storm-generated bed forms show a distinct trend from the beach into deeper water, usually referred to as the normal sediment distribution model. It is here described for six zones (Figure 6.3A):

1. *Backshore and supratidal zone.* Storms frequently erode seaweed and shells of bottom-dwelling fauna and accumulate them landward of the normal beach or barrier zone (supratidal storm beds; Figure 6.3B).
2. *Surfzone and upper shoreface.* These zones are controlled by fair-weather wave action which usually destroys the fingerprints of previous storms. Megaripples, flat swash lamination on offshore bar crests, trough cross-stratification and planar or low-angle swash laminae are the dominant structures.
3. *Middle shoreface.* This zone may preserve some structures formed during storms, such as flat, nearly horizontal bedding or low-angle swaley cross-stratification (Figures 6.3A and C). Mud is sorted out and deposited in deeper water; bottom life is limited to a filter-feeding infauna.
4. *Lower shoreface* (5–20 m water depth). On the lower shoreface the stress imparted by both flow components (oscillating water and geostrophic currents) may stir up both sand and mud, which are redeposited as a graded bed at the same location or nearby (proximal tempestites; Figures 6.2B and C). At the peak of storm action, irregular scours and hollows form at the sea bottom, called pot and gutter casts, which are later filled with the coarsest siliciclastic or bioclastic material available. These casts and tool marks sometimes display bipolar or multi-directional current action. The typical internal sedimentary structure produced by the combined flow regime is thought to be low-angle hummocky cross-stratification (HCS) which normally occurs on top of a graded division with basal lag deposit and subsequent parallel lamination and current ripple cross-stratification (Figure 6.2C). In ideal cases, the hummocky cross-stratification division is overlain by wave ripple cross-stratification and oscillatory ripple marks reflecting the final stage of a waning storm. However, rippled tops may also result from subsequent large waves.
5. *Inner shelf.* On the inner shelf and parts of the outer shelf either hummocky cross-stratification sand layers or thinner graded sandy and silty beds (distal tempestites) with cross-stratified and sometimes rippled tops are typical (Figure 6.2D). Muddy interbeds are produced either by storms which have moved inland and eroded fine-grained material, or by the slow and repeated deposition of suspended river load.
6. *Outer shelf.* At greater water depths on the outer shelf, the current component of the combined storm flow becomes dominant, leading to current-rippled fine sand and silt beds (Figure 6.2D). Storm events affecting this zone are extremely rare, and are followed by long periods of quiescence. In high-energy shelf seas,

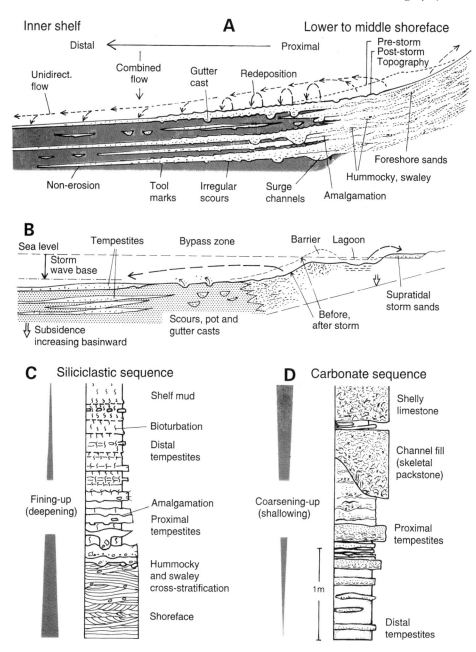

Figure 6.3 *Proximal and distal trends (**A**) without and (**B**) with sediment bypass zone. **C.** Siliciclastic sequence with deepening (fining and thinning) upward trend. **D.** Shallowing (coarsening and thickening) upward calcareous tempestite sequence. Note the amalgamation of proximal sandy to gravelly tempestites. [Modified from: Einsele (1992)]*

distal tempestites occur at water depths up to and in excess of 50 m. They appear to be more discontinuous than proximal ones. In the North Sea, distal tempestites are traceable over tens of kilometres in water depths up to 30 m (Aigner & Reineck 1982).

A somewhat different facies model of tempestites, the sediment bypass model, takes into account that tempestite deposition may occur only on the inner and outer shelf. In shallower water, the material stirred up by storms is either frequently reworked more or less in place (amalgamation) or it is transported basinward through an intermediate zone of non-deposition (Myrow 1992; Figure 6.3B). The typical bypass zone is characterized by isolated pot and gutter casts and by the scarceness of continuous tempestite beds. There may be sequences in which both tempestite models are successively realized. The bypass model appears to be well suited for settings with subsidence rates increasing from the margin toward the centre of the basin.

Characteristics of individual storm layers (tempestites).   Storm layers or tempestites are sheet-like sand, silt and mud beds of considerable lateral extent (e.g. Allen 1984; Seilacher & Aigner 1991). Grain size distribution and composition of tempestites vary greatly. They range from coarse grained (sand and gravel) to silty and muddy types as well as from siliciclastic types with hardly any fossil remains to calcareous bioclastic sandstones, wackestones or packstones (Figure 6.3D). The components of the latter type are derived either from pre-event epibenthic or shallow infaunal populations. Particularly in Proterozoic and Cambrian strata, tempestites frequently consist mainly of reworked muddy intraclasts or reworked microbial mats (flat pebble conglomerates; for example Mount & Kidder 1993).

Common phenomena in the proximal zone are amalgamation and/or cannibalism (Figures 6.3A and C). These terms signify that either pre-existing thick tempestites are truncated, or that thin tempestites are completely reworked by subsequent storms. As a result, the material of older tempestites is incorporated into new ones. This process may occur repeatedly, until a very big storm event ultimately produces a bed, the base of which can be preserved. Multiple reworking promotes abrasion and break-up of mechanically unstable particles and possibly accelerated dissolution of carbonate and other minerals. Typically, amalgamation leads to increasingly mature lag sediments, including placer deposits. In places, skeletal remains of vertebrates, particularly teeth but also coprolites which have been phosphatized prior to reworking, are concentrated in the basal layer of storm beds as bone beds.

Burrows exhumed and partly washed out by storm action are subsequently filled with settling sediment. These sole marks contain a trace fossil association characteristic of shallow-marine environments. Burrow-fill sedimentation can also occur without the formation of a storm deposit (Wanless *et al.* 1988). The post-event community recolonizes either the top of the tempestite, or erosional surfaces that were not covered by a storm bed, for example in somewhat elevated proximal regions. Relatively firm, coarse substrates attract oysters, brachiopods, crinoids, stromatolites, and firm-ground burrowers of the *Glossifungites* association. In distal zones with thin tempestites and rare reworking by successive storms, burrowing by post-event benthic organisms can markedly overprint or completely obscure storm event stratification (Figure 6.3C).

Proximal–distal trends. Tempestites deposited close to the coastline and sediment source are called proximal; those basinward at greater distance from the sediment source are distal. Common proximal–distal trends of tempestites are summarized in Figures 6.2B, C and D and Figures 6.3A and B. The nearshore zone of swaley and large-scale hummocky cross-stratification is usually devoid of muddy interbeds. Basinward, with the presence of muddy intercalations, either a zone of relatively thick and often amalgamated tempestites follows, or there is a zone of sediment bypassing, apart from sediment-filled scours. In any case, tempestites tend to change laterally in thickness and frequently pinch out. Further seaward, the number of individual tempestites over a certain time span first tends to increase, because amalgamation becomes rare, before decreasing due to the different travel distances of individual storm-induced suspension currents. Distal tempestites are thin, fine-grained and show the same inorganic sedimentary structures as distal turbidites, but they differ from turbidites in their faunal characteristics and vertical facies trends (see Section 6.2.3).

Tempestite sequences and recurrence intervals of storm events. Extensive, fairly thick siliciclastic or calcareous tempestite sequences require a substantial continuous source of terrigenous or bioclastic material, which is usually provided by river deltas, rapidly eroding sandy coastal cliffs, or carbonate platforms. With or without sediment bypassing, numerous successive storm events can build up more or less rhythmic tempestite–shale sequences in subsiding basins. These sequences may reflect three different trends of basin evolution:

1. *Steady-state conditions in a foreshore–shelf environment*, i.e. the average sedimentation rate more or less compensates for subsidence. In this case, comparatively thick sequences of alternating tempestites and shelf muds can develop, and the palaeo-water depth at a certain location within the basin will remain constant.
2. *Deepening basin*, that is the average sedimentation rate is lower than subsidence. In this case the vertical sequence will display a trend from thick, relatively coarse-grained proximal tempestites to thin, fine-grained distal tempestites, and finally end up with indistinct mud tempestites or purely autochthonous shelf muds (Figure 6.3C). The transition zone from proximal to distal storm beds may be on the order of 20 to 50 m in vertical section.
3. *Shallowing basin*, with a sedimentation rate higher than subsidence. The resulting tempestite sequence coarsens (thickens) upward (Figure 6.3D), and the transition zone from shelf muds to a siliciclastic or bioclastic foreshore and coastal environment will also be of limited thickness. Such regressive trends favour amalgamation of storm beds.

Storms represented by tempestites in the geological record are less frequent than rare storms in the modern world. Today, a 100-year storm appears to be an exceptionally large storm event (Nummedal 1991). Modern storm layers in seaward-prograding sediments on the inner shelf (20–30 m of water depth) appear at time intervals of some tens to some hundred years (Nelson 1982; Saito 1989). By constrast, ancient tempestite sequences with two to five storm beds per 1 m section indicate a recurrence interval of 1–10 Ka or more, assuming average sedimentation (or subsidence) rates in the range of 20–100 mMa^{-1}. This discrepancy between modern observations and the limited number of preserved tempestites in the ancient record suggests that many storm beds are wiped out or obscured by subsequent very rare, extremely large storm events. The action of earlier weaker storms may be inferred from the occurrence of mechanically abraded, coarse particles and/or the presence of materials from different sources in the ultimately preserved tempestite bed.

Tsunami beds (tsunamiites)

Tsunami waves are generated by submarine earthquakes, large volcanic eruptions, submarine mass movements, and meteorite impacts onto the sea surface. These waves have very large wave lengths (up to 100–200 km), long wave periods (10–20 minutes), and therefore a very deep wave base.

Tsunami waves travel rapidly over large distances; they are barely noticed at the deep ocean surface and do not affect significantly deep-sea sediments (Pickering *et al.* 1991). However, if they approach shallow seas and the coast they 'feel' the sea bottom, are reduced in wave length, and build up extremely high waves (10–30 m) which inundate coastal lowlands. Shortly after, the water flows back to the sea forming strong, seaward-directed return currents (Figure 6.4A). Since the back-flows tend to follow the coastal morphology, they often become channelized and are therefore particularly powerful. These traction currents can erode and transport sediments up to the size of coarse gravel and even blocks. They drop their coarsest load as irregular, often discontinuous beds which can be observed only in part of the available outcrops (Figure 6.4B). Finer-grained, less conspicuous tsunami beds resemble tempestites and are often difficult to discriminate from these event beds. This may be one of the reasons why tsunami deposits have rarely been recorded from ancient sedimentary sequences. In the modern oceans, tsunami waves are by no means rare (e.g. Yamazaki *et al.* 1989).

Distinctive features of tsunami beds.   Besides an erosive base and graded bedding, the most distinctive features of tsunami beds are an unusually coarse grain size in comparison to the underlying and overlying finer grained beds, which may contain either normal storm layers or sediments typical of deeper water beyond the range of storm beds. Gravel or intraclasts of tsunami beds (tsunamiites) are derived from shallower water, the beach zone, and also from the inundated land (e.g. 'exotic' blocks). Therefore, both rock components and faunal and floral remains of tsunamiites tend to indicate mixed sources. Since tsunami waves approach the coastline in several pulses, their return flows also operate in pulses. These may create a section of two or more composite beds, each of them consisting of a coarse-grained base and a finer-grained top layer (Shiki & Yamazaki 1996; Figure 6.4B).

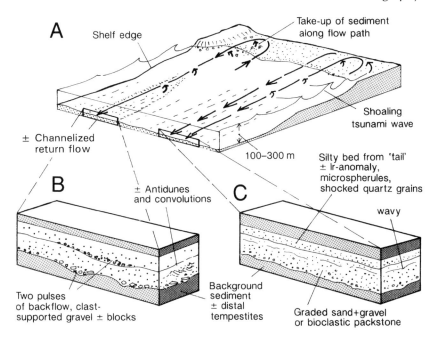

Figure 6.4 **A.** *A tsunami wave approaching the shelf and flooding the coastal zone (based on Minoura & Nakaya 1991). Tsunami deposits may originate from both landward flow and backflow as well as from submarine slope failure (producing mud flows and turbidites).* **B.** *Coarse-grained channelized tsunami deposits showing two pulses of backflow.* **C.** *Meteorite-induced widespread tsunamiite on the outer shelf*

Tsunami beds formed by backflow in relatively deep water have the best preservation potential, whereas tsunami beds deposited on land and partially also those in shallow water are easily eroded, reworked, or overprinted by subsequent storm events. Tsunamiites on land and in the coastal zone contain mixed marine and continental material.

Large meteorite impacts into the sea, as assumed for the Cretaceous-Tertiary boundary, have probably initiated exceptional tsunami waves and return flows which may have formed widely extended tsunami beds at water depths between 100 and 300 m or more (Bourgeois *et al.* 1988; Albertao & Martins 1996). According to these authors, such tsunami beds can be distinguished from normal tempestites or turbidites by wavy internal structures and the presence of some 'exotic' constituents, such as microspherules (microtectites), shock-metamorphosed quartz grains (Figure 6.4C), and a high iridium content (see Section 6.4.2). These constituents seem to be enriched not in the graded, relatively coarse tsunami bed proper, but rather in the following finer-grained bed resulting from the tail of the backflow. 'Normal' tempestites and tsunami beds of medium size are assumed to have similar characteristics.

6.2.2 Sediment Gravity Flow Deposits and Turbidites

Deposits resulting from sediment gravity flows and turbulent suspension currents (turbidity currents) are very common in ancient and modern deep-sea basins. They are mostly well preserved and cover wide areas of the former or present basin floor. In terms of sediment mass, these deposits are the most important group of event deposits. They often provide specific marker beds or successions of beds which can be used for intra-basinal stratigraphical correlation. Since turbidity currents usually evolve from slope failure and gravity mass flows, the deposits of these various processes are genetically related and therefore described under one heading.

The most important prerequisites for the occurrence of mass flow deposits and turbidites in significant quantities are (1) substantial terrigenous sediment influx or high biogenic sediment production; and (2) strong submarine relief. Both groups of event deposits are common on the prodelta slopes and deep-sea fans in front of the mouths of major rivers, at the foot of submarine slopes of deep basins, including deep-sea trenches, and they build up the major portion of thick flysch sequences in orogenic belts. Numerous articles and edited volumes have been published on turbidites and other gravity mass flow deposits (e.g. Walker 1978; Schwarz 1982; Mutti *et al.* 1984; Mutti & Normark 1987; Einsele 1992; Mutti 1992). The flow behaviour of the different types of gravity mass movements is summarized by, for example, Postma (1986) and Stow *et al.* (1996). Here, the most characteristic features and depositional environments of these event deposits in the marine realm are briefly described using a few simplified conceptual models. These can also be applied to lake sediments, though large-scale phenomena are commonly absent in this environment.

Deposits of gravity mass movements

There are two principal types of gravity mass movements which give rise to extensive deposits of particular interest here:

1. *Sliding and slumping.* Mass movement of soft to semi-solid sediments may take place on slopes with angles as gentle as a few degrees. They frequently occur in areas where fine-grained sediments are deposited rapidly, for example in front of deltas. In slides, sediment masses move as a kind of rigid plug without significant internal disturbance. Slumps, however, show considerable internal disturbance such as folding and slip faces, and often evolve into debris or mud flows which are collectively known as slide-debris flows or slideflows.
2. *Sediment gravity flows.* These include mud flows, muddy slide-debris flows, and grain flows. Unconsolidated, quasi-solid masses of meta-stable grain packing may be transformed by cyclic loading into liquefied masses which can move on slopes of less than 1°, excepting pure grain flows. On their way downslope, many mass flows evolve from laminar to fully turbulent systems (e.g. Postma 1986). The presence of mud maintains not only a certain density to the flows, but also provides the flows with cohesive matrix strength, allowing the support of larger particles. Flow masses come to rest if the applied shear stress drops below

the shear strength of the moving material and/or, on land, if their excess pore-water dissipates. The flows 'freeze', which is accomplished either by cohesive freezing, or in the case of a cohesionless sandy matrix, by frictional freezing, or by both processes. In subaqueous environments, individual gravity flows or parts of them can take up additional water from the overlying water body and evolve into sediment masses of lower density and viscosity. These finally generate turbulent suspension currents of high velocity, or turbidity currents.

The principal features of debris flow and mud flow deposits are summarized in Figure 6.5. They reflect the final flow processes immediately before deposition (Lowe 1982), which can be characterized as more or less laminar, cohesive flows of comparatively dense, sediment–fluid mixtures of plastic behaviour.

Debris flow deposits (debrites) and olistostromes (very thick, extensive debrites) consist of a medium- to fine-grained matrix with a varying proportion of matrix-supported clasts. The typical debrite is rich in clasts of different sizes; the clasts may be derived from older sediments and rocks within the basin (intraclasts) or from sources outside the basin (extraclasts, typical for olistostromes). Single clasts or blocks (olistoliths) in olistostromes can reach the size of a house.

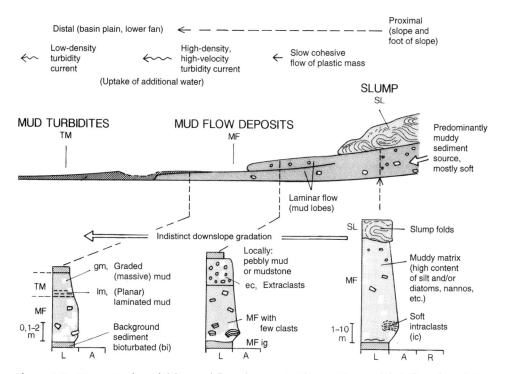

Figure 6.5 *Conceptual model for mud flow deposits (with transition to debris flow deposits) ultimately evolving into mud turbidites. L, lutite (clay and silt); A, arenite (sand); R, rudite (gravel); MF, mud flow deposits; TM, turbidite mud; ig, inverse grading. [Modified from: Einsele (1992)]*

The base of thick and coarse-grained debrites may be scoured. The basal sediments often display a thin sheared zone and inverse grading caused by prograding frictional freezing. The higher portion may exhibit indistinct normal grading. Internally, most debrites lack any bedding phenomena or imbrication of clasts; in some examples, elongate clasts are aligned horizontally and indicate the direction of flow. The top of the bed is either sharp or grades into an overlying turbidite, forming a compound debrite–turbidite couplet (e.g. Stanley 1982; Mutti *et al.* 1984). In places, the top of a debrite may be current-winnowed and therefore transformed to a clast-supported lag deposit. In some cases, a debrite or mud flow deposit is directly overlain by a second debrite or an overlapping lobe of the same mudflow (Figure 6.5). *In situ* traces of burrowing organisms are completely missing within the event bed; they can occur only at the top of mass flow deposits not affected by subsequent winnowing, amalgamation or erosion.

Calcareous debrites resulting from gravity-driven large-scale slope collapse of semi-lithified carbonate buildups often form sheet-like megabreccia beds (see Figure 6.7A) which contain little fine-grained matrix material and therefore are primarily clast-supported.

Mud flow deposits have much in common with debrites; in fact there is no sharp boundary between these two end-members of the same group (Figure 6.5). Mud flow deposits have a muddy matrix with a high silt (or microfossil) content and contain no, or only a small amount of, clasts which are mostly intraclasts that are frequently deformed by the preceding processes of slumping and mass flow. The admixture of gravel or other coarse material from submarine canyons may locally generate pebbly mud or mudstone.

Some workers have introduced additional terms for some mass flow deposits which they ascribed to specific triggering mechanisms: unifites, homogenites, mega-turbidites or seismo-turbidites (e.g. Mutti *et al.* 1984).

Turbidites and deep-sea fan associations

The turbidity current hypothesis as a mechanism for producing graded, sheet-like beds (sandy, silty or muddy turbidites) in marine and lake environments was inferred from the study of ancient rhythmic bed successions, the internal structures of the sand beds, and their allochthonous shallow water fauna (Kuenen & Migliorini 1950). Turbidity (or suspension) currents mostly evolve from slope failures by uptake of water. As long as the density of the suspension is greater than that of the surrounding water body, it tends to move downslope, accelerate, and form a turbulent undercurrent transporting its load into deeper water. The density of suspensions caused by river floods, however, is usually not high enough to produce density undercurrents in the sea. Turbidity current events occur infrequently relative to the human life cycle and therefore large events cannot be directly observed, apart from those observations which can be made in lakes, water reservoirs or artificial flumes. Indirect evidence for the high transport capacity of turbidity currents is gained from reports on the breakage of submarine telegraph cables on continental slopes, as well as in front of submarine canyons at the mouth of some major rivers. The driving force of turbidity currents is primarily a function of (1) the difference in the densities of the suspension and the overlying water body; (2) the submarine relief, i.e. the angle and length of slope; and (3) the thickness of the suspension current.

High-density and thick turbidity currents (Piper & Shor 1988) reach high velocities ($\leqslant 10$ to 20 m s^{-1}) and can carry relatively coarse sand, pebbles and intraclasts. Within the confines of submarine channels, large-scale turbidity currents have the ability to transport gravel ($\leqslant 10$ cm in diameter) as bed load, and may therefore generate lenses of conglomerate at the foot of prodelta slopes. On deep-sea fan lobes and basin plains, they have the capacity to erode the uppermost mud layer in extensive areas (Figure 6.6A). The eroded material feeds the suspension with new sediment which replaces coarser material settling out of suspension in the slackening body and tail of the current. In this way, and by the maintenance of turbulence by gravitational forces (auto-suspension), the current is kept in motion and can travel over long distances. High-velocity suspension currents usually originate from large gravity mass movements on the slopes of deep basins well supplied with sediment.

Low-density turbidity currents flow slowly and can therefore keep only silt and clay-sized material or larger aggregates of fine particles in suspension. Their erosional capacity is very low or non-existent, but weak turbulence maintains such suspension currents for relatively long periods of time. They can attain considerable thicknesses and distribute their suspended load as a thin bed over wide areas. It is assumed that low-density, muddy turbidity currents are often the final stage of sand-bearing suspension currents, which have lost their coarser grain size fraction underway (Figure 6.6A). A more comprehensive treatment of this topic, including a specific nomenclature for the various processes involved in the generation of these event beds, is given by Stow *et al.* (1996).

Types of turbidites and their characteristics. Several types of turbidite, representing the final products of turbidity currents, can be recognized on the basis of their characteristics. For example, texture, internal sedimentary structures, and composition often vary greatly. At least four types of turbidites can be distinguished:

1. *Coarse-grained turbidites.* These are generated by high-density turbidity currents carrying pebbles and clasts as bed load and finer-grained material in suspension. They often show an initial stage of traction sedimentation (coarse-grained conglomeratic sand with plane lamination, cross-bedding, internal scour, and possibly some inverse grading (Figures 6.6A and B). This division is followed by sedimentation from suspension creating either a structureless or a normally graded higher division (Bouma division T_a; Bouma 1962). Typical features are water-escape structures (pillar and dish structures).
2. *Medium-grained sandy turbidites.* Medium-grained siliciclastic and carbonate turbidites reflect deposition from suspension currents of moderate density. The event bed is graded from bottom to top, but grading may become indistinct if the source area does not provide material of a wide range of grain sizes. The lower divisions of the turbidite bed, consisting of plane-laminated and cross-bedded sand (Figures 6.6B and 6.7B; Bouma divisions T_b and T_c) are interpreted as traction structures, while the higher division (T_d) showing laminated mud

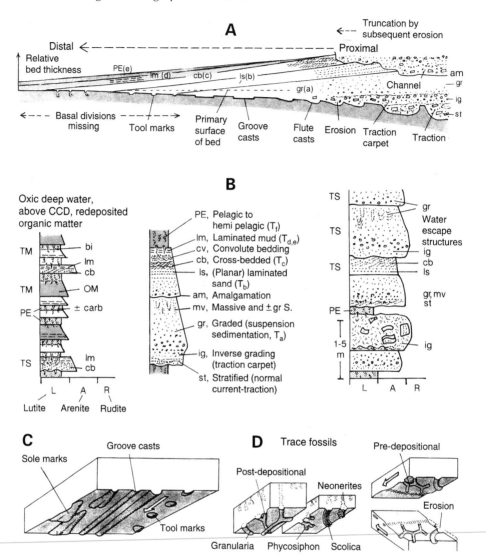

Figure 6.6 **A.** *Proximal–distal trend of siliciclastic sandy turbidites including channel fills. Note the distally decreasing capacity of the turbidity current to erode, to form various sole marks, and to transport coarse particles. Basal divisions of proximal turbidites are distally lost.* **B.** *Individual proximal (right-hand side) to distal turbidites (left) alternating with host sediments; symbols are explained in the intermediate section; TS, turbidite sands; TM, turbidite muds.* **C.** *Sole marks indicating the current direction during erosion.* **D.** *Characteristic trace fossils of sandy turbidites. [Modified from: Einsele (1992)]*

A

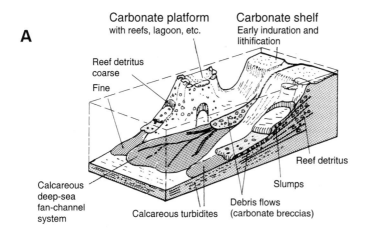

Carbonate platform
with reefs, lagoon, etc.

Carbonate shelf
Early induration and
lithification

Reef detritus
coarse

Fine

Reef detritus

Calcareous
deep-sea
fan-channel
system

Slumps

Calcareous turbidites

Debris flows
(carbonate breccias)

Distal ← — — — — — — — — — — — — — — — — — → Proximal

B

Individual turbidite

cb + cv

lm + gm

ls

gr

Frequently
chert (bands
or nodules)

Erosion

Traction and
traction carpet

Coarse shells
and clasts

C

Fine-grained
carbonate turbidite
(mud turbidite, TM)

PE

bi

gm

lm

Lime ooze
or marl
± plant
remains

1–5
m

L   A   R

Lutite   Arenite   Rudite

Complete ideal
carbonate turbidite

PE

bi

lm+gm

Chert

cb+cv

ls

TS

gr

(± shells
and shell
fragments,
clasts)

0.2–1
m

ig

st

L   A   R

Figure 6.7 *Calcareous debris flow deposits and turbidites.* **A.** *Model of depositional environment.* **B.** *Proximal–distal trend of bioclastic carbonate turbidites.* **C.** *Internal sedimentary structures of individual carbonate turbidites. Symbols are explained in Figure 6.6B. [Modified from: Einsele (1992)]*

may be explained as mixed traction/suspension sedimentation (Lowe 1982). Due to the uptake of eroded deep-sea mud, the amount of autochthonous fauna (nekton, plankton) often increases toward the top of the turbidite bed. The uppermost, structureless and indistinctly graded mud interval (T_e) originates solely from suspension sedimentation.

3. *Carbonate turbidites.* Carbonate sands consisting mainly of skeletal material produced on carbonate shelves and platforms (Figure 6.7A) may hydraulically behave differently from siliciclastic sands. Therefore, carbonate turbidites have often been treated separately from siliciclastic ones, and are sometimes referred to as allodapic limestones (Meischner 1964). Quartz sand in T_a and T_b may be replaced by skeletal particles with diameters much greater than 2 mm, or sand-sized microfossil shells are transported and settle in a manner similar to compact silt grains (T_d). As a result, a graded carbonate turbidite can show distinct jumps in grain size and/or sediment composition in its vertical succession (Figure 6.7C). As with siliciclastic turbidites, the maximum thickness of an individual bed is commonly not attained in the neighbourhood of the source area, but it decreases distally for some distance downcurrent (Figure 6.7B).

4. *Mud turbidites.* The importance of mud turbidites has long been overlooked, despite the fact that mud turbidites form a major part of the sedimentary record of deep-sea fans, such as the Bengal, Indus, Amazon and Mississippi fans. Sediments at the foot of continental slopes and the filling of deep, relatively narrow basins such as the Mediterranean, Black Sea or Gulf of California contain a high proportion of mud turbidites (Einsele *et al.* 1996). Their mode of formation has been discussed by Piper & Stow (1991).

Mud turbidites are regarded either as an end-member of gravity mass flows of mixed granulometry (Figure 6.5), or they are derived solely from muddy sediment sources such as prodelta slopes and other fine-grained slope sediments. In addition to gravity movements, large river floods or muddy sediments stirred up by storms in shallow seas can contribute to the formation of mud turbidites. These beds are deposited from low-density suspension currents and are therefore usually thin; but in the transitional stage from mud flows to muddy suspension currents (hyperconcentrated flow) they may also form thick, indistinctly graded beds (Einsele & Kelts 1982). Taking into account the source and composition of their material, it is possible to distinguish between hemipelagic (Figure 6.8A), pelagic (fine-grained bioclastics; Figure 6.8B), and volcaniclastic (ash) mud turbidites (Kelts & Arthur 1981; Einsele 1992). The redeposition of siliceous ooze, mainly consisting of diatoms and radiolaria, is an important process in the formation of rhythmically bedded marine cherts. The most characteristic features of mud turbidites are a sharp basal contact to the underlying bed, internal normal grading and, under oxic environments, bioturbation at their tops. Proximal mud turbidites may contain a thin laminated sand layer at their base (Figure 6.8A). Upward increasing contents of biogenic opal or carbonate within an individual bed can produce chemical grading. Internally, a laminated division (the E1 of Piper & Stow 1991) is often followed by a structureless, indistinctly graded division (E2). Thin,

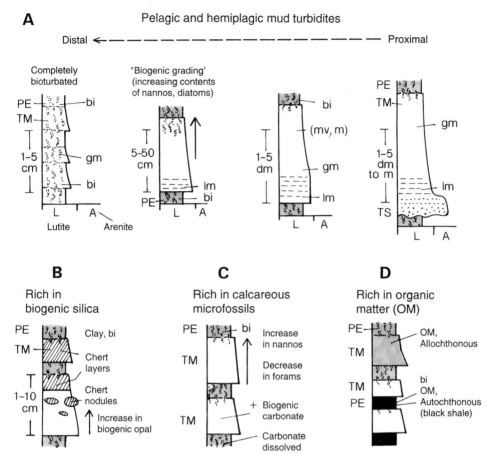

Figure 6.8 *Various types of mud turbidites.* **A.** *Proximal–distal trend of hemipelagic types, predominantly siliciclastic.* **B** and **C.** *Pelagic mud turbidites rich in opaline silica (transformed into chert) or calcareous microfossils; these beds may alternate with pelagic clay devoid of carbonate.* **D.** *Mud turbidites either rich or poor in organic carbon alternating with black shale or oxygenated and bioturbated pelagic deposits. Symbols are explained in Figure 6.6*
[Modified from: Einsele (1992)]

distal mud turbidites tend to become obscured by intense bioturbation and can only be recognized if their material, and/or fauna, differ substantially from the pelagic or hemipelagic background sediment.

Many of these turbidites display pre-event trace fossil associations, characteristic of the pelagic to hemipelagic host sediment, and post-event assemblages which recolonize a freshly deposited turbidite layer (Figure 6.6D; Seilacher 1962). The burrows of the first group are exhumed by the erosive force of the turbidity current and are immediately afterwards filled up with sediment settling out from the slowing suspension current. These *lebensspuren* are preserved on the sole of the turbidite alongside casts produced by current erosion (flute casts) or the imprints of various objects dragged by the current over the sea bottom (groove casts, tool marks; Figures 6.6A and C). The recolonizing trace fossil assemblage has to dig down into the turbidite from a new higher level. If the turbidite is thin, some of the burrowing organisms may reach its base and feed on the background sediments; but if the turbidite is thick, only its top section can be burrowed.

All types of turbidites can contain biogenic carbonate or silica derived from shallow-water environments. If deposited below the calcite compensation depth (CCD) where the background sediments are poor in carbonate or opaline silica, these event deposits produce alternating layers with and without carbonate or biogenic silica (Figure 6.8C). In particular, mud turbidites can form banded sequences. Similarly, mud turbidites containing large amounts of organic matter from their source area, such as those in slope sediments beneath regions of upwelling, can alternate with deep-sea sediment poor in organic carbon (Figure 6.8D). By contrast, slow normal black shale deposition can be interrupted by turbidite inter-beds poorer in organic matter.

A specific problem is the discrimination of distal tempestites induced by storm waves in shallow water and deposited on the middle to outer shelf, from deep-water distal turbidites originating mainly from slope failure. Both types of event deposit result from suspension currents and tend to display some grading as well as the same internal sedimentary structures (i.e. the upper divisions of the Bouma sequence). In the case of tempestites, however, body fossils and trace fossils of both host sediment and event deposit are of shallow-water origin. By contrast, typical turbidites alternate with deep-water host sediments containing corresponding faunal elements, and they show pre-event and post-event trace fossils characteristic of the deep sea (Einsele & Seilacher 1991).

Turbidites in deep-sea fan associations.   Individual large deep-sea fans are normally fed by one single point source, usually a major river; smaller deep-sea fans in a row usually result from sediment input by a number of more or less parallel smaller rivers. In both cases, submarine channel systems on the slope and fans distribute the incoming sediment (Figure 6.9A; Mutti & Normark 1987; Mutti 1992). The channels of modern large submarine fans can be many kilometres wide and have steeper gradients than subaerial rivers on alluvial plains. In the upper and middle fan region, the channels are accompanied by levées (Figure 6.9B) which may rise above the surrounding sea floor by 10–100 m. Large slump and debris flow deposits occasionally bury parts of the pre-existing channel-levée system and thereby cause changes in the sediment distributory system. Relatively dense, fast turbidity currents tend to flow basinward within the confines of the channels and their levées. As a result, the channels of small to medium deep-sea fans evolving close to mountain ranges are mostly filled with coarse-grained and conglomeratic sand beds reflecting traction transport and high-density turbidity currents. Massive channel fills are frequently amalgamated and cut into or pass laterally into thinner bedded turbidites of the levées and fan lobes (Figure 6.9B). Low-velocity, less dense and thicker suspension currents build up channel levées or spill over the levées and form fine-grained, thin-bedded turbidites in interchannel areas which are comparable to the crevasse splay and overbank deposits of subaerial

A Attached fan, channelized

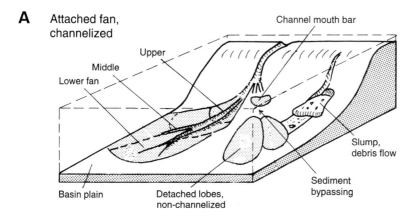

B

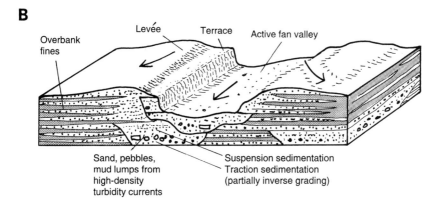

C

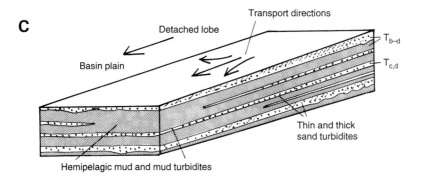

Figure 6.9 **A.** *Model of two contrasting deep-sea fan systems.* **B.** *Channel fill, levée, and overbank deposits of the upper to middle part of the channelized fan.* **C.** *Sheet-like sandy and muddy turbidites on outer (or detached) fan lobe and basin plain. [Modified from: Einsele (1992)]*

meandering rivers. Correlation of such beds becomes difficult even over short distances. The remainder of the suspended load of both thin and dense or thick and less dense turbidity currents is deposited basinward of the channels on unchannelized fan lobes passing into an adjacent basin plain. Both currents leave behind extensive sand and mud layers of the more distal type (Figure 6.9C). It is only the very large gravity mass flows and suspension flows which spread their load as relatively thick beds over large areas of the total fan and basin plain.

The different fan models and the complicated facies architecture of deep-sea fans cannot be discussed here, but are discussed by Shanmugam & Moiola (1985), Mutti & Normark (1987), Mutti (1992) and Reading & Richards (1994).

Current directions, proximal–distal trends and vertical sequences.   Palaeocurrent directions of turbidity currents can be derived from sole marks (Figure 6.6C), internal structures such as cross-bedding, clast and grain orientation, and current ripples (e.g. Collinson & Thompson 1989). Studies on ancient turbidite sequences deposited on lower fan lobes and plains of elongate narrow basins have frequently shown a striking constancy of current directions over large areas. Similarly, current directions on slope aprons fed only by outer shelf and upper slope sediments can be expected to vary moderately. In more proximal, channelized fan associations, as well as in basins supplied with sediment from different sources, however, the palaeocurrent patterns become less regular and sometimes rather complex. When the fan lobes have room to switch (Figure 6.9A), the sediments are dispersed radially. In interchannel areas, the palaeoflow directions are generally deflected from those of the main channel and may show a great variation (Figure 6.9B). Finally, the current patterns of turbidites can occasionally be overprinted by contour currents (Stow *et al.* 1996).

Proximal turbidites including channel fills are often relatively coarse-grained and thick-bedded, and their tops may be truncated by a subsequent turbidity current (amalgamation). Many turbidites display a striking correlation between the dimension of their sole marks, their grain size distribution, and their bed thicknesses. Large flute or groove casts, for example, are often associated with particularly thick and coarse sandy turbidites. Downslope and more distally, the percentage of sand layers in the total fan volume often decreases, a tendency that is also observed in the modern ocean basins (Pilkey *et al.* 1980). Distal turbidites, far from the sediment source, become thinner and finer grained, and they successively lose their basal divisions with sole marks (Figure 6.6A). However, the proximal–distal concept should be applied with caution if deep-sea fan associations with channel systems and overbank deposits are considered.

Taking into account channelized sediment distribution systems and variations in the size of turbidity currents, it is obvious that bed sets and vertical sequences of turbidites can be quite irregular (e.g. alternating bed thickness or grain size). None the less, thickening and coarsening-upward or thinning and fining-upward trends of variable vertical extent can be frequently observed within larger sequences. Minor trends (asymmetric cycles) of some metres up to several tens of metres in thickness can be interpreted as the products of prograding, retrograding, or laterally shifting channel systems and fan lobes (e.g. Mutti 1992; Stow *et al.* 1996).

Thicker depositional sequences with both upward thinning and fining trends are generated by relative sea-level changes which may also result from 'pulses' of tectonic activity. Lowering of sea level accentuates terrigenous sediment input, gravity mass movements, rapid progradation and/or upbuilding of deep-sea fans. In contrast, sea-level rise reduces sediment supply from terrestrial sources and may even cause inactive fan periods, such as those known from the modern Amazon fan where older fan deposits are draped by normal hemipelagic to pelagic sediments. Finally, progress in basin filling will lead to a long-term trend from distal to proximal turbidites and ultimately to shallow-water and continental deposits, as known from fore-arc basins remnant basins, and foreland basins (e.g. Einsele 1992).

6.2.3 Volumes, Travel Distances and Frequency of Mass Flow Deposits and Turbidites

The various types of mass flow deposits and turbidites can be regarded as a family of related phenomena and some of their common characteristics are listed below.

1. *Volumes.* Many mass movements range between 0.001 and several 100 km^3 in volume. The same orders of magnitude are characteristic of the volumes of large mud flows and turbidity currents which evolved from slides and slumps. The famous 1929 Grand Banks earthquake off Newfoundland led to the dislocation of sediment on the order of 100 km^3, including current-induced erosion on the upper slope (Hughes Clarke *et al.* 1990). In addition, a number of extremely large slides and slumps (1000–20 000 km^3) have been reported from the modern ocean margins (see Schwarz 1982). However, many of the mass flow deposits and turbidite beds to be observed in normal field exposures represent smaller sediment displacements, with volumes between 10^3 and 10^6 m^3.

2. *Travel distances.* Debris flows and mud flows can travel distances of several hundreds or thousands of kilometres, as observed in the present-day oceans (Akou 1984; Simm & Kidd 1984). Turbidite flows may redeposit sediments as far as several 1000 km away from their primary location. For example, Elmore *et al.* (1979) described an upper Pleistocene mega-turbidite, 500 km long, more than 100 km wide and up to 4 m thick, from the Hatteras abyssal plain in the western Atlantic. This bed consists predominantly of fluvially derived sand and shelf mud with a large proportion of mollusc shell fragments. In proximal regions, the poorly sorted lower part of the bed ($\geqslant 20\%$ mud) may have been deposited as a sandy debris flow, whereas its upper part and more distal portions reflect deposition from a turbidity current. Another compound, carbonate-bearing debrite-turbidite in the Exuma Sound, Bahamas, is up to 2 m thick and covers an area of more than 6000 km^2 (Crevello & Schlager 1980). In the Ionian abyssal plain, in the Mediterranean, a 12 m thick homogenized Holocene mud layer containing around 50% carbonate, partially derived from intermediate and shallow waters, covers an area of 1100 km^2 in about 4000 m of

water (Hieke 1984). Locally, the layer has a sandy base composed of shell fragments. In ancient rocks (e.g. in the Eastern Alps, Apeninnes, Pyrenees) it is possible to trace specific marker beds across 100–170 km (Hesse 1974; Ricci Lucchi & Valmori 1980; Mutti *et al.* 1984). These observations clearly indicate that gravity flow deposits have a high potential for being widely distributed in relatively large and deep basins.

3. *Frequency.* It is obvious that the frequency of gravity mass flows and turbidite events is a function of sediment supply. Therefore deep-sea fans with average sedimentation rates from 100 to >1000 m Ma^{-1} are locations where event deposits have short recurrence intervals. Thinly bedded, millimetre to centimetre scale, silt and mud turbidites, as observed in interchannel areas or in some modern back-arc basins, have recurrence intervals of tens to hundreds of years. Thicker turbidite sands and muds in middle and lower fan regions as well as in adjacent marine basins (Mediterranean, Gulf of California) were deposited at intervals of several thousands to some ten thousands of years. Turbidites appear to be less frequent in distal fan regions than in more proximal channel and overbank settings (Klein 1985). The longest recurrence times are to be expected for thick mud flows and turbidites (mega-turbidites) which form extensive sheets on submarine fans and basin plains. In ancient rocks, such key or marker beds, often rich in redeposited shallow-water carbonate, occur once every 50 000 to one or several million years, depending on the amount of sediment supply (Mutti *et al.* 1984). They may have been triggered by substantial sea-level fall or very rare, extremely strong earthquakes. In a deep-sea drillhole at the foot of the continental slope off Baja California, only one of these mega-event deposits (of last glacial age) was found in a mud turbidite sequence representing 3–4 Ma (Moore *et al.* 1982). Another well-studied example is the compound slide/mudflow/turbidity current event in the Canary Basin (16–17 Ka; Embley 1980). Here, a 10–20 m thick mud flow deposit covers an area of 30 000 km^2, and the subsequent turbidity current travelled over 1000 km.

6.2.4 Marine Depositional Events Controlled by Relative Sea-Level Changes

In addition to specific marker beds which result from a variety of predominantly local phenomena, depositional events controlled by global and regional sea-level changes (see Chapters 11 and 15) are of particular interest in stratigraphy. This is primarily because sea-level changes are an efficient mechanism for the control of sediment supply. Sediment may be stored during highstand and remobilized later during falling sea level, and this forms the basis for sequence stratigraphy (see Chapter 11). In the following section, third- or higher order sedimentary sequences which appear to be global and the result of eustatic sea-level changes are discussed. However, it must be noted that repeated tectonic activity may also cause third-order sea-level changes.

The position of sandy or bioclastic tempestites within a sea-level controlled sequence is, however, not clear. Tempestite formation requires either sufficient terrigenous sediment influx or significant biogenic carbonate production which is not diluted by siliciclastics. Tempestites probably form preferentially during sea-level fall and are preserved best when deposited during the lowest sea-level stand. In contrast, more is known about the behaviour of sediment gravity flows and turbidites within sea-level-controlled sequences. In shelf/slope settings of passive margins with moderate sediment supply, it is almost exclusively the early to middle lowstand systems tract in which event deposits form (Figure 6.10). Where sea-level fall exceeds subsidence, subaerial valley incision provides material for event deposits and wave-cut erosional surfaces leave behind submarine lag sediments.

By contrast, subsiding deltas receiving very high terrigenous sediment input tend to permanently aggrade and prograde seaward during third-order sea-level variations or tectonic activity. As a consequence, their prodelta and deeper fan sediments reach thicknesses of the order of several hundred to thousand metres per million years, and event deposits may occur in all systems tracts. However, with shorter sea-level variations (periods of ≤100 Ka) and increasing amplitudes of sea-level change, first the transgressive and then the highstand systems tracts become insignificant for the formation of event deposits. Of course, other factors, such as climatic change (e.g. glacials/interglacials) or delta switching, play a role in this respect.

It should be mentioned that some sediment sources respond earlier or later to relative sea-level fall. Pre-existing highstand deltas which have prograded to the shelf edge may undergo large slope failures at the onset of relative sea-level fall, whereas older, drowned lowstand deltas on the shelf or the shelf edge itself may respond later through slope instabilities (Posamentier & Allen 1993). Valley excavation near the coast and its landward continuation will generally lag behind sea-level fall and may continue during rising sea level (Einsele 1996). The predominance of sandy event deposits in small to middle-sized submarine fans indicates nearby mountain ranges and/or the remobilization of coastal and fluvial sediment during relative sea-level fall. Large deep-sea fans fed by major rivers are dominated by turbidite muds.

In settings with rapid differential uplift of the coastal zone, sea-level changes may mimic pulses of uplift and generate step-like valley incision and shoreline regression (Einsele 1996). As a result, sediment influx also varies, but event deposits have a good chance of being formed in all systems tracts of third-order sea-level variations and their parasequences.

Carbonate platforms and nearby slope and basin sediments commonly reflect third-order and higher-frequency sea-level lowstands by karstified platform surfaces and coarse skeletal debris and megabreccias on the lower slopes and basin floors. Exceptions to this rule occur when oversteepened platform margins collapse independently from sea-level changes (Crevello *et al.* 1989). In the case of mixed carbonate–siliciclastic systems, lowstands are frequently characterized by siliciclastic turbidites. Carbonate turbidites tend to be produced during highstands rather than during lowstands, because biogenic carbonate production is higher during times of submerged platforms than during sea-level lowstands (e.g. Schlager *et al.* 1994).

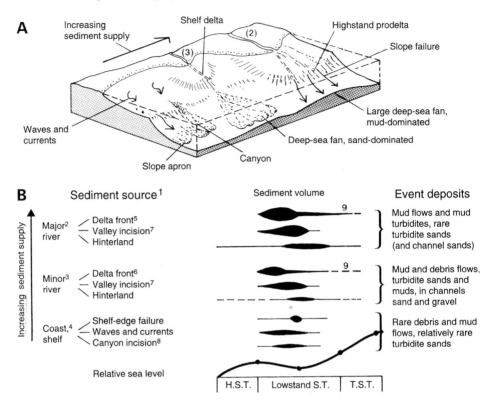

Figure 6.10   *Sediment remobilization and depositional events as related to relative sea-level change in settings of moderate subsidence or uplift.* **A.** *Overview of depositional environments with increasing sediment supply.* **B.** *Sources and remobilization of sediment during specific systems tracts and resulting types of event deposits (partially based on Posamentier and Allen 1993). Notes: 1, mainly for sediment remobilization; 2, from distal mountain range; 3, from mountain range at intermediate distance; 4, little terrigenous input; 5, at shelf edge (highstand delta); 6, on shelf (high to lower sea-level stand); 7, subaerial, close to coastline; 8, submarine; 9, delta front slope failures may continue as a result of high sediment supply. General rule: sea-level control of sediment remobilization and event deposits becomes less distinct with increasing sediment supply and decreasing frequency and amplitude of sea level change. [Reprinted from Sedimentary Geology, 104, Einsele, Event deposits: the role of sediment supply and relative sea-level changes – overview, 11–37, 1996, with kind persmission of Elsevier Science – NL, Sara Burgerhart Straat 25, 1055 KV Amsterdam, The Netherlands.]*

6.2.5 Continental Depositional Events

Depositional events are also known from the continental realm (Table 6.2). However, most of them are of limited extent and therefore useful only for local correlations (e.g. sediment gravity flows on alluvial fans and on fan deltas in lakes). These phenomena commonly occur at intervals of tens to hundreds of years. Overbank fines on flood plains may form either seasonally or at very irregular intervals of tens to hundreds of years. Correlation of individual beds in these settings is difficult without the aid of soil horizons or tephra deposits.

Very rare, catastrophic floods (sheetfloods, flash floods) on alluvial plains, mainly known either from proglacial rivers (e.g. Maizels 1989) or from arid regions (e.g. Karcz 1972; Pflüger & Seilacher 1991), leave behind specific gravel beds as records of the former event. Similarly, exceptionally thick or compositionally unique lacustrine turbidites may provide useful marker beds.

6.2.6 Tephra Events

Volcanic ash layers and other volcaniclastic sediments (tephra deposits) can be of outstanding importance in stratigraphy if they form laterally traceable beds which can be dated. Tephra deposits occur on land as well as in subaqueous environments (Figure 6.11). They originate from volcanic eruptions and can be distributed by wind or current action over wide areas. As long as they contain fresh minerals, tephra layers are well suited to radiometric dating. Individual beds that can be associated with specific large volcanic eruptions are practically isochronous. Isolated tephra beds intercalated between non-volcaniclastic sediments, are often easy to identify and to correlate between distant outcrops and drill cores. For all these reasons, tephra beds frequently represent excellent stratigraphical markers. A thorough description of the great variety of volcaniclastic sediments is given, for example, by Cas & Wright (1987) and Schmincke & Bogaard (1991). Here, two types of tephra deposits are of particular interest:

1. *Fallout tephra deposits.* Fallout deposits or ash falls are produced by pyroclastic activity caused by degassing of magma and phreatomagmatic eruptions, the latter resulting from the contact of hot magma with external water. Hot volcaniclastic fragments and gas are ejected from a volcanic vent and rise as a buoyant plume high into the atmosphere. The ash cloud is directed downwind and may spread over distances of several hundreds to thousands of kilometres. With increasing distance from the source, the distal fallout layer becomes thinner, finer grained and better sorted (Figure 6.11D). The chemical and mineralogical composition of an ash fall bed reflects the nature of the magmatic source. The volcanic fragments, lapilli, ash, dust and lithoclasts derived from the country rock, settle not only on land surfaces and in lakes, but also on the sea bottom, where they may be reworked and redeposited by normal bottom currents, gravity mass flows and suspension currents (Figure 6.11B). In the latter case they form graded ash turbidites. In the ancient record, distal ash fall beds of lake and marine environments have been frequently transformed into bentonite layers or tonsteins.
2. *Pyroclastic flows, volcanic mud and debris flows.* Pyroclastic flows result from the collapse of overloaded volcanic eruption columns, somes and cones, and consist of hot, partially fluidized solid–gas mixtures with high particle concentrations. Under water, the flow material mixes with water and generates tephra-dominated mass flows (Figure 6.11B). The transport distances of these flow types

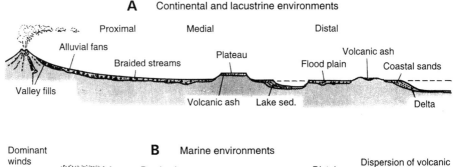

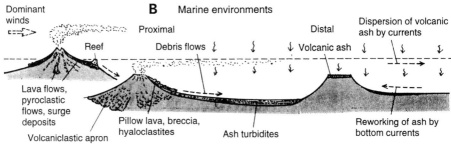

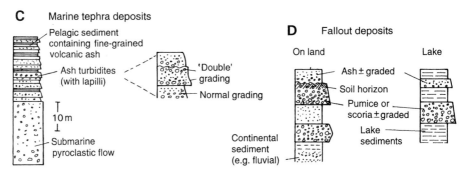

Figure 6.11   *Various tephra events in* **(A)** *continental and* **(B)** *marine environments.* **C.** *Compound section of marine tephra deposits reflecting the transition from maximum to waning stages of submarine eruption.* **D.** *Repeated fallout deposits on land surface and in lake. [Modified from: Einsele (1992)]*

are generally less than those of ash fall deposits; nevertheless, they may reach several tens to hundreds of kilometres. An idealized section of a pyroclastic flow sometimes begins with a high-velocity, low-density gaseous flow producing a thin stratified surge deposit, followed by the main body of the pyroclastic flow producing welded tephra such as ignimbrite, and may end with graded fallout ash (Figure 6.11C). Volcanic mud and debris flows (lahars) result from slope failure of volcanic and non-volcanic material of various ages. They form either during eruptions or as normal gravity mass flows with the aid of water. These types of mud and debris flows carry large proportions of volcaniclastic fragments downslope and over long distances. If they reach the coast they provide material for tephra-dominated submarine mass flow deposits and turbidites.

Volcanic activity produces large amounts of tephra which can be widely distributed both on land and under the sea by the mechanisms mentioned above. In
addition, marine environments provide a higher potential for the volcaniclastic
deposits to be preserved in the sedimentary record. The proportion of volcaniclastic
event deposits varies with tectonic settings. Island arcs and convergent continental
margins associated with an active subduction zone are most effective in producing
large volumes of volcaniclastic materials. The fill of fore-arc and back-arc basins
may consist of volcaniclastic deposits to a large degree. However, as marker beds,
tephra deposits appear to be especially useful in tectonic settings where volcanic
activity is relatively insignificant and rare.

6.2.7 Sediment Redeposition and Event Beds: Concluding Remarks

Most event deposits represent redeposited material that has accumulated earlier
along basin margins. It is possible to distinguish, therefore, four phases in the
generation of event beds (Figure 6.12; Einsele *et al.* 1996):

1. *Pre-event phase of sediment accumulation.* This phase establishes the sediment
   source for the depositional event. Texture and composition of an ultimate
   individual event bed are largely controlled by both the environmental conditions of the primary source region and the area of intermediate pre-event
   sediment storage.
2. *Initiation and triggering of sediment displacement.* These processes include peak
   floods caused by rainstorms or breakage of natural dams and various mechanisms of slope failure. Slope failure may be initiated by rapid sea-level fall, excess
   pore-water pressure, release of water and methane gas from solid gas hydrates
   (e.g. Paull *et al.* 1996), enhanced shear stress and liquefaction caused by extreme
   storms, tsunami waves, earthquakes or volcanic eruptions.
3. *Transport mechanisms.* Transport mechanisms, such as storm and tsunami wave-
   induced currents have a high transport capacity capable of transporting coarse

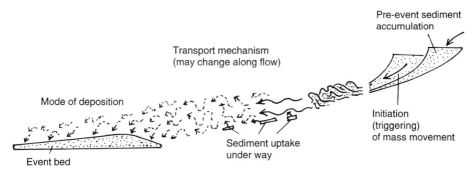

Figure 6.12 *Succession of processes involved in the generation of event deposits demonstrated for gravity-driven mass movements resulting in turbidites*

gravel and blocks in the coastal zone and along submarine channels. Slides and slumps often evolve into various types of mass flow such as plastic flow, liquefied flow, and laminar to turbulent flow, many of which are transformed into high- to low-density suspension flows or turbidity currents. It is important to note that both the loss of sediment or its entrainment can change the composition of the final event bed relative to that of the original sediment.

4. *Mode of final deposition.* Bed type, texture, and sedimentary structures of a depositional event are controlled to a large degree by the mechanism of final deposition. They reveal the character of the flow at its final stage (Postma 1986; Stow *et al.* 1996). The mechanism triggering the depositional event cannot usually be inferred from these or other criteria of the event bed.

6.3 NON-DEPOSITIONAL AND EROSIONAL EVENTS

Non-deposition and erosion do not create distinct beds with sharp lower and upper boundaries, but rather produce specific surfaces or event horizons. For this and other reasons, these horizons constitute a group of stratigraphical features entirely different from depositional events. Non-deposition and erosion can occur on land and below the sea under a range of different circumstances. Many of these generate omission or erosion surfaces of only limited areal extent and are therefore of minor interest here. Of importance are events that affect large areas within a sedimentary basin or leave behind their signature in one or more basins.

Neglecting erosional disconformities caused by prior tectonism, it appears that the sedimentary record of non-deposition and erosion are of interest only if they span, in contrast to depositional events, a considerable period of time. A consequence of this statement is that fossil organisms are important in characterizing these types of event horizons. Both physical and biological processes often play an important part in the formation of these event horizons. In discussing this type of event, we shall first briefly look at some sedimentological phenomena and soils on land, before examining the marine systems that are dominated by biological processes and skeletal remains.

6.3.1 Continental Non-depositional Events

Sediments left as erosional lags in fluvial channels or on winnowed desert pavements are of limited lateral extent or are poorly preserved in the sedimentary record. Of more importance are widespread soil horizons that develop on surfaces either starved of sediment or produced by non-deposition. Typically this may occur on wide fluvial plains, raised river terraces, dried lake basins or on isolated plateaux (Figure 6.13A). If the soils are later buried under younger sediments and become fossilized as paleosols, they can be traced over considerable distances. The characteristics of paleosols depend strongly on the climate and vegetation during soil formation (e.g. Wright 1986; Retallack 1990). Vegetated soils of the humid zone often display traces of rootlets and increased clay mineral content (e.g. illite) and

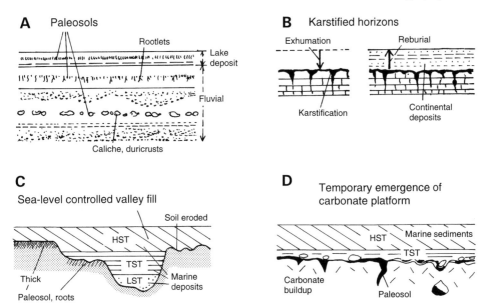

Figure 6.13 *A. Various paleosols in fluvial deposits. **B.** Stratal discontinuity caused by previous period of karstification on land, later exhumed by erosion or reburied under younger continental sediments. **C.** Valley incision and soil formation during sea-level lowstand; the landform is later successively filled and overlain by late lowstand deposits (LST), transgressive (TST) and highstand deposits (HST). **D.** A carbonate platform, which emerged and karstified during sea-level lowstand, is overlain by the sediments of the transgressive and highstand systems tracts (TST and HST)*

organic matter, whereas soils of semi-arid regions tend to be characterized by various types of nodules and duricrusts, such as calcrete, caliche, siliceous crusts (silcrete) and lateritic crusts (ferricrusts). The time period necessary to form mature and therefore distinctive soils is in the order of at least a thousand years, but in many cases may be much longer. Paleosols, in contrast to the event deposits discussed earlier (Figure 6.1), often represent longer time intervals in sedimentary sections than the thick sediments which accompany them (Kraus & Bown 1986). In many fluvial sequences paleosols occur repeatedly and constitute one of the best marker horizons of such deposits. Soil horizons may also be associated with valleys cut during sea-level lowstands (Figure 6.13C).

Similar to paleosols, coal seams produced in swamps represent fairly long periods of siliciclastic non-deposition. Therefore, widespread coal seams in coal-bearing sequences are frequently used for correlation and refinement of the stratigraphical resolution (e.g. Rahmani & Flores 1984; Lyons & Alpern 1989).

The development of karst surfaces on top of exhumed carbonate rocks on land may also provide marker horizons (Figure 6.13B). They are also common on marine carbonate buildups which repeatedly emerged above sea level (Figure 6.13D). Karst surfaces are usually irregular and often overlain by residual soil derived from carbonate dissolution or by deposition. In simple cases, the time span of a 'karstification event' can be constrained by determining the ages of the karst rock and its overlying stratum. In addition, dissolution features and the associated soil formation may give some hints (Ford & Williams 1989). Datable fauna from soil-filled solution pits, shafts and dolines may also help bridge the stratigraphical gap.

6.3.2 Marine Lag Sediments and Shell Concentrations

Non-skeletal lag sediments

Wave action and wave-induced currents in the coastal, foreshore, and inner shelf zones frequently erode and winnow pre-existing sediments. Fine-grained particles are commonly carried away to be deposited in deeper water, whereas the coarse material such as gravel and lithoclasts stays behind to form a lag deposit. Repeated reworking leads to an enrichment of those components which are mechanically and chemically stable (Figure 6.14). In addition, some specific authigenic minerals

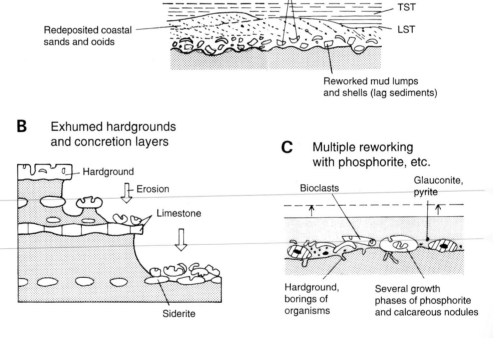

Figure 6.14 *Various types of submarine, mostly non-skeletal, stratal discontinuities.* **A.** *Erosional lag and seaward transported coastal sand and ooids (carbonate or iron ooids; lowstand deposits, LST) overlain by transgressive deposits (TST).* **B.** *Hardgrounds or concretion horizons exhumed by wave and current action.* **C.** *Close-up of B, containing phosphorite nodules and accentuated by multiple reworking. See text for explanation. [Modified from: Einsele (1992)]*

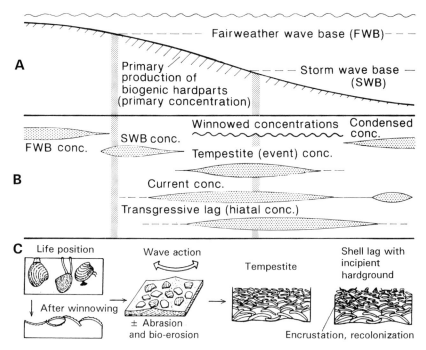

Figure 6.15 *Various modes of skeletal concentrations.* **A.** *Cross-section of shallow-marine basin; primary production of biogenic hardparts changes along onshore–offshore gradient.* **B.** *Processes generating skeletal concentrations along A.* **C.** *Examples of shell (bivalves) concentrations resulting from winnowing, wave action, storms and geostrophic currents, and overprint by life–death interaction (see Figure 6.16). [Modified from: Fürsich & Oschmann (1993)]*

may form under these conditions, such as glauconite in sandy lags. Lag sediments are also known from isolated subaqueous highs affected by waves and currents. Good examples of widespread lag sediments develop under wave action during both transgressive and regressive seas. Concretions of carbonate and phosphorite, formed earlier in deep water within the sediment, may be exhumed and concentrated in this type of lag (Figures 6.14B and C).

Skeletal concentrations (shell beds)

Most large shells are produced by organisms, typically molluscs, that live as infauna within the upper centimetres to sub-metres of soft shallow-water sediments. The density of these infaunal communities varies with water depth (Figure 6.15A) and with other factors, but is commonly not sufficient to produce skeletal concentrations such as observed in reefs. To obtain skeletal and lithoclast concentrations on soft ground, wave and/or current action is usually required as summarized in Figures 6.15B and C. There are two basic genetic types of skeletal concentrations (Kidwell 1991):

1. *Allogenic mode*. Here the event concentration results from a single brief hydraulic or biogenic episode, such as a storm event and the subsequent colonization by epifauna (Figure 6.16A). The storm provides exotic shells and/or lithoclasts eroded in shallower parts of the basin, but may also exhume local shells of soft-bottom dwellers and mix them with the allochthonous shells. After the storm, the new shellground forms a harder substrate than the previous soft ground. This change in environment leads to the colonization of a new gravel-dwelling epifauna on top of or within the dead shell bed. This shift in the faunal composition is referred to as life–death interaction or taphonomic feedback. A special case is the rapid burial and preservation of *in situ* fauna by a depositional event (obrution *lagerstätten*; Brett & Seilacher 1991).

2. *Autogenic mode*. Here the shell (or clast) concentration is caused by reworking of *in situ* skeletal remains and winnowing of the fine-grained matrix (Figure 6.16B). As with the allogenic mode, the shell bed allows colonization by gravel dwellers; that is, the introduction of new species. Normal sedimentation restores the soft bottom and its associated fauna with time.

There may be transitions between these two basic modes of shell concentration and, in addition, many shell beds will be affected by repeated phases of shell

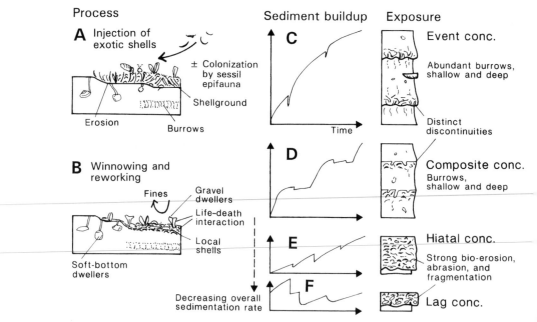

Figure 6.16 *Different modes of shell bed formation (**A** and **C**) with allochthonous (exotic) bioclasts delivered either by normal bottom currents or suspension currents. **B**. Concentration of local shells by reworking and winnowing of the fine matrix. **D**. Shell beds originating from repeated phases of sediment starvation. **E** and **F**. Complex shell bed caused by continuous sediment starvation (or bypassing) and repeated reworking (mainly allochthonous shells).* [Modified from: Kidwell (1991)]

injection and/or *in situ* reworking. As a consequence one can distinguish four types of skeletal bioclast concentrations (Table 6.3) which also reflect a trend from substantial to negligible sedimentation or subsidence rates (Figures 6.16C–F; Kidwell 1993). Hiatal and lag concentrations show indistinct discontinuities and strong bio-erosion, abrasion and fragmentation of bioclasts. The shell orientation or biofabric caused by storms and unidirectional currents is predominantly convex-up (Figure 6.15C), although wave action sometimes leads to edge-wise, fan-shape or chaotic fabrics. The faunal associations involved in the various types of skeletal event beds vary through the Phanerozoic due to the evolution of the organisms that produce the bioclasts. In addition, regional differences in water temperature and nutrient supply play an important part in determining the type of skeletal event bed produced.

6.3.3 Stratal Discontinuities Controlled by Relative Sea-Level Changes

Emergence of the coastal zone or of submarine plateaux may cause valley incision, soil formation and karst formation (see Sections 6.2.5 and 6.3.2; Figures 6.13C and D). However, many of these features tend to become coevally or subsequently removed by subaerial mechanical and chemical erosion. The stratigraphical value of these irregular discontinuities in a repeatedly emerging environment may therefore be limited. Stratal discontinuities, which can be better defined and preserved, form in those parts of a basin which experience either no, or only very brief, periods of emergence. Here, subaqueous erosion or sediment bypassing may generate discontinuities.

   In shallow seas, erosional lags and the various skeletal concentrations characterize certain basinal locations which vary with sea level (Figure 6.17). They occur preferentially at the end of the lowstand systems tract (LST), around the transition from the transgressive (TST) to the highstand systems tract (HST), i.e. along the shallow portion of the maximum flooding surface (MFS) and at the top of the highstand systems tract (HST). Lag concentrations in the LST display an erosional base due to emergence and subaqueous erosion, or alternatively they represent hiatal concentrations which result from sediment bypassing (e.g. shelly shoals). In

Table 6.3 *Genetic classification of shell concentrations. [Modified from Kidwell (1993)]*

Product	Process
Event concentration	Short hydraulic episode (e.g. storm), ± subsequent colonization by epifauna
Composite concentration	Multiple erosional and depositional events, amalgamation, moderate sediment buildup
Hiatal concentration (condensed section)	*In situ* accretion and amalgamation, mixed fauna and slow net sedimentation
Lag concentration	Exhumation and concentration of mechanically and chemically resistant hard parts of different ages, indicating significant stratigraphic truncation

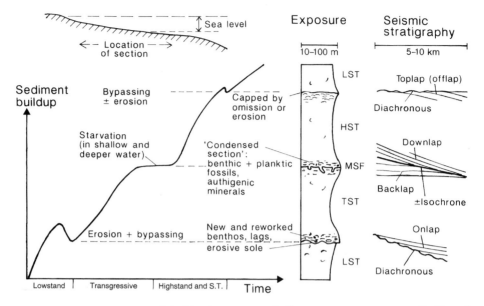

Figure 6.17 *Various types of shell beds in marine shallow-water sections controlled by relative sea-level change. Nature and faunal composition of beds vary with distance of section from shoreline and water depth. Note that seismic stratal patterns such as onlap, downlap, etc., cannot be seen in outcrops of limited lateral extent. [Modified from: Kidwell (1991)]*

seismic records these beds mark an onlap surface and are diachronous. Concentrations of coastal sands and iron ooides on the inner shelf often result from reworking and redeposition of these sediments during lowering sea level (Figure 6.14A). The ooids probably form in the nearshore zone (e.g. Bayer 1989; Einsele 1992).

Shell beds and enrichments of authigenic minerals such as phosphorite and glauconite around the condensed horizon of the LST correspond either to hiatal concentrations or to composite concentrations. They are located at the boundary between seismically backlapping and downlapping surfaces and are isochronous. Finally the HST may be capped by hiatal or lag concentrations which indicate a diachronous downlap surface.

Apart from these well-defined positions in a third-order sequence, shell concentrations can also occur within the individual systems tracts. Their number and mode largely depends on the rates of basin subsidence and sedimentation (Table 6.4; Kidwell 1993). In environments with low subsidence and sedimentation rates, third-order sequences (1–2 Ma) approximately follow the scheme outlined in Figure 6.17. In addition, they may contain a number of composite concentrations and event concentrations. With increasing sedimentation rates, third-order sequences thicken and may exhibit an increasing number of composite and event concentrations, while hiatal and lag concentrations become rare. Shallowing-upward parasequences may be recognized, for example, which are capped by non-depositional flooding surfaces and composite concentrations. High rates of subsidence and deposition commonly prevent the formation of discontinuity

Table 6.4 *Bioclastic concentrations and discontinuity surfaces in shallow-marine basins of different rates of subsidence and sedimentation. See Table 6.3 for types of concentrations; SB, sequence boundary. [Modified and simplified from: Kidwell (1993)]*

	Low subsidence ($\leqslant$10 m Ma^{-1})	Moderate subsidence (100 m Ma^{-1})	High subsidence ($\geqslant$1000 m Ma^{-1})
Sequence stratigraphic characterization	Thin (10–20 m) third-order sequences with distinct SBs; parasequences indistinct	Third-order sequences of medium thickness (100–300 m); shallowing-upward parasequences capped by non-depositional flooding surfaces	Third-order sequences; parasequences; no through-going discontinuity surfaces; downlap surfaces common
Types of concentrations			
Event	±Common	Common	Common
Composite	Common	Common (e.g. at top of parasequences)	Common (± local, on flooding surfaces)
Hiatal	Widespread	Limited (e.g. at maximum transgression surface)	Rare
Lag	Present at SBs	Rare	Rare

surfaces; surfaces of erosion and non-deposition only occur locally. The type and importance of skeletal concentrations may vary from location to location as well as with geological time (Kidwell 1993).

As shown in the transect through a shallow, differentially subsiding basin (Figure 6.18), the erosional unconformities of successive transgressive–regressive cycles merge landward and there form pronounced composite beds which represent hiatuses of increasing time span. Unconformities associated with the regressive phase are frequently overlain directly by transgressive lags. Figure 6.18 also demonstrates that the number of sea-level cycles, which can be identified in vertical sections by means of lags and skeletal concentrations, varies at different localities along the transect.

6.4 OTHER STRATIGRAPHICAL EVENTS

Some specific processes which leave a distinct signature in the stratigraphical record do not fit into the scheme of depositional or non-depositional events outlined above. Two of these processes are briefly described here, because their products are useful in correlating sedimentary successions at both a local and regional scale and have even been used inter-continentally.

6.4.1 *In Situ* Earthquake Structures

Earthquake-induced, *in situ* deformation of sediment layers ('seismites'; Seilacher 1984) shows a number of distinctive sedimentary structures that indicate their

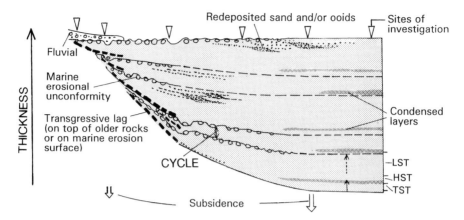

Figure 6.18   *Transect of shallow, differentially subsiding basin which has experienced several cycles of relative sea-level change with differing periods and amplitudes. Regressive and transgressive lags merge and mix landward; the stratigraphic section in the basin centre tends to be complete, but may contain thin condensed sections (e.g. black shales). [Modified from: Einsele (1992)]*

origin. Such structures are known from fluvial, lacustrine and coastal environments, as well as marine environments around island arcs and in orogenic belts (e.g. Sims 1979; Obermeier *et al.* 1990; Kanaori *et al.* 1993). Seismites form, and can be preserved, only on flat surfaces where no large gravity mass movements can be triggered by the earthquake event. In ancient rocks, however, the nature of *in situ* sediment deformation is often not as clear. Principally, one can distinguish between two types of shock-induced *in situ* deformation structures (Allen 1984; Guiraud & Plaziat 1993):

1. *Phenomena of liquefaction and fluidization.* These phenomena are primarily encountered in water-saturated sands and silts with a loose grain fabric (Figure 6.19A–D) and include hydroplastic deformation, loss of primary structures, sand injections (dykes and sills), sand blows, and sandy to muddy surface volcanoes.
2. *Soft-sediment deformation.* Soft-sediment deformation and/or fracturing of cohesive material at or close to the surface may give micro-folds, cracks, brecciation (Figure 6.19E). If the liquefied bed is overlain by a cohesive bed, the latter usually breaks and becomes disorganized.

Historical earthquakes show that individual seismites can be generated at distances of 10–100 km away from the earthquake epicentre (Allen 1986). Consequently, individual seismites can be traced over long distances if the depositional environment and the preservation potential of the beds are favourable.

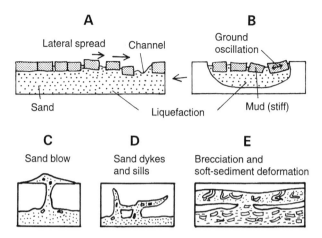

Figure 6.19 *Earthquake-induced seismites displaying* in situ *disturbances of flat-lying sediments. [Based on information in Guiraud & Plaziat (1993) and Greene et al. (1994)]*

6.4.2 Traces of Meteorite Impacts

In geological history meteorites have repeatedly passed the atmosphere and hit the earth's surface. Bolide (meteorite) impacts and their effects on marine and terrestrial organisms and sediments have been vividly discussed in the past 10–20 years (e.g. D'Hondt 1994; Toon & Zahnle 1995). They have created craters, accompanied by laterally ejected rock fragments, and sent large dust clouds into the atmosphere. Other mechanical and chemical signals of such events include burnt forests, shocked minerals (quartz, zircon), high-pressure polymorphs of SiO_2 (stishovite, coesite), and melt-rock (small beads and larger tectites).

Medium-sized impacts with recurrence intervals of between 10^3 and 10^7 years may have damaged large continental areas through blast waves, fires, earthquakes and catastrophic tsunamis along the coasts. Some very large and rare meteorite impacts in the earth's history, with recurrence intervals of tens of million years, appear to have generated signals that may be traced globally. In this case, evaporation of the meteorite and its subsequent fallout as dust and dissolved compounds from the atmosphere may have produced thin sediment layers enriched in rare elements such as iridium. Meteorite impacts on the ocean surface are assumed to have generated large tsunami waves and tsunamiites (see Section 6.2.1). Long-lasting darkness and global cooling caused by atmospheric dust may have affected photosynthesis and caused significant shifts in the marine biochemical system. The most controversial issue is the impact-induced mass extinction of specific groups of organisms on land and in the ocean as postulated, for example, for the Cretaceous–Tertiary boundary.

The question is whether or not stratigraphers and sedimentologists studying ancient rock sequences should look for specific meteorite signatures in their normal fieldwork. As indicated above, there are various distinctive signals, but most of them are not readily observable in the field. New findings of ancient meteorite events are usually made at or close to the impact site where morphological, sedimentological and other consequences are evident. The traces of known and dated bolide impacts can be found elsewhere only by detailed investigations using a variety of methods. If this can be done, bolide impacts are of great importance for the correlation of certain stratigraphical time intervals.

6.5 PRESERVATION OF VARIOUS EVENT HORIZONS

The preservation potential of event beds and event horizons strongly depends on the environment in which they were formed. Depositional events are well preserved in lakes, in the deep sea, and in shallower marine environments below the storm wave base where bottom currents are too weak to rework the event beds. However, there are many exceptions to this general rule. For example, localized bottom currents following the morphological contours of the sea floor are able to erode and redeposit deep-sea sediments. Episodic erosion and redeposition also takes place along the channels of deep-sea fans. Similarly, non-depositional event horizons require an environment where post-event burial under younger sediments can protect them from subsequent erosion. In marine environments this situation occurs at water depths below the storm wave base. Taking into account sea-level changes, the maximum flooding surface seaward of the shelf is an appropriate example (cf. Figure 6.18). Soil horizons on alluvial plains have a good chance of being preserved in settings in which there is considerable subsidence.

Beds produced in environments with permanent or frequent current reworking, such as flood plains, coastal areas and shallow seas (above the storm wave base) have a generally low preservation potential. Those parts of marine basins which emerge during sea-level fall commonly preserve only a limited number of event horizons formed during the proceeding transgressive and highstand systems tracts. Little subsidence and/or slow long-term sediment aggradation promote the destruction of event beds in these environments. Only a limited percentage of the original number of event beds is preserved. This is the reason why, for example, supratidal and foreshore storm- and tsunami-generated sands are less frequent in the geological record than would be expected from the recurrence interval of such events today. Other examples of relatively poor preservation are thin mud turbidites and mud tempestites, because they are more or less obliterated by post-depositional bioturbation in the Phanerozoic, but not in older sediments.

Multi-phase non-depositional event horizons form in environments that are basically unfavourable for their preservation, since sedimentation and burial is unlikely to occur rapidly. However, the accumulation of lithic and skeletal hardparts resistant to current erosion and mechanical, biological (bio-erosion) and chemical destruction allows their preservation until finally subsidence of the basin floor (or relative sea-level rise) causes their burial under normal sediment. Where these lag concentrations survive they record not only the final stage of their development but also indicate, by their fossil content and authigenic minerals, the time span involved in their evolution.

Diagenetic overprint, such as cementation by carbonate, often accentuates primary discontinuities in the sediment build-up and thereby improves the recognition of event horizons and their sedimentary structures in field exposures and drilling cores (Einsele *et al.* 1991).

6.6 SUMMARY

Stratigraphical events result from different processes, the most important being: (1) depositional; and (2) largely non-depositional events. Specific biological events have not been discussed in this chapter. The events described can be summarized as follows:

6.6.1 Large-scale Depositional Events

Large-scale depositional events are abundant in marine basins receiving high terrigenous sediment influx or producing large amounts of biogenic material. Remobilization of sediment is promoted by a variety of processes, including rapidly lowering sea level. The event bed facies is controlled by the mechanism of final deposition. Preservation of event deposits requires a low-energy environment, limited bioturbation, and sufficient subsidence to prevent post-depositional erosion. Individual event deposits with specific properties in terms of thickness, lateral extent, sedimentary structures, composition, and faunal content can be used as isochronous marker beds. They include the following:

1. Storm wave and tsunami wave-induced deposits (tempestites and tsunamiites) which occur mainly in shallow-marine sediments and are usually only partially preserved. They form sheet-like, sometimes discontinuous, basinward thinning and fining beds of considerable lateral extent, and recur in stratigraphical sections at intervals of the order of thousands of years. Sandy siliciclastic and bioclastic tempestites typically display gutter casts and various sole marks, combined flow features (wave and current action), suspension sedimentation (grading), rippled tops, and shallow-water faunal associations. Their components reflect coastal and foreshore erosion and, as a result of frequent reworking and amalgamation, may be mechanically and chemically–mineralogically mature. Tempestites consisting of mud, brecciated mud (including micritic carbonate), and brecciated microbial mats (flat pebble conglomerates) also occur in specific depositional environments. Distinct tsunami deposits appear to be restricted to limited areas, but can prograde into deeper water than tempestites. They contain coarser components of mixed land and coastal-shallow water provenance and may record the pulses of the tsunami waves.
2. Sediment gravity flow deposits and turbidites, including bioclastic and volcaniclastic sediments, constitute a family of genetically related event deposits typical of deep basins. They sometimes show transitional and coupled features. In terms of volume and lateral extent they represent the most important group of event deposits in deep basins. They typically contain displaced shallow-

water fauna and exhibit proximal–distal trends. In mixed siliciclastic/carbonate systems, carbonate turbidites are mostly associated with sea-level highstands, whereas siliciclastic turbidites predominate during lowstands. Proximal bioclastic carbonate turbidites tend to be coarser grained than their siliciclastic counterparts. Apart from textural grading, they frequently display chemical and mineralogical grading, especially in mud turbidites (e.g. chert bands or nodules). Sandy and muddy turbidites frequently deviate in their contents of carbonate and organic carbon from their pelagic or hemipelagic host sediments (e.g. if deposited below the CCD or in well-oxygenated bottom waters). Individual olistostromes, mega-turbidites, and successions of several event beds comprising turbidites of different composition often provide marker horizons for stratigraphical correlation.

Channelized deep-sea fan systems generate a complex facies pattern of gravity mass flow deposits and both sandy and muddy turbidites which show varying palaeocurrent directions. More regular bedding and consistent transport directions are found in lower, detached fan lobes and basin plains. Minor coarsening- or fining-upward cycles are caused by prograding, retrograding, or laterally shifting fan lobes. Sea-level changes and local tectonic activity may create larger cyclic sequences and accentuate or hamper the occurrence and intensity of gravity mass flow deposits and turbidites. The recurrence interval of mass flow and turbidite events varies greatly, from relatively frequent, thin silt and mud turbidites (50 to several 100 years; e.g. in channelized deep-sea fan systems) to thick, extensive key beds ('mega-turbidites', 50 Ka to 1 Ma).

3. Volcanic eruptions produce a variety of event deposits (tephra events) on land and in subaqueous environments, including fallout (ash fall), pyroclastic flows, volcanic mud and debris flows, all of which have the potential to cover a wide area. Comparatively rare tephra beds alternating with non-volcanic sediments can usually be traced extensively and provide, if datable, excellent isochronous stratigraphical marker beds. Submarine tephra (debris flows, ash turbidites) are distributed in a similar way as non-volcaniclastic material.

6.6.2 Condensation, Non-depositional and Erosional Events

These occur in settings which lack sediment supply and/or during periods in which a high-energy hydraulic regime prevents deposition and causes sediment bypassing or erosion. These requirements are met in a variety of depositional environments, including the following:

1. In the continental realm, significant periods of non-deposition enable the formation of soils or duricrusts, which often provide good marker beds in fluvial sequences. Karst surfaces on top of carbonate build-ups are particularly useful in zones of periodic emergence above sea level. Aeolian desert pavements or truncation surfaces may be of more local interest.
2. Shallow-marine erosional lags are concentrations of coarse, stable particles, including lithoclasts, bioclasts, exhumed carbonate concretions, phosphorite,

and authigenic minerals such as glauconite. They usually represent a significant stratigraphical hiatus and are mostly diachronous. In sea-level-controlled systems, erosional lags typically occur at the base of the lowstand and transgressive system tracts as well as at the top of the highstand system tract although they may be less distinct in such locations.

3. Skeletal event concentrations result from the injection of allochthonous bioclasts into muddy environments. If not rapidly buried, these shell beds may be overprinted by recolonization (life–death interaction).

4. Skeletal event concentrations (shell beds) may also show both limited erosion, or non-deposition, and very slow net deposition forming composite concentrations and hiatal concentrations. They may contain both autochthonous and allochthonous bioclasts and often display *in situ* bio-erosion, abrasion and fragmentation. Recolonization and other life–death interactions (taphonomic feedback) are common. Sea-level-controlled condensed sections marking the maximum flooding surface in hydraulically active zones belong to this category. In quiet deeper water, condensed sections are stratigraphically complete. In basins with onshore decreasing subsidence rates, stratal discontinuities merge landward.

6.6.3 Other Events

Large-magnitude earthquakes can produce specific structures in marine and continental sediments at or close to the surface (seismites). Individual seismites are synchronous but cannot normally be traced over large distances. Very rare and large meteorite impacts may also leave specific traces of potential global extent.

REFERENCES

Aigner, T. & Reineck, H.E. 1982. Proximality trends in modern storm sands from the Helgoland Bight (North Sea) and their implications for basin analysis. *Senckenbergiana maritima*, **14**, 183–215.

Akou, A.E. 1984. Subaqueous debris flow deposits in Baffin Bay. *Geomarine Letters*, **4**, 83–90.

Albertao, G.A., Martins, P.P. 1996. A possible tsunami deposit at the Cretaceous–Tertiary boundary in Pernambuco, northeastern Brazil. *Sedimentary Geology*, **104**, 189–201.

Allen, J.R.L. 1984. *Sedimentary structures, their character and physical basis*. Developments in Sedimentology, 30B, Elsevier, Amsterdam.

Allen, J.R.L. 1986, Earthquake magnitude–frequency, epicentral distance, and soft-sediment deformation in sedimentary basins. *Sedimentary Geology*, **46**, 67–75.

Bayer, U. 1989. Stratigraphic and environmental patterns of ironstone deposits. In Young, T.P. & Taylor, W.E. (eds) *Phanerozoic ironstones*. Geological Society Special Publication No. 46, 105–117.

Bouma, A.H. 1962. *Sedimentology of some flysch deposits*. Elsevier, Amsterdam.

Bourgeois, J., Hansen, T.A., Wiberg, P.L. & Kauffman, E.G. 1988. A tsunami deposit at the Cretaceous–Tertiary boundary in Texas. *Science*, **241**, 567–570.

Brett, C.E. & Seilacher, A. 1991. Fossil lagerstätten: a taphonomic consequence of event sedimentation. In Einsele, G., Ricken, W. & Seilacher, A. (eds) *Cycles and events in Stratigraphy*. Springer, Berlin, 283–297.

Cas, R.A.F. & Wright, J.V. 1987. *Volcanic successions: modern and ancient.* Unwin Hyman, London.

Clifton, H.E. (ed.) 1988. *Sedimentologic consequences of convulsive geological events.* Geological Society of America, Special Paper No. 229.

Collinson, J.D. & Thompson, D.B. 1989. *Sedimentary structures*, 2nd edition. Chapman & Hall, London.

Crevello, P.D. & Schlager, W. 1980. Carbonate debris sheets and turbidites, Exuma Sound, Bahamas. *Journal Sedimentary Petrology*, **50**, 1121–1147.

Crevello, P.D., Sarg, J.F. & Read, J.F. (eds) 1989. *Controls on carbonate platforms and basin development.* Society of Economic Paleontologists and Mineralogists, Special Publication No. 44.

D'Hondt, S. 1994. The impact of the Cretaceous–Tertiary boundary. *Palaios*, **9**, 221–223.

Dott, R.H. 1983. Episodic sedimentation – how normal is average? How rare is rare? Does it matter? *Journal of Sedimentary Petrology*, **53**, 5–23.

Einsele, G. 1992. *Sedimentary basins: evolution, facies, and sediment budget.* Springer, Berlin, Heidelberg and New York.

Einsele, G. 1996. Event deposits: the role of sediment supply and relative sea-level changes – overview. *Sedimentary Geology*, **104** (Special Issue), 11–37.

Einsele, G. & Kelts, K. 1982. Pliocene and Quaternary mud turbidites in the Gulf of California: sedimentology, mass physical properties and significance. In Curray, J.R., Moore, D.M. *et al.* (eds) *Initial Reports DSDP 64*, US Government Printing Office, Washington, DC, 511–528.

Einsele, G. & Seilacher, A. 1991. Distinction of tempestites and turbidites. In Einsele, G., Ricken, W. & Seilacher, A. (eds) *Cycles and events in stratigraphy.* Springer, Berlin, Heidelberg and New York, 377–382.

Einsele, G., Ricken, W. & Seilacher, A. 1991. Cycles and events in stratigraphy – basic concepts and terms. In Einsele, G., Ricken, W. & Seilacher, A. (eds) *Cycles and events in stratigraphy.* Springer, Berlin, Heidelberg and New York, 1–19.

Einsele, G., Chough, S.K. & Shiki, T. 1996. Depositional events and their records – an introduction. *Sedimentary Geology*, **104** (Special Issue), 1–9.

Elmore, R.D., Pilkey, O.H., Cleary, W.J. & Curran, H.A. 1979. Black shell turbidite, Hatteras abyssal plain, western Atlantic ocean. *Geological Society of America Bulletin*, **90**, 1165–1176.

Embley, R.W. 1980. The role of mass transport in the distribution and character of deep-ocean sediments with special reference to the North Atlantic. *Marine Geology*, **38**, 28–50.

Ernst, G., Schmid, F. & Seibertz, E. 1983. Eventstratigraphie im Cenoman und Turon NW-Deutschlands. *Zittelina*, **10**, 531–554.

Ford, D. & Williams, P. 1989. *Karst geomorphology and hydrology.* Unwin Hyman, London.

Fürsich, F.T. & Oschmann, W. 1993. Shell beds as tools in basin analysis: the Jurassic of Kachchh, western India. *Journal of the Geological Society of London*, **150**, 169–185.

Greene, M., Power, M. & Youd, T.L. 1994. Liquefaction: what is it and what to do about it. *Earthquake Basics, Brief No. 1*, Earthquake Engineering Research Institute, Oakland, California.

Guiraud, M. & Plaziat, J.-C. 1993. Seismites in the fluviatile Bima sandstones: identification of paleoseisms and discussion of their magnitudes in a Cretaceous synsedimentary strike-slip basin (Upper Benue, Nigeria). *Tectonophysics*, **225**, 493–522.

Hesse, R. 1974. Long-distance continuity of turbidites: possible evidence for an Early-Cretaceous trench–abyssal plain in the East Alps. *Bulletin of the Geological Society of America*, **85**, 859–870.

Hieke, W. 1984. A thick Holocene homogenite from the Ionian abyssal plain (eastern Mediterranean). *Marine Geology*, **55**, 63–78.

Hughes Clarke, J.E., Shor, A.N., Piper, D.J.W. & Mayer, L.A. 1990. Large-scale current-induced erosion and deposition in the path of the 1929 Grand Banks turbidity current. *Sedimentology*, **37**, 613–629.

Kanaori, Y., Kawakami, S., Jairi, K. & Hattori, T. 1993. Liquefaction and flowage at archeological sites in the inner belt of central Japan: tectonic and hazard implications. *Engineering Geology*, **35**, 65–80.

Karcz, I. 1972. Sedimentary structures formed by flash floods in southern Israel. *Sedimentary Geology*, **7**, 161–182.

Kauffman, E.G. 1988. Concepts and methods of high-resolution event stratigraphy. *Annual Review of Earth and Planetary Sciences*, **16**, 605–654.

Kauffman, E.G., Elder, W.P. & Sageman, B.B. 1991. High-resolution correlation: a new tool in chronostratigraphy. In Einsele, G., Ricken, W. & Seilacher, A. (eds) *Cycles and events in stratigraphy*. Springer, Berlin, Heidelberg and New York, 795–819.

Kelts, K. & Arthur, M.A. 1981. Turbidites after ten years of deep-sea drilling – wringing out the mop? In Warme, J.E., Douglas, R.G. & Winterer E.L. (eds) *The Deep Sea Drilling Project: a decade of progress*. Society of Economic Paleontologists and Mineralogists, Special Publication No. 32, 91–127.

Kidwell, S.M. 1991. Taphonomic feedback (live/dead interactions) in the genesis of bioclastic beds: keys to reconstructing sedimentary dynamics. In Einsele, G., Ricken, W. & Seilacher, A. (eds) *Cycles and events in stratigraphy*. Springer, Berlin, Heidelberg and New York, 268–282.

Kidwell, S.M. 1993. Taphonomic expressions of sedimentary hiatuses: field observations on bioclastic concentrations and sequence anatomy in low, moderate and high subsidence settings. *Geologische Rundschau*, **82**, 189–202.

Klein, G.deV. 1985. The frequency and periodicity of preserved turbidites in submarine fans as a quantitative record of tectonic uplift in collision zones. *Tectonophysics*, **119**, 181–193.

Kraus, M.J. & Bown, T.M. 1986. Paleosols and time resolution in alluvial stratigraphy. In Wright, V.P. (ed.) *Paleosols, their recognition and interpretation*. Blackwell, Oxford, 180–207.

Kuenen, P.H. & Migliorini, C.I. 1950. Turbidity currents as a cause of graded bedding. *Journal of Geology*, **58**, 91–127.

Lowe, D.R. 1982. Sediment gravity flows: II. Depositional models with special reference to the deposits of high-density turbidity currents. *Journal Sedimentary Petrology*, **52**, 279–297.

Lyons, W.P. & Alpern, B. (eds) 1989. *Peat and coal: origin, facies, and depositional models*. Elsevier, Amsterdam.

Maizels, J. 1989. Sedimentology, paleoflow dynamics and flood history of Jökulhlaup deposits: paleohydrology of Holocene sediment sequences in southern Iceland sandur deposits. *Journal Sedimentary Petrology*, **59**, 204–223.

Meischner, K.D. 1964. Allodapische Kalke, Turbidite in riff-nahen Sedimentationsbecken. In Bouma, A.H. & Brouwer, A. (eds) *Turbidites*. Elsevier, Amsterdam, 156–191.

Minoura, K. & Nakaya, S. 1991. Traces of tsunami preserved in inter-tidal, lacustrine and marsh deposits: some examples from Northeast Japan. *Journal of Geology*, **99**, 265–287.

Moore, D.G., Curray, J.R. & Einsele, G. 1982. Salado-Vinorama submarine slide and turbidity current off southeast tip of Baja California. In Curray, J.R., Moore, D.G. *et al.* (eds) *Initial Reports DSDP* 64, US Government Printing Office, Washington, DC, 1071–1082.

Mount, J.F. & Kidder, D. 1993. Combined flow origin of edgewise intraclast conglomerates: Sellick Hill Formation (Lower Cambrian), South Australia. *Sedimentology*, **40**, 315–329.

Mutti, E. 1992. *Turbidite sandstones*. Agip, Istituto di Geologia, Università di Parma.

Mutti, E. & Normark, W.R. 1987. Comparing examples of modern and ancient turbidite systems: problems and concepts. In Legget, J.K. & Zuffa, G.G. (eds) *Marine clastic sedimentology*. Graham and Trotman, 1–38.

Mutti, E., Ricci Lucchi, F., Seguret, M. & Zanzucchi, G. 1984. Seismoturbidites: a new group of resedimented deposits. *Marine Geology*, **55**, 103–116.

Myrow, P.M. 1992. Bypass-zone tempestite facies model and proximality trends for an ancient muddy shoreline and shelf. *Journal of Sedimentary Petrology*, **62**, 99–115.

Myrow, P.M. & Southard, J.B. 1996. Tempestite deposition. *Journal of Sedimentary Research*, **66**, 875–887.

Nelson, H. 1982. Modern shallow-water graded sand layers from storm surges, Bering Shelf: a mimic of Bouma sequences and turbidite systems. *Journal of Sedimentary Geology*, **52**, 537–545.

Nummedal, D. 1991. Shallow marine storm sedimentation – the oceanographic perspective. In Einsele, G., Ricken, W. & Seilacher, A. (eds) *Cycles and events in stratigraphy.* Springer, Berlin, Heidelberg and New York, 227–248.

Obermeier, S.F., Jacobson, R.B., Smoot, J.P., Weems, R.E., Gohn, G.S., Monoe, J.E. & Powars, D.S. 1990. Earthquake-induced liquefaction features in coastal setting of South Carolina and in the fluvial setting of the New Madrid seismic zone. *US Geological Survey Professional Paper* 1504.

Paull, C.H., Buelow, W.J., Ussler III, W. & Borowski, W.S. 1996. Increased continental-margin slumping frequency during sea-level lowstands above gas hydrate-bearing sediments. *Geology,* **24,** 143–146.

Pflüger, F. & Seilacher, A. 1991. Flash flood comglomerates. In Einsele, G., Ricken, W. & Seilacher, A. (eds) *Cycles and events in stratigraphy.* Springer, Berlin, Heidelberg and New York, 383–391.

Pickering, K.T., Soh, W. & Taira, A. 1991. Scale of tsunami-generated sedimentary structures in deep water. *Journal of the Geological Society of London,* **148,** 211–214.

Pilkey, O.H., Locker, S.D. & Cleary, W.J. 1980. Comparison of sand layer geometry on flat floors of 10 modern depositional basins. *Bulletin of the American Association of Petroleum Geologists,* **64,** 841–856.

Piper, D.J.W. & Shor, A.N. 1988. The 1929 'Grand Banks' earthquake, slump, and turbidity current. In Clifton, H.E. (ed.) *Sedimentologic consequences of convulsive geologic events.* Geological Society of America, Special Paper No. 229, 77–92.

Piper, D.J.W. & Stow, D.A.V. 1991. Mud turbidites. In Einsele, G., Ricken, W. & Seilacher, A. (eds) *Cycles and events in stratigraphy.* Springer, Berlin, Heidelberg and New York, 360–376.

Posamentier, H.W. & Allen, G.P. 1993. Variability of the sequence stratigraphic model: effects of local basin factors. *Sedimentary Geology,* **86,** 91–109.

Postma, G. 1986. Classification of sediment gravity-flow deposits based on flow conditions during sedimentation. *Geology,* **14,** 291–294.

Rahmani, R.A. & Flores, R.M. (eds) 1984. Sedimentology of coal and coalbearing sequences. *Special Publication of the International Association of Sedimentologists* No. 7.

Reading, H.G. & Richards, M. 1994. Turbidite systems in deep-water basin margins classified by grain size and feeder system. *Bulletin of the American Association of Petroleum Geologists,* **78,** 792–822.

Retallack, G.J. 1990. *Soils of the past.* Unwin Hyman, Boston.

Ricci Lucchi, F. & Valmori, E. 1980. Basin-wide turbidites in a Miocene 'over-supplied' deep-sea plain: a geometrical analysis. *Sedimentology,* **27,** 241–270.

Saito, Y. 1989. Modern storm deposits in the inner shelf and their recurrence intervals, Sendai Bay, northeast Japan. In Taira, A. & Masuda, F. (eds) *Sedimentary facies in the active plate margin.* Terra Scientific Publishers, Tokyo, 331–344.

Schlager, W., Reijmer, J.J.G. & Droxler, A. 1994. Highstand shedding of carbonate platforms. *Journal Sedimentary Research,* **B64,** 270–281.

Schmincke, H.-U. & Bogaard, P. van den, 1991. Tephra layers and tephra events. In Einsele, G., Ricken, W. & Seilacher, A. (eds) *Cycles and events in stratigraphy.* Springer, Berlin, Heidelberg and New York, 392–429.

Schwarz, H.-U. 1982. *Subaqueous slope failures – experiments and modern occurrences.* Contributions to Sedimentology 11, Schweizerbart, Stuttgart.

Seilacher, A. 1962. Paleontological studies on turbidite sedimentation and erosion. *Journal of Geology,* **70,** 227–234.

Seilacher, A. 1984. Sedimentary structures tentatively attributed to seismic events. *Marine Geology,* **55,** 1–12.

Seilacher, A. & Aigner, T. 1991. Storm deposition at the bed, facies, and basin scale: the geologic perspective. In Einsele, G., Ricken, W. & Seilacher, A. (eds) *Cycles and events in stratigraphy.* Springer, Berlin, Heidelberg and New York, 249–267.

Shanmugan, G. & Moiola, R.J. 1985. Submarine fan models: problems and solutions. In Bouma, A.H., Normark, W.R. & Barnes, N.E. (eds) *Submarine fans and related turbidite systems.* Springer, New York, 29–34.

Shiki, T. & Yamazaki, T. 1996. Tsunami-induced conglomerates in Miocene upper bathyal deposits, Chita peninsula, central Japan. *Sedimentary Geology*, **104**, 175–188.

Shiki, T., Chough, S.K. & Einsele, G. (eds) 1996. Marine sedimentary events and their records. *Sedimentary Geology*, **104** (Special Issue).

Simm, R.W. & Kidd, R.B. 1984. Submarine debris flow deposits detected by large-range side-scan sonar 1000 km from source. *Geo-Marine Letters*, **3**, 13–16.

Sims, J.D. 1979. Records of prehistoric earthquakes in sedimentary deposits in lakes. *Earthquake Information Bulletin*, **11**(6), 229–233.

Snedden, J.W., Nummedal, D. & Amos, A.F. 1988. Storm- and fair-weather combined flow on the central Texas continental shelf. *Journal of Sedimentary Petrology*, **58**, 580–595.

Stanley, D.J. 1982. Welded slump-graded sand couplets: evidence for slide generated turbidity currents. *Geo-Marine Letters*, **2**, 149–155.

Stow, D.A.V., Reading, H.G. & Collinson, J.D. 1996. Deep sea. In Reading, H.G. (ed.) *Sedimentary environments: processes, facies and stratigraphy*. Blackwell Science, Oxford, 395–453.

Toon, O.W. & Zahnle, K. 1995. All impacts, great and small. *Geotimes*, March 1995, 21–23.

Walker, R.G. 1978. Deep-water sandstone facies and ancient submarine fans: models for exploration for stratigraphic traps. *Bulletin of the American Association of Petroleum Geologists*, **62**, 932–966.

Walliser, O.H. 1995. *Global events and event stratigraphy*. Springer, Berlin and Heidelberg.

Wanless, H.R., Tedesco, L.P. & Tyrell, K.M. 1988. Production of subtidal and surficial tempestites by hurricane Kate, Caicos platform, British West Indies. *Journal of Sedimentary Petrolology*, **58**, 739–750.

Wright, V.P. (ed.) 1986. *Paleosols, their recognition and interpretation*. Blackwell, Oxford.

Yamazaki, T., Yamaoka, M. & Shiki, T. 1989. Miocene offshore tractive current-worked comglomerates – Tsubetugaura, Chita peninsula, Japan. In Taira, A. & Masuda, F. (eds) *Sedimentary facies in the active plate margin*. Terra Scientific Publications, Tokyo, 483–494.

7
Cyclostratigraphy

Andrew S. Gale

In this chapter cyclostratigraphy is restricted to the study of the sedimentary record produced by climatic cycles of regular frequency, tens to hundreds of thousand years in duration, which are generated by variations in the earth's orbit and are known as Milankovitch Cycles. A range of sedimentary cycles can be recognized in the stratigraphical record, these being the product of a broad range of phenomena. However, it is those cycles driven by global climate which have the most use in stratigraphical correlation. Cyclostratigraphy is concerned particularly with these cycles and the precise identification of their frequency. This has enabled geologists to develop an orbital timescale graduated in tens or hundreds of thousands of years for parts of the geological column. Cyclostratigraphy also investigates the frequently complex way in which orbital cycles have influenced earth's climate, oceans and ice-caps, and attempts to interpret how the cycles seen in the stratigraphical record have formed. The subject is a young one: the term cyclostratigraphy is less than 10 years old, and was first used at a meeting in Perugia in Italy in 1988. However, the idea that orbital cycles might affect the earth's climate was suggested first by Croll (1875), and their potential use in developing geological timescales was discussed in an extraordinarily prescient paper by Gilbert (1895), who identified bedding cycles in the Cretaceous of the Western Interior Basin of the USA, and interpreted them as the product of orbital forcing.

It was a Serbian mathematician, Milutin Milankovitch, who calculated the orbital variations accurately for the first time, and showed quantitatively how these determined the amount of solar radiation reaching the earth (Milankovitch 1941). He proposed that these orbital cycles were responsible for climatic changes

Unlocking the Stratigraphical Record: Advances in Modern Stratigraphy. Edited by P. Doyle and M.R. Bennett.
© 1998 John Wiley & Sons Ltd.

leading to ice-ages. His work was not treated seriously at the time, because the changes in solar energy he calculated seemed much too small to have caused glacial and interglacial periods.

Milankovitch cycles remained unproven in the stratigraphical record until 1976, when Hays *et al.* (1976) demonstrated regular cyclicity of 100 Ka, 41 Ka, 23 Ka and 19 Ka in oxygen isotope and other data, in deep sea cores covering the last 500 Ka. The periodicity exactly matched major orbital cycle frequencies previously calculated by Milankovitch (1941), and demonstrated that orbital cycles acted as a pacemaker to the ice-ages. Furthermore, the small, orbitally moderated changes in solar energy which had previously been considered insufficient to have caused large climate oscillations were shown to be strongly amplified by feedback mechanisms. This work led to the establishment of an orbital timescale which now extends back 20 Ma to the base of the Miocene (e.g. Shackleton *et al.* 1995).

Identification of orbital cycle frequencies in older sediments has not enjoyed such dramatic success. Although regular bedding cycles with frequencies which fall within the Milankovitch Band are a conspicuous feature of many sedimentary successions of Palaeozoic, Mesozoic and Cenozoic age, accurate identification of cycle frequency has often proved difficult to achieve. This is probably because the signal is blurred by the numerous other time-dependent variables which affect sedimentation, such as rapidly changing accumulation rates and hidden hiatuses (Schwarzacher 1993). Only two timescales for pre-Cenozoic intervals have been constructed to date: for the Devonian (Givetian; House 1995a) and the Cretaceous (Cenomanian; Gale 1989, 1995).

The expression of climatic cycles in sediments takes a myriad of forms, because climate change has complex effects on physical, chemical and biological systems. These effects often have significant lag times, and complex interferences and feedback mechanisms are common. Climatic cycles have been identified from numerous parameters, including: mineralogy and geochemistry, which records changing sediment flux or biological productivity, for example; variation in stable isotope ratios of oxygen, reflecting changing temperatures and ice volume; and changes in the relative abundance of fossil species.

In this chapter the main cycle frequencies in the Milankovitch Band which are known to affect the earth's climate are briefly outlined (Section 7.1), before various methods by which individual cycle frequencies can be identified in the sedimentary record are discussed (Section 7.2). In Section 7.3 the processes by which climatic cycles are thought to affect sedimentation and the amplification effects of diagenesis are described. Finally, the relationship between sea levels and cyclostratigraphy is discussed before looking to the time when orbital timescales will form the basis of a high-resolution geochronology for the Phanerozoic (Sections 7.4 and 7.5).

7.1 ORBITAL CYCLES: THE ASTRONOMICAL PERSPECTIVE

Metronomic variations of the earth–moon and earth–sun orbital patterns result in cycles of greatly varying frequencies (Figure 7.1), from the *Calendar Band* (e.g. twice daily tidal cycles, the lunar month, the equinoxes, the annual cycle), to the *Solar*

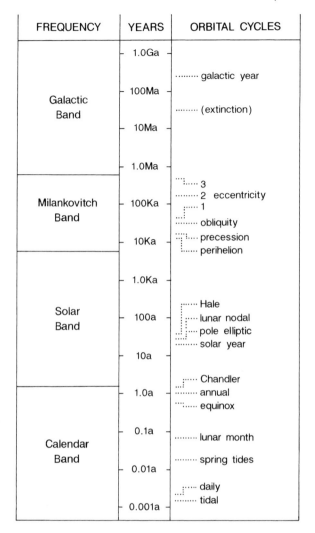

Figure 7.1 *Logarithmic table to show the orbital frequencies which exert an influence on temporal energy reaching the outer atmosphere. [Modified from: House (1995b)]*

Band (e.g. sunspot cycles at 11 years, lunar nodal cycles at 18.61 years), through the *Milankovitch Band* (10 Ka–1 Ma) to the long-term *Galactic Band* (e.g. the cosmic year at 220–250 Ma – the period taken by the solar system to move around the Milky Way Galaxy). Many of the shorter cycles are recorded in the geological record, either as growth bands preserved in fossil animals and plants or as tidal or annual sedimentary events (see the review in House 1995b). It is, however, the lower frequencies of the Milankovitch Band which particularly concern cyclostratigraphy, because bedding in sedimentary rocks often shows a periodicity of tens of thousands of years, as can be demonstrated by simple division of bed numbers into radiometrically dated intervals.

Cycles of the Milankovitch Band fall in the frequency interval of 10 Ka to 1.0 Ma, and are caused by the complex orbital patterns of the sun–moon–earth system. These changes affect both the amount of insolation (solar energy) reaching the earth's surface, and the seasonal distribution of insolation. Three main cycles are found: those of precession, obliquity and eccentricity (Figure 7.2), which combine to produce an intricately detailed curve (Figure 7.3). The actual variations in insolation are of about 5% and affect the earth's climate in a complex manner; they include various feedback mechanisms which augment the effects of the eccentricity cycle in particular. The effects of different cycle frequencies vary latitudinally, such that precessional effects are dominant at low latitudes and obliquity at higher ones.

7.1.1 The Precession Cycle (19–23 Ka)

Precession is the combined effect of the precession of the equinoxes and the movement of the perihelion, expressed as the movement of the axial projection of

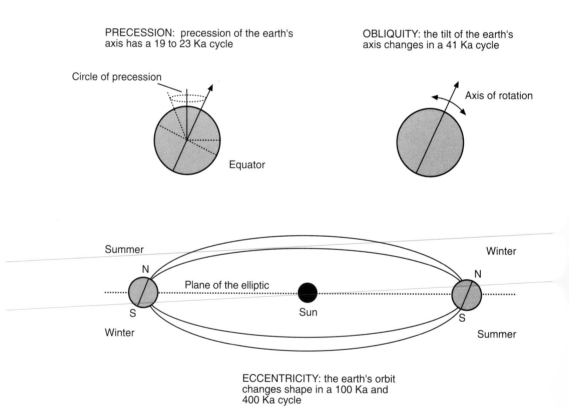

Figure 7.2 *Diagram of the earth–moon–sun system and the oscillations that produce changes in insolation. These may lead to orbitally forced signatures in the sedimentary record. [Modified from: House (1995b)]*

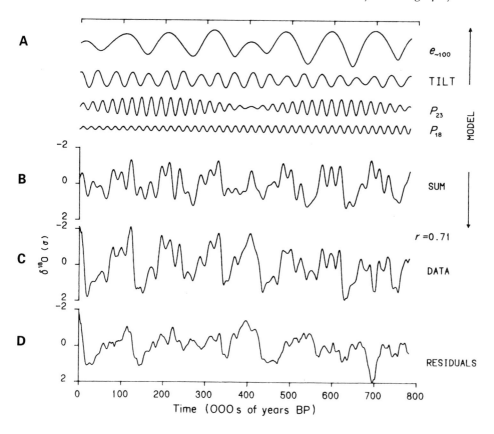

Figure 7.3 *Variations in oscillations of the earth–sun system. **A.** Oscillations caused by eccentricity (e_{-100}), obliquity (tilt) and precession (p_{23} and p_{18}). **B.** The sum of the four signals, a measure of energy received by the outer atmosphere. **C.** The oxygen isotope record, the manifestation of the Milankovitch signal. **D.** Residual products after deducting **B** from **C.***

the earth's rotational axis relative to the stars. At the present time, two peaks with periodicities of 19 and 23 Ka are dominant.

7.1.2 The Obliquity Cycle (41 Ka)

The angle between the earth's celestial equator (projection of the equator onto the sky) and the plane of the earth's orbit – the ecliptic – varies by about 3.5°, fluctuating between 21.5° and 24.4° with a periodicity of 41 Ka. This affects the insolation received by the earth by changing the intensity of the seasonal cycle, and affecting the latitudinal insolation gradient.

7.1.3 The Eccentricity Cycle (106, 410 Ka)

There is considerable variation in the orbit of the earth–moon system around the sun, which results in more and less strongly elliptical pathways of orbit. The most important of these are the 106 and 410 Ka cycles.

There is good evidence that duration of the lunar day and the lunar month have changed with time, and it is requisite that the periods of precession and obliquity must also have changed. The periods of these frequencies have increased slowly through time, as calculated by Berger *et al.* (1989; Figure 7.4). It is important to note that the duration of eccentricity cycles has not changed.

7.2 IDENTIFICATION OF CYCLE FREQUENCY

The combination of the main Milankovitch cycle frequencies (19, 23, 41, 106 Ka) produces a complex insolation curve (Figure 7.3), which is matched closely by oxygen isotope data from the deep-sea Pleistocene record (Imbrie 1985). The component frequencies can be identified as discrete peaks by the use of spectral analysis (Figure 7.5), which have consistent peak ratios. Superficially, the identification of cycle frequency appears to be a simple process; in reality it often is not, for reasons which are only partly understood.

Firstly, the relative contribution of different Milankovitch frequencies to a composite curve varies with both time and latitudinal position. For example, the short eccentricity cycle varies in amplitude through time (Figure 7.6): the obliquity

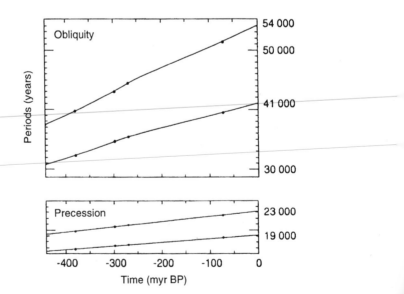

Figure 7.4   *Changing periods of obliquity and precession through time. [Modified from House (1995b)]*

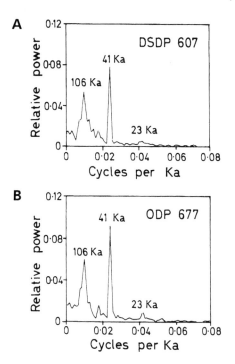

Figure 7.5 *Fourier spectrum of oxygen isotope data for the past 2.5 Ma for two deep ocean cores.* **A**. *DSDP 607.* **B**. *ODP 677. The major peaks of precession (23 Ka), obliquity (41 Ka) and eccentricity (106 Ka) are indicated. [Reproduced with permission from Blackwell Science, from: Weedon (1993)]*

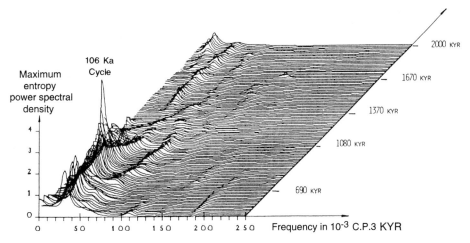

Figure 7.6 *Evolutive maximum entropy spectral analysis of a deep-sea Atlantic core, to show in three-dimensions the changing amplitude of the 106 Ka eccentricity cycle over the past 2–3 Ma. [Modified from: Berger (1988)]*

signal is predicted to be dominant at higher latitudes (particularly north and south of 60°), and the precession signal at low latitudes. The idealized composite curves of the four main Milankovitch frequencies are therefore found only rather infrequently. Secondly, bedding cycles in sedimentary rocks are a product of various geological processes, some of which, like variable accumulation rate, can obscure the orbital signals.

7.2.1 Cycle Counting

The presence of cycles of regular thickness in sediment is evidence that a regular cycle in time was responsible for their formation (Schwarzacher 1975). Establishing that the cyclicity falls within the Milankovitch Band can be achieved simply by dividing the number of bedding couplets (e.g. limestone–shale alternations) into the duration of an interval dated by radiometric methods. Of course, this does not necessarily mean that the cycles in question were formed by the slow climatic changes of orbital forcing; they may be rapidly but infrequently deposited event beds related, for example, to tectonic events or storms. However, it is now being recognized that orbitally induced changes in climate and sea level can control the deposition of rapidly formed deposits like turbidites. Several studies have shown convincingly that some turbidite sequences display Milankovitch Band frequencies and that individual turbidite events were orbitally forced (see Section 7.3; Foucault *et al.* 1987; Weltje & De Boer 1993).

Cycle counting can only give an approximation of the frequency of the cycles in any given succession for two reasons. Firstly, gaps in the succession can significantly alter the calculated frequencies; this is a problem particularly when the biostratigraphy is of low resolution. Secondly, changes in the dominant high-frequency cycle affecting sedimentation can take place, notably switchover between precession and obliquity signals, which can be predicted to have the effect of systematic doubling or halving of bed thickness (21–41 Ka). However, identical changes in bed thickness can be generated also by changes in accumulation rate (subsidence and/or supply) without changing cycle frequency. Sorting out the processes involved is not easy and probably is never conclusive (e.g. Ten Kate & Springer 1993; Giraud *et al.* 1995).

Cycle counting is a worthwhile exercise when it can be shown that a succession does not contain gaps, and good radiometric dates are available. Gale (1995) counted 107 couplets (chalk–marl alternations) of very even thickness in the Middle and Upper Cenomanian of western European basins, which he identified as precession cycles, and which gave a duration for that interval of 2.24 Ma. Radiometric dates from bentonites for the same biostratigraphically defined interval give a duration of 2.2 Ma.

7.2.2 Time-Series Analysis

The standard method of identifying peak frequencies related to orbital signals is spectral analysis. The most widely used methods are based on Fourier analysis of time-series in terms of component sine and cosine waves (Weedon 1989, 1993). The power spectrum plots the squared average amplitude or 'power' of each regular component sine and cosine wave against frequency. The size of a spectral peak indicates the average importance of that component in the whole series. It is also necessary to identify 'red noise' generated by low-frequency oscillations. A spectral analysis of Pleistocene oxygen isotope data for the past 800 Ka (Imbrie 1985) is shown in Figure 7.3 and the various orbital peaks are marked. The distinctive ratios of the spectral peaks (5:1 precession:eccentricity; 2:1 obliquity:precession) aids their identification.

Application of time-series analysis to geological data runs up against various inherent features of the geological record which can distort and obscure environmental signals. Hiatuses generate additional noise and can actually alter the wavelength ratios of pairs of regular cycles (Weedon 1993). Varying coefficients of compaction and changing sedimentation rates dramatically affect time-series by generating harmonic and tone peaks, which can be identified because these lie at integer multiples of the fundamental peak frequency. Varying short-term sedimentation rates lead to broadening and loss of spectral peaks (Weedon 1993), and systematic (monotonic) change in sedimentation rates can distort a cyclic signal such that no distinct spectral peak is produced. An example of this was given by Fischer *et al.* (1991) where the high-frequency bedding cycle, obvious to the eye, could not be recognized by spectral analysis. Identification of systematically changing sedimentation rate is aided by the use of spectra for short, overlapping sub-sections of the data, which are compared to see if distinctive spectral peaks progressively change in frequency through the succession as sedimentation rate changes – the 'moving window' see Melnyk & Smith 1989). The presence of significant spectral peaks only demonstrates that regular cycles in time are present, which can also be achieved by inspection of field sections, cores or time-series.

Filtering of time-series allows isolation of the regular cycles. Low pass filtering involves removal of all high-frequency components from time-series, and band pass allows examination of oscillations with restricted wavelengths. Weedon & Jenkyns (1990) used filtered time-series to examine data from the Belemnite Marls of the Dorset coast (Figure 7.7). This succession contains a high-frequency bedding cycle (mean thickness of 35–40 cm), grouped into visually discernible bundles in the lower part of the section with a mean wavelength of 300 cm, widely attributed to the combined precession–short eccentricity cycles. However, Weedon & Jenkyns (1990) argue that if this was in fact the case, then amplitude of the precession cycles should vary *within* the eccentricity cycle. Filtering suggests that the bundles have an irregular duration, and therefore these authors attribute the bundling to long-term irregular variations in climate.

A series of studies have attempted to identify the orbital frequencies controlling bedding in the conspicuously rhythmic basinal limestone–marl–shale successions of the Vocontian Basin in south-east France (Figure 7.8; Huang *et al.* 1993; Giraud *et al.* 1995). These studies are of particular interest because they used different methods of study and arrived at similar conclusions. Huang *et al.* (1993) used square-wave Walsh power spectra on logged sections from the Berriasian–Valanginian interval, and the analyses exhibit various frequency peaks, which they

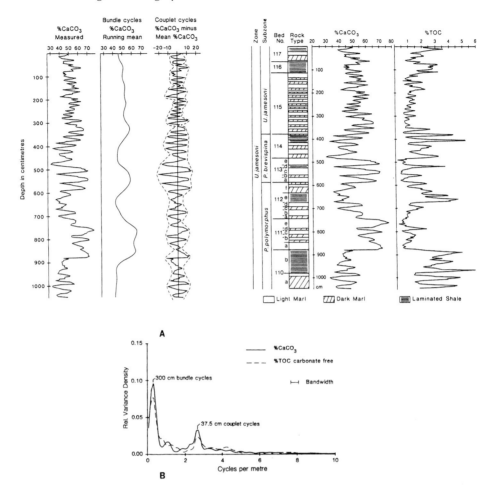

Figure 7.7   A. *Time-series from the Belemnite Marls (Lower Jurassic) on the Dorset coast, based on carbonate analysis and Total Organic Carbon. Note the conspicuous bundling.* **B.** *Time-series analysis of these data shows clearly both high-frequency cyclicity and bundling.* **C.** *Isolated bundle and couplet cycles, produced by smoothing the data; the envelope around the couplet cycles is out of phase with the bundle cycles. Weedon & Jenkyns (1990) interpreted the bundling as an irregular long-term climatic cycle, unrelated to eccentricity. [Reproduced with permission from the Geological Society, from: Weedon & Jenkyns (1990)]*

identified as Milankovitch periodicities from a consideration of sedimentation rates. They tried to identify the frequency peaks of the spectra by comparing wavelength ratios to the periodicity ratios of the Milankovitch orbital elements, based on frequencies calculated for the Early Cretaceous by Berger *et al.* (1989). This works well when multiple peaks are present, and they were able to confidently identify a strong obliquity peak, and a less well developed precession peak. When a single peak was present, as in the Hauterivian, it could only be identified with uncertainty from mean sedimentation rate as representing obliquity.

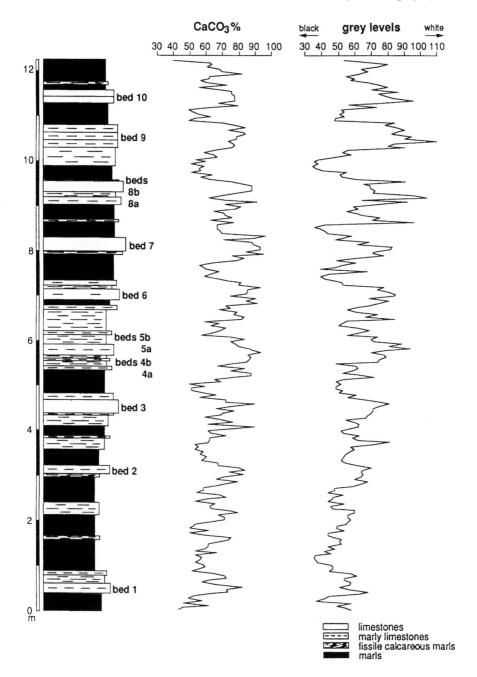

Figure 7.8 *The Valanginian late S. verrucosum Zone in the Angles core, Vocontian Basin, Vergons, south-east France. From spectral analysis and band-pass filtering, Giraud et al. (1995) identify a 119 cm signal as the obliquity cycle, and a 71 cm signal as precession (Figure 7.9). [Reproduced with permission from the Geological Society, from: Giraud et al. (1995)]*

Giraud *et al.* (1995) studied the Valanginian only (Figure 7.8) and used closely spaced samples to investigate serial variation in carbonate content and colour (greyness). They used two spectral analyses to process the data, and discovered a single major peak at 119 cm which they identified from mean sedimentation rate (based on radiometric durations of the length of the Valanginian) as an obliquity signal. Band-pass filtering indicated a minor peak at 71 cm which they tentatively identified as a precession signal. This is expressed in the succession as a minor cyclicity which splits the major carbonate beds into two (Figure 7.9). Again, peak ratios proved to be an important aid in identification of frequencies.

Time-series analysis does have limitations, particularly when dealing with rapid and irregular variation in the thickness of high-frequency beds (Fischer *et al.* 1991); the high-frequency bedding signal is clearly a most important feature in the succession even if it does not appear as a spectral peak. Spectral analysis is, however, certainly a powerful tool when the presence of multiple peaks allows

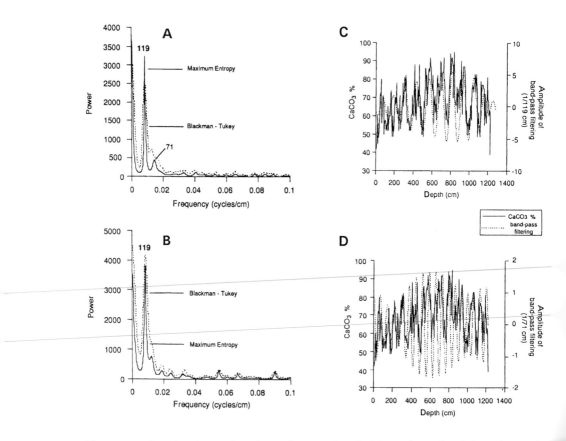

Figure 7.9   **A** and **B**. *Spectral analysis of time-series (CaCO₃ and grey level) from the Angles core (see Figure 7.8).* **C.** *and* **D**. *Band-pass filtering of data centred on a frequency of 119 cm (C) and 71 cm (D). Note the doubling of the 119 cm peaks by smaller 71 cm peaks. [Modified from: Giraud et al. (1995)]*

ratios to be compared with those of known Milankovitch periodicities. Identification of the frequency of a single, strong, spectral peak ultimately must rely upon mean sedimentation rates determined from radiometric data and is subject to the same constraints as cycle counting. In short, spectral analysis is a valuable aid to frequency identification, but does not always provide the answers.

7.2.3 Cycle Ratios: The Precession–Eccentricity Syndrome

The consistent ratios between the mean precession index (21 Ka) and the short (E1: 106 Ka) eccentricity cycle can result in a highly distinctive grouping of five precession couplets in one E1 eccentricity bundle. Furthermore, the short eccentricity bundles are themselves commonly enveloped in superbundles of four, interpreted as the long eccentricity cycle (E2: 400 Ka). This association has been called the precession–eccentricity syndrome, and is beautifully illustrated by the Albian Scisti a Fucoidi in the Piobbico core from central Italy (Herbert & Fischer 1986; Fischer *et al.* 1991; Figure 7.10). The Scisti a Fucoidi were deposited in deep water (1–2 km) and comprise hemipelagic shales, marls and limestones, which display striking rhythmic variation in carbonate content, bioturbation and organic content on several scales. The highest frequency bedding couplets have a mean thickness of 0.1 m, and are redox cycles reflecting varying sea-floor oxygenation. They are bundled into groups of five, defined by dark laminated marls.

Herbert & Fischer (1986) and Fischer *et al.* (1991) used calcimetry (measurement of carbonate) and densitometry (measurement of darkness) to quantify stratigraphical variation through the Piobicco core. Their results (Figure 7.10) demonstrate the cycle hierarchies of the precession–eccentricity syndrome; the conspicuous E1 bundles of five precession couplets are grouped in fours within E2 superbundles. In Fourier plots, the precession couplets signal fails to rise above background noise, even though this is the most conspicuous feature to the eye. The bundle cycles (E2) are most dominant, but the long eccentricity (E2) cycles do not show up, presumably because these are alternately long and short. Fischer *et al.* (1991) comment upon the shortfalls of applying mathematical tools to geological data, and conclude that high-frequency cycles can easily be obscured by variations in accumulation rate.

The precession–eccentricity syndrome is extremely valuable in the identification of hiatuses and, when calibrated with biostratigraphy, even allows a highly accurate estimation of their duration. For the Middle Cenomanian, Gale (1995) demonstrated that two chalk–marl alternations representing precession cycles (B44, B45) present in the section at Rheine in north Germany were missing in the most complete Anglo-Paris sections. High-resolution oxygen isotope data from Folkestone (Figure 7.11) show bundling of the precession signal by both E1 and E2 through this part of the section. Two of the precession peaks are missing from an E1 bundle at the level of the hiatus, confirming that the gap represents about 40 Ka at Folkestone.

In the absence of obliquity 'doubling' the precession signal, or short eccentricity creating bundles of five, it does not seem to be possible mathematically to identify whether a single bedding frequency represents a precession or an obliquity signal.

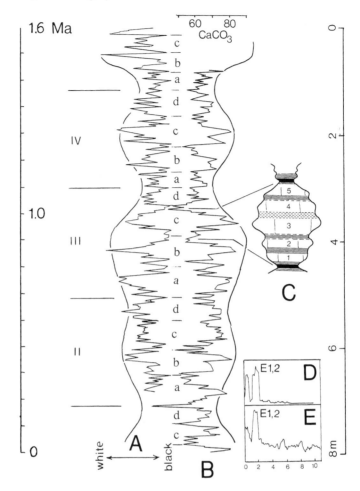

Figure 7.10   *An 8 m segment of (Albian) Scisti a Fucoidi in the Piobbico core from Umbria, Italy, to illustrate the precession–eccentricity syndrome. The bundles II–IV represent long eccentricity (E2: 400 Ka) cycles. Time-series are based on darkness values obtained by microdensitometry (**A**), and carbonate values are based on 2 cm spaced samples (**B**). **C**. An enlarged 100 Ka segment of the core to show its fivefold precessional structure. **D** and **E**. Spectral analysis of the carbonate data; note the 100 and 400 Ka peaks, and the lack of a high-frequency peak. [Modified from: Fischer et al. (1985)]*

7.3 SEDIMENTOLOGICAL EXPRESSION AND INTERPRETATION OF CYCLES

Since the effects of Milankovitch cycles on climate are global in extent, sedimentary responses to the changes they cause should be identifiable in all depositional environments. In practice, ancient cycles are more readily interpreted in marine environments, partly because they are commoner, and also because the biostratigraphy for marine facies is generally of higher resolution than that develop-

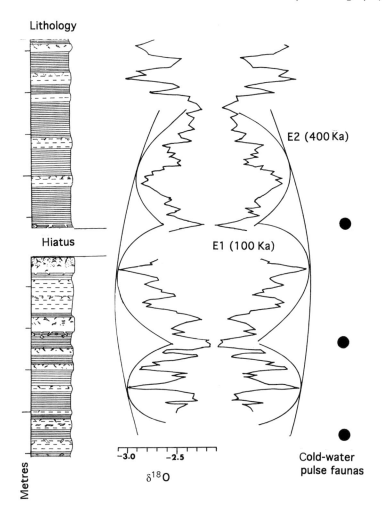

Figure 7.11 *Bundling of precession cycles in the mid-Cenomanian of Folkestone Kent, based on oxygen isotope data in Paul et al. (1994). The high-frequency cycle represents precession (21 Ka), bundled into fives by short eccentricity (E1: 106 Ka) and a superbundle of four (E2: 400 Ka). The hiatus results in two precession couplets missing from a bundle*

ed for continental facies. Strong precessional signals have, however, been identified from Lake Playa complexes in the Triassic (Olsen 1986) and the Eocene (Fischer & Roberts 1991), from North America, where they control alternate wet–dry periods. Most of the examples of cycle interpretation described below are from marine facies. The mechanisms outlined below are believed to have been important in the sedimentary expression of climatic cycles in the Milankovitch Band.

7.3.1 Productivity and Dilution Cycles

Productivity cycles are generated when carbonate supply varies against a background of constant fine clastic deposition, sometimes called a 'clay clock'. Carbonate supply is determined by productivity, which is dependent upon upwelling and surface mixing to supply nutrients to phytoplankton. Dilution cycles are the product of fluctuating clastic supply against a background of constant carbonate productivity, and are controlled by climatic factors, at least in pelagic and hemipelagic settings. In hemipelagic settings, clay is originally supplied in runoff, which is highly responsive to source-area climate (Fischer *et al.* 1985). In pelagic facies, clay is mostly carried in by wind.

Both types of cycle produce similar alternations of carbonate-dominated beds with clay-rich calcareous shales or marls; one of the commonest outer-shelf facies of the Mesozoic (Figure 7.12). The cycles can theoretically be distinguished because in productivity couplets the shale component of a couplet should maintain an even thickness over a wide area, and thickness variation is consequently controlled by variation in carbonate content (Einsele & Ricken 1991). The reverse should be true for dilution cycles but, in practice, these criteria do not seem to work. Dilution and productivity cycles are set at opposite ends of a continuous spectrum (Figure 7.13), and the real problem is identification of the dominant process.

Figure 7.12   *Evenly spaced alternations of limestone and shale in the Barremian of Vergons, south east France. The cycles have a Milankovitch frequency, but do they represent obliquity or precession?*

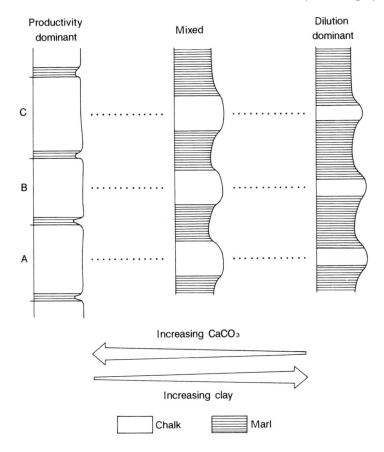

Figure 7.13 *Productivity and dilution cycles: extremes of a continuum*

In the chalk–marl cycles of the Cenomanian of the Anglo-Paris Basin (Figure 7.14), there is evidence that the carbonate-rich part of the couplet was generated under higher productivity conditions. In particular, nannofossils from the chalks include taxa which have been used as productivity markers, an abundance of planktic foraminifera, and show evidence of more rapid deposition. In contrast, the marls register slightly slower deposition and lower productivity, but also an increase in the input of clay.

An extreme form of dilution cycles in a low carbonate system are the turbidites from the Pliocene of Corfu described by Weltje & De Boer (1993). Fan lobe development was found to precisely match the precession signal, which probably controlled sediment supply to a delta-fed turbidite system. It is likely that other ancient turbidite systems were also under Milankovitch control.

Figure 7.14   *The Lower Chalk (Cretaceous, Cenomanian) at Culver cliff, Isle of Wight, England. The conspicuous chalk–marl rhythmicity corresponds to the 21 Ka precession signal*

7.3.2 Redox Cycles

Redox cycles are caused by variations in either organic carbon or oxygen supply to deep marine areas (De Boer & Smith 1993), and appear characteristically as an alternation of darker, organic-rich, often laminated marl or clay and lighter bioturbated carbonate-rich marl. They represent movement of the redox layer (oxic–anoxic boundary) from beneath the sediment–water interface to a level in the water column above it. Beneath the carbonate compensation depth, redox cycles may be represented by alternations of organic-rich mudstones and radiolarites. The two most probable controls on formation of redox cycles are oxygen supply, which is most common in restricted, periodically stagnant basins, and productivity, which controls organic flux. In the view of Pedersen & Calvert (1990), productivity is almost invariably the main cause of anoxia. Productivity is itself controlled by nutrient supply from upwelling or surface mixing, which are both related to climatic changes.

Varying oxygen levels also control the trace fossil content of sediments deposited around the redox boundary. As bottom water oxygen decreases, the depth of penetration, size and diversity of burrows all gradually decline, until finally a laminated organic-rich sediment is formed (Savrda & Bottjer 1989). The deep burrow *Chondrites* is tolerant of and abundant in low oxygen conditions, and is characteristically the last trace fossil to be seen before anoxia (Bromley & Ekdale 1984; Figure 7.15).

Figure 7.15   *Redox cycle (limestone–laminated shales) in the Blue Lias (Jurassic, Sinemurian), west of Lyme Regis. Note the lower laminated organic-rich (anoxic) shale, with Chondrites in the more oxic upper part, overlain by a limestone representing fully oxygenated conditions*

7.3.3 Dissolution Cycles

The accumulation of carbonate in deep water is moderated by dissolution at the sediment–water interface, and at the Carbonate Compensation Depth (CCD) total carbonate dissolution occurs. The CCD will vary in its depth depending on many factors, such as temperature, the supply of carbonate, and the concentration of dissolved CO_2, all of which can be controlled by climate (Einsele & Ricken 1991). Dissolution cycles produce thick limestones and thinner marls, rather similar to productivity cycles in appearance. It is likely that dissolution cycles were common during the Cretaceous, because the CCD was located at shallow depths (2–3 km).

7.3.4 Diagenesis and Cyclicity

After deposition, all sediments undergo changes at relatively low temperatures and low pressures, which are referred to collectively as diagenesis. Much early diagenesis is bacterially moderated, and commonly involves the precipitation of carbonate and silica cements, often as concretions. The precise stratigraphical distribution of concretionary layers is therefore often controlled by primary depositional features such as original concentration of carbonate, or presence of minor hiatuses above concretions (Raiswell 1988). Diagenesis commonly augments primary cyclic sedimentary features by, for example, cementing limestone beds, or the formation of rhythmically bedded flint nodules in pelagic carbonates (Figure 7.16) which in Cretaceous chalks dominantly have a precession frequency. However, whilst cyclicity is often made more conspicuous by diagenesis, some primary signals such as those of oxygen isotopes are altered by the same processes. It is most unlikely that early diagenesis can create rhythmic bedding out of originally undifferentiated primary sediment (see Hallam 1986).

7.4 EUSTASY AND ORBITAL CYCLES

In times of glaciation, as in the Late Pleistocene, major sea-level fluctuations are the direct consequence of Milankovitch-driven changes in ice-volume. In the Pleistocene, sea levels were affected mostly by the 41 Ka obliquity cycle and the 106 Ka eccentricity cycle, with some evidence of the 21 Ka (mode) precession cycles. In the Carboniferous glaciation, there is much less evidence of high-frequency cyclicity in sea level and the main control may have been the 400 Ka long eccentricity cycle. There is a consensus that eustatic cycles caused by Milankovitch Band orbital effects correspond roughly with the fourth, fifth and possibly third order cycles of Vail.

Figure 7.16 *Beds of flints (silica concretions) in the Coniacian–Santonian Chalk of Scratchell's Bay, Isle of Wight, England. Although entirely diagenetic, these pick out a climatic signal that was controlled by an orbital frequency – probably precession*

The earth during the Mesozoic is usually considered to have been a greenhouse world in which ice-caps were not present. Under these conditions, was there any relationship between Milankovitch Band cycles and sea level, and if there was, what process brought it about? Triassic platform carbonates in Hungary and Austria contain well-developed cyclicity, which displays the characteristic 5:1 bundling ratios of the precession–eccentricity syndrome (Goldhammer *et al.* 1987). Each minor cycle has a thin vadose diagenetic cap, caused by emergence and exposure to freshwater. Peritidal facies are absent, so the emergence is not a consequence of autocyclic progradation, and the succession records an asymmetric eustatic signal indicating sea-level changes of a few metres. There is no evidence of polar ice in the Triassic, so what was responsible for formation of these cycles? Land ice controls ocean water volume, so a requirement for glacioeustatic fluctuation is that a significant area of land must lie within 30–40 ° of the poles. This was true for the Triassic; perhaps changes in mountain glacier development was sufficient to account for the small sea-level changes required.

7.5 CORRELATION OF ORBITAL CYCLES AND CYCLOCHRONOLOGY

A detailed chronology based on identified orbital cycles now exists for the Quaternary, based on the SPECMAP oxygen-isotope curve. This curve is constructed from analyses of foraminiferal tests derived from deep-sea sediments, and is anchored to the Present; it serves as a standard for Quaternary chronology, and it records the growth and decay of ice-sheets. In addition, an astronomically calculated insolation curve extending back far into the Neogene has been used to correct radiometric dates for the Late Miocene to Pleistocene interval (Hilgen 1991). Ice-sheet meltwater (Roof *et al.* 1991), aeolian wind transport (Tineke *et al.* 1991) and carbonate productivity show strong signals, particularly of precession, which imply a direct response to insolation (Fischer 1995). The SPECMAP orbital timescale is of course anchored at the Present, and calculations of insolation values must decrease in accuracy into the more distant past. It follows that orbital timescales for much of the geological past will be 'floating' intervals defined by biostratigraphical or magnetostratigraphical markers (Fischer 1995).

The correlation of bedding cycles at outcrop or in cores enables a very detailed regional and interbasinal understanding of cyclostratigraphy to be attained. High-resolution correlation allows condensation to be understood and the duration of hiatuses to be followed precisely. This type of study has been undertaken in the Valanginian and Aptian of north-west Germany (Schneider 1964), the Berriasian–Barremian interval of the Cretaceous Vocontian Basin in south-east France (Cotillon 1991), and in the Cenomanian of western Europe (Gale 1989, 1995) (Figure 7.17). The cyclostratigraphy can be integrated with other schemes, including biostratigraphy, magnetostratigraphy and isotope stratigraphy, which allow basinal sedimentary processes to be examined in detail.

Identification of orbital frequencies in long rhythmically bedded successions is often more difficult than establishing the correlations themselves. The ideal situ-

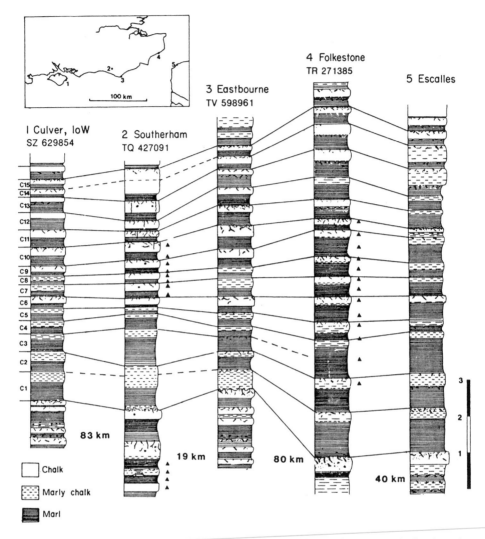

Figure 7.17 Correlation of marl–chalk productivity cycles in the Lower Chalk of southern England and northern France. [Modified from: Gale (1989)]

ation occurs when the precession–eccentricity syndrome is developed within the 400 Ka cycle, because this not only enables a cyclochronology to be obtained readily, but also provides an instant check on completeness within two orders of bundling. This method was elegantly demonstrated by Herbert & Fischer (1986) and Fischer *et al.* (1991). The identification of cycle ratios can be verified by spectral analysis, although thickness variation in the highest frequency signal causes problems (Fischer *et al.* 1991).

When an obvious hierarchy is not developed, but the succession is dominated by high-frequency bedding cycles of even thickness, things are more problematic. The situation is a common one, because the eccentricity cycle does not maintain a constant amplitude, but rather waxes and wanes (Berger 1988). The common high-frequency signals are precession (mode at 21 Ka) and obliquity (41 Ka); the question therefore is how to distinguish between these. Rarely, the two interpenetrate to produce a double bedding cycle (Giraud *et al.* 1995), but more usually it is necessary to use mean sedimentation rates to identify cycle frequency. This can be achieved by either cycle counting or spectral analysis, and both are subject to the imprecisions of most pre-Neogene radiometric dates; for example, calculated durations of Cretaceous stages can vary by 50–80%. If the precession and obliquity cycle frequencies are unevenly mixed in a succession where there is variation in sedimentation rate, it is probably impossible to separate them. Gale (1995) used cycle counting on a composite Cenomanian succession to produce a timescale for the stage in western Europe, based on the presumption that the ubiquitous high-frequency bedding signal represented precession. Statistical attempts to recognize cycle hierarchy in this succession have been unsuccessful (P. Ditchfield, pers. comm. 1997); if some obliquity cycles do go unrecognized, the timescale developed is still two orders of magnitude more accurate than that based on radiometric data.

The ultimate control on the accuracy of ancient Milankovitch timescales is their reproducibility in different facies and regions (Fischer 1995). To compare successions, however, it is necessary to have high-resolution independent correlation based on other stratigraphical means – biostratigraphy, magnetostratigraphy and chemostratigraphy, for example. The best means of correlation for this purpose are independent of facies, and perhaps include high-resolution carbon isotope curves (e.g. Gale *et al.* 1993) and magnetic fingerprints (e.g. Napoleone & Ripepe 1989). It is important to remember that without the assistance of these, even a thoroughly investigated orbital timescale will just comprise a series of cycles floating in time and space.

7.6 SUMMARY

The geological proof of Milutin Milankovitch's theory that orbital cycles have controlled the earth's climate is one of the major scientific discoveries of the latter part of the twentieth century. The identification of cycles of Milankovitch frequencies in deep-sea cores and elsewhere has greatly advanced our understanding of the Quaternary, and has led to the construction of a high-resolution orbital timescale graduated in precession units of 21 Ka mean duration which extends back to the base of the Miocene.

In the more distant geological past, Milankovitch cycles are conspicuous features of many sedimentary successions, but identifying their frequency has not always been straightforward or conclusive. The construction of timescales for the pre-Neogene will be greatly aided by the distinctive ratios of the precession–eccentricity syndrome which produces bundling of four to five precession cycles within a single short (E1: 100 Ka) eccentricity cycle, and four short eccentricity cycles within a single long (E2: 400 Ka) eccentricity cycle. Hemipelagic facies in particular have considerable sensitivity to climatic signals and afford some of the best opportunities to develop ancient timescales.

REFERENCES

Berger, A. 1988. Milankovitch theory and climate. *Reviews of Geophysics*, **26**, 624–657.

Berger, A., Loutre, M.F. & Dehant, V. 1989. Pre-quaternary Milankovitch frequencies. *Nature*, **323**, 133.

Bromley, R.G. & Ekdale, A.A. 1984. *Chondrites*: a trace fossil indicator of anoxia in sediments. *Science*, **224**, 872–874.

Cotillon, P. 1991. Varves, beds and bundles in pelagic sequences and their correlation. In Einsele, G., Ricken, W. & Seilacher, A. (eds) *Cycles and events in stratigraphy*. Springer-Verlag, Berlin and Heidelberg, 820–839.

Croll, J. 1875. *Climate and time in their geological relations*. Appleton, New York.

De Boer, P. & Smith, D.G. 1994. *Orbital forcing and cyclic sequences*. International Association of Sedimentologists Special Publication No. 19, Blackwells, Oxford.

Ditchfield, P. & Marshall, J.D. 1989. Isotopic variation in rhythmically bedded chalks: palaeotemperature variation in the Upper Cretaceous. *Geology*, **17**, 842–845.

Einsele, G. & Ricken, W. 1991. Limestone–marl alternation – an overview. In Einsele, G., Ricken, W. & Seilacher, A. (eds) *Cycles and events in stratigraphy*. Springer-Verlag, Berlin and Heidelberg, 23–47.

Fischer, A.G. 1995. Cyclostratigraphy, quo vadis? In House, M.R. & Gale, A.S. (eds) *Orbital forcing timescales and cyclostratigraphy*. Geological Society Special Publication No. 85, 199–204.

Fischer, A.G. & Roberts, L.T. 1991. Cyclicity in the Green River Formation (lacustrine Eocene) in Wyoming. *Journal of Sedimentary Petrology*, **61**, 1146–1154.

Fischer, A.G., Herbert, T.D. & Premoli Silva, I. 1985. Carbonate bedding cycles in Cretaceous pelagic and hemi-pelagic sequences. In Pratt, L.M., Kauffman, E.G. & Zelt, F.B. (eds) *Fine-grained deposits and biofacies of the Cretaceous western interior seaway: evidence of cyclic sedimentary processes*. Society of Economic Paleontologists and Mineralogists Second Annual Midyear Meeting, Golden, Colorado Field trip No. 9, 1–37.

Fischer, A.G., Herbert, T.D., Napoleone, G., Premoli Silva, I. & Ripepe, M. 1991. Albian pelagic rhythms (Piobbico core). *Journal of Sedimentary Petrology*, **61**, 1164–1172.

Foucault, A., Powichrowski, L. & Prud'Homme, A. 1987. Le control astronomique de la sedimentation turbiditique: example du Flysch a Helminthoides des Alpes Ligures. *Academie des Sciences Comptes Rendus, Serie II*, **305**, 1007–1011.

Gale, A.S. 1989. A Milankovitch timescale for Cenomanian time. *Terra Nova*, **1**, 420–425.

Gale, A.S. 1995. Cyclostratigraphy and correlation of the Cenomanian Stage in Western Europe. In House, M.R. & Gale, A.S. (eds) *Orbital forcing timescales and cyclostratigraphy*. Geological Society Special Publication No. 85, 177–197.

Gale, A.S., Jenkyns, H.C., Kennedy, W.J. & Corfield, R. 1993. Chemostratigraphy versus biostratigraphy: data from around the Cenomanian–Turonian boundary. *Journal of the Geological Society of London*, **150**, 29–32.

Gilbert, G.K. 1895. Sedimentary measurement of geological time. *Journal of Geology*, **3**, 121–125.

Giraud, F., Beaufort, L. & Cotillon, P. 1995. Periodicities of carbonate cycles in the Valanginian of the Vocontian Trough: a strong obliquity control. In House, M.R. & Gale, A.S. (eds) *Orbital forcing timescales and cyclostratigraphy*. Geological Society Special Publication No. 85, 143–164.

Goldhammer, R.K., Dunn, P.A. & Hardie, L.A. 1987. High frequency glacio-eustatic oscillations with Milankovitch characteristics recorded in northern Italy. *American Journal of Science*, **287**, 853–892.

Hallam, A. 1986. Origin of minor limestone–shale cycles: climatically induced or diagenetic? *Geology*, **14**, 609–612.

Hays, J.D., Imbrie, J. & Shackleton, N.J. 1976. Variations in the Earth's orbit: pacemaker of the Ice-Ages. *Science*, **194**, 1121–1132.

Herbert, T.D. & Fischer, A.G. 1986. Milankovitch climatic origin of Mid-Cretaceous black shale rhythms in central Italy. *Nature,* **321**, 739–743.

Hilgen, F.J. 1991. Extension of the astronomically calculated (polarity) time scale to the Miocene/Pliocene boundary. *Earth and Planetary Science Letters,* **107**, 349–368.

House, M.R. 1995a. Devonian precessional and other signatures for establishing a Givetian time-scale. In House, M.R. & Gale, A.S. (eds) *Orbital forcing timescales and cyclostratigraphy*. Geological Society Special Publication No. 85, 37–51.

House, M.R. 1995b. Orbital forcing timescales: an introduction. In House, M.R. & Gale, A.S. (eds) *Orbital forcing timescales and cyclostratigraphy*. Geological Society Special Publication No. 85, 1–18.

Huang, Z., Ogg, J.G. & Gradstein, F.M. 1993. A quantitive study of Lower Cretaceous cyclic sequences from the Atlantic Ocean and the Vocontian Basin (SE France). *Paleoceanography,* **8**, 275–291.

Imbrie, J. 1985. A theoretical framework for the Pleistocene ice ages. *Journal of the Geological Society of London,* **142**, 417–432.

Melnyk, D.H. & Smith, D.G. 1989. Outcrop to subsurface cycle correlation in the Milankovitch frequency band: Middle Cretaceous, Central Italy. *Terra Nova,* **1**, 432–436.

Milankovitch, M. 1941. *Kanton der Erdbestrahlung und seine Anwendung auf das Eiszeitenproblem*. Serbian Academy of Science, Belgrade, 133.

Napoleone, G. & Ripepe, M. 1989. Cyclic geomagnetic changes in mid-Cretaceous rhythmites, Italy. *Terra Nova,* **1**, 437–442.

Olsen, P. 1986. A 40-million-year lake record of early Mesozoic orbital forcing. *Science,* **234**, 842–848.

Paul, C.R.C., Mitchell, S.F., Marshall, J.D. Leary, P.N., Gale, A.S., Duane, A.M. & Ditchfield, P.W. 1994. Palaeoceanographic events in the Middle Cenomanian of Northwest Europe. *Cretaceous Research,* **15**, 707–738.

Pedersen, T.F. & Calvert, S.E. (1990) Anoxia vs. productivity: what controls the formation of organic-carbon-rich sediments and sedimentary rocks? *Bulletin of the American Association of Petroleum Geologists,* **74**, 454–466.

Raiswell, R. 1988. A chemical model for the origin of minor limestone–shale cycles by anaerobic methane oxidation. *Geology,* **16**, 641–644.

Roof, S.R., Mullins, H.T., Gartner, S., Huang, T.C., Joyce, E., Prutzman, J. & Tjalsma, L. 1991. Climatic forcing of cyclic carbonate sedimentation during the last 5.4 million years along the West Florida continental margin. *Journal of Sedimentary Petrology,* **61**, 1070–1088.

Savrda, C.E. & Bottjer, D.J. 1989. Trace-fossil model for reconstructing oxygenation histories of ancient marine bottom waters: application to Upper Cretaceous Niobara Formation, Colorado. *Palaegeography, Palaeoecology, Paleoclimatology* **74**, 49–74.

Schneider, F.K. 1964. Erscheinungsbild und Entstehung der rhythmischen Bankung der altkretazischen Tongesteine Nordwestfalens und der Braunschweiger Bucht. *Fortschrift Geologischen der Rheinland und Westfalen,* **7**, 353–382.

Schwarzacher, W. 1975. *Sedimentation models and quantitative stratigraphy*. Developments in Sedimentology 19, Elsevier, Amsterdam.

Schwarzacher, W. 1993. *Cyclostratigraphy and the Milankovitch theory*. Developments in Sedimentology 52, Elsevier, Amsterdam.

Shackleton, N.J., Crowhurst, S., Hagelberg, T., Pisias, N.G. & Schneider, D.A. 1995. A late Neogene time scale: application to leg 138 sites. In Pisias, N.G., Mayer, L.A., Janacek, T.R., Palmer-Julson, A. & van Andel, T.H. (eds) *Proceedings of the Ocean Drilling Program, Scientific Results,* **138**, 73–101.

Ten Kate, W.D. & Springer, A. 1993. Orbital cyclicities above and below the Cretaceous–Tertiary boundary at Zumaya (N Spain), Agost and Relleu (SE Spain). *Sedimentary Geology,* **87**, 69–101.

Tineke, N.F., Kroon, D., Ten Kate, W.D. & Sprenger, A. 1991. Late Pleistocene periodicities of oxygen isotope ratios, calcium carbonate contents, and magnetic susceptibilities of western Arabian Sea margin Hole 728A (ODP Leg 117). *Ocean Drilling Program Science Research Proceedings,* **117**, 309–320.

Weedon, G.P. 1989. The detection and illustration of regular sedimentary cycles using Walsh power spectra and filtering with examples from the Lias of Switzerland. *Journal of the Geological Society of London,* **146**, 133–144.

Weedon, G.P. 1993. The recognition and stratigraphic implications of orbital-forcing of climate and sedimentary cycles. *Sedimentary Review,* **1**, 31–48.

Weedon, G.P. & Jenkyns, H.C. 1990. Regular and irregular climatic cycles and the Belemnite Marls (Pleinsbachian, Lower Jurassic, Wessex Basin). *Journal of the Geological Society of London,* **147**, 915–918.

Weltje, G. & De Boer, P.L. 1993. Astronomically induced paleoclimatic oscillations reflected in Pliocene turbidite deposits on Corfu (Greece): implications for the interpretation of higher-order cyclicity in ancient turbidite system. *Geology,* **21**, 307–310.

8
Strontium Isotope Stratigraphy

John M. McArthur

Two of the most important and difficult tasks in geology are the precise determination of the relative age of rocks (whether rocks in one locality are older, younger or the same age as rocks in another) and the precise quantification of age, in terms of millions of years. Without both, the geological history of the earth cannot be understood.

The numerical age of igneous and metamorphic rocks has traditionally been measured using isotopic systems that are based on radioactive decay; for example, the decay of rubidium to strontium, samarium to neodymium, potassium to argon, and uranium to lead (Chapter 12). Such traditional methods are rarely applicable to sedimentary rocks, although recent developments in ultramicro-methods of dating single grains of feldspar and zircon are revolutionizing this state of affairs. Traditionally, relative ages of sediments have been determined by studying the fossils they contain; the determination of numerical age was possible only where volcanic or intrusive rocks datable by radio-decay methods provided fortuitous intercalations within a sequence of sediments. Within the past 30 years it has become common to determine both the relative and numerical age of sediments through a study of the magnetic orientation of minerals within them (magnetostratigraphy). Still more recently this method has been supplemented by isotope stratigraphy, a methodology for determining both relative and absolute ages based on the measurement of the stable isotopic composition of oxygen and strontium in fossils. The latter method, known as strontium isotope stratigraphy and commonly abbreviated to SIS, has developed only within the last 10 years but looks set to take an important place in stratigraphy beside magnetostratigraphy as a powerful way to date sediments.

Unlocking the Stratigraphical Record: Advances in Modern Stratigraphy. Edited by P. Doyle and M.R. Bennett.
© 1998 John Wiley & Sons Ltd.

Dating and correlation with strontium isotopes relies on the measurement of an $^{87}Sr/^{86}Sr$ value in a sample; this ratio expresses simply the relative numbers of atoms of the isotopes of strontium that have mass numbers 86 and 87. On precipitation of a marine mineral, such as biogenic calcite, it incorporates some strontium from seawater. That strontium will have an $^{87}Sr/^{86}Sr$ equal to that of strontium in the oceans at that time. As the $^{87}Sr/^{86}Sr$ of strontium in the world's oceans has changed with time in a known way (Figure 8.1), comparing the $^{87}Sr/^{86}Sr$ value of a fossil with a standard curve permits an age for the mineral to be deduced.

Any mineral that incorporates marine strontium can be dated with SIS, provided the strontium can be isolated without being contaminated by strontium from other sediment components, such as detrital clays, which contain strontium with a different isotopic ratio. Foraminifera, belemnites, brachiopods, and whole-rocks comprised of fairly pure chalk, have been used commonly both to construct calibration curves (Figure 8.1) and to date rocks of unknown age. A few attempts have been made to use other materials, such as barite, and apatite in fish teeth, bones and phosphorite deposits; these attempts have met with mixed success. Most marine sedimentary rocks contain enough carbonate (or apatite or barite) to permit dating and correlation using SIS. Since the $^{87}Sr/^{86}Sr$ of continental waters reflects local sources of strontium, SIS does not work for non-marine deposits except, perhaps, on a very local scale for lacustrine sediments.

Strontium isotope stratigraphy, however, has a use wider than simply dating and correlating marine sediments. The fluctuations of the marine $^{87}Sr/^{86}Sr$ curve (Figure 8.1) reflect the operation of geological processes on a global scale; the positions of maxima and minima, and the rates at which change in $^{87}Sr/^{86}Sr$ occurs, tells us about the rate and scale of global weathering and seafloor spreading, especially when $^{87}Sr/^{86}Sr$ data are combined with other isotopic and geological information.

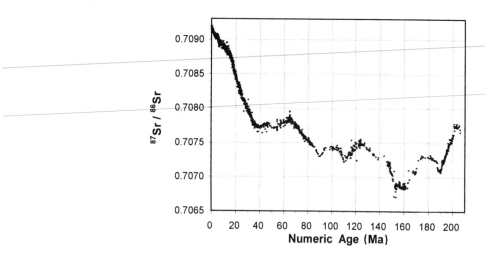

Figure 8.1   *Variation of $^{87}Sr/^{86}Sr$ during the past 206 Ma*

8.1 HISTORY OF STRONTIUM ISOTOPE STRATIGRAPHY

The simple underlying principles of strontium isotope stratigraphy (SIS) were enunciated first by Wickman (1948), who suggested that the $^{87}Sr/^{86}Sr$ of strontium in seawater should increase as 'a one-valued function of time' because the oceans are the repository of the products of continental weathering, which include strontium with an $^{87}Sr/^{86}Sr$ greater than that of marine strontium. He postulated that a record of this increase should be preserved in marine limestone, gypsum and anhydrite, as these sediments incorporate marine strontium on precipitation from seawater. Measurement of their $^{87}Sr/^{86}Sr$ would therefore allow determination of the time of mineral formation. Wickman's idea was tested by Peterman *et al.* (1970), Dasch & Biscaye (1971), Veizer & Compston (1974), Burke *et al.* (1982) and others, who measured $^{87}Sr/^{86}Sr$ of marine carbonates of known age and so revealed variations in $^{87}Sr/^{86}Sr$ between 0.707 and 0.709 during parts of the Phanerozoic. This was not the monotonic increase that Wickman predicted, for reasons given later.

Since the paper by Burke *et al.* (1982), SIS has been refined to the point where it is a viable method for dating and correlating marine sediments. Improvements continue through improving methods for measuring for $^{87}Sr/^{86}Sr$; by a growing understanding of how diagenesis affects the $^{87}Sr/^{86}Sr$ value preserved in a rock; and by applying the method to new materials, e.g. abiotic marine carbonate cements.

Calibration curves of an accuracy sufficient for useful stratigraphy have now been defined for much of the past 206 Ma (Figure 8.1; McArthur 1994; Howarth & McArthur 1997). Calibration curves available for other time periods are of variable, but mostly lesser, quality than those available for the past 206 Ma. This is partly because of the difficulty of finding well-preserved material for analysis (a difficulty that increases with numerical age), and partly because the difficulty of assigning numerical ages to rocks increases with age.

8.2 FUNDAMENTALS OF STRONTIUM ISOTOPE STRATIGRAPHY

8.2.1 Trends of $^{87}Sr/^{86}Sr$ with Time

As Figure 8.1 shows, the $^{87}Sr/^{86}Sr$ of strontium in the oceans has varied with time; during the Phanerozoic values have fluctuated between 0.7092 and 0.7068. Specific values of $^{87}Sr/^{86}Sr$ may occur more than once, but this seldom presents a problem as other criteria (magnetostratigraphy, lithostratigraphy and biostratigraphy) are usually available for deciding which part of the isotope curve is relevant to a particular problem. More restrictive is the fact that during some periods of time $^{87}Sr/^{86}Sr$ changed too little for the method to have much temporal resolution, e.g. for much of the Aalenian (176.5 to 180.1 Ma) (Figure 8.1).

8.2.2 Data Presentation

Strontium isotope data are reported as $^{87}Sr/^{86}Sr$ values (i.e. the number of ^{87}Sr atoms relative to the number of ^{86}Sr atoms) because that is what mass spectrometers actually measure. In addition, many users also report the arithmetic difference between a measured value ($^{87}Sr/^{86}Sr_{Sample}$) and an accepted datum ($^{87}Sr/^{86}Sr_{Datum}$).

$$\Delta^{87}Sr = [^{87}Sr/^{86}Sr_{Datum} - {}^{87}Sr/^{86}Sr_{Sample}] \times 10^5$$

The factor 10^5 (some workers use 10^6) is introduced in order to make the numbers more manageable, e.g. $\Delta^{87}Sr = -7.5$ (or -75) rather than -0.000075. Two common data are the $^{87}Sr/^{86}Sr$ for strontium in today's oceans (0.709175), a value termed Modern Seawater Strontium and abbreviated to MSS, and the $^{87}Sr/^{86}Sr$ for a widely available standard called NIST 987 (formally SRM 987), which has a value of 0.710248.

8.3 THE GEOCHEMISTRY OF STRONTIUM

8.3.1 What Controls the $^{87}Sr/^{86}Sr$ of Marine Strontium?

The strontium in the ocean comes from three sources (Figure 8.2), and each has a distinct value of $^{87}Sr/^{86}Sr$. Hydrothermal circulation at mid-ocean ridges supplies strontium leached from ridge basalts (i.e. from the mantle); rivers supply strontium weathered from continental crust; while recrystallizing carbonate sediments provide, in comparison, a small supply by advection and diffusion of strontium from sediment pore waters. The temporal variation of marine $^{87}Sr/^{86}Sr$ seen in Figure 8.1 results from a variation with time in the amount of strontium being supplied to the ocean from these three sources, and in the $^{87}Sr/^{86}Sr$ values of strontium in rivers and strontium from pore waters. The $^{87}Sr/^{86}Sr$ from mid-ocean ridge circulation has changed little during the Phanerozoic.

The mantle source of strontium occurs prominantly at mid-ocean ridges. Seawater is drawn into hydrothermal circulation systems and loses most of its strontium into anhydrite, which precipitates at shallow depths in the oceanic crust. Deeper in the oceanic crust strontium is leached from basalt by the hot seawater. Fluids venting to the seafloor therefore have an $^{87}Sr/^{86}Sr$ that is determined by the proportion of basalt-strontium and residual seawater; the value is usually about 0.703, and the flux of strontium about 1.0×10^{12} g a^{-1}. As water in the open oceans has always had an $^{87}Sr/^{86}Sr > 0.703$, hydrothermal circulation at mid-ocean ridges has always worked to decrease the $^{87}Sr/^{86}Sr$ of marine strontium.

The crustal source of strontium derives from the weathering of continental rocks, the soluble products of which are carried to the oceans *via* rivers and groundwater outflow onto continental shelves. The $^{87}Sr/^{86}Sr$ of this strontium depends upon the age and type of rock being weathered. River supply is dominated by the world's larger rivers, which have a rather restricted range of $^{87}Sr/^{86}Sr$ values (0.705–0.735) because they average the $^{87}Sr/^{86}Sr$ from a wide variety of crustal rocks. The global average $^{87}Sr/^{86}Sr$ for modern rivers is 0.712 ± 0.001, and the flux is about 3.3×10^{12} g a^{-1}. The flux and $^{87}Sr/^{86}Sr$ of groundwater outflow is all but unknown, but its $^{87}Sr/^{86}Sr$ is not likely to differ much from riverine values. The addition of

Source of Sr	Mid-oceanic ridges	Recrystallization of carbonate sediment	Rivers
Amount per year (metric tonnes)	1×10^6	0.3×10^6	3×10^6
$^{87}Sr / ^{86}Sr$	0.704	0.708	0.712

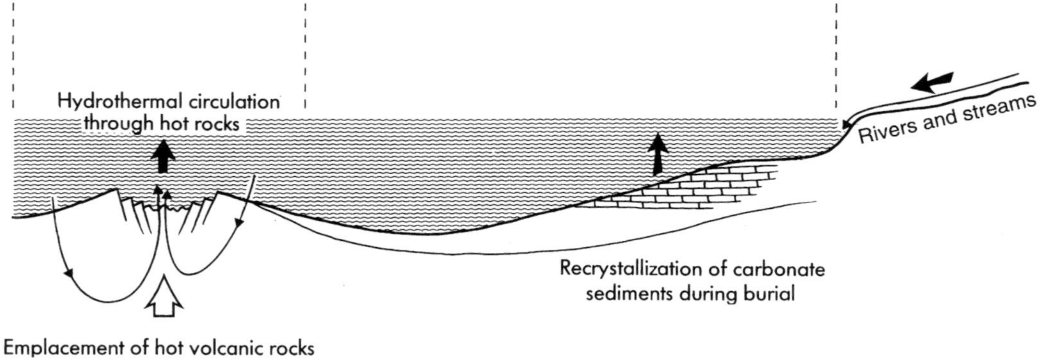

Figure 8.2 Schematic cross-section through the earth's crust showing the major sources of supply of Sr to the oceans. The amount supplied each year, and the ratio of each source, is shown in the boxed section [Reproduced with permission of Academic Press, Inc, from: McArthur (1992)]

crustal strontium to the ocean *via* rivers and groundwater outflow therefore increases the $^{87}Sr/^{86}Sr$ of marine strontium.

In comparison with the above sources, recrystallizing carbonate sediments provide a very small supply of strontium to the ocean. The strontium content of marine carbonates ranges from a few hundred to a few thousand micrograms of strontium per gram of sediment (parts per million, ppm) and some of this is released to pore water during recrystallization, mostly within the first few tens of metres of burial, and diffuses back into overlying seawater. Today, this strontium has an $^{87}Sr/^{86}Sr$ value of 0.708 and a flux of 0.3×10^{12} g a^{-1}.

Of these three inputs of strontium to the oceans, the $^{87}Sr/^{86}Sr$ of strontium from mantle sources at mid-ocean ridges has probably changed, only very gradually, by less than 0.001 during geological time. The other quantities (ratios and fluxes) have varied in response to changes in the amount of mid-ocean ridge volcanism, and the effects of changing world climates on weathering and river flow. The complex interplay of these five influences has led to the variations in the $^{87}Sr/^{86}Sr$ of marine strontium that are shown in Figure 8.1.

8.3.2 Homogeneity of Marine $^{87}Sr/^{86}Sr$

Reason for homogeneity

The global applicability of SIS rests on the fact that today's oceans are isotopically homogenous with respect to $^{87}Sr/^{86}Sr$ (with a present analytical precision of $^{87}Sr/^{86}Sr$ measurement of about ± 0.00001). The homogeneity extends throughout the world's oceans, into restricted seas such as Hudson's Bay, the Mediterranean Sea, the Gulf of Corinth, much of the Baltic Sea, and into the seaward end of estuaries: sea water must be diluted by more than 50% with river water before the effect on $^{87}Sr/^{86}Sr$ is measurable, except for rare cases where a river's strontium concentration or $^{87}Sr/^{86}Sr$ value is extreme (e.g. the Swan River of Western Australia).

Uniformity of $^{87}Sr/^{86}Sr$ is expected because the residence time of strontium in the oceans ($>10^6$ years) is far longer than the time it takes currents to mix the oceans ($>10^3$ years), so the oceans are thoroughly mixed on times scales that are short relative to the rates of gain and loss of strontium. Put another way, an atom of strontium reaching the ocean participates in thousands of mixing cycles before permanently becoming part of the sediment. This gives the isotopes of strontium time to homogenize. Strontium-isotope stratigraphers assume that the ocean has always been well mixed on a time scale of a few thousand years and therefore isotopically uniform with respect to $^{87}Sr/^{86}Sr$. Additionally, they assume that the oceans have always had about the same amount of dissolved strontium as they do now. All available evidence suggests that both assumptions are good for the Phanerozoic at least. Nevertheless, neither assumption is proven and were either (or both) not true, the effect on SIS would be appreciable.

Consequences of homogeneity

The isotopic uniformity of marine $^{87}Sr/^{86}Sr$ means that strontium in marine minerals will have identical $^{87}Sr/^{86}Sr$ values in sediments deposited simultaneously in any part of the world. A particular $^{87}Sr/^{86}Sr$ value therefore may be viewed as being characteristic, everywhere in the world, of a precise point in time (even though it may not be possible to say which point, in terms of millions of years), and correlation with strontium isotopes is equivalent to correlation with time surfaces and is therefore chronostratigraphy. Biostratigraphers commonly adopt the working assumption that the appearance and disappearance of fossils used in correlation are everywhere simultaneous, but this assumption is extraordinarily difficult to prove. Time and practice have determined which fossils seem to have synchronous boundaries and which do not. Biostratigraphy is therefore potentially diachronous whereas SIS is not. Stratigraphy with strontium isotopes is therefore *potentially* more accurate than biostratigraphy; in practice, biostratigraphy is almost always superior because it is much more developed than is SIS.

8.4 CORRELATION AND NUMERICAL DATING

8.4.1 Correlation

Sedimentary rocks have traditionally been correlated with fossils. Strontium isotope stratigraphy can correlate equally well. This ability is illustrated in Figure 8.3, where the temporal equivalence of geographically distant rocks is demonstrated in physical terms. In the imaginary sections illustrated, rocks formed at level A in each section have identical $^{87}Sr/^{86}Sr$. If there is good reason to believe both sequences are approximately equivalent in age (say from the regional geological context) then rocks at level A in each sequence are equivalent in age, and are said to correlate; what that age is in numerical terms is not revealed by this correlation.

8.4.2 Establishing Numerical Ages with $^{87}Sr/^{86}Sr$

The ability to correlate with strontium isotopes is useful. More useful is the ability to determine numerical ages in millions of years. Before this is possible, calibration curves of $^{87}Sr/^{86}Sr$ against stratigraphical level (e.g. Figure 8.3) must be given a numerical scale (Figure 8.1); to do this, the $^{87}Sr/^{86}Sr$ value and numerical age for a sample must be determined. Establishing the numerical age of sedimentary rocks is difficult because directly dating them with radioactive isotopes (to provide numerical tie-points) is possible only where they contain glauconite, or primary feldspars or zircons from volcanic eruptions intercalated into the sequence in volcanic rocks (e.g. lavas, ash horizons). Most numerical ages assigned to sedimentary rocks are therefore determined by interpolation between such rare tie-points, which are dated in specific localities. Numerical age for other localities requires that the stratigraphical level of the tie-point be correlated

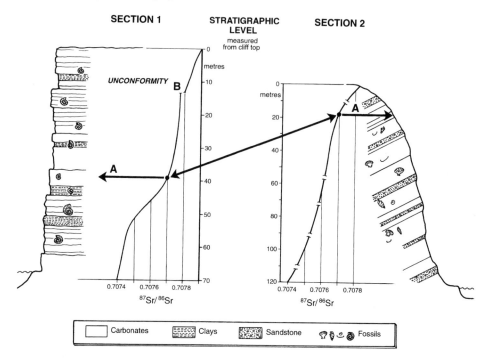

Figure 8.3   *Correlating rocks with Sr isotopes: using $^{87}Sr/^{86}Sr$ curves for two widely separated imaginary cliff sections. Levels in the sequences with identical $^{87}Sr/^{86}Sr$ formed at the same time and therefore correlate; for example, the rocks at level A*

to other localities by biostratigraphy or magnetostratigraphy – a procedure that introduces uncertainty.

Most of the data used to compile Figure 8.1 derive from measurement of $^{87}Sr/^{86}Sr$ in samples from carbonate sequences for which a detailed biostratigraphy exists and has been numerically calibrated often by high-order biostratigraphical correlation from tie-points, or by magnetostratigraphy. The final numerical scale therefore includes all the uncertainties associated with magnetostratigraphy and/or biostratigraphy. For the latter, these include uncertainties in fossil recognition, plus those inherent in the numerical calibration of the biostratigraphical scheme.

8.5 SAMPLES FOR STRONTIUM ISOTOPE STRATIGRAPHY

8.5.1 Sampling Media

The minerals used so far for SIS are biogenic carbonate (chalk and carbonate ooze, macro and microfossils), the mineral apatite (francolite) in fish teeth, bones and phosphate deposits, barite, and abiotic marine carbonate cement. Pure minerals give the most reliable ages but bulk materials, such as chalk and lithified limestone, give good ages if analysed in a manner that avoids contributing contaminant strontium from extraneous minerals.

Calcitic macrofossils (e.g. brachiopods and belemnites) frequently yield good ages. Being large, they are easy to sample, simple to clean of contaminant material, and they resists diagenesis well. In contrast, aragonitic macrofossils sometimes give aberrant data for no apparent reason. Being small, foraminifera are more difficult to sample; being hollow, their interiors often contain contaminants, such as clays and diagenetic calcite. When taken from carbonate-rich sediments, these problems seem resolvable, as contaminant clays are present in low abundance and strontium in diagenetic calcite is generally derived very locally from recrystallizing carbonate. In clastic sediments, contaminant clays are abundant; moreover, they supply diagenetic carbonate to strontium with a very high $^{87}Sr/^{86}Sr$. This can severely alter the $^{87}Sr/^{86}Sr$ value of a foraminifera, even when secondary calcite is present in amounts of only 1%.

Bulk rock materials, such as chalk, carbonate ooze and phosphorite, are more susceptible than are fossils to contamination from strontium in extraneous phases, such as sea-salt and clay minerals. By careful preparation, such contamination can virtually be eliminated in many instances. Lithification of carbonate and apatite alters $^{87}Sr/^{86}Sr$ but proper preparation of whole-rock samples can minimize the effect and reveal original, or near-original, values of $^{87}Sr/^{86}Sr$. Early marine cements in lithified rocks may record the original $^{87}Sr/^{86}Sr$ of seawater, but their use relies on the assumption that the early diagenetic pore water from which they form had an $^{87}Sr/^{86}Sr$ indistinguishable from that in overlying seawater. The limited data available on this point suggest that this may be true for depths in sediments of at least a few tens of centimetres, but perhaps not for greater depths.

Barite ($BaSO_4$) should be particularly useful for dating abyssal red clay sequences as it seems to be the only material likely to preserve an $^{87}Sr/^{86}Sr$ record at oceanic depths beneath which carbonate is not preserved. Diagenetic reactions ensure that barite is seldom preserved in sediments that accumulate at a depth of less than a few hundred metres, so its use for dating sequences on continental margins is limited.

8.5.2 Sample Preservation Criteria

All samples taken from rocks have undergone diagenetic alteration so, before analysing a material for $^{87}Sr/^{86}Sr$, the degree to which it has preserved its original value must be assessed. SIS is most valuable where preservation of a sample is good and where alteration can be shown to have had a negligible effect on $^{87}Sr/^{86}Sr$. There are several ways of assessing the preservational state of a sample and these are briefly discussed below, although a more comprehensive summary can be found in McArthur (1994). Usually, several methods need to be combined in order to arrive at a realistic assessment of a sample's preservational state.

Scanning electron microscopy is good for revealing alteration in materials that have a systematic structure, such as biogenic carbonate, in which it is assumed that good preservation of ultrastructural morphology proves a sample's integrity. A few examples of preservational assessments made by SEM are shown in Figure 8.4.

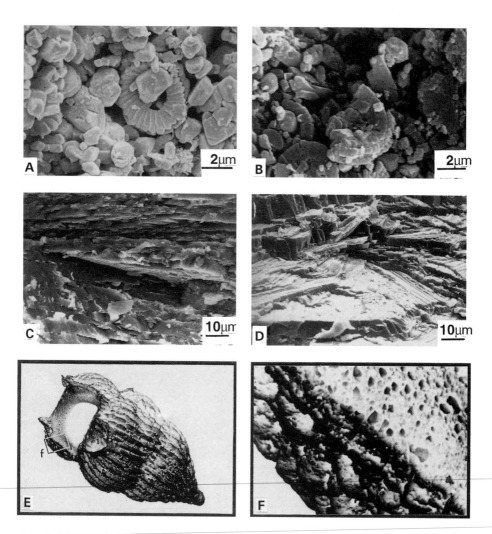

Figure 8.4   *SEM photomicrographs demonstrating good preservation for SIS:* **A.** *Chalk from Lagerdorf, Germany; some overgrowth is seen in both but spar is rare.* **B.** *Chalk, from Kjølby Gaard, Denmark;* **C.** *and* **D.** *different specimens of* Pycnodonte *form K/T boundary sediments on Seymour Island, Antarctica, showing unaltered compact layering in calcite.* **E.** *good preservation of a specimen of* Pseudotextularia deformis *(200 mm long axis) from 1 m below the K/T boundary in the Danish chalk, Kjølby Gaard.* **F.** *detail of aperture in* **E.** *showing only minor nucleation of calcite on the specimen's surface.*

X-ray diffraction (XRD) is a useful way of detecting alteration in biogenic carbonates, as it can detect the presence of as little as 0.5% of a crystalline contaminant, such as calcite introduced into aragonite by diagenesis. Unfortunately, XRD is insensitive for poorly crystalline phases, such as illite, which may host contaminant strontium, and as much as 10% of these phases may escape detection.

Thin-sections and cathodoluminescence are powerful ways of assessing the suitability of material for analysis and have been used mostly on carbonates. Those that luminesce are generally regarded as recrystallized, but some modern carbonate shells luminesce and some diagenetic cements do not, so this criteria must be used with caution.

Chemical analysis may reveal the presence of contaminants. Detectable concentrations of aluminium may warn of contamination from clay minerals that can affect $^{87}Sr/^{86}Sr$ even when present at very low abundance as they usually contain strontium with a very high $^{87}Sr/^{86}Sr$ value. High concentrations of iron or magnesium (>400 ppm and >50 ppm respectively) may indicate alteration of calcite, but lower concentrations do not necessarily prove good preservation. The concentration of magnesium in biogenic calcite varies from 10^2 to 10^5 as a result of biological discrimination and so magnesium concentrations reveal little about diagenesis, although for macrofossil aragonite, concentrations of magnesium greater than 100 ppm suggest it is altered.

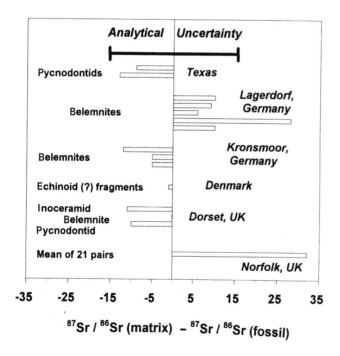

Figure 8.5 *Comparison of $^{87}Sr/^{86}Sr$ in macrofossils and closely associated samples of nannofossil chalk. $\Delta^{87}Sr = (^{87}Sr/^{86}Sr_{macrofossil} - ^{87}Sr/^{86}Sr_{chalk}) \times 10^6$. Differences of less than 30 in $\Delta^{87}Sr$ would commonly be thought to be within analytical error*

For biogenic carbonates, Sr/Ca ratios lower than those found in modern equivalents are often taken to indicate that diagenetic recrystallization has occurred, because inorganic recrystallization discriminates against re-incorporation of strontium into precipitating abiotic carbonate. Nevertheless, the Sr/Ca in biogenic carbonate is governed by biological discrimination, the Sr/Ca in seawater, and the temperature and rate of calcite precipitation. Such effects are difficult to separate.

If a sample has an unusual stable isotopic composition it is unlikely to have retained its original $^{87}Sr/^{86}Sr$. The converse is not necessarily true, however; samples may recrystallize without altering their stable isotopic composition in a noticeable way.

A particularly valuable indicator of good preservation is that well-preserved samples show a consistency of $^{87}Sr/^{86}Sr$ amongst sub-samples and a similar concordance of values amongst different materials from a single site or stratigraphical level. This is illustrated in Figure 8.5, which shows a comparison of $^{87}Sr/^{86}Sr$ in macrofossils and bulk chalk from chalk strata (cleansed of contaminants by pre-leaching in acid prior to analysis). Most differences are within the limits of analytical uncertainty, but a few are not (e.g. the difference within the English chalk beneath Norfolk, and for one sample pair from Germany). Clearly, such tests are a good way of detecting subtle changes in $^{87}Sr/^{86}Sr$.

8.5.3 Sampling Methods

Sub-sampling of material for analysis is best done by hand-picking fragmented samples under the microscope, by micro-drilling, or by using a laser to vaporize microscopic areas of a polished block, or thin section, prior to isotopic analysis of the vapour. If the strontium isotope stratigrapher lacks the tools to isolate a sub-sample by physical means, isolating the target phase may be possible using a selective attack that dissolves the target mineral without also dissolving recrystallized derivatives or contaminant components, such as overgrowths and clays. Once a sub-sample is obtained, the strontium within it is isolated from other elements by well-established techniques of ion-exchange chromatography. The separated strontium is then analysed for its $^{87}Sr/^{86}Sr$ value using thermal-ionization mass spectrometry.

Physical methods of separation

Most workers select material for analysis using a microscope, with the aim of selecting that part of a bulk sample that is as pristine as can be found. Visual appraisal of material in this way is, in the hands of the experienced, a remarkably good way of selecting the best-preserved material. Foraminifera, fragments of macrofossil shell and grains of peloidal apatite, are typical materials separated in this fashion. For lithified samples, micro-drills can be used to isolate the unaltered, or least altered, parts of the rocks as seen in thin section, or to sample from cut surfaces. In this way, sample weights of as little as 2 mg can be obtained. Finally, a recent development that takes micro-sampling to its ultimate is to use a laser beam under vacuum to vaporize a small part (a few micrograms or less) of a sample and to measure the $^{87}Sr/^{86}Sr$ of the vapour. In this way, unaltered primary biogenic calcite can be separated from diagenetic overgrowths and cements and analysed for $^{87}Sr/^{86}Sr$. Such equipment is, in 1997, almost prohibitively expensive; in due time, it may become commonplace and so revolutionize SIS.

Chemical methods of separation

In favourable circumstances, strontium in target minerals can be isolated without significant contamination from strontium in accessory minerals by using selective cleaning and dissolution methods. For example, a partial dissolution of the target mineral will remove much of the contaminant strontium on, and in, accessory minerals (e.g. clay minerals). A subsequent step that dissolves more, but not all, of the target phase mineral, will then yield a sample with minimal contributions from contaminant phases and so a near-original $^{87}Sr/^{86}Sr$. For such strategies it is best to use a concentrated *weak* acid (one partially ionized in aqueous solution), such as 90% acetic acid, rather than a dilute *strong* acid (one completely ionized in aqueous solution), such as 10% hydrochloric acid, as the latter reagent is much more aggressive than the former. For carbonates and apatites, other useful dissolution reagents are acetic acid/ammonium acetate buffered to pH 4.5, ion-exchange dissolution at pH 5 (Kralik 1984), and tri-ammonium citrate buffered to pH 8.

Figure 8.6 shows how $^{87}Sr/^{86}Sr$ of a carbonate sample changes on incremental dissolution in acid as a result of contributions from contaminants, and so empha-

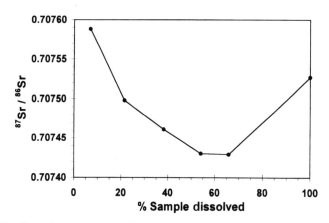

Figure 8.6 $^{87}Sr/^{86}Sr$ of an impure chalk (75% CaCO₃) sample as a function of sequential dissolution in dilute acetic acid solution. The early fractions contain much contaminant Sr which gives them a ratio higher than in later fractions; the latter more accurately reflect the real $^{87}Sr/^{86}Sr$ of the sample. The high value at 100% dissolution arose through the near-complete dissolution of the sample residue in concentrated hydrochloric acid, a process that dissolved much of the clay minerals in the rock

sizes the importance of a thorough cleaning step if contamination is to be minimized.

8.6 ACCURACY AND PRECISION OF STRONTIUM-ISOTOPE STRATIGRAPHY

The quality of a date derived by SIS from well-preserved samples is defined by its precision and accuracy. Precision refers to how reproducible the age is; that is, determined on 10 different samples from the same stratigraphical level, what is the spread of numerical ages that result? Accuracy refers to how closely the resulting age approaches the true age, which can never be known, only more closely approximated as we get better at numerical calibration.

Three things affect the quality of a date obtained by SIS: (1) the slope of the curve of $^{87}Sr/^{86}Sr$ against numerical age; (2) the analytical uncertainty of the $^{87}Sr/^{86}Sr$ analysis of a sample; and (3) the accuracy of the age model used to calibrate the isotope curve. Each is discussed below.

The slope of the curve

The precision of a numerical age derived by SIS depends upon the gradient of the $^{87}Sr/^{86}Sr$ curve: where the gradient is steep the precision of an age determination (<0.3 Ma) may better that available from biostratigraphical study. The rate of change of $^{87}Sr/^{86}Sr$ as a function of numerical age is shown in Figure 8.7. The fastest rate of change in $^{87}Sr/^{86}Sr$ we know of, about 66×10^{-6} per Ma, occurred during the Early Miocene. Owing to the sinuosity of the strontium-isotope calibration curve (Figure 8.1), a value of $^{87}Sr/^{86}Sr$ may represent more than one numerical age. The approximate age of a sample can be determined by lithostratigraphy, biostratigraphy or magnetostratigraphy, so this complication seldom presents a problem.

Analytical quality

Analysis of a sample in different laboratories should give the same $^{87}Sr/^{86}Sr$ (within analytical uncertainty). Unfortunately, this does not happen, so three standards are used to correct measurements to a common datum. They have the following values for $^{87}Sr/^{86}Sr$: 0.710248 for NIST 987 (known until recently as SRM 987); 0.708022 for Eimer and Amend (E & A); 0.709175 for EN-1, a carbonate standard prepared from a modern *Tridachna* clam from Enewetak Atoll and presumed to be representative of Modern Seawater Strontium (MSS). Many laboratories reported $^{87}Sr/^{86}Sr$ for other modern shells, or modern seawater, or both, as proxies for EN-1.

All $^{87}Sr/^{86}Sr$ measurements involve some analytical uncertainty. Modern instruments can measure $^{87}Sr/^{86}Sr$ to a precision of about $10–20 \times 10^{-6}$; that is, any single measurement will, when repeated, give a value that differs from its predecessors by about that amount or less. Replication improves precision through calculation of a mean of replicates, but not to much below a precision of 10×10^{-6} (Thirlwall 1991).

235

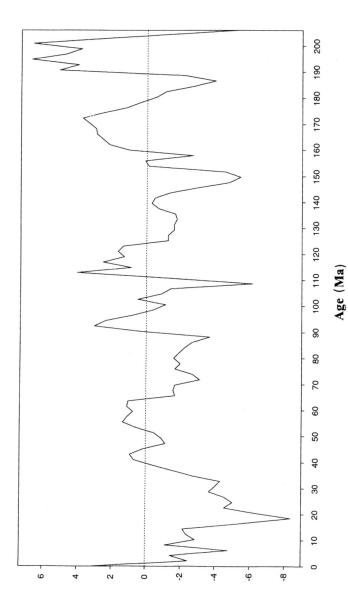

Figure 8.7 Rate of change of marine $^{87}Sr/^{86}Sr$ during the past 206 Ma. The line is the first differential of a line of best fit to the data in Figure 8.1

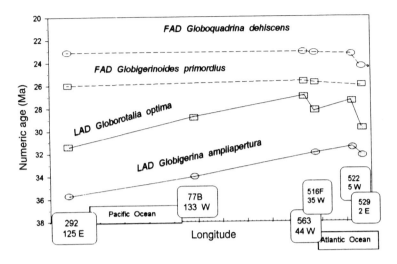

Figure 8.8 *Diachroneity and synchroneity of biostratigraphical boundaries revealed by SIS. Last appearances of* Globigerina ampliapertura *and* Globorotalia optima *young eastward, whilst the first appearances of* Globigerinoides primordius *and* Globoquadrina dehiscens *are synchronous. Site localities: 292, western Philippine Basin; 593, Tasman Sea; 77B, equatorial Pacific; 563, central North Atlantic; 516, 516F, western South Atlantic; 522, 529, eastern South Atlantic. [Reproduced with permission of the American Geophysical Union, from Hess* et al. *(1989)]*

The numerical calibration of standard curves

Standard curves such as Figure 8.1, where $^{87}Sr/^{86}Sr$ is plotted against numerical age, are readily understandable, but disguise the fact that they are based on plots of $^{87}Sr/^{86}Sr$ against stratigraphical level; that is, the position of a sample in a rock sequence, or a borehole core. To convert stratigraphical level into numerical age is seldom easy and generally must be done by correlating to the sample site levels of known age using biostratigraphy or magnetostratigraphy; rarely, it can be done by radiometric dating of rocks intercalated within the sediments to be dated.

The magnitude of the inaccuracy of such correlation depends upon the method used and can be large. For example, inaccuracies of several million years arise from uncertainties in biostratigraphically based age assignments used for the Cenozoic part of the standard curve. The uncertainties arise from many sources. Biostratigraphical datum levels may be diachronous (Figure 8.8) or restricted geographically. Numerical ages assigned using biostratigraphy require correct recognition of boundaries and their defining taxa, interpolation, extrapolation, high-order correlations, and assumptions about sedimentation rate, all of which introduce non-systematic and unquantifiable uncertainty into biostratigraphical ages used to construct the standard curve. For these reasons the emphasis in age calibration for SIS is shifting from a reliance on biostratigraphy towards a greater reliance on magnetostratigraphy. Problems exist for this method as well; principally in the correct identification of magnetic reversals.

8.7 STATISTICS IN STRONTIUM ISOTOPE STRATIGRAPHY

Calibration curves, such as Figure 8.1, have been defined by plotting $^{87}Sr/^{86}Sr$ data against numerical age. It is useful to be able to convert pictorial trends such as Figure 8.1 into quantitative relations, such as an equation or a table, from which numerical age can be computed or read, given a specific $^{87}Sr/^{86}Sr$. There are many ways to derive such equations and tables but all rely initially on computing a line of best fit to data pairs comprising $^{87}Sr/^{86}Sr$ and numerical age.

Regression techniques

For short time periods (<5 Ma), $^{87}Sr/^{86}Sr$ often seems to vary linearly with time, so it is possible to fit linear regressions to trends of $^{87}Sr/^{86}Sr$ against numerical age. Where non-linearity exists, curves can be fitted using polynomial regressions functions, such as power laws or exponential curves. Many statistical packages available for personal computers, and many hand-held calculators, will fit such polynomials but, despite their simplicity, both linear and non-linear regressions have severe shortcomings.

Firstly, linear regression can obscure useful details such as minor inflexions that might be valuable, either for correlating rocks and integrating different stratigraphical schemes or because they reflect the operation of global shifts in the geochemical cycle of strontium. Polynomial regression requires the prior identification of turning points in the data in order that an appropriate order of polynomial curve can be chosen. More importantly, there is no reason to suppose that nature conforms to such numerical relationships, however tidy they appear to us. Moreover, these regression methods treat one variable (the dependent variable, y) as dependent on the other (the independent variable, x) which is assumed to be free of uncertainty. As neither $^{87}Sr/^{86}Sr$ nor numerical age are dependent variables, nor is either free of uncertainty, simple regressions of $^{87}Sr/^{86}Sr$ on age, or of age on $^{87}Sr/^{86}Sr$, are inappropriate.

Statistical techniques

A better way to fit lines to data is to use a statistical method that makes no assumptions about how each datum relates to any other. Such fits do not conform to predetermined notions of how the curve should look; for example, that it should look linear, or have three turning points. The shape of the fit is controlled by the data alone. Details of one such method, termed LOWESS regression, and its application to SIS are given in Howarth & McArthur (1997). Figure 8.9 shows such a LOWESS fit to the $^{87}Sr/^{86}Sr$ and numerical age data for the interval 0 to 10 Ma. The fit cannot be reduced to a formula with which to calculate numerical age from $^{87}Sr/^{86}Sr$, but it can be used to derive a look-up table predicting numerical age from $^{87}Sr/^{86}Sr$, an extract from which is shown in Table 8.1.

Table 8.1   *Numerical ages predicted for $^{87}Sr/^{86}Sr$ values 0.707790 to 0.707984; extract from the look-up table of Howarth & McArthur (1997). The table was generated from a statistical LOWESS fit to the data shown in Figure 8.1*

$^{87}Sr/^{86}Sr$	Minimum age	Mean age	Maximum age
0.707970	>30.20	30.49	<30.79
0.707971	>30.18	30.46	<30.77
0.707972	>30.15	30.44	<30.74
0.707973	>30.13	30.41	<30.71
0.707974	>30.11	30.38	<30.68
0.707975	>30.08	30.36	<30.65
0.707976	>30.06	30.33	<30.63
0.707977	>30.03	30.31	<30.60
0.707978	>30.01	30.28	<30.57
0.707979	>29.98	30.26	<30.55
0.707980	>29.96	30.23	<30.52
0.707981	>29.93	30.20	<30.50
0.707982	>29.90	30.18	<30.47
0.707983	>29.88	30.15	<30.44
0.707984	>29.86	30.13	<30.42

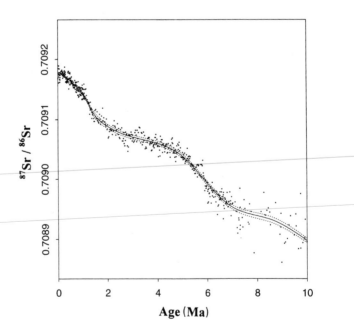

Figure 8.9   *A robust LOWESS fit for the interval from 0 to 10 Ma. The central solid line denotes the mean of the data. Confidence limits of 2 s.e. about the mean value are shown as dotted lines. These limits represent the confidence with which the position of the mean is known, and say little about the distribution of the data*

Dealing with data uncertainty when fitting trends to data

Currently, samples used to compile standard curves carry uncertainties on their numerical ages (derived from magnetostratigraphy or from numerically calibrated biostratigraphy) that are extraordinarily difficult to quantify. As a result, no SIS calibration curve has yet been prepared that includes uncertainties on the numeric ages used to compile it.

8.8 USE OF STRONTIUM ISOTOPES IN DATING

Strontium isotope stratigraphy has been used widely to (1) determine subsidence histories for atolls; (2) refine the depositional history of shelf sequences; (3) predict stratigraphical level in sedimentary rocks lacking age-diagnostic fossils (e.g. Darai Limestone of Papua New Guinea) as an aid to oil exploration; (4) estimate the duration of stratigraphical gaps; and (5) date phosphogenesis. Two important applications are reviewed below.

Estimating the duration of hiatuses and dating unconformities

Miller *et al.* (1988) used $^{87}Sr/^{86}Sr$ to estimate the duration of hiatuses in sedimentation at two DSDP Sites (548, 549) on the Goban Spur by dating material above and below suspected breaks in the sequence. The duration of the breaks was estimated to be 4.8 ± 1.8 Ma at Site 548, and 3.6 ± 1.8 Ma at Site 549, and these estimates were similar to those derived from biostratigraphy.

Sequence stratigraphical boundaries (i.e. unconformities) are sometimes marked by the formation of hardgrounds, phosphorite, and by boring and attached faunas. Hardgrounds may be datable by SIS if microsampling is undertaken and this potential for dating needs to be explored. Precise dating may reveal the degree of diachronicity along such surfaces, and quantify rates of transgression and regression. Measurement of $^{87}Sr/^{86}Sr$ in material overlying a hardground will date the time at which sedimentation was resumed.

Dating episodes of phosphogenesis

Phosphorites comprise two main types: replacement phosphorite, formed by phosphatization of (usually) precursor carbonate; and authigenic phosphorite, formed by precipitation of apatite, a P-rich mineral, directly from pore water. Apatite contains up to 2700 ppm of strontium, making the mineral very suitable for dating by SIS. In both phosphorite types, apatite appears to record the value of marine $^{87}Sr/^{86}Sr$ during formation (McArthur *et al.* 1990; Stille *et al.* 1994). Nevertheless, much more validation of this conclusion is needed by $^{87}Sr/^{86}Sr$ analysis of Recent replacement phosphorites that have been dated by independent means (e.g. U-series dating), and by analysis of $^{87}Sr/^{86}Sr$ of modern pore waters, in order to establish to what extent their $^{87}Sr/^{86}Sr$ reflects the $^{87}Sr/^{86}Sr$ value of sea.

8.9 STRONTIUM-ISOTOPES AND GLOBAL CHEMICAL CYCLES

The fluvial flux of strontium to the oceans is about three times greater than the mid-ocean ridge flux (see Section 8.3.1), so the ups and downs in the global strontium isotope curve (Figure 8.1) reflect mainly changes in the $^{87}Sr/^{86}Sr$ value and amount of strontium supplied to the oceans each year by rivers; this strontium comes from the weathering of continental rocks. Old basement rocks have a high $^{87}Sr/^{86}Sr$ (>0.72), whilst young basaltic rocks have a low $^{87}Sr/^{86}Sr$ (0.704). Phanerozoic sediments have intermediate values. The periods when $^{87}Sr/^{86}Sr$ increased or decreased greatly with time therefore reflect periods when the long-term equilibrium between continental weathering and mid-ocean ridge volcanism was out of balance. Such periods might reflect times of enhanced weathering or mid-ocean ridge volcanism, or periods of unroofing of rocks with high or low $^{87}Sr/^{86}Sr$. Consequently, some effort has been put into trying to relate variations in marine $^{87}Sr/^{86}Sr$ to global tectonic events. For example, the driving force for the rapid increase in $^{87}Sr/^{86}Sr$ during the period from 42 Ma to present has commonly been ascribed to uplift of the Himalayas and its enhanced denudation, or the development of major ice sheets in Antarctica. The Himalayas were not significant mountains much before the Miocene, yet the rapid rise in $^{87}Sr/^{86}Sr$ started in the latest Eocene, so the former proposal seems unlikely. Neither proposal explains why marine $^{87}Sr/^{86}Sr$ appears to have peaked about 150 Ka ago and is now decreasing (Howarth & McArthur 1997).

Small but abrupt changes in the marine $^{87}Sr/^{86}Sr$ record (20×10^{-6} over time scales of <0.5 Ma) might occur in response to sudden dumping into the ocean of unusually large amounts of strontium with extreme $^{87}Sr/^{86}Sr$ values, for example in response to rapid climate change; whether this has really occurred is uncertain. What is certain is that *cyclic* fluctuations in $^{87}Sr/^{86}Sr$ do not occur on time scales shorter than a million years, owing to the dampening affects of the long residence time of strontium in the oceans (2–4 Ma) and the large size of the oceanic strontium reservoir (1.3×10^{17} mol).

ACKNOWLEDGEMENTS

Thanks to R. J. Howarth for providing Figure 8.9 and to M. F. Thirlwall in whose laboratory much of the data presented here were obtained.

REFERENCES

Burke, W.H., Denison, R.E., Hetherington, E.A., Koepnick, R.B., Nelson, H.F. & Otto, J.B. 1982. Variation of $^{87}Sr/^{86}Sr$ throughout Phanerozoic time. *Geology*, **10**, 516–519.

Dasch, D.J. & Biscaye, P.E. 1971. Isotopic composition of Cretaceous-to-Recent pelagic foraminifera. *Earth and Planetary Science Letters*, **11**, 201–204.

Hess, J., Stott, L.D., Bender, M.L., Kennet, J.P. & Schilling, J-G. 1989. The Oligocene marine microfossil record: age assessments using strontium isotopes. *Paleoceanography*, **4**, 655–679.

Howarth, R.J. & McArthur, J.M. 1997. Statistics for strontium isotope stratigraphy. A robust LOWESS fit to the Sr-isotope curve for 0–206 Ma, with look-up table for the derivation of numerical age. *Journal of Geology*, **105**, 441–456.

Kralik, K. 1984. Effects of cation-exchange treatment and acid leaching on the Rb–Sr system of illite from Fithian, Illinois. *Geochimica Cosmochimica Acta*, **48**, 527–533.

McArthur, J.M. 1994. Recent trends in Sr isotope stratigraphy. *Terra Nova*, **6**, 331–358.

McArthur, J.M., Sahami, A.R., Thirlwall, M.F., Osborn, A.T. & Hamilton, P.J. 1990. Dating phosphogenesis with Sr isotopes. *Geochimica Cosmochimica Acta*, **54**, 1343–1351.

Miller, K.G., Feigenson, M.D., Kent, D.V. & Olson, R.K. 1988 Upper Eocene to Oligocene isotope ($^{87}Sr/^{86}Sr$, $\delta 18O$, $\delta 13C$) standard section, Deep Sea Drilling Project Site 522. *Paleoceanography*, **3**, 223–233.

Paytan, A., Kastner, M., Martin, E.E., Macdougall, J.D. & Herbert, T. 1993. Marine barite as a monitor of seawater strontium isotope composition. *Nature*, **366**, 445–449.

Peterman, Z.E., Hedge, C.E. & Tourtelot, H.A. 1970. Isotopic composition of strontium in sea water throughout Phanerozoic time. *Geochimica Cosmochimica Acta*, **34**, 105–120.

Stille, P., Riggs, R., Clauer, N., Ames, D., Crowson, R & Snyder, S. 1994. Sr and Nd isotopic analysis of phosphorite sedimentation through one Miocene high-frequency depositional cycle on the North Carolina continental shelf. *Marine Geology*, **117**, 253–273.

Thirlwall, M.F. 1991. Long-term reproducibility of multicollector Sr and Nd isotope ratio analysis. *Chemical Geology (Isotope Geosciences Section)*, **94**, 85–104.

Veizer, J. & Compston, W. 1974. $^{87}Sr/^{86}Sr$ composition of seawater during the Phanerozoic. *Geochimica Cosmochimica Acta*, **38**, 1461–1484.

Wickman, F.E. 1948. Isotope ratios: a clue to the age of certain marine sediments. *Journal of Geology*, **56**, 61–66.

9
Borehole Data and Geophysical Log Stratigraphy

Alf Whittaker

Traditionally, stratigraphical studies have used observations and data derived from surface exposures or from human excavations. Increasingly over the last two decades, however, the earth's sedimentary basins and basement rocks have been probed by thousands of boreholes and wells, whose description, analysis and interpretation necessitate some changes to stratigraphical procedure and methodology. It is fundamentally important to realize that geophysical wireline logs, or 'electric' logs as they are sometimes known, can be used either as supplementary data in routine subsurface stratigraphical description of borehole or well sections, or in a 'stand alone' way as original prime data sources in the absence of direct lithological samples.

The basic procedures for establishing formal stratigraphical units with data observed at the earth's surface also apply to data acquired in the subsurface, whether onshore or offshore (Whittaker *et al.* 1991). Despite some differences in approach, which are necessary because of the methods used and their attendant problems, the same generic principles used when dealing with surface exposures are applied when dealing with data and material from the subsurface. For example, complete lithological and palaeontological descriptions (or logs) of core samples or cuttings derived from boreholes are required in written and graphical form. Accurate geographical locational information is also necessary. Similarly, boundaries of stratigraphical units should be indicated with their depths from an established reference datum, and the naming of stratigraphical units should be taken where possible from nearby geographical features. To ensure the successful use of

Unlocking the Stratigraphical Record: Advances in Modern Stratigraphy. Edited by P. Doyle and M.R. Bennett.
© 1998 John Wiley & Sons Ltd.

this data source it is vital to understand the methods of drilling and logging processes so as to be aware of both the strengths and pitfalls of the techniques used.

This chapter covers aspects of the drilling process, its associated mud system, and the acquisition of rock samples from the subsurface. This basic information is needed to complement the description of the *in situ* subsurface conditions in which geophysical wireline log data are acquired and which have an important impact on interpretation procedures. The chapter also describes the types of open-hole logs in common use and how they are used in lithological description, the interpretation of sedimentary environments, the recognition of sequence boundaries, and in stratigraphical correlation.

9.1 ACQUISITION OF ROCK SAMPLE MATERIAL FROM BOREHOLES

The drilling process was initially a means of tapping underground water supplies or of proving and providing access to mineral deposits concealed in the subsurface. Until fairly recently it was common practice for geoscientific research organizations and geological surveys to carry out research drilling programmes which retrieved 100% of core sample material from the top of the borehole to the bottom. Nowadays, however, much of the deeper drilling around the world is carried out by the oil industry. Coring is only carried out at a few critical points in the drilling programme of an oil well because of the lengthy procedures involved and associated expense.

9.1.1 The Drilling Process

Modern drilling rigs use methods which rotate a cutting bit at the bottom of the hole. In this system a motor-driven rotary table on the derrick floor drives a 10-m-long, square device called a kelly bar to rotate via a square-holed bushing (Figure 9.1). As the rotary table turns, the kelly is forced to rotate, taking the drill string and bit with it. The kelly serves three purposes: (1) it is attached to the drill string and transmits torque from the rotary table; (2) it permits the drill string to be moved vertically as the string is lowered during penetration; and (3) it is also the unit through which the drilling mud is pumped down the drill pipe. In operation, the rotating part of the equipment begins at the swivel above the kelly bar and extends down the drill string to the point where the drill bit comes into contact with the bottom of the hole. The swivel assembly itself sustains the weight of the drill string, allows its rotation, and provides a rotating pressure seal, and a conduit for circulating the drilling mud to the top of the string.

9.1.2 The Mud System

The drilling process, of course, necessitates the destruction of rock material by the drill bit at depth, the return of the cuttings to the surface, and their eventual

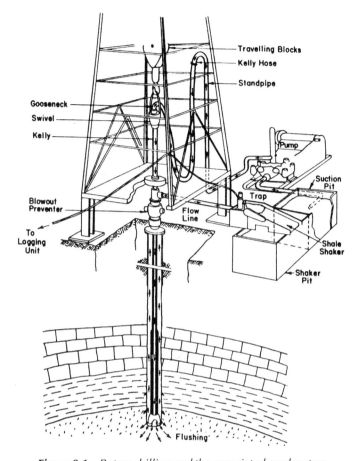

Figure 9.1 *Rotary drilling and the associated mud system*

disposal. To achieve this, the drilling process is assisted by the borehole's circulatory fluid or mud system. The drilling mud serves several purposes: it cools and lubricates the drill bit and the drill string; it removes the drilled solids, allowing release at the surface; it forms a gel to suspend the cuttings when the fluid column is static; and it controls the subsurface pressure.

9.1.3 Acquisition of Rock Samples

Even in the earliest days of drilling it was common practice for the driller to make a record of the strata or rocks passed through while drilling. Nowadays, the wellsite geologist is responsible for recording and evaluating geological and geophysical data while drilling ahead. His/her main duties include the description or logging of sample material produced from the borehole, maintaining an up-to-date lithological log, and recommending coring at significant intervals, such as formation boundaries, correlation horizons, porous intervals, or potential hydro-carbon horizons.

Cuttings samples are taken continuously during the drilling process at 3-m intervals or less, depending partly upon the speed of drilling and partly on the amount of detail required. The basic cuttings log is prepared and kept up to date using samples collected from the shale shaker and washed before description. Often, the only information that the geologist has about the newly drilled ground at depth is through observation of the penetration rate of the drill string and bit. A change in the rate of penetration can indicate a change in rock type, so that if there is a target horizon to be cored, circulation samples are obtained by temporarily terminating drilling, lifting the drill pipe a short way off the bottom of the hole but continuing the mud circulation so as to pump bottom hole cuttings to the surface for examination. Naturally there is a time interval between when the formation is drilled and when cuttings return to the surface. This is known as lag time and depends upon pump capacity, mud viscosity, hole diameter and the depth of hole. A knowledge of these variables allows an approximation of sample depth to be calculated. Normally there are a few cuttings or a small percentage of the newly penetrated lithologies present in the washed and prepared sample, allowing a judgement to be made on whether or not to start coring.

With both increasing depth and increased speed of penetration, the quality and usefulness of cuttings samples decreases. Sample mixing and contamination from various levels in the well can be great in soft formations and in caved portions of the borehole, and may provide difficulties in making accurate depth determinations. To help with the problem of bed or formation boundary definition it is found useful to provide a percentage lithology log of the cuttings to help at a later stage in the interpretation process with the precise determination and depth of formation boundaries by using lithological data in combination with wireline logs. By contrast, if drilling is relatively slow in hard rock types such as quartzite and if the borehole wall is consequently relatively stable with little caving, the cuttings are less contaminated, more representative of their true *in situ* formations, and their boundary and depth definitions are more reliable.

Geologists are keen to maximize the amount of sampling, and especially coring. Conventional coring employs a core barrel of varying length depending upon the precise requirements. The coring apparatus consists of an outer barrel, a 'floating' inner barrel, a spring box catcher which retains the core in the barrel when the assembly is retrieved from the hole, and a cutter head or diamond bit sometimes called a 'crown'. Mud circulates from the drill pipe between the two barrels to the bit, and factors such as weight on bit, rotary speed and circulation rate are determined by local conditions. It is important to note that it is necessary to pull the entire drill string out of the hole to recover the core using a conventional coring system. This, in turn, increases the amount of time required for the operation and therefore the expense. Nevertheless, the advantages of such a system include the obtaining of a large diameter (greater volume) core, usually with good recovery. During sporadic spot coring, and even during continuous coring operations, it is good practice to collect cuttings samples so as to provide some representative material in the event of incomplete or even unsuccessful recovery caused by poor formation conditions.

Worthy of brief mention is the practice of wireline coring, used nowadays in deep and ultradeep continental scientific drilling, as well as in drilling for minerals and hydrocarbons. The great advantage with this technique is that continuous coring can be achieved without pulling the drill pipe from the hole to recover each core. To achieve this the core barrel sits inside the drill pipe and can be retrieved, when full, through the hollow centre of the drill string cylinder. Other advantages include the fact that drilling ahead ('rock-bitting') without coring can be alternated with the coring process until the bit has worn out. The cost reduction achieved in this way can be substantial. Disadvantages of wireline coring include the space requirement of additional surface equipment, the fact that the cores are considerably smaller than conventional cores, and the possibility of a smaller percentage core recovery.

9.1.4 Sidewall Coring

Sidewall coring devices are used to take small core plugs from the side of a borehole. One advantage of sidewall cores is that they permit the sampling of the formation after cessation of the main drilling operation. Such samples are usually acquired as part of the geophysical wireline logging process, although rotating sidewall mechanical methods are also sometimes used. In the wireline process a cable gun carrying 36 very small core barrels or tubes, each attached by thin wires to a metal stem, is lowered down the hole on an electric logging line (wireline) to the required depth and then driven or shot into the formation by a detonated charge. The tiny core barrels are then retrieved by hoisting or pulling the assembly.

Advantages of sidewall coring techniques are that samples can be obtained from any desired depth in a previously drilled hole, and that the method can be an invaluable aid in assisting geophysical wireline log interpretation. The disadvantages are as follows:

1. the cores are of small size (*c.* 10 mm diameter and 15 mm length);
2. they have been subjected to considerable *in situ* flushing by the drilling mud filtrate; and
3. they are commonly severely compacted by the explosive impact of the small core barrel, which may affect observations on such properties as porosity.

9.1.5 Stratigraphical Practice using Subsurface Rock Samples

Described above are the sorts of methods used to acquire geological material and data directly from the subsurface. Each sample is described in as much detail as possible using standard descriptors for rock type, colour, sedimentological features, palaeontology, structural features, porosity and so on (Whittaker *et al.* 1991). Core data obviously provide the best geological information from the subsurface so that in continuously cored sections it is only the quantity of material available that makes stratigraphical practice different from that employed at surface exposures.

Nevertheless, although lithological and lithostratigraphical data from continuous cores can match those from surface exposures, it will be apparent that macrofossils are much less abundant in cores than in equivalent exposures. It follows that biostratigraphical data based on surface macrofossil studies are not necessarily easily related to core or cuttings sample material. Furthermore, many of the current biostratigraphical schemes are based originally on macrofossils distributed in a sequence at outcrop near basin margins. However, fossils from cuttings samples, which provide the bulk of subsurface material available to stratigraphers, tend to consist of microfossils, causing problems of comparison with existing surface-defined macrofossil biostratigraphical schemes. Consequently, if the more typical and abundant subsurface sample material obtained from cuttings (microfossils) is used it can adversely affect biostratigraphical correlation. Similarly, if cuttings samples are used alone, details of lithostratigraphy, sequence, sedimentology and the depth of beds (i.e. lithostratigraphical boundaries) cannot be defined as precisely as they can when working with good surface exposures.

Examination of cores and cuttings provides direct information on the physical properties of concealed subsurface formations. The rock data recorded in this way form the basis of, or serve to confirm, geophysical ('electric') wireline log interpretations which may also provide a great deal, if not all, of the missing lithostratigraphical detail unavailable to the subsurface geologist.

9.2 GEOPHYSICAL WIRELINE LOGGING

Geophysical downhole logging measures and records data in open or cased boreholes. A sonde or measuring instrument, up to 18 m long and 10 cm diameter, is lowered downhole via a cable (wireline) connected to surface recorders and computers (Figure 9.2). The geophysical well log itself is a continuous recording of a geophysical parameter down or along a borehole to produce a geophysical well log. The measurements recorded are logged or plotted continuously against depth in the well. A good quality geophysical log, produced under good subsurface conditions at the correct logging speed, can form a 'high fidelity' record of the stratigraphy traversed by the logging tool in the borehole (Whittaker *et al.* 1985; Rider 1986).

The methods of geophysical downhole logging were invented by the Schlumberger brothers in France during the early part of the twentieth century (Allaud & Martin 1977) by carrying out electrical resistivity measurements in exploratory boreholes. These techniques became known as *carrotage électrique* or 'electrical coring' because of the analogy with 'mechanical coring' which referred to the analysis of rock samples taken while drilling. Since 1933 the process has become known colloquially as 'electric logging'. In the first electrical logging operation undertaken, in 1927 at Pechelbronn in Alsace, the measurements were not carried out continuously but at 1-m stations downhole. The procedure initially was slow, with the loggers measuring about 50 stations per hour. However, after plotting the

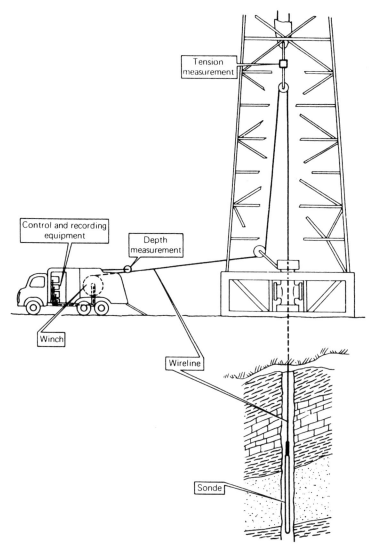

Figure 9.2 *Schematic diagram of geophysical wireline log measurement in a borehole. [Reproduced with permission from Editions Technip, from: Desbrandes (1985)]*

data it soon became clear that the variations in resistivity corresponded to important lithological changes. Not long after this it became possible to correlate peaks and troughs of electrical resistivity in boreholes over a wide area. Comparison of log data with cores from several boreholes showed that accurate lithological identification could be achieved, and electric logging gradually began to replace coring.

Although some of the geophysical log data are primarily of interest to petroleum geologists for the direct, downhole location and evaluation of hydrocarbon deposits, the purely geological or stratigraphical uses of downhole geophysical logs are of prime importance to geologists in the fields of stratigraphical correlation, lithological description, recognition of sedimentary cycles, subsurface sedimentology and in obtaining compaction and burial history.

Many different modern geophysical wireline logs are available, utilizing (1) mechanical measurements (e.g. caliper tool, recording hole diameter and geometry); (2) *in situ* 'spontaneous' phenomena (e.g. borehole temperature, spontaneous electrical currents, natural radioactivity); and (3) induced measurements (e.g. resistivity, induction, velocity, density, neutron). The logs are recorded immediately after cessation of drilling but before inserting casing in the newly drilled section of hole, and, of course, after the drilling tools have been removed from the hole.

9.2.1 Geophysical Data Acquisition

The data are acquired from the borehole using specialized equipment quite separate from that used in the drilling process. In onshore drilling, a specially adapted, heavy-duty vehicle (logging truck) is used, containing sophisticated electronic equipment for data recording plus the necessary computers, logging instruments, cable and associated logging drum. Offshore, similar equipment is located in a small logging cabin installed permanently on the platform or drilling rig. The logging instruments (tools or sondes as they are variously known) are lowered into the borehole at the end of the wireline. Most modern logs are recorded digitally with a sampling rate sometimes about once every 150 mm, but also as low as once every 30 mm. The huge quantities of data acquired in a well or borehole which may be several thousands of metres deep are stored in the computer, enabling a full printout as soon as the logging run has finished.

With the exception of temperature logging, logs are normally recorded while pulling the sonde from the bottom of the hole upwards. The cable, of course, acts as a data transmission medium as well as providing support for the sonde. The cable, which is guided manually during logging, is wound around the drum which pulls it at speeds of between 300 m per hour to 1800 m per hour depending on the type of sonde. It is necessary to be able to check logging depths mechanically, which is achieved by the wireline cable having magnetic markers placed at regular intervals. This is because it is sometimes found that drillers' depths and loggers' depths are discrepant, and wireline log apparent depths have to be corrected for tensional stretching.

Rig time is very costly and consequently methods have been devised to combine several different logging devices in one combination tool for simultaneous acquisition of different types of data. One other advantage to the user from this practice is the positive benefit to depth correlation between depths as measured by logging runs and depths measured by drilling. An example of combination open-hole tools is the Schlumberger DIL–Sonic–GR sonde which is over 18 m long and only 85 mm in diameter; it gives a simultaneous recording of the dual induction laterolog (DIL) (a conductivity measuring device), plus the sonic velocity log, plus the gamma ray log measuring the natural radioactivity of the formation. Such combinations are only possible by using modern computers to memorize and depth-match the different readings. Geophysical logging of a deep borehole can take many hours or days even using these combination instruments.

9.3 THE SUBSURFACE LOGGING ENVIRONMENT

Before discussing the applications of logs to stratigraphical work we need to consider the subsurface environment in which the recordings are made, and the limitations they impose on the logs. The sonde itself should be capable of (1) giving an accurate and repeatable measurement; and (2) making a measurement of a representative *'in situ'* sample of the formation. The words *'in situ'* are shown in inverted commas because the drilling process itself has dramatically altered the real *in situ* formation environment by the very process of drilling the hole, which during that process is full of mud. Accordingly, the formation has acquired massive, artificial 'porosity', at least locally, as a result of the drilling process, plus unnatural fluid content.

9.3.1 Subsurface Pressure Conditions

The in-hole (subsurface) pressure conditions are determined by a combination of what might be called (1) the formation pressure, i.e. the pressure under which the subsurface formation fluids (including gases) are confined; and (2) the pressure of the drilling mud column. The latter is hydrostatic and depends solely on the depth of the borehole, the height of the mud column and the density of the mud itself. Under normal circumstances, drillers and mud engineers have the difficult task of ensuring that the pressure of the drilling mud is fractionally higher than that of the subsurface formations drilled through. Accordingly, the mud is forced into porous and permeable formations.

9.3.2 Formation Pressures

In geological basins it is found that pore fluid pressures increase with depth. They increase from the hydrostatic pressure (best thought of as the 'normal' pressure) to an overpressured state. Overpressure itself is defined as any pressure above the hydrostatic for a particular depth; it means that formation fluids are being pressed by the containing rocks. Shallow formations tend to show a 'normal' (hydrostatic) pressure, although with increasing depth overpressure increases and the formation fluids bear more of the rock overburden pressure (Figure 9.3).

Overpressures can increase up to a maximum called the lithostatic (geostatic or overburden) gradient, which is taken to represent the probable maximum pressure likely to be found in a well at any depth. True lithostatic gradient is in fact very variable, depending upon the densities of formations and rock successions, which

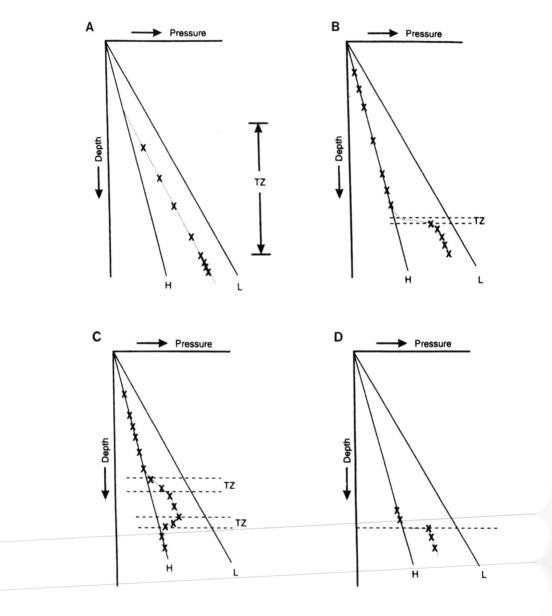

Figure 9.3   *Examples of pressure gradients (pressure versus depth) for wells in various basin settings. TZ, Transition zone; H, hydrostatic gradient; L, lithostatic gradient. [Modified from: Swarbrick & Osborne (1996)]*

do not necessarily increase uniformly, nor linearly, down a well but may decrease over a specific interval going down a succession, e.g. in salt beds sandwiched between cemented ('tight') sandstones. Most geophysical logging instruments will withstand pressures up to a maximum of 1000–1800 kg cm^{-2}, that is pressures much greater than those normally found in logging sedimentary basins. The trend towards even deeper drilling of the continental crust has led to the development of the so-called hostile environment logs, that is logs to cope with the higher pressures and temperatures expected at these greater depths.

9.3.3 Invasion

An important pressure-related aspect of deep drilling is that under normal circumstances the pressure of the mud column will be greater than the formation pressure, so that as the drill bit enters into a porous formation, the drilling mud will be forced into it. This process is known as invasion (Figure 9.4). The porous formation acts as a filter separating the mud into fluid and solid parts. The liquid used to mix the mud (known as the mud filtrate) flows into the formation while the solids are deposited around the borehole wall. This mud cake can build up to form an impermeable barrier. Importantly, during this process the original formation fluids can be swept laterally away from the hole thereby affecting the acquisition of wireline log data. Since the logging takes place in open hole immediately following the drilling, the process of invasion clearly has a profound impact on interpretation procedures. The amount of penetration into the formation by the filtrate is described by its 'depth' or 'diameter' of invasion. Interestingly, in very porous and permeable rocks the mud cake forms quickly and invasion does not penetrate the formation very deeply, whereas in poorly porous and impermeable zones, and in fractured formations, mud cake deposition is slow and invasion can be very deep.

9.3.4 Subsurface Temperatures

The regular increase of temperature with depth in the earth's crust is a well-documented phenomenon although the increase is not linear. Temperature inversions occur because of variations in the thermal conductivity of rocks. The increase of temperature with depth is known as the geothermal gradient and in sedimentary basins ranges between 20 °C and 35 °C per kilometre (Figure 9.5). In the UK average bottom hole temperatures (BHT) are about 26 °C per km, though UK basement rocks have an average of less than 20 °C per km. It is worth noting that *in situ* subsurface temperatures are disturbed by the drilling process. Cold mud entering a warm or hot formation lowers the temperature, and it takes at least 5–10 days for equilibrium to be approached, after the cessation of drilling operations. Borehole logging temperatures, by contrast, are usually recorded only a few hours after mud circulation has ceased.

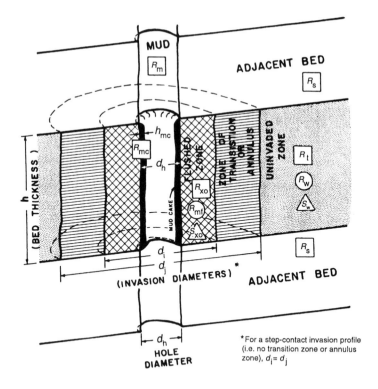

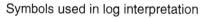

Symbols used in log interpretation

☐   Resistivity of the zone
○   Resistivity of the
    water in the zone
△   Water saturation
    in the zone

Figure 9.4   *Diagram to show the principle of invasion, invasion diameters and the associated mud cake, flushed zone, transition zone and the uninvaded formation. The diagram
also shows the standard symbols used in petrophysical log interpretation. [Reproduced with
permission from: Schlumberger (1977)]*

9.4 DEPTH OF INVESTIGATION AND BED BOUNDARIES

Practical limitations include lateral depth of investigation, absolute vertical resolution (thin bed resolution), and possible problems of bed boundary definition.

As noted above, the logging tools themselves are not operating in an undisturbed
formation so it is not surprising that the acquired data need careful consideration
and a full understanding of the techniques used and principles involved. In the
drilling of porous and permeable beds the invaded zone rather than the unaltered

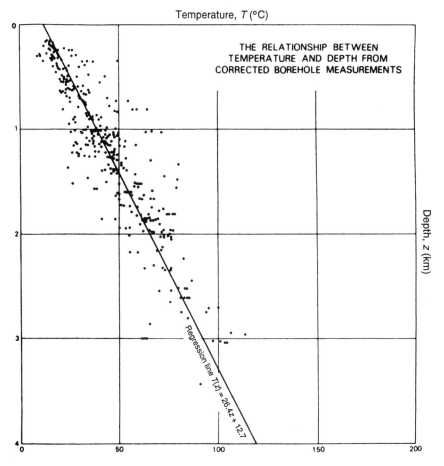

Figure 9.5 *The geothermal gradient for the UK. [Reproduced with permission from the British Geological Survey, from: Downing & Gray (1986)]*

formation and its natural contained fluids is nearest to the logging instrument. Logging tools with a deep, lateral 'penetration' into the formation beyond the shallow invaded zone have been developed for certain circumstances. This depth of lateral 'penetration' into the formation is called the depth of investigation. The logging instruments have been designed for shallow or deep investigation (laterally into the formation) to examine what might be called 'near-hole' and 'further-field' properties of the formation in the vicinity of the borehole.

Most geophysical logs have a very local (i.e. near-borehole-wall) area or 'depth' of investigation. The area of investigation by the sonde itself reaches from the borehole wall immediately adjacent to the sonde (shallow investigation) up to a distance of about 5 m from the borehole wall (deep investigation). It is generally true that the depth of investigation of the logging instrument depends on the separation of the transmitter and receiver devices within the sonde actually producing the bombarding signal. In the case of the resistivity devices, for example, the micro-inverse resistivity sonde (Microlog) has electrodes only 2.54 cm apart, and investigates only the resistivity of the mud cake on the borehole wall where that is present; in contrast, the induction conductivity sonde (Induction log) with transmitter and receiver 1 m apart investigates about 5 m into the formation.

9.4.1 Minimum Bed Resolution

In general, the logging tool is only capable of making a true value ('absolute') measurement of a bed's physical property if the bed is thicker than the transmitter–receiver distance of the tool. In the case of the Microlog, the resistivity of a bed down to 2.54 cm thickness can be resolved, while the Induction log can resolve true tool resistivities only in beds thicker than 1.2 m.

Of great importance to stratigraphical studies, as distinct from the requirement to measure absolute values of a particular geophysical property, is the fact that beds much thinner than the sonde's transmitter–receiver distance are identifiable, with the proviso that the value indicated by the log will only be a percentage of the true reading. In practical terms, and on an empirical basis, the thinnest beds identifiable are about 10 cm or so thick, although comparison of properly recorded wireline log data with good core sample material suggests that thinner beds than this can be recognized under good conditions.

9.4.2 Bed Boundaries

In lithostratigraphy, a bed is the smallest formal unit recognized. Like the basic lithostratigraphical unit, the formation, it has lithological characteristics that distinguish it from adjacent beds. At outcrop, bed boundaries are often readily recognized and are commonly abrupt, or sharp. Geophysical wireline logs cannot show this abruptness because the logging tool is moving during the recording of data. The gamma ray log, for example, is a record of the formation's radioactivity which is effectively counted by the sonde. An averaging effect of the counting caused by the tool's movement up the hole during logging occurs and causes distortion at bed boundaries. Over bed boundaries, half of the gamma ray count will be from one bed and half from the adjacent bed, so that the average value obtained has no real formation equivalent. This is caused by the gamma ray tool travelling too quickly up the hole. In this case the logging speed is critical and the interpreter must be aware of its possible effects.

During examination of the data, and when interpreting the detailed borehole stratigraphy, the logs present a series of peaks and troughs. It is usual practice to take the bed boundaries at the point of maximum change in a given value (the point of maximum slope). Naturally, other data such as cuttings description or sidewall core information will be incorporated into the final interpretation.

9.5 TYPES OF OPEN-HOLE GEOPHYSICAL LOGS IN COMMON USE

The following types of geophysical logs are useful to the geologist in making stratigraphical interpretations: caliper logs, resistivity logs, spontaneous potential logs, gamma ray and spectral logs, neutron logs, density logs, and finally sonic or acoustic logs. These are discussed below.

9.5.1 Caliper Logs

Caliper logs measure borehole diameter, size and shape. The more simple mechanical tools have two arms and measure the diameter in one direction only. Four-arm calipers, or dual caliper tools such as the borehole geometry tool, measure hole diameter in two perpendicular directions and give more information about hole rugosity, size and volume. Caliper logs provide information on rock properties.

Where the hole is on-gauge (i.e. where it has the same diameter as the drill bit), the formation is usually hard or competent, such as in massive limestones or older, dense, non-porous rocks. By contrast, areas of over-gauge hole with a larger diameter than the bit size are said to be 'caved' or washed out and might indicate the presence of unconsolidated sediment, fractured or soluble rocks. Areas of under-gauge hole, where the hole diameter is smaller than the bit size, are often known as 'tight spots' to the drilling crew, and may be caused by smectite (swelling clay) in clay and mudstone formations. Perhaps more commonly, however, the under-gauge hole is caused by the build up of mud cake opposite permeable strata. This can indicate a porous and permeable sandstone body and therefore a potential reservoir horizon in the sequence.

9.5.2 Resistivity Logs

In electrical resistivity logging, electric currents are passed through the formation under investigation. Formation resistivity, which is the reciprocal of conductivity, can be defined as the degree to which a substance resists the flow of electric current. Most rock and rock matrix constituents are essentially insulators, while their contained fluids are conductors. The exceptions to this fluid conductivity are hydrocarbons, which are infinitely resistive. If a formation is porous and contains brine, then the resistivity will be low; if the same formation contains hydrocarbons, then its resistivity will be very high. Current flow in a formation is through the water contained in the pore spaces, and formation resistivity therefore provides information on pore fluid content, a property of vital importance to the oil industry.

Conventional resistivity devices give a so-called normal curve which is produced from the recorded voltage drop between two sampling electrodes. A short normal electrode with a spacing of between 50 and 60 cm is useful in correlation, in the definition of bed boundaries, and in measuring resistivity near the borehole wall. The lateral device has three effective electrodes giving a zone of investigation approximately equal to the electrode spacing. Lateral curves are asymmetrical and distorted by thin beds and adjacent beds. Conventional devices like these may be affected by the borehole itself as well as the adjacent formations. Other resistivity instruments using focusing electrode tools can provide better results for thin bed resolution and can investigate the formation to near, medium or far distances laterally.

Induction logs measure the formation conductivity and will operate in non-conductive muds or in empty holes. Relatively thin (*c.* 1 m) beds can be defined and good determinations of true resistivity are possible. Microresistivity tools can allow determination of permeable zones by detecting the presence of mud cake on the borehole wall; they can also measure the resistivity of the so-called flushed zone immediately adjacent to the borehole wall.

Resistivity logs may also provide information on gross lithology, bedding, stratigraphical correlation, rock texture, facies, overpressure and source rock characteristics. The abundance of shale in sedimentary sequences permits the use of resistivity logs as an indicator of gross lithology, and although the logs do not allow direct lithological identification as such, they can apparently be very sensitive indicators of grain size, chemical content, density and water content. Resistivity logs are frequently also used in correlation because they are the only logs available, although they would not normally be regarded as the best logs for such purposes as the data are influenced by changes in formation pressure and water salinity, neither of which is necessarily related to the stratigraphy.

Resistivity variations in oil zones can be related to grain size changes because a coarse-grained sand will generally have a low (irreducible) water saturation and hence higher resistivity, whereas a fine-grained sand with higher water saturation will show lower resistivity. Fining-upwards sandstones filled with hydrocarbons should show a regular upwards decrease in resistivity. Resistivity logs can also reveal cyclic trends caused by varying sand and mudstone content in a sequence.

Compaction of mudstone with depth can be recorded on electric logs. As compaction increases with depth so does resistivity, although it is the conductivity logs which commonly display this phenomenon best. Both conductivity logs and sonic logs behave in this way, apparently as a result of a relationship between conductivity (or acoustic travel time) and mudstone or shale porosity. In some basin sections a reversal of shale conductivity with depth is observed and this is attributed to overpressure. The abrupt change is thought to be related to an increase in shale porosity which happens as an overpressured zone is entered. It is important to establish that these types of changes are not caused by lithological variations.

If the mudstone or shale is rich in organic matter, shale resistivity can be increased. This is because organic matter is usually highly resistive, which may be increased still further on maturation and production of free oil in the shale voids. However, the presence of a source rock cannot be deduced from resistivity data alone as many factors can give rise to high resistivity values.

Of particular importance to the geologist is the dipmeter log. This comprises a number of orientated microresistivity logs recorded via four-arm dipmeter tools generally, although both three- and six-arm dipmeters are available. The curves obtained from each arm are correlated by computer, and stratal dips are calculated. The correlation interval within which the computer will attempt log correlation can be specified, as can the step distance and search angle over which the computer examines for each correlation. For detailed sedimentological work the correlation interval and step distance should be small ideally, with a search angle of 35° or 45°. The most usual presentation of the computer data is the so-called tadpole plot.

9.5.3 Spontaneous Potential Logs

Spontaneous potential, self potential, or SP logs measure the electrical currents that occur naturally in boreholes as a result of salinity differences between formation water and the borehole fluid (or mud filtrate). These natural currents or 'spontaneous potentials' originate from the electrical disequilibrium caused by connecting (normally unconnected) formations vertically as a result of drilling a hole in the ground. They are useful for detecting permeable beds in certain situations, for locating bed boundaries, for correlation purposes and for giving an indication of shaliness in a sequence.

All deflections on the SP log indicate the presence of a permeable bed but not the amount of permeability. In the early days of geophysical logging the SP log was used routinely in correlation but has now been superseded as a correlatory tool by other log data types. Nevertheless, it is still useful for correlation in areas of varied water salinity. It was one of the first tools to allow recognition of shale–sand distinction in clastic sequences. There are no absolute values in SP logging, so the amount of deflection of the curve to left or right which occur in thick shale intervals is defined as zero. It is called the shale base line and all SP values are related to this line. In siliciclastic sand–shale sequences the SP log curve is related to shale amount, with a full SP deflection in clean sand zones and a reduced deflection in shaly zones. This makes the log useful as a grain size and facies indicator, despite the fact that the SP log has been largely superseded as a facies indicator by the gamma ray log.

9.5.4 Gamma Ray and Spectral Gamma Logs

The gamma ray log is a record of the formation's natural radioactivity. The gamma rays come from the naturally occurring elements uranium, thorium and potassium. The single gamma ray log alone gives the radioactivity of the three elements combined together, while a spectral gamma ray log shows the different amounts of each of the above-mentioned radioactive elements present in the section being logged.

Most rocks are radioactive to a greater or lesser extent and the gamma ray log is therefore useful in finding and assessing radioactive mineral deposits. However, shales have a high gamma radiation output relative to other sedimentary rocks. On this basis the gamma ray log is a useful lithological indicator and is normally part of the basic logging suite.

Additionally, like the SP log, the gamma ray log can be used for facies studies since, although it is shale content that the gamma ray log indicates, it can be interpreted in terms of grain size. A coarse-grained sand will contain very little mud, a medium-grained sand will contain some mud, while a fine-grained sand might contain a fair amount of mud – changes which can be picked out in the gamma log. The gamma ray log is perhaps the best-known correlation tool. It gives some indication of lithology, has plenty of 'character', and can be resolved down to thicknesses of 10 cm or less. Gamma log values change in the vertical sense but are consistent laterally and are therefore of great use in correlation.

In the case of unconformities, high gamma ray values in places occur as long 'spikes' or peaks which are often associated with uranium concentrations. These concentrations can indicate unusual conditions of deposition. Empirically they are seen to be associated with gaps in the sequence which represent long time intervals and therefore unconformities.

The spectral gamma ray log can be used in the identification of detrital minerals such as feldspar and mica on the assumption that only potassium radioactivity is present in these minerals. Subtraction of the total shale volume obtained from the thorium log will give the volume of potassium radioactive minerals present. Combined with other logs, the potassium radioactivity values will indicate either feldspars or micas. Depositional environment may be interpreted by considering thorium, potassium and uranium as indicators. The association of uranium with marine shales and of thorium with terrestrial deposits has led to the suggestion that study of such concentrations may give an indication of the amount of marine influence on the depositional environment.

In the past, gamma ray logs were measured in counts per second, but are nowadays calibrated in API (American Petroleum Institute) units. The radioactivities observed range from a few API units in salt (halite) or anhydrite to 200 or more in shales. A high gamma ray reading in a sedimentary section usually indicates shales although the interpreter should be aware that potassium, as well as occurring in the clay silicate mineral structure of mudstones and shales, also occurs chemically in evaporites and in some rock-forming minerals such as feldspars.

Worthy of brief mention is the recently developed, so-called geochemical logging tool (GLT) containing gamma ray spectroscopy sensors which analyse the concentrations of 10 elements in the rock. In addition to the measurements of natural radioactivity for thorium, uranium and potassium concentrations, there is a delayed activation measurement for determining aluminium concentrations, plus a system for the measurement of prompt gamma rays following thermal neutron capture which gives information about silicon, calcium, iron, sulphur, titanium and gadolinium content. Recent research suggests that log-derived geochemical properties from elemental concentrations in sedimentary environments can provide quantitative analyses that could only previously be achieved via core analysis.

9.5.5 Neutron Logs

Neutron logs are used for the definition of porous formations and for a quantitative estimate of porosity. Neutrons are emitted from a radioactive source located in the sonde; in the formation the neutrons collide with materials so that some energy is lost. The amount of energy lost depends on the mass of the nucleus with which the neutron collides so that the greatest loss occurs when the neutron strikes a hydrogen nucleus, because neutrons have almost identical mass to that of a hydrogen atom. In effect therefore, the neutron log responds to the amount of hydrogen present in the formation (the so-called *hydrogen index*). In clean formations it reflects the amount of liquid-filled porosity and is principally a measure of the formation's water content, whether that water is free pore-water, bound water, or water of crystallization. The quantity of hydrogen per unit volume is converted directly to neutron porosity units, although different lithologies need different conversion factors. The log is normally calibrated to limestone so that it is sometimes known colloquially as the 'limestone curve'.

Qualitatively, the log is particularly useful in discriminating between gas and oil, since gas has a relatively low hydrogen content. Used together with a gamma ray log it helps to define shales, tight formations and porous sections. In combination with density and/or sonic logs it is useful in determining lithology. The linear limestone porosity units are calibrated via the API (American Petroleum Institute) Neutron pit in which the response of a logging tool in 19% porosity, water-filled limestone is defined as 1000 API units.

Neutron logs, despite being primarily a porosity tool, are also able to distinguish shales (high porosity) from limestones or sandstones (moderate to low porosity) especially in situations where sonic and density logs are not available. In particular, the neutron log is useful in the recognition of complex multi-mineral rocks.

9.5.6 Density Logs

The formation density log is used mostly as a porosity logging device. It is a continuous record of the formation's bulk density, which includes its solid matrix and the fluid enclosed in its pore spaces. In geological terms, a sandstone with no porosity at all will have a bulk density of 2.65 g cm^{-3}, the density of pure quartz.

The formation is bombarded with gamma rays from a radioactive source in the sonde applied to the borehole wall. These gamma rays collide with electrons in the formation, and energy is lost from them (Compton scattering). These scattered gamma rays are eventually counted by a detector to give a measure of the formation density. Information is obtained on:

1. porosity;
2. the identification of certain minerals (especially in evaporite sequences);
3. the location of coal seams;
4. the detection of gas;
5. source rock evaluation; and
6. the identification of lithology in combination with other tools.

Furthermore, the density log is valuable in calculating seismic reflection coefficients for the production of reliable synthetic seismograms, and can augment the interpretation of gravity data acquired from surface measurements.

As well as its use as a porosity indicator and in the determination of the pore fluids, the density log is also a powerful lithological tool because of the wide range of rock densities (g cm^{-3}) encountered; for example, coal 1.2–1.8, halite 2.05, sandstone up to 2.65, limestone up to 2.71, dolomite up to 2.87, anhydrite 2.98.

9.5.7 Sonic or Acoustic Logs

Sonic or acoustic logs measure, in effect, the speed of sound in the formation. The *interval transit time* shown on the log and known as Δt (delta-*t*) is actually the reciprocal of the sonic velocity in the formation and is the time required for a compressional sound wave to traverse 1 foot of the formation (it is conventional to use imperial units in sonic log work). The interval transit time for a formation depends upon the formation's lithology and porosity, so that when the lithology is known, the log becomes very useful for determining porosity. The sonic tool is borehole compensated and thus reduces unwanted effects at places in the borehole where rugosity, or departure from cylindrical shape, might be present.

Sonic devices use two transmitters located in the sonde, one above and one below two pairs of receivers (Figure 9.6). When one transmitter is triggered, a sound wave enters the formation and the elapse time between detection of the first arrival at two receivers is measured. The speed of sound in the sonde and in the drilling mud is less than that in the formation and therefore the first arrivals at the receivers correspond to sound travel paths in the formation very near to the borehole wall. Delta-*t* (Δt) is usually measured in microseconds per foot and because it is the reciprocal of the velocity of the compressional sound wave, it is useful for determining formation velocities. Integrated transit times are shown on the sonic log as a series of ticks and this makes it an invaluable tool for interpretation of seismic reflection data.

In addition to these uses and the determination of porosity, the sonic log is useful for: (1) stratigraphical correlation; (2) identification of lithology; (3) evaluation of shaly sands; and (4) compaction and burial history studies.

Sonic log values in microseconds per foot for common rock types range from sandstone 51–56, limestone 47.5, dolomite 43.5, anhydrite 50 to halite 67. Shales and coals give variable responses generally higher than the above, although it must be remembered that sandstones and carbonates give increased values when they have significant pore space. Nevertheless, a combination of the gamma ray log and the sonic log provides a very sensitive indicator.

9.6 LITHOLOGICAL DETERMINATION

Logs useful in lithological determination are the gamma ray, sonic, formation density and SP. It is also apparent from the above discussion that if rock forma-

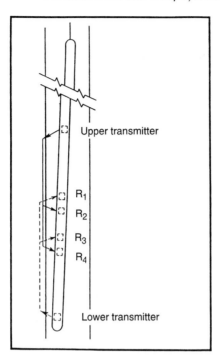

Figure 9.6 *Schematic diagram of the Schlumberger BHC Sonic sonde (BHC = borehole compensated) showing the positions of upper and lower transmitters and the corresponding receivers (R_2 and R_4 for the upper transmitter; R_1 and R_3 for the lower transmitter). [Reproduced with permission from: Schlumberger (1972)]*

tions were comprised solely of monomineralic layers then it would be relatively simple to compare values obtained from the wireline logs with the same properties determined experimentally, and so to deduce lithology. However, because of the overlap of many log parameters for different, commonly occurring rock types, it is good practice, indeed necessary in many situations, to use log data in combination rather than to consider each log separately.

By plotting interval transit time from the sonic log against bulk density, or the latter against neutron 'porosity', it is possible to locate the position of different minerals uniquely. By varying the porosity of each and assuming a standard fluid filling and saturation, it is possible to translate the mineral points into unique curves which describe both mineralogy and porosity. Lithology expressed as mineral combinations may be estimated by interpolation between the curves, and water-filled porosity by interpolation along their length. However, such two-dimensional cross-plots do not necessarily produce unique solutions by themselves. The more complex the lithology (and the more complex any contained fluids), the more log data are needed to evaluate the formation; it may in certain situations be necessary to overlay various standardized log traces, for example by making the zero neutron porosity value correspond with the bulk density value, in order to evaluate the formation and maximize the interpretation.

There is a limit to the number of logs that can be handled conveniently by visual examination and cross-plotting techniques. However, it is possible to consider three or more logs (multi-log analysis) simultaneously by using computer methods and computer-derived interpretations. Many of these methods use statistical techniques to cope with the huge amounts of data available. Bayesian methods, for example, are used to characterize and sort depth points into populations that represent type lithologies. In many of the methods used, some preparation of the logs precedes the ultimate statistical analysis, and care must be exercised to ensure that the input data and results are geologically meaningful. At the end of the day, it is the responsibility of the geologist to ensure that the ensuing interpretations are geologically sound.

In the case of subsurface lithostratigraphy, it is important to remember that drilling boreholes and wells results in two independent but related lithological data sources. These are: (1) the direct lithological data from the drilling process itself, that is cuttings, cores, mud logs and drilling parameters data; and (2) data coming from the wireline geophysical logging process comprising mainly geophysical log data, but also including sidewall lithology sampling acquired via the wireline. For the most reliable lithological interpretation, it is essential that all the data are integrated so as to maximize the interpretation.

9.7 SEDIMENTOLOGY, ENVIRONMENTS, FACIES AND SEQUENCES

Integrated lithological information, geophysical log data and analysis of micropalaeontological biota from cuttings can provide useful sedimentological, environmental and facies information. This is especially the case when the strata in question are known from cores obtained in adjacent boreholes, or from surface outcrops. The petroleum industry utilizes such data routinely in the exploration, development and production of oil fields, and particularly in the study of basin development and evolution.

Petrographical description of rock cuttings from the well reveal much about the sedimentology to the geologist, but literally only on the microscopic scale. Cores, of course, will reveal some of the meso- and perhaps macroscopic features of at least some of the rock sequence drilled through, but regrettably these are not taken regularly or routinely. Much important sedimentological, facies and sequence data can be deduced from the routine examination of geophysical log suites in the determination of bed thickness, nature of lithological boundaries and in the recognition of rhythmic sequences such as cyclothems. A tool of particular importance to the interpretation of the structural disposition, scale, nature and orientation of strata and cross-bedding is the dipmeter. The combination of all these data sets can enable and facilitate detailed palaeoenvironmental reconstruction, as well as the recognition of sequence boundaries.

Log shapes in sand bodies have been studied in depth, with the conclusion that except in very poorly sorted rocks the SP and the gamma ray log in clastic sequences may correspond closely to grain size profiles. Various log shapes have been recognized, including the bell shape (a narrowing- or decreasing-upwards profile), the cylinder (a block, uniform or parallel-sided profile), and the funnel (a widening- or increasing-upwards profile) (Figure 9.7). Also recognized is the curve character, which can be smooth, complex or serrated, and the sand body upper and lower boundary contacts, which can be abrupt or gradational. In such cases bed thickness is readily apparent, as is the nature of the bed contact and fining- or coarsening-upwards characteristics (Figure 9.8). The parameter measured by the particular tool must always be borne in mind as the gamma ray log, if used on its own, can be quite misleading. For example, a channel sandstone with an abrupt base might normally be expected to have a lower natural level of radioactivity than an underlying mudstone. However, sandstones quite commonly contain radioactive potash feldspars, heavy minerals or basal mudstone intraclasts, making it less readily distinguishable from the mudstone below, so that the gamma ray log may appear gradational rather than abrupt.

9.7.1 Dipmeter Interpretation

Dipmeter data nowadays routinely provide both structural and sedimentological data useful in the interpretation of facies and sequence. The interpretation stems from the recognition of four, basic, easily recognized, tadpole plot motifs (Figure 9.9). By convention, the various motifs are marked using a standard colour code as

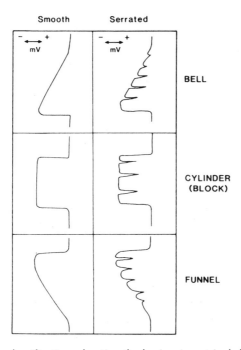

Figure 9.7 *Log shape classification, showing the basic geometrical shapes used to describe SP and gamma ray logs. [Reproduced with permission from Blackie & Sons, from: Rider (1986)]*

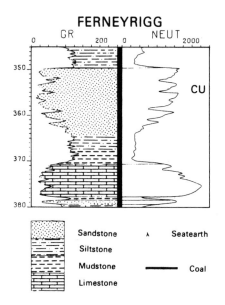
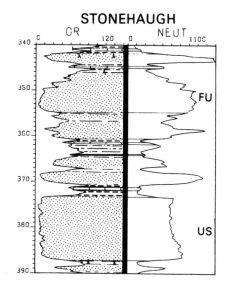

Figure 9.8   *Examples of log shapes from Carboniferous sequences in boreholes from northern England. A coarsening-upwards (funnel-shaped) unit is seen in the Ferneyrigg borehole (confirmed from core data), while fining-upwards (bell-shaped) and uniform (cylinder-shaped) sandstone with a sharp base and top are seen in the Stonehaugh borehole (confirmed from core data). [Reproduced with permission from the Geological Society, from: Whittaker* et al. *(1985)]*

follows. Green patterns are successions of dips of similar orientations and magnitudes, and are mostly indicative of structural dips; they are common in argillaceous rocks or other thinly bedded or laminated sediments laid down in low-energy environments. Red (upward-decreasing dips) and blue (upward-increasing dips) patterns are commonly indicative of both faults and cross-bedding. In fact, where a sedimentary origin can be demonstrated, the size of cosets and the nature and orientation can be used in environmental interpretation or for prediction of the subsurface extent of reservoir bodies. In mixed sequences it is possible by using a computer to determine the structural dip from green patterns and then to tilt them back to their original orientations to give the direction of sediment transport.

The recognition of faults can also have important sedimentological implications. Obviously it is necessary to be able to judge where strata may be missing as in the case of normal faulting, or repeated in the case of reverse faulting, so as to be aware of the effects of structure and tectonics on sedimentological models being proposed. It is also possible to infer from dipmeter data that the faults themselves may display evidence of syn-sedimentary activity (growth faults).

Clusters of randomly orientated dipmeter tadpoles are known as 'bag o' nails' and can result from a variety of causes. In many cases such patterns result from poor data or from the inability of the computer to make good correlation between the microresistivity traces. However, the random pattern may indicate structurally disturbed zones related to faulting or relatively intense folding or flexuring. In

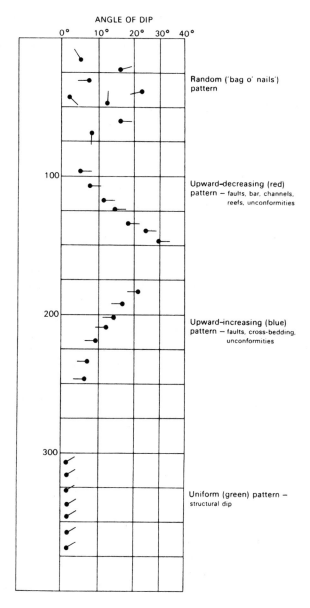

Figure 9.9 *The four main dipmeter tadpole plot patterns. [Reproduced with permission from the Geological Society, from: Whittaker et al. (1985)]*

some instances the random pattern may also result from sedimentological causes, for example in thick massive bodies with little internal orientation such as reefs, or in contemporaneously slumped sediments where bedding has been locally intensely deformed.

9.8 STRATIGRAPHICAL CORRELATION AND SEQUENCE ANALYSIS

One of the prime geological applications of geophysical wireline logs is in stratigraphical correlation and in sequence analysis. In connection with this it is important to realize that although subsurface stratigraphical description and nomenclature use lithostratigraphy, biostratigraphy and chronostratigraphy, the scarcity of sample material from borehole sources and even factors such as the problems of precise sample depth location, commonly mean that the geophysical log data assume great importance.

The geological succession derived from description of available sample material and from associated log analysis routines can be compared with established standard sequences and rock types and a correlation made. Log shapes depict various patterns, which can be thought of in terms of curvature and arrangement, and degrees of serration or smoothness, that is whether the log patterns display rapidly alternating peaks and troughs of substantial amplitude. Such observations show what are best thought of as characteristic signatures which apply at various scales.

An example of regional correlation and large-scale application is shown in Figure 9.10, which is from the Mercia Mudstone Group of the Irish Sea Basin. The characteristic log signatures of many of the well-established lithostratigraphic units are readily recognizable from geophysical log data alone, although all the correlations shown are actually confirmed by sample material and some of them by micropalaeontological data. From such wireline geophysical data it is possible to recognize lithostratigraphical divisions such as groups, formations, members and even individual beds originally described at outcrop.

An example of recognition of lithostratigraphical units on wireline log data at a lower level in the hierarchy than the group level is provided by subdivisions in boreholes drilled through the lower part of the Lias Group, the lowermost division of the Jurassic System. The lithostratigraphical units in Table 9.1 have been recognized in the field for many years and are traceable over large tracts of country. The equivalent units as recognized on the wireline log data are shown in Figure 9.11 as units LL1 (Blue Lias) to LL5 (Green Ammonite Beds). The log units are constrained by the biostratigraphy, especially the ammonite biostratigraphy, of the fully-cored British Geological Survey Burton Row borehole. Lower Lias unit LL1 is seen to be characterized by an extremely serrated (spiky) log signature reflecting the rapid limestone–mudstone alternations typical of the Blue Lias. Also worthy of note, however, are longer wavelength components of the signal which can be seen by producing an 'average' line through the spikes. These sinusoidal shapes indicate the presence of further lithostratigraphical subdivisions in many of the larger

269

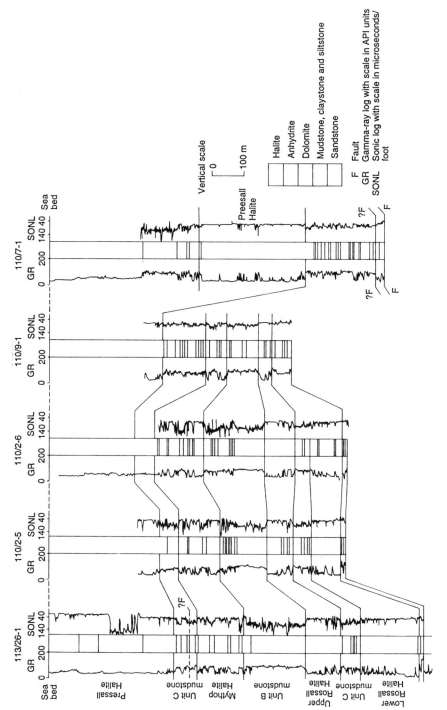

Figure 9.10 *Large-scale regional lithostratigraphical and log correlation of the Mercia Mudstone Group (Triassic) in selected boreholes in the East Irish Sea Basin. [Modified from Jackson et al. (1995)]*

Table 9.1 *Lithostratigraphical units within the lower part of the Lias Group of the Jurassic System (see Figure 9.11)*

Lithostratigraphical unit in outcrop	Wireline unit
Green Ammonite Beds	LL5
Belemnite Marl	LL4
Black Ven Marl	LL3
Shales with beef	LL2
Blue Lias	LL1

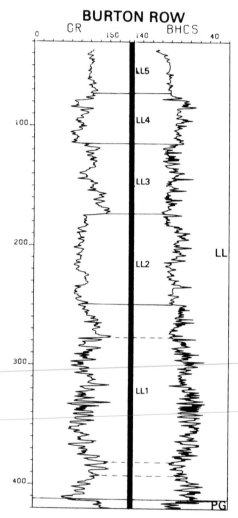

Figure 9.11 *Geophysical wireline log data used to delineate Lower Jurassic lithostratigraphical divisions at formation level in the Burton Row borehole, Somerset, UK. PG = Penarth Group (Triassic), LL1 to LL5 represent Jurassic lithostratigraphical divisions in upward sequence as follows: Blue Lias, Shales with Beef, Black Ven Marl, Belemnite Marl, Green Ammonite Beds. [Reproduced with permission from the Geological Society, from: Whittaker et al. (1985)]*

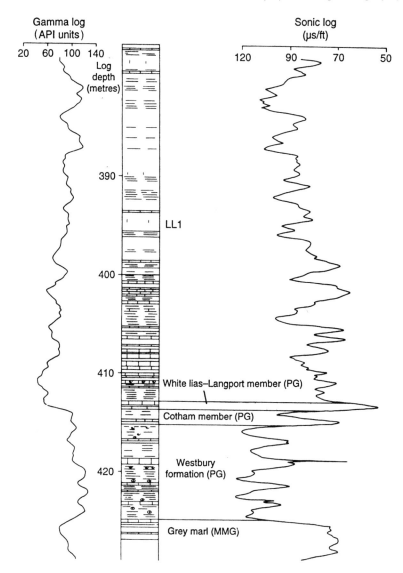

Figure 9.12 *Detailed correlation and thin bed recognition of uppermost Triassic and lowermost Jurassic strata in the Burton Row borehole, Brent Knoll, Somerset, UK, using wireline log data and the detailed lithological description of the fully-cored section. The borehole core strata log can be correlated on a bed-by-bed basis with identical lithostratigraphy in the nearby north Somerset coastal exposures. MMG = Mercia Mudstone Group, Triassic; PG = Penarth Group, Triassic; LL1 = lowest lithostratigraphical division of Lower Jurassic = Blue Lias. [Reproduced with permission from the Geological Society, from: Whittaker et al. (1985)]*

divisions already recognized. One example of this is the occurrence of a predominantly mudstone lithological division near the base of the Blue Lias (LL1 unit), which has been described in detail in the field over a wide area as the Lavernock Shales in South Wales and the Saltford Shales of the Bristol area; they are equivalent to the 'Angulata' Clays of the north of England and occur in the *Alsatites liasicus* Biozone of the Hettangian Stage (Jurassic System). The short wavelength part of the signal in successions like these (alternating limestone–mudstone) can also be used to demonstrate the cyclicity of sequences at relatively high frequencies.

Thin bed recognition and high-resolution stratigraphy can be achieved using good-quality log data because the lowest vertical bed resolution is about 10 cm or even less. Figure 9.12 is also part of the fully-cored section penetrated by the Burton Row borehole and demonstrates the detailed lithological record of strata near the base of the Jurassic. It is seen that the lithological log produced from description of the core correlates precisely with the gamma ray and sonic logs run over the same interval. The bed-by-bed correlation can be achieved from gamma ray log data to core data, and from core data to outcrop sections on the Somerset coast.

9.9 SUMMARY

This chapter has discussed how stratigraphical studies are carried out using borehole sample material and downhole geophysical wireline log data, ideally in combination but separately in a 'stand alone' way if necessary. In order to appreciate some of the strengths and weaknesses of the methods, data acquisition processes have been outlined together with the *in situ* physical conditions present in the deep subsurface environment where the drilling and logging actually take place.

It has been shown how borehole sample and geophysical wireline log data are used in combination to achieve lithological identification and the establishment of a lithostratigraphical description of concealed, sedimentary rocks. These interpretations are augmented by biostratigraphical information and seismic reflection data where available, to present as full a picture as possible of crustal structure and basin evolution.

The geophysical wireline log data themselves, even when compared with visually described core material from the same borehole interval, have properties and reveal features that cannot be detected by eye. For these reasons the log data greatly augment and enrich geological studies in general and stratigraphical research in detail. In the subsurface environment in particular, the log data enable more objective definition of stratigraphy and the elucidation of a more detailed geological history; they allow repeatable objective description, and are amenable to mathematical treatment, signal analysis and automatic correlation.

ACKNOWLEDGEMENT

This chapter is published with the permission of the Director, British Geological Survey, NERC.

REFERENCES

Allaud, L.A. & Martin, M. 1977. *Schlumberger, the history of a technique.* John Wiley, Chichester.

Desbrandes, R. 1985. *Encyclopedia of well logging.* Editions Technip, Paris.

Downing, R.A. & Gray, D.A. (eds) 1986. *Geothermal energy – the potential in the United Kingdom.* British Geological Survey, HMSO, London.

Jackson, D.I., Jackson, A.A., Evans, D., Wingfield, R.T.R., Barnes, R.P. & Arthur, M.J. 1995. *United Kingdom offshore regional report: the geology of the Irish Sea.* British Geological Survey, HMSO, London.

Rider, M.H. 1986. *The geological interpretation of well logs.* Blackie, Glasgow.

Schlumberger 1972. *Log interpretation. Volume 1 – Principles.* Schlumberger, New York.

Schlumberger 1977. *Log interpretation charts.* Schlumberger, Ridgefield.

Swarbrick, R.E. & Osborne, M.J. 1996. The nature and diversity of pressure transition zones. *Petroleum Geoscience,* **2,** 111–116.

Whittaker, A., Holliday, D.W. & Penn, I.E. 1985. *Geophysical logs in British stratigraphy.* Geological Society of London, Special Report No. 18.

Whittaker, A., Cope, J.C.W., Cowie, J.W. *et al.* 1991. A guide to stratigraphical procedure. *Journal of the Geological Society of London,* **148,** 813–824.

10
Principles of Seismic Stratigraphy

Alf Whittaker

Chapter 9 reviewed the importance and relevance to stratigraphy of subsurface geophysical wireline log data. The value of that sort of data to geologists is augmented greatly when it is realized that geophysical logs form the direct link between borehole sample data and seismic reflection data. In fact seismic reflection data are effectively a transform of sonic (acoustic) log data, the former being presented in two-way reflection time, and shown as linear time values on the seismic profile's vertical axis, the latter being presented on geophysical logs vertically in terms of linear depth. The geophysical log data therefore provide an important method of geologically calibrating seismic results.

It is also important to note that the vertical (stratigraphical) resolution of down-hole geophysical log data is about two orders of magnitude better than that of conventional seismic reflection data. It was seen in Chapter 9 that high-quality log data can be resolved down to about 10 cm or so. Typical oil company seismic reflection data, by contrast, can be resolved vertically at best down to thicknesses between 10 m (in shallow crust), and 100 m (in deeper crust) depending upon the wavelengths involved. The relatively coarse vertical resolution of seismic reflection data is an essential point to keep in mind when considering stratigraphical resolution (in terms of actual bed thicknesses) during seismic and sequence stratigraphical analyses. Seismic sections are excellent for showing continuity of reflectors and therefore gross lateral stratal relationships, but they are much less useful for high-resolution stratigraphical studies involving thin beds and laminae. However, the geophysical log data are capable of adding to the seismic data some of the missing detail and assisting in obtaining a high-resolution stratigraphy. In this

Unlocking the Stratigraphical Record: Advances in Modern Stratigraphy. Edited by P. Doyle and M.R. Bennett.

respect, therefore, the log data and the seismic data are very much complementary to each other.

In the context of stratigraphical applications, the reverse of this situation, where the seismic data have the advantage over log data, can be demonstrated using the assessment of breaks in the sequence as an example. Lateral continuity and parallelism of reflectors suggest continuity of deposition without break in sedimentation, whereas truncation, onlap, downlap and other observable reflector characteristics indicate breaks in sedimentation caused by either non-deposition or erosion, or both. Sequence stratigraphy in particular treats packages of continuous and conformable strata bounded by unconformities as representing discrete depositional episodes in the history of a basin (see Chapter 11).

Concepts that were eventually to impact significantly on what has become known as seismic stratigraphy (and later, sequence stratigraphy; Chapter 11) began with the work of Sloss (1963), who, basing his ideas on outcrop evidence from the cratonic interior of North America, concluded that depositional sequences were present which were unconformity-bounded. These unconformity-bounded records of sedimentation were regarded as major tectonic cycles punctuating the Palaeozoic succession. They were also seen as sizeable units or packages of strata which, in terms of the stratigraphical classification, were higher than Group or Supergroup, and were believed to be traceable over large areas of the North American continent. In addition, in the 1960s and later, there was an extended phase of research into sedimentary processes and modern depositional systems, which provided the analogues with which to interpret ancient depositional systems. This type of facies analysis was developed in parallel with, but quite independently of, a great improvement in the quality of seismic reflection data which ultimately permitted modelling of the expected seismic reflection response of different facies associations. These seismic facies units could then be used as a predictive tool to assess certain rock types and therefore aid the identification of different lithologies in the concealed subsurface.

This chapter will consider briefly the techniques and methodology of the acquisition and processing of seismic reflection data before turning to the interpretation of the results with special reference to stratigraphy.

10.1   SEISMIC REFLECTION TECHNIQUES AND METHODOLOGY

10.1.1   Seismic Waves

The seismic reflection method is analogous to echo sounding. It relies on the reflection of sound waves from geological boundaries present at depth in the subsurface (Figure 10.1). Data acquisition commences with an artificial energy source being introduced into the earth: on land, by using an explosive charge such as dynamite, or vibrators; at sea, by using air guns. The seismic waves themselves are of two types, body waves and surface waves. The former actually propagate through the body of a medium, while the latter propagate along the interface between media of different properties. Of the two main types of body waves

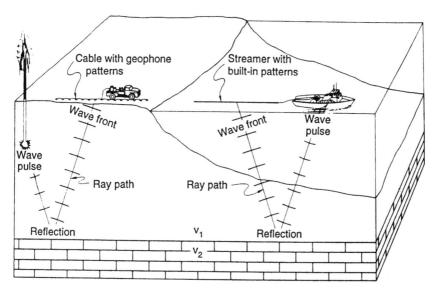

Figure 10.1 *Principles of seismic reflection surveying onshore and offshore. [Modified from: Tucker (1974)]*

recognized, P- or longitudinal waves are ones in which the particle motion is in a direction parallel to the direction of propagation, while S- or shear (or transverse) waves have particle motion in a direction transverse to the direction of propagation. P-waves are the most important in seismic reflection work.

Having introduced energy into the earth, the seismic (or sound) waves travel down into the crust until they meet a geological boundary where there are significant changes in physical properties (velocity and density). A small amount of the energy from the sound wave is then reflected back to the surface, while the remaining sound wave continues its downward travel (becoming weaker) to the next boundary, where the process is repeated. These changes of physical properties are known as acoustic impedance contrasts and are caused by changes in lithology, porosity and fluid content. The deeper the boundary or junction, the further the reflected wave has to travel back to the surface and therefore the later its arrival there.

Reflected back to the surface, the wave front meets a geophone or hydrophone receiver array which produces a high-fidelity transform of the earth movement into a varying electrical voltage, eventually to be converted into the familiar seismic signal. The time taken for the reflected seismic waves to rebound to the surface since the original energy input is known as the two-way travel time. The layout of arrays and geophone patterns is important and depends upon the target or aims of the seismic survey as well as on the structural and tectonic trends in the geology of the region under examination. After this the data undergo a complicated series of processing procedures including demultiplexing, deconvolution, band-pass filtering, velocity corrections, stacking and migration.

10.1.2 Seismic Velocity

The velocity, or the rate of propagation of a seismic wave through a medium, is equal to the distance travelled by the wave divided by the time taken ($V = D/T$, where V is velocity, D is distance, and T is time). Seismic velocities are of great importance to geologists for a variety of reasons, not least because the seismic section, given in terms of time, requires knowledge of the velocity of the rock column in order to allow calculation of depth or thickness which, for geological work in general and stratigraphical applications in particular, are of prime importance. Seismic velocity nowadays is usually measured in metres or kilometres per second, despite the fact that sonic or acoustic geophysical logs are still mostly calibrated in feet per second. Seismic time is measured in seconds and milliseconds (with 1000 milliseconds in a second). Seismic velocities depend on various things, e.g. the type of seismic wave (in most consolidated rocks P-waves propagate more quickly than S-waves), or the composition and physical properties of the rock (usually, the harder and more compact a rock, the quicker the velocity). A typical range of P-wave velocities is shown in Figure 10.2. It will be seen that most rock types typical of sedimentary basins have a wide range of velocities rather than a unique value. Generally speaking, low velocities are associated with clastic rocks and high velocities with carbonates and evaporites. However, velocity alone is insufficient to distinguish rock type.

As well as the relationship of velocity to rock type, porosity and fluid content, sediments and rocks also compress or compact with depth of burial so that velocities in sedimentary rocks (especially mudstone/clay sequences) increase steadily, but not necessarily uniformly, with depth to give a velocity gradient. Commonly used in seismic studies is the expression the *interval velocity* of a sequence, which is the average velocity of a particular succession of rocks and is related to the velocity gradient of that succession. Each interval is defined by the *interval transit time*, or the time taken for the acoustic pulse to traverse that succession in milliseconds. As suggested above, the density of rocks also affects the speed of seismic waves, with denser rocks usually having higher velocities (Figure 10. 2). Despite the wide range of velocities noted above, bulk densities display a lesser range.

10.1.3 Acoustic Impedance and Reflection Coefficients

Acoustic impedance is the product of a rock's velocity multiplied by its density, and this relationship is important in determining what are known as reflection coefficients, or reflectivity. The relatively insignificant role of density, or sometimes a complete lack of density data, necessitates that density is ignored so that often only velocity data are used in calculating reflection coefficients. These can be positive or negative depending on velocity relationships, that is whether there are increases or decreases of velocity at a geological interface. If there is no velocity change at a lithological interface there will be no reflection. However, a reflection on a seismic section rarely represents a single bedding plane but is commonly the

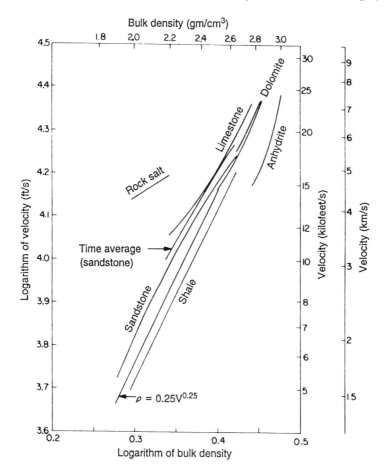

Figure 10.2 *Velocity–density relationships in rocks of different lithology, using time-average and Gardner equations. (The time-average equation relates rock velocity to its porosity and fluid content, while Gardner's equation relates density to velocity.) [Reproduced with permission from the Society of Exploration Geophysics, from: Gardner et al. (1974)]*

sum, or composite, from several bedding surfaces. Thin beds with rapid velocity changes, which is a common situation in the sedimentary column, produce complicated, composite reflections (Figure 10.3).

Both the acoustic log data and the bulk density log data combined are used to study acoustic impedance contrasts in the earth's subsurface and to calculate seismic reflection coefficients in determining the accurate geological identification of individual seismic reflectors. The same log data are also used to geologically calibrate and interpret the whole seismic reflection section and thereby facilitate accurate subsurface geological mapping which is fundamentally important to a proper understanding of the three-dimensional disposition of geological sequence and structure.

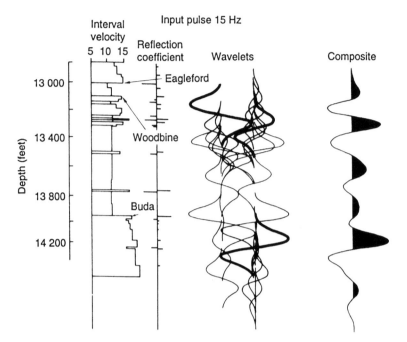

Figure 10.3 *Production of a synthetic seismogram from a geological well section (on the left, with interval velocities in kilofeet per second) via its positive and negative reflection coefficients and an input wavelet of 15 Hz frequency. The composite synthetic seismic trace of the geological section is shown on the right. [P.R. Vail, R.G. Todd and J.B. Sangree, © 1977, reprinted with permission of the American Association of Petroleum Geologists.]*

10.1.4 Data Processing

Processing of the seismic data is necessary to convert the data recorded in the field into interpretable seismic sections. The idea is to produce a high-quality seismic display (with what is known as a high signal-to-noise ratio) that matches as closely as possible a geological cross-section of the subsurface. Table 10.1 lists the main stages in a processing procedure (and the actual functions) in the order in which they occur, bearing in mind that the number of processes and applications varies between individual surveys and sometimes even between individual profiles. The Header Sheet, displayed at one end of the seismic section, shows the various processes applied to the data.

10.2 INTERPRETATION OF SEISMIC REFLECTION DATA

10.2.1 Geological Identification of Reflectors

It is possible to pick seismic reflectors as events on a seismic section without specifically identifying geological horizons, and to produce a structural interpreta-

Table 10.1 *Stages in a seismic reflection data processing procedure (CMP = common mid point; NMO = normal move out)*

Process	Function
Demultiplex and edit	Rearranges data. Edits out poor data
Statics correction	Corrects for elevation and near-surface velocity
Deconvolution before stack	Sharpens seismic signature
Collect into CMP gathers	Rearranges traces
Velocity analysis	Establish velocity function to give best stack
NMO correction	Time shift to traces to make stack
Muting	Suppresses unwanted signals
Stack	Improves signal-to-noise ratio
Deconvolution after stack	Sharpens seismic signature
Filter	Removes unwanted frequencies
Migration Display	Places seismic events in true position

tion of the region. It is even possible under favourable circumstances, or with experience of the basin being examined, to recognize characteristic seismic signatures of various stratigraphical units. An example is shown in Figure 10.4 where, in the region concerned, the named stratigraphical intervals each have a characteristic seismic signature. However, for many purposes it is important to be able to identify individual reflectors geologically so that depths to key horizons such as source rocks or reservoir rocks can be determined. Although there may be some rough indication from the character of the reflections to allow a guess at possible lithologies, this is unlikely to provide sufficient detail for specialist work. Given the range of velocities (Figure 10.2) of sedimentary rocks, it is not possible to identify lithology directly from velocity data alone, although familiarity of the sequence in a well-known basin may serve to identify unusual velocities with particular rock types and therefore aid correlation. Nevertheless, there are three principal sources of data for the geological identification of reflectors: borehole information, local outcrop data, and regional geological knowledge.

10.2.2 Borehole Data

Borehole data provide a prime source of information about the concealed subsurface geology, and especially the stratigraphy. However, before attempting to tie in the borehole stratigraphy to the seismic section it is vital to be as certain as possible that boundaries in the boreholes are correlated consistently and placed as accurately as can be ascertained.

It was stressed in Chapter 9 that all relevant borehole data should be brought into play when identifying formation boundaries. This is achieved by using, in combination, both cuttings information and geophysical log data, especially the data from the sonic, gamma, density and neutron logs. Since important stratigraphical horizons do not necessarily have a change in acoustic impedance associated with them, it is important also to remember that there will only be a seismic

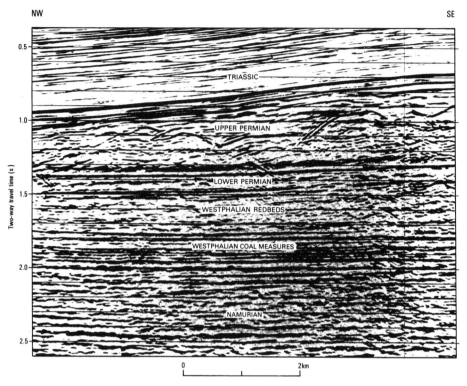

Figure 10.4   *An example of the seismic reflection character or signature of various strati-*
graphical intervals present in parts of the southern North Sea region. [Reproduced with per-
mission from the British Geological Survey, from: Cameron et al. (1992)]

response across boundaries that do display such a change. Consequently, the most
important geophysical log with which to achieve geological calibration, and ultimately stratigraphical correlation, is the sonic log.

Sonic logging measures the formation's interval transit time (Δt), which is the
inverse of velocity. Δt is recorded in microseconds per foot and ranges from 140 or
100 (low velocity) on the left-hand side of the sonic log track to 40 (high velocity)
on the right-hand side. The interval transit time values are integrated with depth
to produce the integrated travel time, presented as a series of small horizontal
peaks or ticks on the depth axis (Figure 10.5). Every 10 ms, the display shows one
larger tick. The spacing of these individual (1 ms) ticks against the depth measurement shows how far sound would travel through that sequence in 1 ms. Consequently, if the depths of the formation boundaries are known, it is necessary only
to count the number of ticks between successive formation tops to give the
one-way travel time spacing of reflectors on the seismic section. Of course, these
values need to be multiplied by two to obtain the two-way travel time seen on the
seismic profile.

One way of achieving this is to produce what is known as a stick diagram by
drawing a table of formation tops, depths, one-way travel time and two-way travel

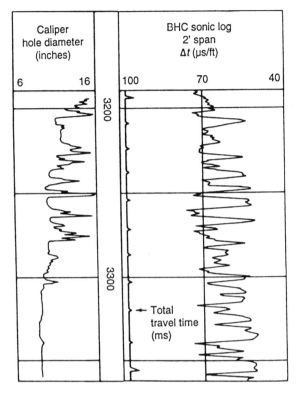

Figure 10.5 *Schematic presentation of sonic log data (right-hand track) showing interval transit time (Δt) in microseconds per foot. Also shown are the integrated travel-time ticks (shown in the diagram as 'total travel time') in milliseconds. [Reproduced with permission from: Schlumberger (1972)]*

time. The stick diagram vertically shows two-way travel time, with ticks marking the relevant geological levels at the relevant two-way travel times, so that ticks projecting to the left indicate a negative reflection coefficient (a change from hard to soft) while those to the right indicate a positive reflection coefficient (a change from soft to hard). As most sonic logs do not commence at the top of the well, the zero value in the table will not coincide with the zero or sea level of the seismic section. The stick diagram will show therefore the relative time spacing of reflectors but not the absolute values of two-way travel time.

This can be achieved by calibration from a well velocity survey in which a series of controlled detonations in a pit near the well top are exploded and recorded by a series of geophones that have been lowered down the hole to previously determined levels, usually close to important geological horizons. The time taken for the sound waves to travel from the surface to these levels can be measured and the two-way travel time calculated.

10.2.3 Vertical Resolution of Seismic Data

It is also important for geologists, and especially stratigraphers, to understand something of seismic frequencies and seismic wavelengths. This is because such information is critical in determining reflector, bed and sequence thicknesses. In order to be capable of resolution, reflector thicknesses must be of comparable dimensions to the seismic wavelength. A knowledge of the wavelength of the input seismic signal is necessary therefore in determining the relevant thicknesses. However, in the case of thin beds, reflections from the top and from the bottom of the same bed will interfere with each other.

The frequency spectrum of the acoustic signal generated varies according to the energy source (Figure 10.6). Division of the velocity by the frequency enables the wavelength to be calculated by means of the following formula:

$$\text{Wavelength} = \text{Velocity} / \text{Frequency}$$

The higher the frequency of the waveform, the greater or better will be the vertical resolution. Given a bed of intermediate velocity, in order to completely resolve reflections from top and base, the bed thickness must be greater than one-half the dominant wavelength. This allows time for the bed-top reflection to be completely clear of the interface, before the bed-base reflector can meet it in its upward path and cause interference. It is possible to resolve bed thickness down to one-quarter of the dominant wavelength, generally regarded as the limit of resolution.

There is a loss of energy caused by absorption as the seismic wave travels deeper into the earth's crust. This gives rise to attenuation or decrease of the higher frequencies, resulting in turn in an increase in the wavelength and resolvable bed thickness with depth. Consequently, thin beds, small faults and structures are not

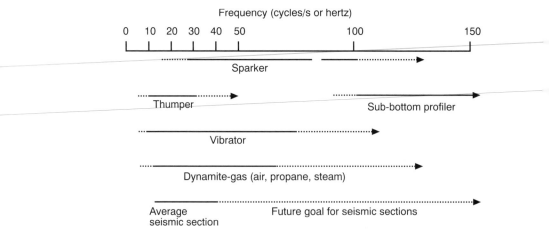

Figure 10.6   *A frequency spectrum acoustic signal from 0 to 150 Hertz (cycles per second) showing frequency ranges for different energy sources. [Modified from: Tucker (1974)]*

recorded at great depth. Velocities of 1.5–2.0 km s^{-1} and frequencies of about 50 Hz (hertz, or cycles per second) are typical in the shallow crust and give wavelengths of 30–40 m. The best vertical resolution is consequently between 7 and 15 m (i.e. one-quarter to one-half a wavelength). In contrast, velocities in the deeper crust are commonly between 5.0 and 6.0 km s^{-1} with frequencies of about 20 Hz, giving wavelengths of 250–300 m and best vertical resolutions of 75–150 m.

It is worth noting that what are known as high-resolution seismic surveys are carried out, but they only provide useful geological information at shallow depths. The method is used a great deal in coal exploration where attempts are made to map individual coal seams, which in turn throws light on the subsurface structure and especially faults of relatively small throw. The energy sources used (small dynamite charges or high-frequency vibrating methods) provide dominant frequencies of several hundred hertz. Such methods generating approximately 200 Hz in rock sequences with velocities of 1.5 km s^{-1}, give wavelengths of about 7.5 m which corresponds to a resolution of about 1.3 m (i.e. a quarter of the wavelength).

An example of the importance of frequency in geological resolution of seismic data is shown in Figure 10.7. In this modelled example a laterally continuous higher sandstone bed produces a continuous reflector whether the frequency is 20 Hz or 50 Hz; the higher sandstone bed overlies three, laterally discontinuous sandstone lenses two of which die out successively towards the left. The laterally discontinuous, lower sandstone lenses can be resolved by the 50 Hz pulse, but not by the 20 Hz pulse, which suggests only a thinning of the main mudstone horizon separating the upper sandstone from the lower, lensing beds.

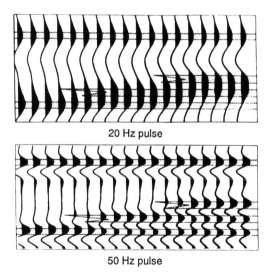

20 Hz pulse

50 Hz pulse

Figure 10.7 *Comparison of the effect of higher and lower frequency pulses on the vertical resolution of geology. See text for detail. [P.R. Vail, R.G. Todd and J.B. Sangree, © 1977, reprinted with permission of the American Association of Petroleum Geologists.]*

10.2.4 Horizontal Resolution of Seismic Data

As well as problems of vertical resolution, it is important also to be aware of lateral or horizontal resolution problems. When reflected energy returns to the surface from depth it comes not from a point source but from a circular area where the wave front has impinged on a geological boundary. This is called the Fresnel Zone and is defined as the area of the geological reflector which has met the first quarter of the wavelet. It limits the horizontal resolution of the seismic data effectively to one-quarter of the dominant frequency. It can also be responsible for reflection events present on a section that are analogous to sideswipe, i.e. features that are derived from geology that is off the seismic line.

10.3 SEISMIC STRATIGRAPHY

Seismic stratigraphy is the interpretation of stratigraphy and depositional facies from seismic reflection data. Seismic reflection terminations and configurations are used to deduce stratification patterns, depositional sequences, environmental conditions and lithofacies estimation.

10.3.1 Seismic Signature or Character

The field geologist has at his/her disposal various characters or attributes with which he/she can describe rocks, such as colour, rock type, hardness, grain size and fossil content. By noting the relevant data from outcrop to outcrop, the

Table 10.2 *Seismic reflection parameters used in seismic stratigraphy, and their geological significance. [Modified from: Mitchum et al. (1977b)]*

Seismic facies parameters	Geological interpretation
Reflection configuration	Bedding patterns
	Depositional processes
	Erosion and palaeotopography
	Fluid contacts
Reflection continuity	Bedding continuity
	Depositional processes
Reflection amplitude	Velocity–density contrasts
	Bed spacing
	Fluid content
Reflection frequency	Bed thickness
	Fluid content
Interval velocity	Estimation of lithology
	Estimation of porosity
	Fluid content
External form and areal association of seismic facies units	Gross depositional environment
	Sediment source
	Geological setting

geologist can eventually arrive at a correlation which can be refined as more data become available. As in the case of geophysical log data, various geological features or properties have a seismic expression too. The range of such seismic features is smaller than those available to the field geologist but can still be of use in correlating reflections along and between seismic profiles and in distinguishing stratal or geological horizons.

There are two important basic attributes of seismic data, known as reflection configuration and reflection character. *Reflection configuration* is the general geometry of the individual reflections and their angular relationships with each other, a feature which has proved of use in seismic sequence stratigraphy and which records onlapping and downlapping relationships (Table 10.2). *Reflection character* describes the look of the individual reflections, and includes the following important attributes: reflection amplitude, polarity and continuity.

Reflection amplitude describes the measure of reflection peaks and troughs and in theory relates to the relative value of the acoustic impedance contrasts across the reflector boundary. Care needs to be exercised with this feature, however, because the processing of data can result in equalizing of the amplitudes. Changes in the vertical sense may locate major changes in rock type or unconformities, whereas lateral changes may indicate facies differences. Polarity describes how a positive or negative reflection coefficient is shown on a seismic trace. The normal convention is that a positive reflection coefficient will show as a trough (white) while a negative reflection coefficient will show as a black peak. Continuity describes the lateral continuity of reflections. A continuous reflection maintains itself for a considerable distance (several kilometres) whereas a discontinuous reflection manifests itself as an alignment but with gaps in it. The amount of lateral continuity can be interpreted in terms of lateral lithological continuity, whereas discontinuous reflector continuity can suggest lateral discontinuity of strata. It is important to remember that pitfalls in interpretation can arise from non-geological sources such as noise and diffractions.

10.3.2 Reflections and Geological Boundaries

Reflections are produced where there are vertical changes in acoustic impedance, which commonly are places where there are vertical changes in rock type. Vertical changes in rock type are fundamental to stratigraphical studies and, discounting faulting, are often associated with three basic geological phenomena: bedding surfaces, unconformities and diachronous facies boundaries.

Bedding surfaces or stratal surfaces usually represent periods of non-deposition or a rapid vertical change in the type of sediment being deposited. Bedding planes in vertically uniform lithologies may represent pauses in sedimentation possibly accompanied by some local erosion. Such pauses are commonly described as diastems or non-sequences. These relatively short interruptions in sedimentation, involving only a brief interval of time with little or no erosion before resumption of deposition, are, of course, gaps in the sequence and may include the down dip, basinward, conformable equivalents of unconformities. However, these very short breaks are ones that would occur within a particular sedimentary environment and are unlike those which relate to unconformities associated with major changes in environment.

An unconformity is a gap in the geological record formed when deposition of sediment ceases for a considerable time. The most spectacular and best known examples are called angular unconformities and illustrate the folding and faulting of an older sedimentary record, its planing down by erosion, and the deposition of a younger sequence over the old one. All strata above the unconformity surface are younger and all strata below are older, although the unconformity surface itself is not a time plane because its age varies laterally. Angular relationships invariably show up well on seismic sections, assuming there to be a contrast in acoustic impedance across the boundary (Figure 10.8). Given that there will be lateral changes in rock types overlying each other, the reflection amplitude, character and phase may also change laterally. Disconformities or non-angular unconformities will also only be observed if there is a change in acoustic impedance contrast across the boundary. The great advantage of seismic reflection data is that angular unconformities displaying acoustic impedance contrasts up dip can be traced laterally, and stratigraphically reasonably precisely, into equivalent conformable basinward sequences.

10.3.3 Seismic Sequence Analysis

Seismic sequence analysis subdivides the seismic profile into packages of concordant reflections separated by discontinuity surfaces, which are defined in turn by systematic reflection terminations. These packages of concordant reflections (seismic sequences) are interpreted as depositional sequences made up of genetically related strata bounded at top and base by unconformities and their correlative conformities (see Chapter 11). Reflection terminations are interpreted to include erosional truncation, toplap, onlap and downlap.

Sequence boundaries are commonly unconformities that are associated with differences in the angularity of seismic reflectors above and below an unconformable surface. This angularity makes the sequence boundary recognizable, but although there may often be a good velocity–density (acoustic impedance) contrast across the boundary, the amplitude and polarity of the reflections generated at the unconformable surface may be quite variable. The reason for this is the presence of variable lithologies at the unconformable surface and the associated rapid horizontal and lateral variations of physical properties.

One of the first steps in seismic stratigraphical analysis is to search the seismic data for unconformities simply because these, together with their correlative conformities, separate seismic sequences. The cessation of sedimentary deposition can indicate either erosion of a previously existing section, or non-deposition. These breaks can have important implications; for example, that the sediment source was no longer available, that tectonic activity had taken place, or that the environment had changed significantly.

Of particular significance in seismic stratigraphy is the recognition of the nature

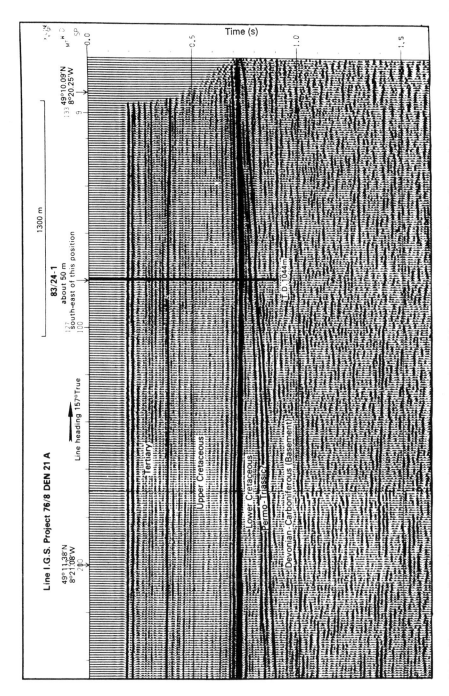

Figure 10.8 *Two unconformities are seen on this seismic section from offshore south-west England. The angular (Late Kimmerian) unconformity between Upper and Lower Cretaceous follows the flat-lying, high-amplitude reflector at just over 700 ms two-way travel time. Less obvious is the onlapping relationship shown at the dipping unconformity surface between the ?Permo-Triassic and basement (Devonian–Carboniferous) sequences. [Reproduced with permission from the British Geological Survey, from: Evans et al. (1981)]*

of reflector angularity at the tops and bases of depositional sequences. Some common seismic angularity patterns are shown in Figure 10.9. Erosional truncation is an important reflector termination, which is observed at the top of a seismic sequence. Another reflector pattern, known as toplap, is caused by non-deposition across the top of a prograding set of sediments. Other angularity patterns can be observed at the bases of sequences, one of the most common being that of onlap. This occurs when reflectors progressively terminate against an inclined surface.

Central to the concept of seismic sequence stratigraphy is the idea that primary seismic reflections follow chronostratigraphical (time stratigraphical) correlation patterns rather than time transgressive lithostratigraphical units (Vail *et al.* 1977; Chapter 11). This means that the seismic reflections essentially represent 'time lines'. Bearing this principle in mind, the basic concepts of depositional sequences as recognized on seismic reflection profiles and the particular importance of unconformities are illustrated in Figure 10.10 (Mitchum *et al.* 1977a). Mitchum *et al.* defined a depositional sequence as a stratigraphical unit composed of relatively conformable successions of genetically related strata and bounded at top and bottom by unconformities or their correlative conformities.

Figure 10.10A shows a generalized diagrammatic stratigraphical section through a typical seismic succession. Seismic sequence boundaries are defined by surfaces A and B which pass laterally from unconformities to correlative conformities. Individual stratal units are traced by following stratification surfaces and they are assumed to be conformable where successive strata are present. Hiatuses are seen where stratal units are missing.

Figure 10.10B, with geological time plotted as the vertical axis, shows a generalized chronostratigraphical section derived from the seismic stratigraphical section given in Figure 10.10A. The geological time range of the seismic stratigraphical sequence between surfaces A and B varies from place to place, but this variation is restricted within synchronous limits which are determined by those parts of the sequence boundaries that are unconformities. With regard to this sequence, limits occur at the beginning of stratal unit 11 and at the end of stratal unit 19. It is also apparent from Figure 10.10B that seismic sequence boundary B in its unconformable part (left-hand part of the diagram) represents a non-depositional hiatus, whereas seismic sequence boundary A, which has two unconformable parts, displays a non-depositional hiatus in the right-hand part of the diagram while the

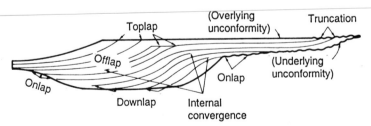

Figure 10.9 *Seismic stratigraphic reflection terminations within an idealized seismic sequence. [R.M. Mitchum, P.R. Vail and J.B. Sangree, © 1977, reprinted with permission of the American Association of Petroleum Geologists.]*

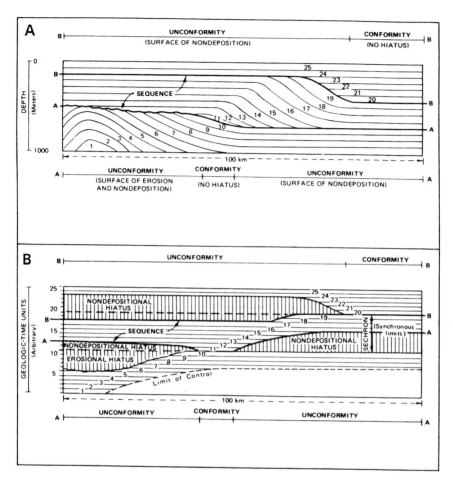

Figure 10.10 *Basic concepts of depositional sequences. See text for details. [R.M. Mitchum, P.R. Vail and S. Thomspon, © 1977, reprinted with permission of the American Association of Petroleum Geologists.]*

unconformity at the left-hand side represents both a non-depositional hiatus and an erosional hiatus.

10.3.4 Seismic Facies Analysis

Having determined seismic sequence boundaries by using reflection termination patterns, further seismic facies analysis, using seismic reflection data, can provide information on palaeoenvironments and lithofacies (Table 10.3).

Facies boundaries, or lateral changes in lithology, result from lateral changes in the environment of deposition or sediment supply. Roughly planar facies boundaries come about by the stacking of successive sediment layers which exhibit

Table 10.3   *Geological interpretation of seismic facies parameters. [Modified from: Mitchum et al. (1977b)]*

Reflection terminations (at sequence boundaries)	Reflection configurations (within sequences)	External forms (of sequences and seismic facies units)
Lapout	Principal stratal configuration	Sheet
Baselap	Parallel	Sheet drape
Onlap	Subparallel	Wedge
Downlap	Divergent	Bank
Toplap	Prograding clinoforms	Lens
Truncation	Sigmoid	Mound
Erosional	Oblique	Fill
Structural	Complex sigmoid-oblique	
Concordance (no	Shingled	
termination)	Hummocky clinoforms	
	Chaotic	
	Reflection-free	
	Modifying terms: even, wavy regular, irregular, uniform, variable, hummocky, lenticular disrupted, contorted	

lateral facies changes. Facies analysis using seismic data is the analysis of reflection configurations and various seismic parameters including amplitude, continuity, frequency and interval velocity. Together they provide information on gross lithology, stratification and depositional features of the sediments. The seismic parameters are mapped as seismic units, or three-dimensional packets of reflections which differ from those of adjacent units (Figure 10.10). Despite the fact that there is no unique relationship between reflection configuration and, for example, specific rock types, in practice the seismic analysis often provides a reasonable framework of the depositional environment and lithology when combined with non-seismic data.

A commonly recognized feature on good-quality seismic reflection data is the seismic facies unit. This is a traceable (i.e. mappable) three-dimensional seismic unit made up of groups of reflectors whose character and parameters differ from those of adjacent facies units. Detailed consideration and delineation of these units can be useful in terms of deducing the environmental setting, depositional processes and possible rock types present.

Various reflection configurations are shown in Figure 10.11. Descriptive terms for parallel or subparallel patterns can be modified by words such as even or wavy. This particular type can occur in several external forms, commonly as sheet, drape and fill, and suggests uniform rates of deposition. Divergent patterns are wedge-shaped with lateral or down-dip thickening occurring within individual reflection cycles rather than by onlap or toplap. These patterns suggest lateral variation in the rate of deposition, or progressive tilting of depositional surfaces. Progradational patterns are interpreted as strata which reflect significant deposition due to lateral outbuilding or prograding. Various types of these progradational patterns are called clinoforms because they have developed on gently sloping depositional

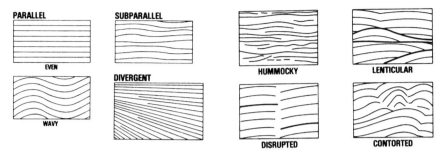

Figure 10.11 *Various seismic reflection configurations and modifications. [R.M. Mitchum, P.R. Vail and J.B. Sangree, © 1977, reprinted with permission of the American Association of Petroleum Geologists.]*

surfaces (Figure 10.12). The sigmoidal form shows S-shaped reflections with gently dipping upper and lower parts but with more steeply dipping and thicker middle parts. The parallelism of the 'topset' reflections suggests upbuilding of the uppermost sediment set and implies a rising sea-level during deposition. The oblique form shows dipping reflections which terminate updip by toplap and downdip by downlap, and may be the product of a combination of high sediment supply and either slow or no basin subsidence during a period of sea-level stillstand. The complex sigmoidal oblique form consists of an alternation between sigmoidal and oblique types of prograding reflectors. This pattern is indicative of alternating upbuilding and depositional bypass of the topset sediments, perhaps within a progradational phase during a general sea-level rise. The shingled form is the same as the oblique parallel form except for being very thin and suggesting

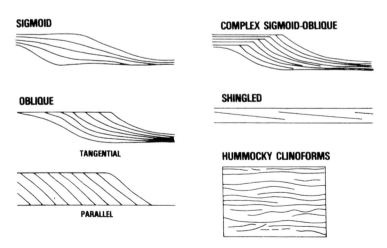

Figure 10.12 *Seismic reflection patterns interpreted as prograding clinoforms. [R.M. Mitchum, P.R. Vail and J.B. Sangree, © 1977, reprinted with permission of the American Association of Petroleum Geologists.]*

progradation into very shallow water. The hummocky clinoform type comprises irregular and discontinuous subparallel reflectors which form a random, low-relief, hummocky pattern with random terminations. This type is thought to result from the occurrence of small interfingering sediment lobes building out into shallow water in a prodelta or interdelta position.

Chaotic configurations show discontinuous and discordant reflections which result from sediments that were initially ordered but have subsequently been deformed, or alternatively they result from deposition in a very high energy environment. Areas of the seismic data that are reflection-free result from sediments which are homogeneous, non-stratified, highly contorted or steeply dipping (Figure 10.13).

10.3.5 External Forms of Seismic Facies Units

An appreciation of the three-dimensional external form of seismic facies units is important in maximizing the interpretative potential of seismic sections. Examples of these are shown in Figure 10.14. Some of the types such as mounds and fills can be subdivided further, depending on factors such as origin, and internal reflection configuration. However, sheets, wedges and banks are mainly of a large size and are common in shelf areas. They display various parallel, divergent and progradational internal reflection patterns. Sheet drapes are interpreted in terms of sediment deposition over a surface showing relief, but one which has not affected the characteristic uniform thickness. This suggests, perhaps, a deep water environment. Lenses are present in various seismic facies associations but are especially common as the external form of progradational clinoforms. Mounds and fills have diverse origins and are interpreted to derive from sediments forming promontories or filling depression features.

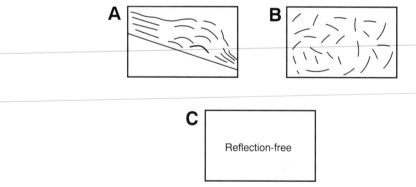

Figure 10.13 *Drawings of chaotic and reflection-free seismic reflection patterns.* **A.** *A chaotic pattern which may be interpreted as original stratal features that are still recognizable after deformation.* **B.** *Not recognizable as any stratal pattern.* **C.** *A reflection-free zone.* [R.M. Mitchum, P.R. Vail and J.B. Sangree, © 1977, reprinted with permission of the American Association of Petroleum Geologists.]*

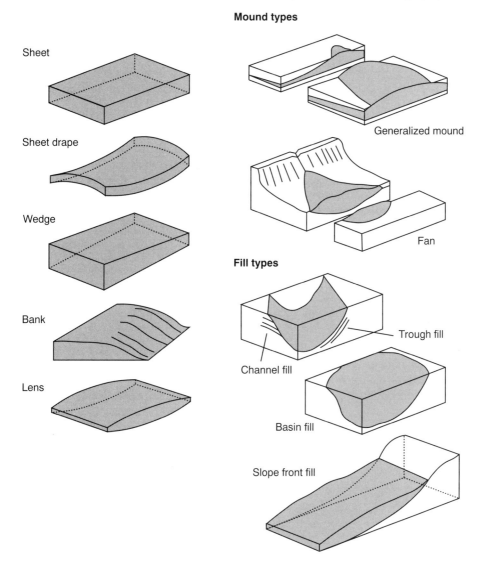

Figure 10.14 *External forms of some seismic facies units. [R.M. Mitchum, P.R. Vail and J.B. Sangree, © 1977, reprinted with permission of the American Association of Petroleum Geologists.]*

10.3.6 Forward Modelling of the Geology

Forward modelling of the detailed geology from sonic and density logs to produce a synthetic seismogram is one of the recommended ways to tie a seismic line to a well section. Digital log data are used to calculate reflection coefficients for all the important geological boundaries. The computer then models the effects of an input downgoing wave of known form which calculates primary reflector positions and possible multiple reflector positions. It is usual for the geophysical logs to be re-plotted with time rather than depth on the vertical axis so that the logs can be placed next to the anticipated seismic response. This is then compared with the seismic section requiring calibration to see if a correlation with the borehole geology or stratigraphy is obtained.

Synthetic seismics are limited by the fact that the input wavelet is always a best guess. A much better and more realistic correlation between the borehole data and the seismic data is possible by running a vertical seismic profile (VSP) in the well concerned. A VSP is similar in many ways to a well velocity survey except that more geophones are used and instead of just recording the first arrivals, the whole wave train is recorded. The plot of this wave train is displayed with time on a horizontal axis and depth on a vertical axis. As well as the direct wave, reflected arrivals from horizons at greater depths than the geophones are recorded. Downhole reflections can then be matched to the seismic section and tied to the well.

10.3.7 Inverse Modelling

Inverse modelling of seismic data proceeds from the seismic trace to the geology, or more correctly, lithology. This is achieved by inputting reflection coefficient data to the seismic trace which provides information on the velocity of the crustal geology. Consequently, in an area where stratigraphical interval velocity data are known, velocity logs can be produced which can be transformed into lithological or geological logs. Such logs are sometimes known as seislogs and can be of great value to the stratigrapher. Effectively the process can be considered as transforming each seismic trace of a survey or profile into a borehole record.

10.3.8 Subsurface Mapping using Seismic Data

Conventional seismic data are usually acquired on a grid basis so that individual profiles are crossed by others. Having identified the reflectors geologically, individual events are mapped out and tied from section to section, until a loop which does not mis-tie is obtained. By this stage of the interpretation it should be possible to decide which events are mappable regionally and how many ought to be traced or mapped. It is normal practice to pick several horizons to help constrain the interpretation and to obtain a three-dimensional regional picture from these. The interpreted sections are then digitized and processed to produce reflection time values at each shot point. These are contoured either manually or automatically to produce isochron or two-way reflection time maps. Isochrons, or lines of equal time value, and facies maps can also be produced to give a generalized view of basin development and palaeogeography. These, in turn, can be helpful in further and more detailed interpretation.

10.4 SUMMARY

This chapter has described how stratigraphical studies are carried out using seismic reflection data, especially when the interpretations from these data are augmented by well-documented subsurface geological information and by geophysical log data. Although the seismic data can be used in a stand-alone way, the results which are achievable are greatly enhanced by input of detailed stratigraphical information from all sources. Consequently, seismic reflection data provide additional techniques and methods in terms of stratigraphical applications, and also lend themselves to a more holistic approach to stratigraphy in general.

ACKNOWLEDGEMENTS

This chapter is published with the approval of the Director, British Geological Survey, NERC. Selected diagrams were redrawn by Matthew R. Bennett.

REFERENCES

Berg, O.R. & Woolverton, D.G. 1985. Seismic stratigraphy II: an integrated approach to hydrocarbon exploration. *American Association of Petroleum Geologists Memoir*, **39**.

Cameron, T.D.J., Crosby, A., Balson, P.S., Jeffery, D.H., Lott, G.K., Bulat, J. & Harrison, D.J. 1992. *United Kingdom offshore regional report: the geology of the southern North Sea*. British Geological Survey, HMSO, London.

Evans, C.D.R., Lott, G.K. & Warrington, G. 1981. *The Zephyr (1977) Wells, South-Western Approaches and Western English Channel*. Report of the Institute of Geological Sciences, No. 81/8.

Gardner, G.H.F., Gardner, L.W. & Gregory, A.R. 1974. Formation velocity and density – the diagnostic basics of stratigraphic traps. *Geophysics*, **39**, 770–780.

Gregory, A.R. 1977. Aspects of rock physics from laboratory and log data that are important to seismic interpretation. *American Association of Petroleum Geologists Memoir*, **26**, 15–46.

Hardage, B.A. 1987. *Seismic stratigraphy. Handbook of geophysical exploration. Seismic exploration, Vol. 9*. Geophysical Press Limited, London and Amsterdam.

Mitchum, R. M., Vail, P.R. & Thompson, S. 1977a. Seismic stratigraphy and global changes of sea level, Part 2: The depositional sequence as a basic unit for stratigraphic analysis. *American Association of Petroleum Geologists Memoir*, **26**, 53–62.

Mitchum, R.M., Vail, P.R. & Sangree, J.B. 1977b. Seismic stratigraphy and global changes of sea level, Part 6: Stratigraphic interpretation of seismic reflection patterns in depositional sequences. *American Association of Petroleum Geologists Memoir*, **26**, 117–133.

Schlumberger, 1972. *Log interpolation. Volume 1 – principles*. Schlumberger, New York.

Sheriff, R.E. 1977. Limitations on resolution of seismic reflections and geologic detail derived from them. *American Association of Petroleum Geologists. Memoir*, **26**, 3–14.

Sloss, L.L. 1963. Sequences in the cratonic interior of North America. *Bulletin of the Geological Society of America*, **74**, 93–114.

Tucker, P.M. 1974. *Seismic interpretation for geologists manual*. Oil and Gas Consultants International, Inc., Tulsa.

Vail, P.R., Todd, R.G. & Sangree, J.B. 1977. Seismic stratigraphy and global changes of sea level, Part 5: Chronostratigraphic significance of seismic reflections. *American Association of Petroleum Geologists Memoir*, **26**, 99–116.

Whittaker, A., Holliday, D.W. & Penn, I.E. 1985. *Geophysical logs in British stratigraphy.* Geological Society of London, Special Report No. 18.

Whittaker, A., Cope, J.C.W., Cowie, J.W. *et al.* 1991. A guide to stratigraphical procedure. *Journal of the Geological Society of London*, **148**, 813–824.

11
Sequence Stratigraphy

Stephen J. Vincent, David I. M. Macdonald and Peter Gutteridge

Sequence stratigraphy is presently regarded as one of the most important unifying concepts in sedimentary geology. It seeks to explain the depositional patterns of sediments on a basinal scale, with reference to changing sea level and tectonic subsidence. The sequence stratigraphical model is actually a series of four linked models (Carter *et al.* 1991).

1. In its simplest form, sequence stratigraphy is a descriptive discipline, which uses unconformities and their correlative conformities to split sedimentary successions into unconformity-bounded sequences. Sequence boundaries have been used successfully in regional stratigraphical correlation, and, debatably, in inter-regional and global correlation.
2. Additionally, the growth of these sequences can be modelled mathematically in terms of eustatic sea-level change and tectonic subsidence or uplift, using accommodation models. In marine clastic systems, the relative changes in these two parameters create or add accommodation space, where sediment may be deposited, between the 'floor' of the sediment–water interface and the 'ceiling' of the sea surface.
3. A knowledge of the gross geometry of a sequence can be used to predict the distribution of sediment types, without prior knowledge of the lithologies. This predictive capability of the model has made it a powerful tool in the oil industry, where it is of considerable use in the interpretation of seismic sections (see Chapter 10).
4. Lastly, and most controversially, the accommodation model can be used to derive a eustatic sea-level curve from an original coastal onlap curve. In the

Unlocking the Stratigraphical Record: Advances in Modern Stratigraphy. Edited by P. Doyle and M.R. Bennett.
© 1998 John Wiley & Sons Ltd.

extreme form, it has been claimed that it is possible to construct a global eustatic sea-level curve (Haq *et al.* 1987): the global sea-level model. This has been a hotly debated topic (see Chapter 15), and the third- and higher-order global eustatic sea-level changes predicted by the chart are no longer accepted by most workers. It must be stressed, however, that the global sea-level model alone is not sequence stratigraphy, but an interpretative derivation; discrediting the sea-level chart does not discredit sequence stratigraphy.

11.1 HISTORICAL BACKGROUND

Although it has been greeted as a totally new approach to the description and correlation of bodies of rock, sequence stratigraphy does have a past. From the time of James Hutton, unconformities have been used to separate major rock units; in some cases, the differences between rocks on either side of the unconformities have been so profound, that they have been used as the division between systems; as, for example, in the distinction between the Silurian and the Old Red Sandstone in north-west Europe. Sloss (1963) used widespread unconformities to divide the Palaeozoic platformal sequences of the North American interior into 'sequences' *sensu lato*. There has also been a long history of study of cyclic sedimentation, particularly in relatively stable settings such as the Carboniferous of North America and Eurasia. Much of the long range correlation that was achieved with these rocks (e.g. Ramsbottom 1973, 1978) was by the correlation of marine bands. Such work provides an historical context for the advent of seismic stratigraphy and sequence stratigraphy in the 1970s and 1980s.

Sequence stratigraphy grew from seismic stratigraphy (see Chapter 10), and was almost entirely the product of a small team working at the Exxon Production Research Company, under the leadership of Peter Vail. The key realizations that came from seismic stratigraphy were that (1) seismic reflectors are time lines, rather than lithological boundaries; (2) unconformities can be correlated with conformities in a basinward direction; and (3) sequence-bounding unconformities define lenticular sediment bodies tens to hundreds of metres thick and kilometres to tens of kilometres wide. These observations, combined with outcrop and sub-surface geological data, form the basis of sequence stratigraphy; the study of rock relationships within a chronostratigraphical framework of repetitive genetically related strata bounded by surfaces of erosion or non-deposition, or their correlative conformities. The seminal work on the topic was the linked series of papers by Vail and co-workers (Vail *et al.* 1977a), published in AAPG Memoir 26 (Payton 1977). Since then, the concepts of sequence stratigraphy have been widely applied, and almost as widely misunderstood. In particular, there has been confusion between the sequence stratigraphical model and the global sea-level model.

The sequence stratigraphical model was originally derived from work on marine siliciclastic systems, where the depositional systems are the simplest and most amenable to numerical modelling. This simple case is used to illustrate the way the model works. The strengths and weaknesses of the original model are then discussed before we go on to explore two settings where there are more degrees of freedom, and the model is necessarily more complex. In marine carbonate systems material can be produced *in situ*, rather than just being reworked in the case of siliciclastic systems. Additional factors, such as light and nutrient availability, also come into play. Non-marine and, in particular, fluvial depositional systems are more complex still, with a free surface as the 'ceiling' to deposition, as well as a greater lateral variability in facies. The sequence stratigraphy of these rocks, which forms the current research frontier in sequence stratigraphy, is covered in the final part of the chapter.

11.2 THE SEQUENCE STRATIGRAPHICAL MODEL

The idea behind sequence stratigraphy is simple: sedimentary successions can be divided into unconformity-bound units (sequences) which form during a single cycle of sea-level change. Sequences can be split into smaller units (systems tracts) which are genetically linked, and represent different stages in a single sea-level cycle. Systems tracts are formed of one or more parasequences, which are asymmetrical sedimentary cycles representing growth of the sedimentary succession during (at most) a few hundred thousand years. Parasequences are arranged into parasequence sets which show either progradational, aggradational or retrogradational stacking patterns. Sequence stratigraphy sets out to place the generation of these stratal patterns within a predictable, chronostratigraphical framework through an understanding of the controls on the availability of accommodation space, i.e. the space beneath a conceptual equilibrium surface that separates erosion from deposition available for potential sediment to accumulate (Jervey 1988). Within the marine realm, this equilibrium surface, termed base level, is effectively sea level, although some processes will erode to slightly below this. The availability of accommodation space can be modelled using a simple algorithm that links eustatic sea-level change, subsidence (or uplift), sediment added and water depth. The accommodation model is expressed as follows:

$$NS = \Delta E + S \qquad WD2 = A - SED \qquad A = WD1 + NS$$

where NS = new space added between time 1 and time 2; ΔE = eustatic sea-level change between time 1 and time 2; S = subsidence between time 1 and time 2; WD = water depth; A = accommodation space available between time 1 and time 2; and SED = sediment added between time 1 and time 2. These various factors are illustrated and defined in Figure 11.1.

11.3 SEQUENCE STRATIGRAPHY IN MARINE SILICICLASTIC SYSTEMS

11.3.1 Geometric and Genetic Terms

Figure 11.2A shows a cross-section through a hypothetical basin margin during 29 Ma of its history. This illustration has been constructed using a sinusoidal

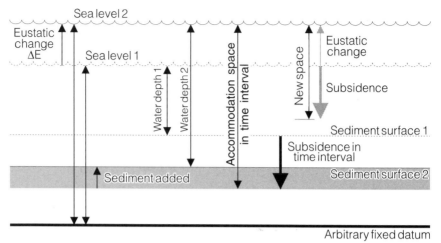

Figure 11.1   *Definitions of key parameters used in the accommodation model*

eustatic sea-level curve with subsidence increasing with distance from a tectonic hinge point, similar to that of a passive margin. Data for control points A and B are tabulated in Figure 11.2B, and plotted on Figures 11.2C and 11.2D; note how relative sea level forms a combination of the subsidence and eustatic curves. A chronostratigraphical chart can be created showing the extent of sedimentation for each 1 Ma iteration of the model (Figure 11.2E). This chart is also known as a Wheeler diagram.

This is the sequence stratigraphical model in its simplest form, generated as a function of subsidence rate and eustatic sea-level change, with sedimentation and subsidence rates held constant through the 29 Ma. Using the illustration in Figure 11.2, five main features should be identifiable.

1. *Stratal terminations*. These are where individual units end, either by onlap, toplap or downlap (see Figure 10.9).
2. *Two sequence boundaries* (SB). Note that the lower sequence boundary (at 5 Ma) is a Type 1 sequence boundary, created when relative sea level falls below the offlap break (the break in slope of the previous clinoform). This results in emergence and erosion updip, a significant basinward shift in facies and coastal onlap, and the development of a lowstand fan in the basin. The upper sequence boundary (at 21 Ma) is more subtle; this is a Type 2 sequence boundary, which occurs when, despite a drop in eustatic sea level, the rate of this fall is outpaced by subsidence at the offlap break (Figure 11.2D). Relative sea level does not, therefore, fall below the offlap break so that the eustatic fall in sea level is not associated with significant incision or a basinward shift in facies. A Type 2 sequence boundary is recognized on seismic data by a basinward shift in coastal onlap, landward of the offlap break (Figure 11.2E). Sequence-bounding unconformity surfaces can be traced into a basin to a point, basinward of which, sedimentation was continuous throughout the time that the sequence boundary

was forming, and the sequences become entirely conformable. In most basins, this correlative conformity occurs in deep water facies.

3. *Systems tract boundaries.* One can recognize four system tracts in this example:
    (i) During the intervals 0–5 Ma and 14–21 Ma highstand systems tracts (HST) are developed. Sediments both aggrade and prograde, with the updip onlap point moving progressively landward, and the downdip downlap point moving progressively basinward (Figure 11.2E). As each successive stratal unit covers a larger area than the one before, and the sedimentation rate is constant, topset beds exhibit a thinning-upward trend (Figure 11.2A). Highstand systems tracts may develop at any time between the rising (R) and falling (F) inflection points of the eustatic curve, depending on the rates of subsidence and sediment supply (Figures 11.2C and 11.2D).
    (ii) From 5 to 11 Ma there is a lowstand systems tract (LST). This contains the F inflection point of the eustatic curve and is made up of two parts: firstly, there is a lowstand fan (5–9 Ma) comprising basin-floor and slope fans, which have a characteristic mounded topography. The coastal onlap point is well within the basin at this time. Secondly, there is a lowstand wedge (9–11 Ma), developed during relative sea-level rise at the shoreline. In this model the lowstand wedge displays aggradational geometries, although on seismic sections progradation is commonly also observed. It is important to note that no time-equivalent sediments are being deposited updip of the offlap break; instead active erosion occurs in this region.
    (iii) In the intervals 11–14 Ma and 27–29 Ma transgressive systems tracts (TST) are developed. Sediments have a strongly retrogradational pattern, with the updip onlap point and the downdip downlap point moving progressively landward (Figure 11.2E). Stratal patterns are dominated by topset beds and contain the R inflection point of the eustatic curve, when the rate of eustatic sea-level rise is at its greatest (Figure 11.2D). The lowstand and transgressive systems tracts are separated by the transgressive surface above which flooding occurs across previously formed topsets. This is commonly referred to as the initial flooding surface (IFS). The transgressive and highstand systems tracts are separated by the maximum flooding surface (MFS) which represents the most landward position of the shoreline. This occurs when the provision of new space for topset accumulation has slowed to a point where it is matched by sediment supply, and marks a switch from retrogradational to progradational parasequence stacking patterns. In a basinward position, the toes of the highstand parasequences downlap onto the maximum flooding surface, hence its geometric name of downlap surface. Here the surface is starved of coarse clastic sedimentation for a considerable period of time (Figure 11.2E) and is usually marked by a condensed section. This is important in biostratigraphical correlation, as it represents the greatest concentration and most landward development of marine fauna.
    (iv) The 21–27 Ma interval is a shelf-margin systems tract (SMST), developed above a Type 2 sequence boundary. Sedimentation is displaced strongly basinward across this boundary (Figure 11.2E), and strata prograde and aggrade slightly.

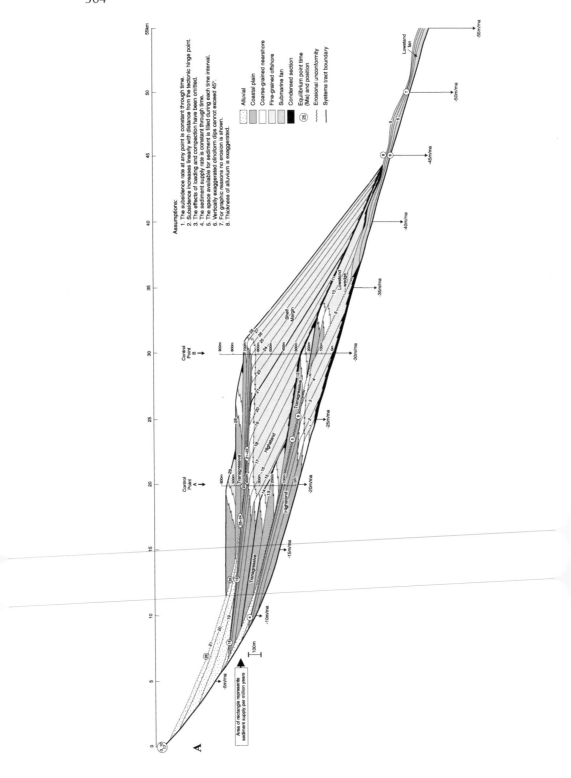

Assumptions:
1. The subsidence rate at any point is constant through time.
2. Subsidence increases linearly with distance from the tectonic hinge point.
3. The effects of loading and compaction have been omitted.
4. The sediment supply rate is constant through time.
5. The space available for sediment is filled during each time interval.
6. Vertically exaggerated clinoform dips cannot exceed 45°.
7. For graphic reasons no erosion is shown.
8. Thickness of alluvium is exaggerated.

Alluvial
Coastal plain
Coarse-grained nearshore
Fine-grained offshore
Submarine fan
Condensed section
25 Equilibrium point time (Ma) and position
Erosional unconformity
Systems tract boundary

Area of rectangle represents sediment supply per million years

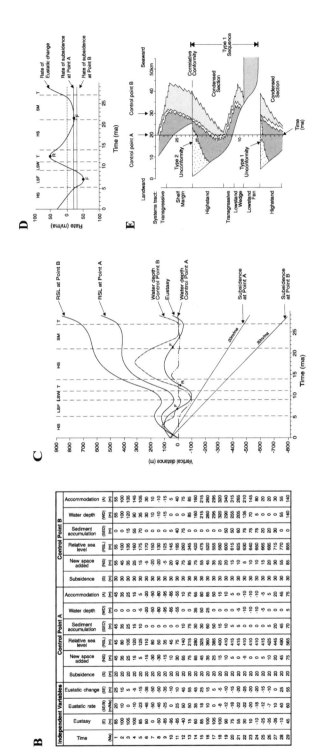

Figure 11.2 *Sequence stratigraphical depositional model showing depositional systems tracts and their bounding surfaces.* **A.** *Geometry of the model with depth (sometimes referred to as the 'Exxon slug').* **B.** *Data table for control points A and B.* **C.** *Relative sea-level change and water depth through time for control points A and B.* **D.** *Rates of eustasy and subsidence for control points A and B.* **E.** *Geometry of the model with time (known as a Wheeler diagram). The assumptions used to construct this model are listed in the figure. [Reproduced by kind permission of Exxon Production Research Company]*

4. *Facies*. Note the position of alluvial, coastal plain, nearshore, offshore shelf and submarine fan facies on the cross-section. One of the great strengths of the sequence stratigraphical model is that it explains the factors governing the distribution of sedimentary environments in a unified way. This point is expanded upon in the following section.

5. *The relationship between systems tracts and eustasy*. Note that the positions of systems tracts boundaries are controlled by the rates and not the absolute values of eustasy, with the transgressive systems tract always containing the maximum rate of sea-level rise (the R inflection point; Figure 11.2D). The lowstand or shelf margin systems tract contains the maximum rate of sea-level fall (the F inflection point), with the former being developed when this rate of fall is greater than the rate of subsidence at the offlap break. Note also that water depths at control point A are significantly out of phase with eustasy (Figure 11.2C); maximum water depths occur prior to eustatic maxima. At control point B the curves are in phase, whilst in more seaward positions the curves would again be out of phase, with maximum water depths occurring after the eustatic maxima. Therefore, palaeobathymetry measured at a single well is not a measure of eustasy, but of relative sea level. Also compare the coastal onlap curve with the eustatic curve (Figures 11.2C and 11.2E).

11.3.2 Step-by-Step Through a Sea-Level Cycle

Figure 11.3 shows a single cycle of sea-level change and its effect on the development of a sequence. This should be studied in conjunction with Figure 11.2, in order to understand the mechanics of the model.

1. As relative sea level falls below the offlap break, upper reaches of the basin become emergent, and a Type 1 sequence boundary is created (Figure 11.3A). Former areas of marine shelf become exposed, and are incised by fluvial systems, creating incised valleys. Type 1 sequence boundary formation initiates lowstand fan deposition, which initially comprises sand-prone basin-floor fan units detached from the foot of slope. A major sea-level fall may also result in the development of submarine canyons, which will focus these deposits via a point source. Note that the maximum incision corresponds to the point on the eustatic curve when rate of fall is greatest (Figure 11.2D); this is a very important point which will become more apparent in the succeeding diagrams. Basin-floor fan deposition, canyon formation and incised-valley erosion are all interpreted to occur during a relative fall in sea level (Van Wagoner *et al.* 1988). Slope fans, characterized by turbidite and debris-flow deposits formed in channel–levée complexes, may also be initiated during lowstand fan deposition. The recognition of the lowstand fan stage is the most effective theory to date to explain the distribution of submarine fan facies. It implies that the overriding control on turbidite deposition is eustatic.

2. When relative sea level is at a minimum, depositional systems stabilize beyond the previous offlap break and a topset-clinoform system, termed the lowstand

wedge, is established (Figure 11.3B). This comprises a wedge-shaped body of sediment which contains progradational to aggradational parasequences developed during initial relative sea-level rise (Van Wagoner *et al.* 1988). Lowstand wedge deposits are characterized by shallowing-upward deltas and shingled turbidites, and display landward onlap and basinward downlap (Figure 11.2E). Slope fan deposition may occur during the early part of the lowstand wedge stage, and their upper surface may form a downlap surface for its middle and upper portions (Figure 11.3B). Initial relative sea-level rise will also result in the infilling of incised valleys, which may continue during transgression. Experimental work by Wood *et al.* (1993) concludes that the morphology and subsequent infilling of incised valley systems are highly dependent on the rates of base-level change.

3. As the rate of sea-level rise increases, transgression occurs across the shelf (Figure 11.3C). Incised valleys may be flooded to form estuaries and subsequently filled. Ravinement, by both wave and tidal processes, may accompany transgression, resulting in a number of erosional surfaces, which may develop both at the transgressive surface and at the top of individual parasequences. Topset deposition, including paralic, coastal plain and shelfal systems, will predominate; tidal processes may be widespread, and due to the undersupply of sediment, the transgressive systems tract will be relatively sand-poor.

4. As the rate of sea-level rise slows, sedimentation is able to keep pace and then outpace this rise, resulting in the aggradation to strong progradation of depositional systems during the highstand systems tract (Figure 11.3D). Sediments are dominated by coastal plain and deltaic facies which may prograde out over underlying lowstand deposits.

5. If when eustatic sea level falls again, the rate of fall is not sufficient to create a relative fall in sea level at the offlap break, the proximal parts of the underlying topsets will be exposed, forming a Type 2 sequence boundary, without fluvial incision or lowstand fan deposition (Figure 11.3E). A Type 2 sequence boundary may be very difficult to identify at outcrop, being represented by a subtle unconformity or disconformity and a possible switch from progradational to aggradational parasequence sets.

For a full understanding of the concept of sequence stratigraphy in its purest form, readers are referred to SEPM Special Publication 42 (Wilgus *et al.* 1988), and in particular to the paper by Van Wagoner *et al.* (1988) which outlines all of the most important terms used, and to AAPG Methods in Exploration Series, No. 7 (Van Wagoner *et al.* 1990). Emery & Myers (1996) also provide a comprehensive recent review of the subject.

11.3.3 Strengths and Weaknesses of the Marine Siliciclastic Sequence Stratigraphical Model

The sequence stratigraphical model is a powerful tool, impacting on the fields of correlation, dating, palaeoenvironmental analysis, and on the analysis of tectonic

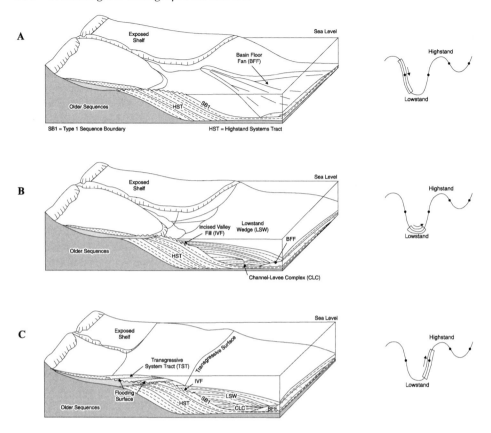

Figure 11.3 *Schematic diagrams illustrating the sequence stratigraphical development of siliciclastic systems during: **A,** early type 1 lowstand (lowstand fan); **B,** late type 1 lowstand (lowstand wedge); **C,** transgression; **D,** highstand; **E,** type 2 lowstand. [Reproduced with permission from Blackwell Science, from: Haq (1991)]*

effects on sedimentary basins. Its greatest strength is that for the first time there is a model which explains all major facies shifts in a sedimentary basin. The core of this theory is the simple, elegant accommodation model which incorporates subsidence (or uplift) and sea-level change in a quantifiable and predictive way.

The accommodation model can be modified to incorporate changes in sedimentation rate, non-linear subsidence rates, and secondary factors such as compaction. Whatever the complexity of the parameters, the basic idea remains unchanged: in marine siliciclastic systems, sediment is deposited in the closest accommodation space to the shoreline. However, because one is dealing with complex natural systems, variations from the predicted behaviour do occur. There is also some confusion between the sequence stratigraphical model itself and other models, such as the global sea-level model, which are derivatives. These real or imagined 'failures' of the sequence stratigraphical model are discussed below.

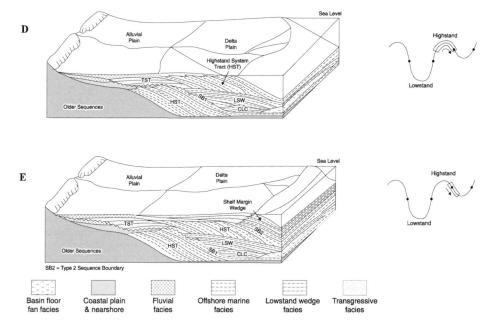

Figure 11.3 *continued*

Sequences in correlation

Sequence stratigraphy arose from the new ways of correlating seismic sections that were developed in the late 1970s (see Chapter 10). As a tool for local to sub-regional correlation, it has proved invaluable. In the oil industry in particular, application of sequence correlation rather than lithostratigraphical correlation successfully explained problems of apparently compartmentalized reservoirs and revolutionized the view of hydrocarbon migration pathways (Figure 11.4).

From these practical beginnings, the Exxon group took their research on in a number of directions which included the suggestion that sequence stratigraphy could be used for inter-regional and global correlation: the global sea-level model. It is at this point that the various strands of their argument usually become confused, and it is useful to break the discussion into five steps.

1. The initial suggestion that sequence boundaries would be correlated world-wide was made by Vail *et al.* (1977b). At this stage, the suggestion was merely a practical observation, based on large numbers of seismic sections, together with biostratigraphical control from wells and key outcrop sections. The biostrati-graphical basis for this suggestion has never been published.
2. If sequence boundaries are indeed global, then there must be a global driving mechanism, independent of the behaviour of individual basins. The original team at Exxon suggested that this mechanism was global eustatic sea-level change resulting from geotectonic and glacial phenomena.

310

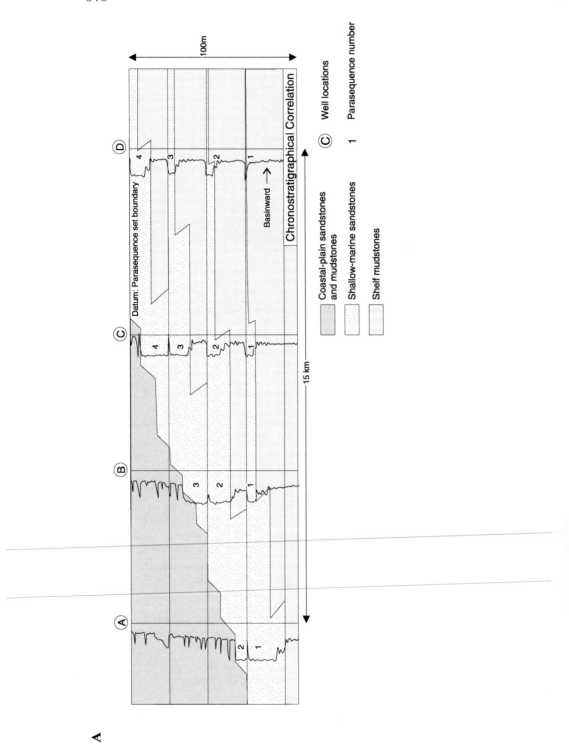

100m

Datum: Parasequence set boundary

4

3

2

1

D

4

3

2

1

C

3

2

1

B

2

1

A

Basinward →

Chronostratigraphical Correlation

15 km

© Well locations

1 Parasequence number

Coastal-plain sandstones
and mudstones

Shallow-marine sandstones

Shelf mudstones

311

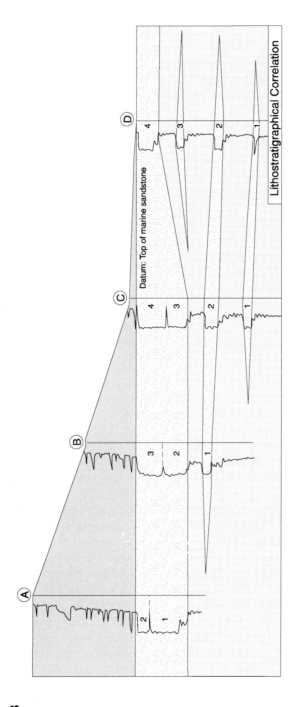

Figure 11.4 Comparison of (**A**) the chronostratigraphical and (**B**) the lithostratigraphical correlation of a progradational parasequence set of paralic to shallow-marine siliciclastic sediments. [Reproduced with permission from the American Association of Petroleum Geologists, from: Van Wagoner et al. (1990)]

3. Following from this, they constructed a chart showing their proposed cycles of global eustatic sea-level change. The first version of this was a coastal onlap chart, with an asymmetrical pattern showing the average relative position of onlap through time, relative to a datum point; in later versions (e.g. Haq *et al.* 1987), they derived a sinusoidal eustatic sea-level curve using the accommodation model (see Chapter 15).
4. They applied dates to sequence boundaries, using both well and outcrop material. It was proposed that this dated series of events could form the basis of a new stratigraphical framework, with, potentially, a very high resolution.
5. The theory that global eustatic sea-level change drove the observed cyclicity became almost accepted as a fact and many authors published apparent correlations with the global standard. At one time it looked as though there would be enough tempting correlations to the 'Haq' curve to make the global sea-level model a self-fulfilling prophecy. However, publication of significant mismatches to the prediction (e.g. Carter *et al.* 1991) and the careful demonstrations of a tectonic cause for some sequences (e.g. Underhill 1991; see also Chapter 15) have resulted in most earth scientists rejecting the global sea-level model as a global norm.

Scale problems and alternative correlation methods

The sequence stratigraphical model is predicted to work at any scale and for any tectonic setting, and can be used both at outcrop and in the subsurface. However, there are problems of scale which diminish the usefulness of the technique.

In the subsurface, sequences are first recognized on seismic sections, which have a vertical resolution of no better than about 10 m, and commonly more than that (see Chapter 10). In contrast, in outcrop one can easily recognize features on the scale of metres to tens of metres, but larger units are impossible to recognize unless in areas of exceptional exposure such as the Book Cliffs of Utah (e.g. Van Wagoner 1995).

From the very first, it was recognized that sequences were at different scales. First-order cycles (>50 Ma) were the large-scale changes occupying more than a geological period, and generally too large to image even on seismic sections. Progressively smaller cycles were assigned to second (3–50 Ma), third (0.5–3 Ma) and fourth (0.1–0.5 Ma) orders. Third-order cycles form the foundation of seismic stratigraphy as they are commonly of a scale well resolved by seismic. In field-based studies, fourth-order sequence boundaries can be recognized with spacings of no more than a few tens of metres (e.g. Plint 1996) which would be at or below the limit of seismic resolution. Higher-order cycles have also been proposed, although there has been much debate over their validity, with this argument being complicated by the fact that fifth- and higher-order cycles are not adequately defined.

The mismatch between field and seismic scale observations has complicated the dating of sequence boundaries. A sequence boundary is created almost instantaneously in geological time, hence their chronostratigraphical value. However, in certain circumstances they can be time-transgressive. In the Holocene Gironde estuary, for example, lowstand fluvial sediments form a thin layer of coarse thalweg deposits which are progressively onlapped by tidal estuarine muds and sands of the transgressive systems tract, such that the lowstand systems tract deposits are diachronous and locally coeval with those of the transgressive system (Allen & Posamentier 1993). In addition, in basins with differing rates of subsidence undergoing the same eustatic sea-level fall, sequence boundary formation will not be synchronous or may not even occur, as it is the combination of subsidence and eustasy (assuming constant sedimentation) which go to make up relative sea level (Parkinson & Summerhayes 1985; Figure 11.5). In fact, the age of sequence boundaries will only be synchronous (within the limits of biostratigraphical resolution) if rates of sea-level fall are much greater than the rates of subsidence in all the basins being considered (Lawrence 1994). Furthermore, since there is hardly ever zero sedimentation on a surface in the ocean, there is a tendency to pick sequence boundaries too high in seismic sections (Cartwright *et al.* 1993). When a sequence boundary is penetrated by a well correlated by a synthetic seismogram, the boundary will be assigned too young an age. This problem does not arise at outcrop. The most precise estimate of the age of a sequence boundary is at the point where it meets the correlative conformity. Dating of this junction is difficult since the unconformity is at its most subtle and the marine fauna is at its most sparse due to the combined effects of very low sea level and clastic dilution.

There is an additional problem with high-order sequences. Mostly there are several of these per stage and the apparent resolution of sequences is better than biostratigraphy. This means that it is impossible to independently test the dating of these boundaries (Miall 1991).

For these reasons, Galloway (1989) has proposed that maximum flooding surfaces should be used to correlate and define stratigraphical units, which he terms genetic stratigraphical sequences. This approach has the advantage that maximum flooding surfaces: (1) represent a continuous section rather than a partial hiatus; (2) contain the richest, and hence most readily datable fauna; and (3) are commonly associated with peaks in radioactive mineral concentrations identifiable on gamma ray electrical logs. This approach has many practical applications (e.g. Underhill & Partington 1994), and it is common practice to initially pick maximum flooding surfaces, particularly on wireline logs, prior to sequence boundary identification. However, whilst the use of maximum flooding surfaces for correlation should be emphasized, genetic sequence stratigraphy is not likely to be widely adopted, partly due to entrenched opinion, because the term 'sequence' implies a wholly conformable succession, and partly because of the economic importance of the recognition of unconformably based lowstand sandstone accumulations.

Sequences and global sea level

Despite the fact that the global sea-level model has been largely discredited as a world-wide standard, there is a fundamental problem in identifying the driving mechanism for sequences. It is generally agreed that first- and second-order cycles

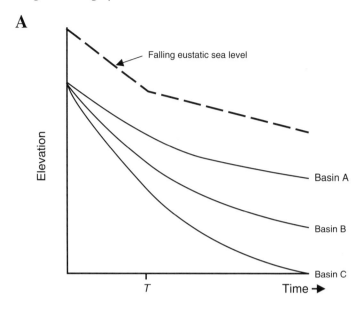

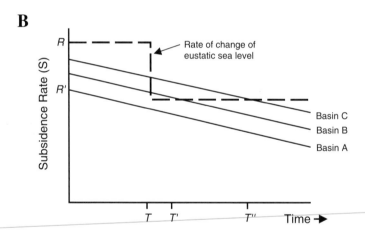

Figure 11.5   *The development of non-synchronous sequence boundaries during eustatic fall in basins with different subsidence histories and assuming constant sedimentation.* **A.** *Subsidence history of three idealized extensional basins (A–C) showing an exponential decay in subsidence with time, and a falling eustatic curve with an instantaneous decrease in the rate of fall at time T.* **B.** *Rates of subsidence and eustatic fall with time. Eustatic sea level initially falls at a rate, R, which is greater than the subsidence of basins A–C, such that each basin will undergo a fall in relative sea level. At time T, there is a decrease in the rate of sea-level fall instantaneously to R'. This is slower than the rate of subsidence in basins B and C, so that a switch to a relative sea-level rise and transgressive shoreline development occurs in these basins. In contrast, relative sea-level fall and regression continues in the most slowly subsiding basin, A. Due to an exponential decrease in subsidence with time, there becomes a point when for basins B and C, subsidence will again fall below the rate of eustatic fall (R'), and will result in a switch from relative sea-level rise to relative sea-level fall and sequence boundary formation. This will occur at times T' and T", respectively. Therefore, in this scenario, sequence boundary formation is non-universal (occurring only in basins B and C), non-synchronous, and occurs during a period when there is no change in the rate of sea-level fall. [Reproduced with permission from the Association of American Petroleum Geologists, from: Parkinson & Summerhayes (1985)]*

are driven by global tectonism, such as changes in continental configuration and changes in the volume of mid-ocean ridges. These processes are slow, with periodicities of 10 Ma or more. Higher-order cycles, with periodicities of 1 Ma or less, are more problematical. The only known parameter which could change sea levels world-wide with this sort of frequency are Milankovitch-driven changes in global ice volume (Pitman 1978). During glacial periods this obviously occurs and there is a fair match between the high-order sea-level curve and the oxygen isotope curve, which is a proxy for ice volume (Williams 1988; see Chapter 7).

There is a problem, however, in identifying a driving mechanism in geological periods without glaciation. Cloetingh (1988) proposed that intraplate stresses could be transmitted rapidly through the continental lithosphere, causing the continental margins to oscillate with an amplitude and frequency sufficient to generate pene-contemporaneous third-order cycles. This theory has not been widely accepted, although Cathles & Hallam (1991) have produced data which apparently support it. White & McKenzie (1988) demonstrated how two-phase lithospheric stretching could produce an onlap cycle at the margins of a sedimentary basin (the 'steers-head' geometry). Episodic fault movement can also produce third-order cycles (Underhill 1991). None of these mechanisms has a global effect.

11.4 SEQUENCE STRATIGRAPHY IN MARINE CARBONATE SYSTEMS

As with siliciclastic systems, the evolution of carbonate systems can be considered in terms of how the interplay between sediment supply and accommodation space affects the distribution and nature of stratal patterns within the stratigraphical record. A fundamental difference between them, however, is the ability of carbonate systems to supply sediment by *in situ* carbonate production. The controls on the rate of carbonate productivity and its response to variations in accommodation space are also quite different to the dynamics of sediment supply in siliciclastic systems. Furthermore, there is a significant environmental and evolutionary control on the structure and diversity of carbonate-producing communities of different ages. This has important implications for the rates of carbonate production, controls on sequence geometry and the response of carbonate platforms to changes in accommodation space through geological time – complications not found in siliciclastic systems.

A carbonate platform is a general term for a sub-basin-scale body of carbonate sediment that was constructed largely as a result of *in situ* carbonate production (Read 1982, 1985). These can either be attached to a continental area or be isolated and surrounded by deep water. A carbonate shelf is a flat-topped carbonate platform which is surrounded by an abrupt break in slope with steep slopes down to basinal areas. A carbonate ramp is a carbonate platform with a gently dipping surface on which deposition ranges from emergent or very shallow to basinal or pelagic (Figure 11.6).

Carbonate platforms commonly show a complex evolution which includes

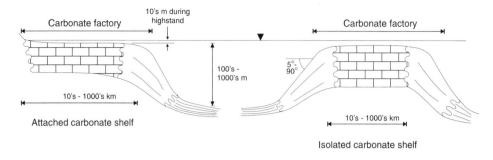

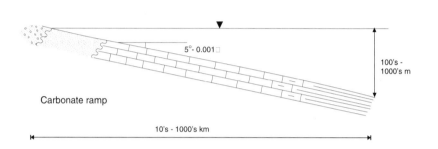

Figure 11.6   *The geometry of carbonate platforms*

both ramp and shelf stages. Tucker (1985) showed that many carbonate platforms are initiated as carbonate ramps which later evolved into flat-topped carbonate shelves. Schlager (1992) showed that the most prolific carbonate production in many tropical carbonate systems takes place in water depths shallower than 15 m (Figure 11.7A). The higher rate of carbonate production in the shallower part of the carbonate ramp leads to its differential aggradation to form a carbonate shelf.

Wright & Faulkner (1990), Burchette & Wright (1992) and Wright (1994) showed that carbonate ramps were the dominant platform type following times of mass extinction or crisis in 'reef' building communities (Figure 11.8). They argued that special conditions of carbonate production prevailed at these times, when the rate of carbonate production was constant with depth (Figure 11.7B). The Early Carboniferous, when the marine ecosystem was recovering from the demise of early Palaeozoic reef-building communities, was such a time of widespread carbonate ramp development (Ahr 1989; Wright & Falkner 1990). Bridges *et al.* (1995) suggested that the palaeo-ocean was enriched with nutrients as a result of several global anoxic events during the Late Devonian which affected the structure of the carbonate-producing communities and, as a side-effect, stimulated the development of carbonate mud moulds. Carbonate ramps were also common in the Tertiary and in modern temperate carbonates whose production versus depth profiles differ from tropical carbonates (e.g. the south-east Australian shelf; Boreen & James 1993,

A Late Cenozoic

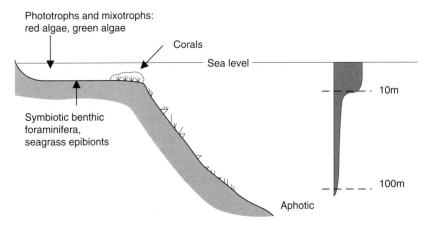

Phototrophs and mixotrophs:
red algae, green algae

Corals

Sea level

10m

Symbiotic benthic
foraminifera,
seagrass epibionts

100m

Aphotic

B Early Carboniferous

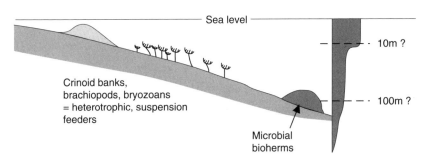

Minor carbonate production
by cyanobacteria and molluscs

Sea level

10m ?

Crinoid banks,
brachiopods, bryozoans
= heterotrophic, suspension
feeders

100m ?

Microbial
bioherms

Figure 11.7 *Contrasting carbonate-producing communities and depth versus production profiles for (A) the late Cenozoic and (B) the Early Carboniferous. Note the increased production rate between 10–100 m [Reproduced with permission from Blackwell Science, from: Wright & Burchette (1996)]*

1995). Carbonate communities on the south-east Australian shelf include bryozoans, crinoids, brachiopods and algae, all of which were important members of late Palaeozoic carbonate communities.

The fact that environmental and ecological conditions can control the nature of carbonate platforms both through time and in space has important implications for carbonate sequence stratigraphy as shelf and ramp systems respond differently to relative sea-level change and produce quite different stratal geometries.

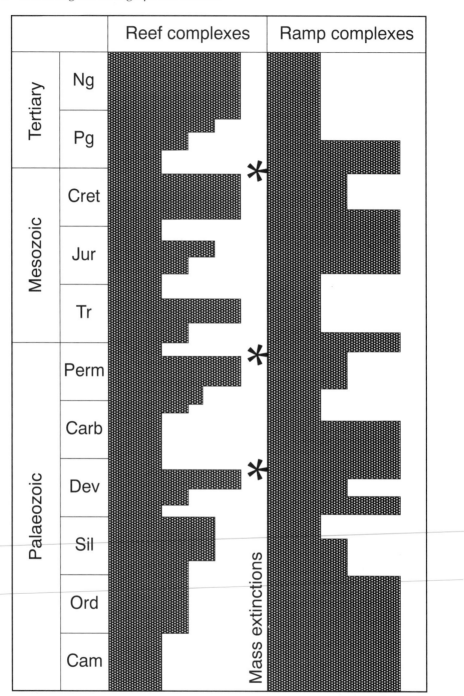

Figure 11.8 *Occurrence of carbonate ramps, reefs and mass extinction episodes through the Phanerozoic. Note that carbonate ramps are most abundant following mass extinction events and/or during periods of restricted reef development. [Modified from: Burchette & Wright (1992)]*

11.4.1 Carbonate Shelves

Lowstand sedimentation

Most carbonate production on shelves takes place on the large shallow subtidal area on the shelf top which is prone to be exposed during third- and higher-order lowstands in sea level. As a consequence, the sequence stratigraphical response of carbonate shelves can be understood in terms of switching carbonate production off during lowstands and on again during highstands. A Type 1 lowstand results in the exposure of the shelf and foreslope (Figure 11.9A). The basin is starved of periplatform carbonate sand and mud because none is being produced on the shelf. Aprons of slumps and lithoclastic shelf and shelf margin material may be deposited over the slope and base of slope. This differs from siliciclastic systems which commonly experience increased sand supply to the basin during lowstands resulting from the rejuvenation and bypassing of the shelf by fluvial systems. A Type 2 lowstand results in the exposure of the outer shelf while the fore slope is still submerged (Figure 11.9D). In this case, carbonate production can continue in a narrow zone adjacent to the shelf margin and a shelf margin wedge may develop. Sediment supply to the basin is reduced but not entirely cut off.

Processes on the exposed shelf depend on palaeoclimate and run-off. In a semi-arid to humid setting these may include karstification and lithification of the carbonates. Dolomitization may also take place during lowstands as a result of meteoric influx.

Transgressive sedimentation

During transgression, carbonate production starts up again once the shelf is flooded (Figure 11.9B). There is commonly a time lag between flooding and re-establishment of maximum carbonate production which results in an initial phase during which sedimentation 'catches-up' with the increase in accommodation space. This is often marked by a short-lived phase of condensed or relatively deep-water subtidal sedimentation. In the basin, the transgressive systems tract may be marked by a change over from the deposition of lithoclasts to the deposition of periplatform carbonate mud and sand so that the maximum flooding surface rarely finds an expression in carbonate systems; this is in contrast to siliciclastic systems where it is typically marked by a condensed surface.

The diagenesis of transgressive systems tracts is characterized by the pumping of marine pore fluid through the carbonate shelf. This results in early marine cementation and, in arid climates, dolomitization of the carbonate shelf (e.g. Tucker & Hollingworth 1986).

Highstand sedimentation

In the highstand systems tract, carbonate production 'keeps up' with relative sea-level rise, and excess carbonate sand and mud are reworked and deposited in surrounding basins (e.g. Gawthorpe & Gutteridge 1990; Figure 11.9C). This differs

A Type 1 lowstand

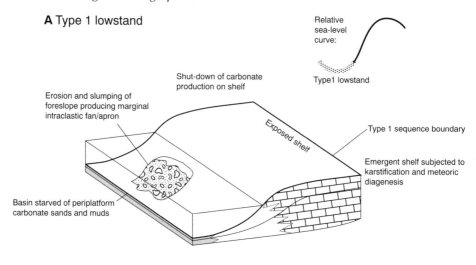

B Transgressive systems tract

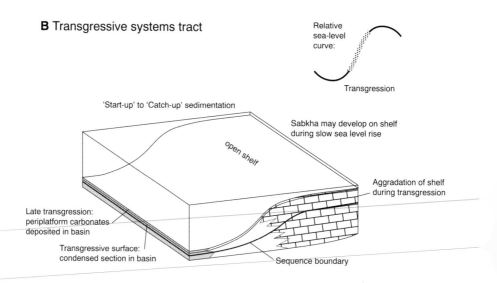

Figure 11.9 *Schematic diagrams illustrating the sequence stratigraphical development of carbonate shelves during (**A**) a Type 1 lowstand, (**B**) transgression, (**C**) highstand, (**D**) a Type 2 lowstand. [Modified from: Sarg (1988)]*

C Highstand systems tract

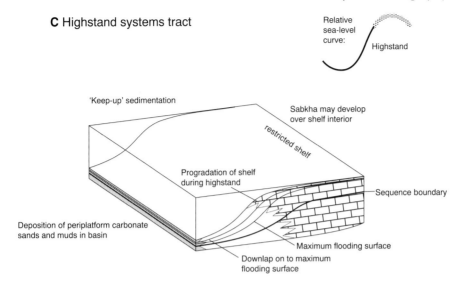

D Type 2 lowstand

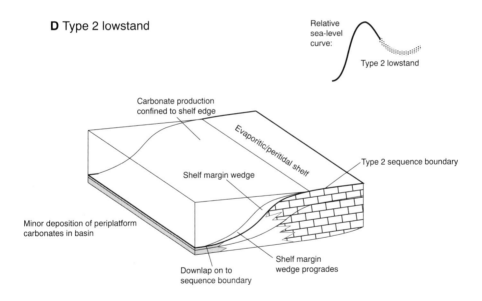

Figure 11.9 *continued*

from siliciclastic systems where sediment is typically trapped on the shelf during early highstands, leading to basin starvation. The export of carbonate sediment gives carbonate platforms the potential to prograde, although this is also controlled by pre-existing topography. Leeder & Gawthorpe (1987) show that carbonate platform margins controlled by active faults may not be able to prograde over the fault scarp because the topography between shelf and basin is maintained by fault movement. Mullins & Neumann (1979) and Eberli & Ginsburg (1989) show that the Bahama Banks are prograding in a north-west direction. This is controlled by deep-water ocean currents which are constructing an apron of periplatform carbonate sand and mud that drapes the base of slope, building a foundation over which progradation can take place.

Drowning of carbonate platforms

Drowned carbonate platforms are a common feature of continental margin successions. Cored wells and outcrop show that shelf carbonates are overlain by a highly condensed, winnowed unit which is commonly glauconitic or phosphatic and contains a concentrated pelagic and nektonic fauna. This is typically overlain by a thick succession of pelagic or distal pro-deltaic sediment. This stratigraphical relationship frequently gives rise to a high-amplitude 'booming' seismic reflector which may be mis-interpreted as a sequence boundary produced during a lowstand. Examples of drowned shelves include Miocene carbonate shelves of the Far East (Epting 1989) and the Early and Middle Jurassic carbonate shelves of the Mediterranean region which are overlain by the highly condensed pelagic Ammonitico Rosso deposit (Jenkyns 1974).

Drowned carbonate shelves deposited during global greenhouse conditions represent a paradox because, under optimum conditions, carbonate production rates can exceed third- and second-order rates of sea-level rise and tectonic subsidence (Schlager 1992). Drowning may be explained if the carbonate system becomes 'ill' prior to drowning. Epting (1989) showed that Miocene carbonate shelves of the Far East retreated from prograding distal siliciclastic systems prior to shut down and drowning. The demise of Cretaceous carbonate shelves in Yugoslavia and Venezuela has been attributed to poisoning of the carbonate system by oceanic anoxic events prior to drowning (Jenkyns 1991; Martinez & Hernandez 1992).

11.4.2 Carbonate Ramps

In contrast to carbonate shelves, carbonate production on carbonate ramps is continuous through the cycle of sea-level change. Carbonate ramps respond to changes in accommodation space by the up- and down-dip migration of facies belts (Figure 11.10). Sedimentary sequences are often difficult to identify owing to the subdued topography and very low depositional slopes of carbonate ramps. At a seismic scale, downlap, onlap and progradation may be difficult to resolve because of the sub-parallel nature of the reflectors. Vertically exaggerated seismic displays may help. The high-resolution sequence stratigraphy of carbonate ramps at an outcrop- or core-scale is also difficult to interpret. In shallow ramp settings several sequence boundaries may be amalgamated to form one composite exposure surface. In the middle and outer parts of the carbonate ramp, sequence boundaries may only form during third- or lower-order sea-level changes. Higher-order sea-level changes may not produce sequence boundaries in this setting but may instead be represented by condensed surfaces or abrupt facies shifts. The sequence stratigraphy of carbonate ramps is reviewed by Burchette & Wright (1992) and Wright & Burchette (1996).

11.4.3 Carbonate–Siliciclastic Systems

Mixed carbonate–siliciclastic sedimentation may occur on carbonate platforms in continental margin settings, surrounding intracratonic basins or fringing basement highs where there is an episodic supply of siliciclastic sediment to the carbonate depocentre. Carbonate productivity is adversely affected by the introduction of siliciclastic sediment, with carbonate production and the deposition of siliciclastic sediments often being mutually exclusive and partitioned in either space or time. Two types of mixed carbonate–siliciclastic systems have been recognized, caused by the following processes.

1. *Spatial variations in siliciclastic supply.* The present south Florida carbonate platform was established during the early Eocene (McKinney 1984; Figure 11.11). Siliciclastic sediment derived from the American continent formed a clastic wedge which passed laterally into the platform. Suspended sediment in the distal part of the siliciclastic wedge suppresses carbonate production. Since the rate of deposition of distal siliciclastic sediment is much lower than that of proximal siliciclastic sediment and shelf carbonates, a deep water channel (the 'Suwannee Channel') formed between the two depocentres. The carbonate system progrades towards the siliciclastic source when siliciclastic input is reduced, and retreats towards the carbonate depocentre when siliciclastic supply is increased. Sequence development in this type of system is influenced by siliciclastic supply which is controlled by climate and tectonism in the hinterland, in addition to sea-level variations.
2. *Reciprocal sedimentation.* This is produced by temporal variations in siliciclastic supply, which often results in the development of alternating siliciclastic- and carbonate-dominated systems at the same location in a basin. This may result from alternate drowning and emergence of a siliciclastic source attached to a carbonate shelf. The late Dinantian carbonate shelf of North Wales was attached to an area of exposed Lower Palaeozoic basement which acted as a siliciclastic source (Figure 11.12). The carbonate shelf succession contains interbedded shallow marine sandstones infilling karstic channels and pipes (Walkden & Davies 1983). During lowstands, the carbonate shelf was karstified and siliciclastic sediment was transported across the shelf in incised channels to supply shoreface sand bodies on the outer shelf (Figure 11.12A). During transgressions,

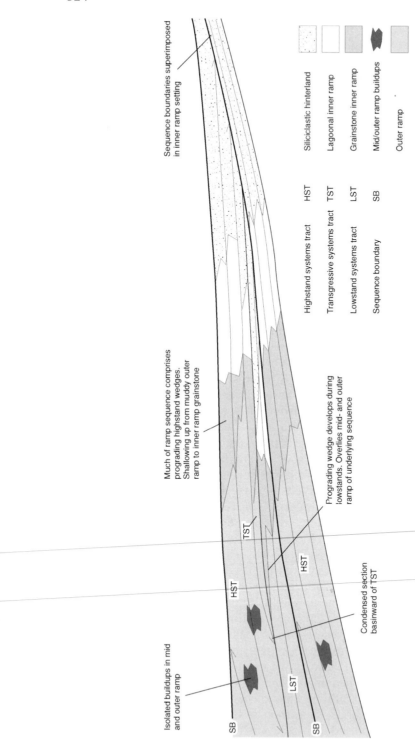

Figure 11.10 Sequence development on a carbonate ramp. [Modified from: Burchette & Wright (1992)]

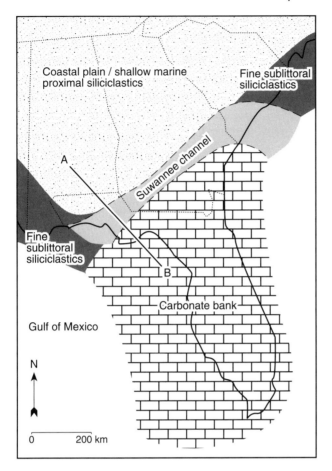

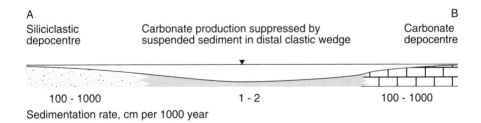

• Siliciclastic and carbonate depocentre separated by shale basin.
• Shale basin migrates towards the carbonate depocentre with an increasing influx of clastics.
• Shale basin migrates away from the carbonate depocentre with a decreasing influx of clastics.

Figure 11.11 *An example of a mixed carbonate–siliciclastic system resulting from spatial variations in sediment supply: the middle Eocene of the south Florida carbonate platform / coastal plain. [Modified from: McKinney (1984)]*

A

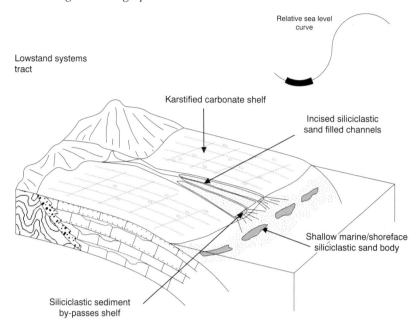

Relative sea level curve

Lowstand systems tract

Karstified carbonate shelf

Incised siliciclastic sand filled channels

Shallow marine/shoreface siliciclastic sand body

Siliciclastic sediment by-passes shelf

B

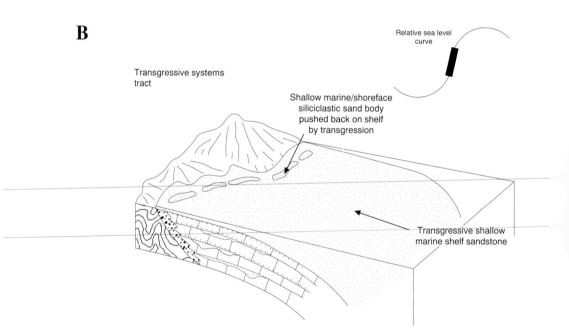

Relative sea level curve

Transgressive systems tract

Shallow marine/shoreface siliciclastic sand body pushed back on shelf by transgression

Transgressive shallow marine shelf sandstone

Figure 11.12 *An example of a mixed carbonate–siliciclastic system resulting from temporal variations in sediment supply: the Late Dinantian carbonate shelf of North Wales during (**A**) lowstand, (**B**) transgression, (**C**) highstand. [Modified from Walkden & Davies (1983) and Davies (1983)]*

C

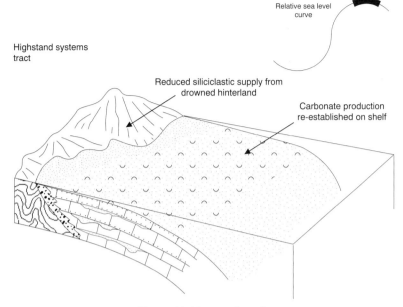

Figure 11.12 *continued*

these sands were reworked shelfward to produce sheet sand bodies (Figure 11.12B). Carbonate sedimentation was re-established on the carbonate shelf during highstands (Figure 11.12C). These can be regarded as carbonate-dominated cycles. Equivalent sediments in northern England are cyclic silici-clastic-dominated deltaic sediments known as Yoredale cycles. Carbonates in these cycles are confined to thin subtidal units which were deposited in shallow marine settings kept free of siliciclastic sediment by the switching of delta lobes. Leeder & Strudwick (1987) recognized tectonic, eustatic and autocyclic controls on sequence development in these cycles (Figure 11.13). A further example of reciprocal carbonate–siliciclastic sedimentation, controlled by variations in sea level, has been documented by Southgate *et al.* (1994) from Late Devonian reef complexes of the Canning Basin, Western Australia.

11.4.4 Mixed Carbonate–Evaporite Systems

Accommodation space changes in evaporitic basins

Evaporitic basins respond to falls in sea level by evaporative draw down and the precipitation of evaporites. Associated carbonate production is generally shut down because either surrounding platforms are exposed or carbonate-producing communities are killed off by hypersaline conditions. Siliciclastic sedimentation is often also limited in this setting because of arid conditions and low run-off.

Evaporite sediments have some of the highest accumulation rates of all sediment types (Kendall 1988), such that sequence boundaries on surrounding carbonate platforms may be equivalent to thick evaporite successions in basins. Evaporite deposition requires an arid climate and a basin which can be readily isolated from the 'world ocean' by a sill or barrier (e.g. Tucker 1991; Kendall 1992). Examples include the Late Permian Zechstein Basin of north-west Europe which was a major intracratonic basin connected to the Boreal Ocean by a seaway (Tucker 1991), and various enclosed intrashelf basins present on the Barents Shelf during the late Palaeozoic (Cecchi 1992).

Ideally, the height of the barrier is such that it is exposed by third- or higher-order sea-level changes. Once the basin is isolated, water-level changes are controlled by a combination of evaporation, in some cases to dryness, run-off from the hinterland and seepage through the barrier if it is permeable; changes often take place rapidly and may bear no relation to global sea-level changes. Evaporite deposition takes place in various shallow- to deep-water environments at times of basin isolation. Carbonate platforms may develop around the edges of the basins during highstands when connection with the ocean is re-established.

As in other systems, sequences are defined by sequence boundaries. A sequence boundary may be expressed as a palaeokarst or evaporite residue over the marginal carbonate platform. Correlative conformities are represented by the base of a lowstand gypsum wedge at the basin margin and at, or just below, the contact between carbonate or carbonaceous laminites and laminated gypsum in the basin centre (Tucker 1991).

Incomplete draw down

The amount of draw down within an evaporitic basin is determined by the degree of aridity, period of isolation and seepage into the basin. Initially, sabkha evaporites are deposited around the basin margin with the precipitation of subaqueous gypsum in shallow hypersaline water. This builds up to form a marginal gypsum wedge such as the Hartlepool Anhydrite (ZS3 in Figure 11.14), which formed during an episode of draw down of the Zechstein Basin and is banked up against the reef margin of the underlying Ford Formation (Tucker 1991; ZS2 in Figure 11.14). If the water column is saturated with gypsum, graded beds and slumps of reworked gypsum sediment will be preserved in the distal part of the marginal wedge and basin. Basin floor deposits will include laminated and bottom-growth gypsum (e.g. Kendall & Harwood 1989). The water column is likely to become stratified which will enhance the preservation of organic matter, leading to the accumulation of source rocks.

The surrounding carbonate platforms will be exposed during draw down and may be subjected to subaerial erosion. The scarcity of meteoric water in arid settings often results in minimal diagenetic alteration of the carbonates during lowstands (Read & Horbury 1993).

Complete draw down

Halite will be precipitated on the basin floor if the relative humidity is less than 76% and there is no surface run-off or seepage from the adjacent ocean (Kendall 1988). Marginal gypsum wedges and laminites may also form during draw down and potash salts may form in conditions of extreme desiccation. Where halite has completely filled the basin subsequent carbonate deposition will take place in an extensive shallow sea (ZS4 Figure 11.14).

Transgressive and highstand sedimentation

When the basin is connected to the open ocean, carbonate platforms may develop around the basin margins. Sabkha and marginal marine evaporites may be deposited as part of the transgressive systems tract if the basin flooding is relatively slow; alternatively, progradational sabkhas may be present in the highstand systems tract. In the basin centre, a thin, condensed, commonly finely laminated succession of pelagic and periplatform carbonates will be deposited. Smith (1989) and Tucker (1991) showed that the morphology of the marginal carbonate ramp or shelf in Zechstein carbonate sequences is determined by the pre-existing topography of the underlying carbonate platform or evaporite basin fill (Figure 11.14).

11.5 FRONTIERS IN SEQUENCE STRATIGRAPHY: NON-MARINE SYSTEMS

Non-marine sequence stratigraphy is a rapidly expanding branch of stratigraphical analysis. Like the application of sequence stratigraphy to marine strata, it is based on the concept that variations in the rates of accommodation space creation or removal and sediment supply will control the generation and arrangement of stratigraphical packages and their bounding surfaces, and that this will vary in a systematic and, hence, predictive manner. Non-marine sequence stratigraphical concepts are, however, far less advanced than those within the marine realm, due to the relative youth of the subject, the difficulty in applying some of the fundamental principles to this setting and the complex nature of many of the depositional systems involved. As a result, the subject is the focus of both significant research and fierce debate within the geological community, and is explored to a greater depth than the more established fields of marine sequence stratigraphy outlined above. An excellent recent review of non-marine sequence stratigraphy has been provided by Shanley & McCabe (1994).

Fluvial systems represent the most volumetrically and economically significant non-marine environment within the stratigraphical record, and the following discussion focuses on these deposits. Although not discussed here, an understanding of the distribution of other non-marine strata-bound mineral resources within sequence stratigraphical framework, such as coal and placer deposits, also form economically important fields.

330

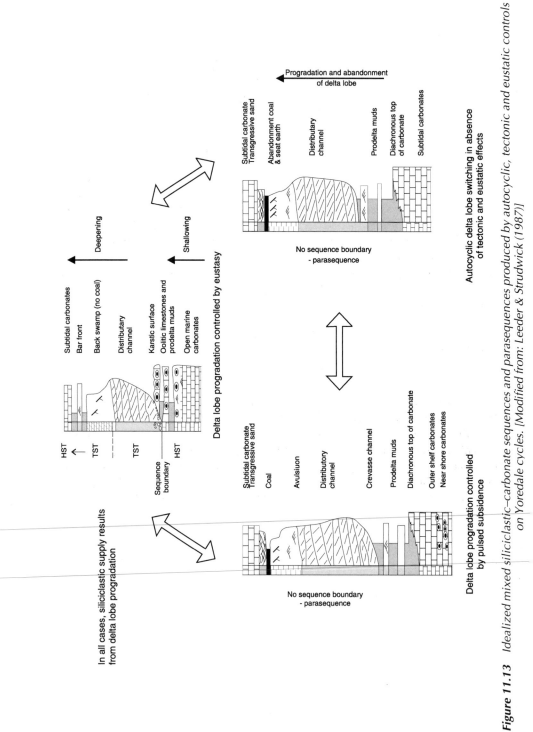

Figure 11.13 Idealized mixed siliciclastic–carbonate sequences and parasequences produced by autocyclic, tectonic and eustatic controls on Yoredale cycles. [Modified from: Leeder & Strudwick (1987)]

ZS4:

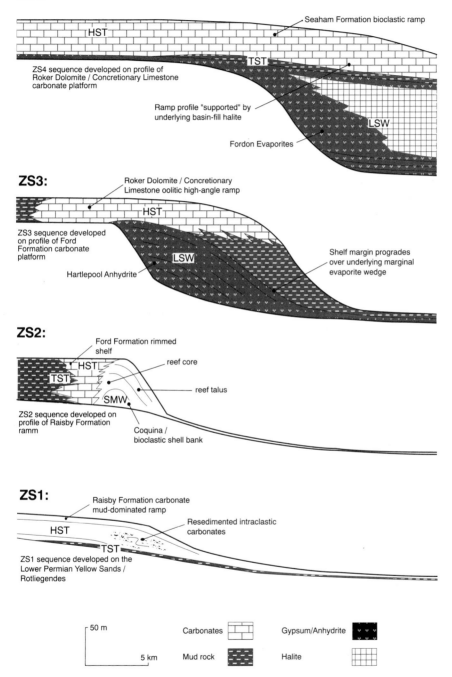

ZS4 sequence developed on profile of
Roker Dolomite / Concretionary Limestone
carbonate platform

Seaham Formation bioclastic ramp

HST

TST

Ramp profile "supported" by
underlying basin-fill halite

LSW

Fordon Evaporites

ZS3:

Roker Dolomite / Concretionary
Limestone oolitic high-angle ramp

HST

ZS3 sequence developed
on profile of Ford
Formation carbonate
platform

LSW

Hartlepool Anhydrite

Shelf margin progrades
over underlying marginal
evaporite wedge

ZS2:

Ford Formation rimmed
shelf

reef core

HST

TST

reef talus

SMW

ZS2 sequence developed on
profile of Raisby Formation
ramm

Coquina /
bioclastic shell bank

ZS1:

Raisby Formation carbonate
mud-dominated ramp

Resedimented intraclastic
carbonates

HST

TST

ZS1 sequence developed on the
Lower Permian Yellow Sands /
Rotliegendes

50 m

5 km

Carbonates

Mud rock

Gypsum/Anhydrite

Halite

Figure 11.14 *Evolving sequence geometry and facies types in successive Zechstein carbonate–evaporite sequences (ZS1 to ZS4) of the Late Permian of north-east England and the adjoining North Sea. [Modified from: Smith (1989) and Tucker (1991)]*

11.5.1 Accommodation Space, Base Level and Equilibrium Surfaces

In non-marine settings, the definition of the conceptual equilibrium surface defining the upper limit of accommodation space is more complex than that defined by base level in the marine realm. In aeolian settings it may be represented by the water table (Kocurek & Havholm 1994; Shanley & McCabe 1994), whilst lacustrine systems create their own base level independent of sea level. In alluvial systems, the concept of the graded river (Mackin 1948) has most commonly been used to approximate this equilibrium surface (e.g. Posamentier *et al.* 1988; Posamentier & Vail 1988; Ross 1989; Shanley & McCabe 1991; Emery & Myers 1996), although some of these authors, in applying the term 'base level' to it, have confused cause (relative sea-level change) with effect (a change in the position of the non-marine equilibrium surface; Schumm 1993). To avoid this confusion, Shanley & McCabe (1994) and Quirk (1996) have adopted the terms 'stratigraphic base level' and 'base profile', respectively.

Base-level controls on changes in accommodation in the fluvial realm

Changes in the shape of a river's graded profile will result in the creation or removal of accommodation space and aggradation or degradation. Numerous factors influence the shape of a graded fluvial system, including changes in discharge, sediment supply, channel form, uplift and the position to which the system grades, the bayline. These may result from changes in climate, tectonics, fluvial system evolution and base level; a detailed discussion of all the factors involved is beyond the scope of this section, but have been addressed by Schumm (1977) and Quirk (1996). First-generation sequence stratigraphical models have concentrated on modelling the effects of base-level change, with fixed rates of sediment supply, due to its regional effect and, hence, predictive nature. The interplay of base level and sediment supply are considered in isolation in the following section.

Whether accommodation space is created or destroyed during base-level change will depend on the relative translation of the bayline with respect to the initial graded stream profile. Where the bayline migrates to a position above the original graded profile or its seaward-projected continuation, accommodation space creation and aggradation will occur (Figures 11.15C, E, F and G). Where the bayline falls below the seaward-projected continuation of the graded profile, accommodation space destruction and degradation will occur (Figure 11.15A), whilst if the bayline migrates along the graded profile or its seaward continuation, equilibrium will be maintained (Figures 11.15B and D).

The response of a fluvial system to base-level fall and basinward bayline migration (regression) will depend on the relative gradient of the downstream (coastal plain) component of the fluvial system and its emergent shelf (Miall 1991; Posamentier *et al.* 1992). Where the shelf is at a lower gradient than the adjacent fluvial profile, the competence of the fluvial system to carry its sediment load will decrease. If the reduction in slope is too large for the river to be able to compensate by a reduction in sinuosity, then aggradation will occur (Figure 11.15C). Where

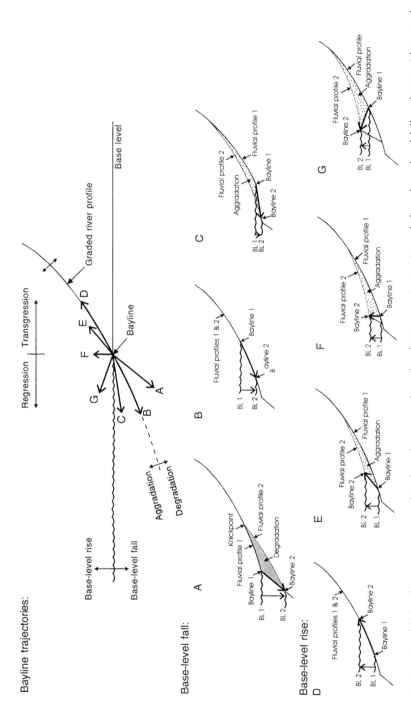

Figure 11.15 Possible bayline trajectories related to changes in base level. Trajectories A–C result from base-level fall and may either result in degradation (A), no change (B), or aggradation (C) of the fluvial system, depending on the position of the new bayline with respect to the projected graded river profile. A base-level fall will always results in regression. A base-level rise may result in either no change (D) or aggradation (E–G) of the fluvial system. This will depend upon the relative rates of base-level rise versus sedimentation rate. Increasing rates of sedimentation with respect to base-level rise will result in a shift from a retrogradation of facies and transgression (E), through aggradation (G), to a progradation of facies and regression of the bayline (F). For further discussion, see the text. [Based on information in: Nummedal et al. (1993)]

there is bathymetric continuity between the graded fluvial profile and the emergent shelf, there is likely to be little or no change in the shape of the fluvial profile and, therefore, no variation in accommodation space (Figure 11.15B). In both these instances, incision, degradation and Type 1 sequence boundary formation will occur only if bayline migration continues beyond the offlap break. If the emergent shelf is at a higher gradient than the fluvial profile, however, and the system is unable to reduce the gradient of its fluvial channel through an increase in sinuosity or by deltaic progradation, incision and degradation will occur immediately (Figure 11.15A).

During base-level rise, the provision of accommodation space will depend on the relative rate of this rise with respect to sediment supply. Where base-level rise is rapid relative to sediment supply, landward bayline migration (transgression) will result in no significant variations in accommodation space, as the bayline simply migrates back along the already established graded river profile (Figure 11.15D; Leopold & Bull 1979; Posamentier & Vail 1988; Allen & Posamentier 1993). Such changes have been observed following dam construction and in some Holocene examples. Accommodation space will only increase during rising base level when there is a component of fluvial progradation and graded profile extension. This will occur when the area available for sediment deposition is restricted, such as within incised valleys, or when rates of base-level rise are slow relative to sediment supply. When sedimentation rates are moderate, but are not able to fill the available accommodation space, a series of backstepping lobes will develop during transgression (Figure 11.15E; Koss *et al.* 1994). However, if sedimentation rates are high, aggradational to progradational bayline stacking patterns may be developed during rising base level (Figures 11.15F and G).

Early sequence stratigraphical models included a shallow gradient marine shelf (Posamentier *et al.* 1988; Posamentier & Vail 1988). As a result, they predict fluvial aggradation during basinward bayline migration, and conclude that apart from minor amounts of aggradation within incised valley systems during late lowstand, alluvial aggradation will be restricted to the late highstand, as depicted in Figure 11.2A, when rapid progradation and basinward bayline migration is anticipated (Posamentier & Vail 1988; Posamentier 1988). Miall (1991) pointed out, however, that many modern marine shelves have a higher gradient than their respective fluvial systems and has called into question the assumptions behind these earlier geometric models. Furthermore, the preservation potential of late highstand deposits is likely to be limited due to erosion during the subsequent base-level fall. For ease of discussion during the remainder of this section, it will be assumed that the shallow marine shelf will have a higher gradient than the lower reaches of its adjacent fluvial system, as in Figure 11.15A, such that a fall in base level will result in an increase in fluvial gradients.

An understanding of the controls on the geometrical arrangement and character of fluvial strata suggests that these may be placed within a sequence stratigraphical framework. Numerical models of alluvial architecture predict that as aggradation rates increase relative to avulsion frequency, the proportion and interconnectedness of channel bodies within an alluvial succession decrease (Allen 1978; Leeder 1978; Bridge & Leeder 1979). Sequence stratigraphical models, as a result, propose that during periods with low aggradation rates, such as during late highstand progradation or lowstand aggradation, laterally extensive stacked sand bodies will develop, whilst during periods with high aggradation rates, such as during early highstand and possibly transgression, more isolated channel bodies will occur (e.g. Posamentier & Vail 1988). Implicit in these latter models, however, is the assumption that avulsion frequency is held constant or at least increases at a lesser rate than sedimentation rate. Recent experimental work suggests that this relationship may not be valid (Bryant *et al.* 1995), and therefore caution must be displayed when applying this simple relationship until more work on the interaction between avulsion and sedimentation rate has been carried out.

Studies of fluvial geomorphology indicate that with increasing gradient, sediment grade and bed load proportion will increase (Lane 1955), and meandering channels will display an increase in sinuosity and channel width:depth ratio (Schumm 1977, 1993). Ultimately, a switch from mixed or suspended load (meandering) to bed load (braided) channel types may also occur (Schumm & Khan 1972). Some or all of these changes may manifest themselves in the stratigraphical record during base-level change, with coarser braided fluvial facies being more prevalent in lowstand and early transgressive systems tracts, and finer-grained meandering fluvial facies being developed in late transgressive and highstand systems tracts. Variations in base level are also likely to influence the pattern of paleosol and coal seam development (Cross 1988; Wright & Marriott 1993; Aitken 1994). Well-drained mature paleosols are anticipated in lowstand systems tracts, although these will have a low preservation potential. Poorly drained conditions during transgression will result in ponded (hydromorphic) soils and the maximum development of coals. Better drained soils and only thin coals are expected in a typical highstand systems tract.

An increasing number of case studies of interbedded fluvial and shallow marine strata demonstrate the applicability of sequence stratigraphical concepts within the estuarine to delta plain environment (e.g. Van Wagoner *et al.* 1990; Dalrymple *et al.* 1994; Van Wagoner 1995; Howell & Aitken 1996). In exceptional circumstances, laterally continuous exposure allows these strata to be traced back into the exclusively fluvial realm, allowing the theoretical considerations of a fluvial system's response to base-level change, outlined above, to be tested. The most notable example of this comes from the work of Shanley & McCabe (1991, 1993; Shanley *et al.* 1992) within Upper Cretaceous strata of the Kaiparowits Plateau, southern Utah, USA. They were able to successfully correlate sequence boundaries between coeval shallow-marine and alluvial and coal-bearing strata, almost 80 km up depositional dip, to enable them to directly compare changes in alluvial architecture and facies with the stacking patterns of equivalent shoreface strata (Figure 11.16). This study forms the largest single influence on current thinking about fluvial sedimentation within a sequence stratigraphical framework in the coastal plain environment.

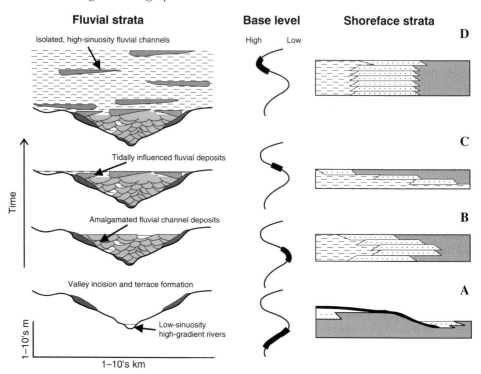

Figure 11.16   *Summary diagram illustrating the relationship between shoreface and alluvial architecture as a function of base-level change in coeval shallow marine and fluvial Upper Cretaceous strata of the Kaiparowits Plateau, southern Utah, USA.* **A.** *Slow rates of base-level rise and base-level fall result in progradational shoreface geometries, sequence boundary formation and lowstand deposition in the near-shore setting. In the non-marine setting, fluvial incision and terrace formation occurs.* **B.** *During slowed rates of base-level fall and initial base-level rise, shoreface geometries switch from progradational to aggradational and/or retrogradational. Incised valleys aggrade with amalgamated fluvial deposits.* **C.** *Increased rates of base-level rise result in retrogradational shoreface parasequences capped by a maximum flooding surface. These are temporally equivalent to tidally influenced fluvial strata.* **D.** *Reduced rates of base-level rise, in this example, result in aggradational shoreface parasequences as this is balanced by sedimentation rates. Isolated meanderbelt sandstones develop in the fluvial realm.* [Reproduced with permission from Blackwell Science, from: Shanley & McCabe (1993)]

11.5.2 Fluvial Sedimentation within a Sequence Stratigraphical Framework

Recognition of sequence boundaries

Channel scour and incision are processes inherent within fluvial systems. The discrimination of incisional events related to a fall in base level below the offlap break from localized autocyclic channel scour is key to the recognition of sequence boundaries within fluvial strata. Best & Ashworth (1997), from observations of the magnitude of autocyclic scour at the confluence of modern large braided rivers, suggest that an erosive boundary with relief greater than five times mean channel depth that extends for distances greater than the floodplain width would provide unambiguous evidence for sequence boundary recognition. Abrupt changes in alluvial architecture, with an increase in channel stacking, amalgamation and grain size, and changes in sediment composition may also be useful criteria, although a number of other independent factors may result in similar stratigraphical features. In interfluve areas, the identification of sequence boundaries becomes even more cryptic, and at present most studies involve the lateral tracing of these surfaces away from incised valley units. Theoretical considerations would suggest these should be marked by mature, well-drained paleosols marking prolonged periods of subaerial exposure (Van Wagoner *et al.* 1990; Wright & Marriott 1993), and abundant rootlet development is commonly observed. Rather than being well drained, however, Aitken & Flint (1996) have identified interfluve sequence boundaries which are characterized by gley paleosols developed under poorly drained conditions. Geochemical studies indicate that these originated under freely drained conditions, but were subsequently overprinted by gley processes as a result of rising water tables during late lowstand to early transgression. This may prove a powerful tool for the identification of interfluvial paleosols, although the preservation potential of these surfaces is low as they are often cannibalized during transgression.

Recognition of maximum flooding surfaces

Maximum flooding within alluvial strata is most clearly defined by the invasion of tidal processes into areas formerly dominated by fluvial processes. Shanley *et al.* (1992) recognized a tidal influence within fluvial sediments up to 65 km inland from their coeval shoreline, although the distance of maximum incursion is dependent on the fluvial dynamics, rate of base-level rise, physiography and tidal range of that particular example. Where base-level rise is sufficient to flood incised valley interfluves, extensive poorly drained swamp and lacustrine facies will also be deposited; and in areas lacking terrigenous input, maximum flooding may be represented by the maximum development of lacustrine carbonates or extensive coals (Cross 1988; Aitken 1994; Aitken & Flint 1995).

Systems tracts

The successful assignment of alluvial strata to their appropriate systems tracts is dependent on the identification of their related bounding surfaces and their position within a sequence; parasequences have not been successfully identified in fluvial strata. A lowstand systems tract, by definition, is bound by a sequence boundary below and the first widespread transgressive surface (initial flooding surface) above. Marine flooding surfaces are commonly absent in the updip portions of incised valleys. Instead, extensive carbonaceous shales and coals which cap incised valley fill and interfluve surfaces have been interpreted as the product of initial flooding (Aitken & Flint 1995). By this definition, an incised valley fill will be assigned to the lowstand systems tract. In many instances, however, apparent initial interfluve flooding is simply the geomorphic expression of final incised valley flooding, when the switch from a previously confined depositional system to one that is free to migrate across the whole coastal plain, results in a dramatic reduction in local sedimentation rates. In addition, Shanley & McCabe (1993, 1994) point out that initial flooding is likely to onlap in a landward direction, such that true fluvial lowstand deposits will be absent in up-dip regions.

In their study of the Kaiparowits Plateau, Shanley & McCabe (1991, 1993), on the basis of regional correlation and architecture, interpreted laterally amalgamated fluvial channel sandstones which immediately overlie sequence boundaries to form during the early stages of transgression (Figure 11.16B). Where developed, these laterally amalgamated channel bodies are likely to grade vertically into thin, typically isolated high-sinuosity channels contained within fine-grained overbank sediments, reflecting an increase in the provision of accommodation space over sedimentation (Figure 11.16D). These channels may contain heterolithic tidally influenced strata, the maximum development of which represents the landward equivalent of the maximum flooding surface which separates the transgressive systems tract from the overlying highstand systems tract (Shanley & McCabe 1991; Shanley *et al.* 1992; Figure 11.16C). In fluvial settings either unaffected by, or upstream of the influence of tidal processes, it is difficult to distinguish the top of the transgressive systems tract from the base of the highstand systems tract as the alluvial architecture of these deposits are similar. In coal-bearing successions, the transition between well developed and more poorly developed coals has been interpreted as the point of maximum accommodation and the maximum flooding surface (Aitken 1994; Aitken & Flint 1995). Aitken & Flint (1995) recognize an increase in amalgamation and net: gross of isolated channel bodies upwards in the highstand systems tract, although late highstand deposits are commonly removed by subsequent sequence boundary formation.

11.5.3 The Applicability of Sequence Stratigraphy to Fluvial Strata

Alluvial successions in the stratigraphical record represent the direct interplay between upstream tectonic, climate and intrinsic controls, and basinal base-level controls on accommodation space and sediment supply. Most work to date has concentrated on the role of base-level changes on the subdivision and correlation of strata, due to its regional and predictive nature, to the exclusion of these other factors. Furthermore, in order to test this end-member control, most studies have been carried out on low-gradient coastal plain fluvial systems in tectonically quiescent settings, where base-level changes are likely to be the most pronounced. There is concern, therefore, that in most instances the stratigraphical record of alluvial strata is far more complex than current fluvial sequence stratigraphical models and geological case studies would suggest (Schumm 1993; Westcott 1993; Shanley & McCabe 1994). These concerns are centred around two points: whether changes in base level are effectively transmitted up a fluvial system to result in predictable changes in accommodation space; and, if so, whether it is possible to distinguish these from the numerous other extrinsic and intrinsic factors which affect the relative interplay between sedimentation and accommodation space in fluvial systems. Both concerns increase, the greater the distance upstream from the bayline.

The transmission of base-level changes upstream

A study of the Gulf Coast fluvial systems' response to the last eustatic lowstand at 18 Ka, when sea level was *c.* 120 m below present levels, and the subsequent rise in sea level, indicate that the influence of base-level change extends approximately 370 km up the Mississippi River (Autin *et al.* 1991), 150 km up the Trinity and Sabine rivers (Thomas & Anderson 1994), and 90 km up the Colorado River (Blum 1994). Although these are significant distances, they are only a fraction of their river lengths, with the Colorado River being 2333 km long, for instance. Numerous factors are thought to be important in influencing the distance to which base-level changes are transmitted up a fluvial system's reach. These include the direction, magnitude, rate and duration of base-level change; the lithology, structure and nature of valley alluvium; and the valley morphology and inclination, and river morphology and adjustability (Schumm 1993). A river system will initially respond to a change in valley gradient, induced by a base-level rise or fall, by changing the sinuosity and efficiency of its channel in order that the system may continue to carry the supplied load with the available discharge (Mackin 1948; Schumm 1993). Where an increase in valley slope is the result of base-level change, the fluvial system will increase its sinuosity, thereby increasing its channel length and decreasing its gradient. Experimental studies indicate that the resultant change in channel pattern may not totally compensate for the change in valley slope, but that channel width: depth and roughness will also increase in order to decrease channel efficiency. Similarly, a decrease in valley slope will result in a decrease in sinuosity, and a decrease in channel width: depth and roughness. The rate of base-level change is also important in this context, as when base-level fall is rapid, vertical channel incision is likely, preventing lateral channel migration and a reduction in channel slope through an increase in sinuosity. In addition, a narrow deep valley will concentrate flow in a confined area, increasing its ability to incise. Braided rivers do not have the ability to increase their channel length through an increase in sinuosity and are, therefore, more susceptible to incision. It is only when the magnitude of base-level change is sufficiently large that the resultant variation in valley slope cannot be absorbed by adjustments in channel form, or when other factors prevent effective channel adjustments, that the graded profile will alter and aggradation or degradation will occur.

Variations in the nature of the river substrate will affect knickpoint migration; where a resistant bedrock lithology crops out, a local base level will be established. The cohesiveness of the valley alluvium will also influence knickpoint characteristics. If this is cohesive, rapid upstream migration is likely, with the knickpoint maintaining its definition, whilst with non-cohesive alluvium, the effects of a base-level change are likely to be rapidly dissipated.

Geometric considerations of a graded river profile indicate that a fall in base level will be most strongly felt in its downstream reaches. Here the gradient of the fluvial system is at its lowest and therefore any drop in base level will result in the most dramatic increase in gradient. The relative change of slope will decrease upstream, however, because of the convexity of the graded river, such that the ability of channel adjustments to absorb this change become increasingly likely. In addition, the magnitude of potential incision between the original and new profiles will decrease upstream, such that it will become increasingly difficult to distinguish between erosional events resulting from base-level drop and localized, autocyclic scouring. Koss *et al.* (1994), from experimental work, have suggested that base-level changes have little effect on the upper reaches of a drainage basin. Base-level changes will be most keenly felt in low-gradient systems.

Schumm (1993: 289) concluded that 'the propagation of the effects of base-level change along a large alluvial river probably will be moderate, moving upstream, but not for long distances. Total rejuvenation of the drainage system is not expected, although the effect will be greatest where base-level change is great, incision rapid, and the rivers are confined.'

Distinguishing between base level and other controls on alluvial strata

In addition to the potential inability of base-level changes to be transferred efficiently upstream within a fluvial system, the degree to which upstream controls on sediment supply and accommodation space effect the genesis and architecture of alluvial strata, independent of base-level change, will determine the ability of sequence stratigraphical analysis to provide a correlative and predictive template for alluvial strata.

The relative importance of upstream versus downstream controls on alluvial strata have been explored in detail during an examination of the evolution of the Colorado drainage during the last 20 Ka by Blum (1990, 1994). During the last glacial lowstand (18 Ka), the former shoreline was approximately 100 km offshore from the modern Colorado River mouth (Frazier 1974), and an incised valley was cut by the Colorado River as it extended across the continental shelf. Rapid transgression followed, which slowed at 6 Ka until a relative stillstand developed between 3 Ka and the present day (Figure 11.17E).

The Colorado drainage is divided into two main components (Figure 11.17A): the upper Colorado drainage, to the west of the Balcones Escarpment, from which sediment is predominantly supplied; and the lower Colorado drainage, which comprises an upper, erosional bedrock confined valley which acts essentially as a sediment conduit to a lower depositional, largely undissected Quaternary alluvial–deltaic complex and the Gulf Coast Basin. Four unconformity bound, radiometrically dated, late Pleistocene to Recent stratigraphical units are recognized throughout much of the Colorado drainage. Upstream of the apex of the Quaternary alluvial–deltaic complex, younger stratigraphical units occur at progressively lower elevations due to protracted bedrock valley deepening (Figures 11.17B and C). Superimposed on this long-term trend, periods of valley migration, widening and sediment storage, followed by incision, sediment removal and flood plain abandonment, produced the observed stratigraphical units. These episodes of alluvial aggradation and degradation are unrelated to changes in base level (Figure 11.17E), but instead, represent the response to fluctuations in the relationship between discharge and sediment concentration driven by upstream climatic and environmental changes.

Downstream of the Quaternary alluvial–deltaic plain apex, approximately 90 km upstream from the present shoreline, older stratigraphical units are progressively buried, with the depth of burial increasing downstream (Figures 11.17B and D). This arrangement is related to base-level change, with the base of the lowermost stratigraphical unit, the Eagle Lake Alloformation, representing incision into deposits of the underlying last interglacial highstand during glacial base-level fall. Transgression followed during the present interglacial, with sediment storage on the predominantly subaerial alluvial–deltaic plain occurring mainly during the final stages of base-level rise. This last point fits with models of increasing alluvial sedimentation during periods of slowing base-level rise. It is important to note, however, that within the available accommodation space created by the rise in base-level, the development of unconformity-bound stratigraphical packages was controlled by climatic variations that were unconnected with the apparent changes in base level. In addition, due to the higher frequency of these events, there is no unique surface to which the eustatically controlled sequence boundary, as defined in coastal and marine strata, can be correlated.

Like climatic controls, tectonic variations in accommodation space and sediment supply have an important impact on the alluvial architecture of many fluvial successions, which may overwhelm any evidence for base-level change, or may superficially resemble the effects of base-level change itself. Intrinsic controls may also have a significant effect on variations in sediment supply within a fluvial system (Schumm 1977; Westcott 1993).

Summary

Fluvial sequence stratigraphy is an exciting and expanding field of stratigraphical study; one in which concepts are still rapidly evolving and a diversity of opinions are held. Generally accepted, however, is the view that it is the lower reaches of a fluvial system that are most likely to be affected by changes in base level (Figure 11.18). Shanley & McCabe (1993), for instance, suggest that it is low-gradient fluvial systems within perhaps 100–150 km of their bayline which have the potential to be greatly affected by base-level change, and that it is this portion of the fluvial system which is most likely to be preserved in the stratigraphical record. Indeed, in many instances the application of sequence stratigraphical principles to fluvial strata has proved highly successful (and profitable; e.g. the Prudhoe Bay Field, North Slope, Alaska; Emery & Myers 1996). However, in moderate- to high-gradient systems, in tectonically complex regions and in the upstream portions of all fluvial systems, caution must be taken when ascribing variations in facies and alluvial architecture to changes in base level (Figure 11.18). Instead, other extrinsic and intrinsic factors which control the relative interplay between sedimentation and accommodation space are likely to be paramount. This does not invalidate the application of sequence stratigraphical principles to fluvial environments, as sequence stratigraphical analysis still forms an important tool

342

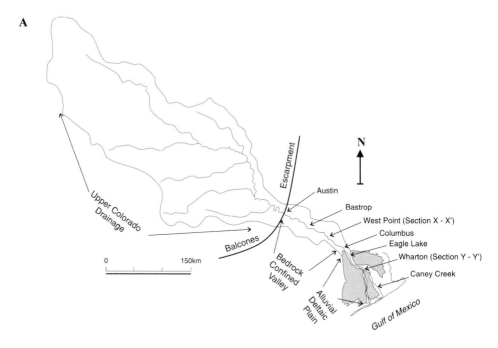

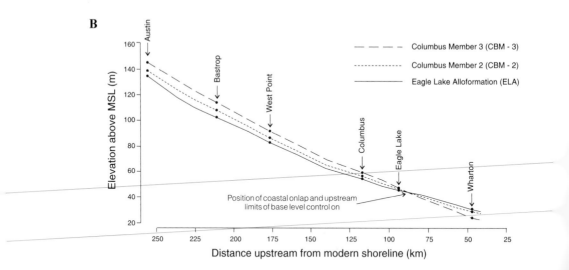

Figure 11.17 *The alluvial architecture of the lower Colorado valley, Texas, USA.* **A.** *Location of the Colorado River.* **B.** *Longitudinal profiles of the differing allostratigraphical units o the lower Colorado valley, illustrating the upstream limit of base-level controls, with progress ive incision upstream of this point and aggradation below it. Upstream of Eagle Lake, longi tudinal profiles for Columbus Bend members 1 and 2 (CBM-1 and CBM-2) are essentially th same, whilst downstream of this point, those for the Eagle Lake Alloformation (ELA) and Col umbus Bend Member 1 are essentially the same.* **C.** *Schematic cross-section of the lowe*

C

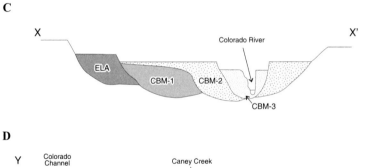

D

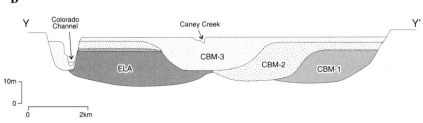

E

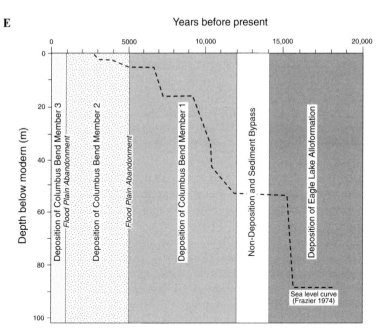

Figure 11.17 *(continued) Colorado valley, near West Point, Texas, in the bedrock-confined Inner Coastal Plain.* **D.** *Schematic cross-section of the lower Colorado valley, near Wharton, Texas, on the alluvial–deltaic plain.* **E.** *Timing of alluvial depositional and erosional events on the alluvial–deltaic plain of the lower Colorado River in relation to the Quaternary sea-level curve for the Gulf of Mexico of Frazier (1974). [Reproduced with permission from the Association of American Petroleum Geology, from: Blum (1994)]*

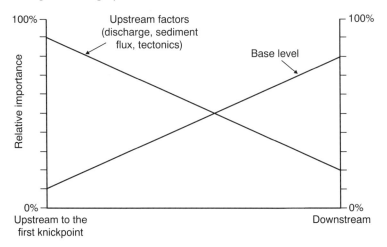

Figure 11.18   *Schematic depiction of the relative importance of upstream and downstream factors on changes in fluvial systems. [Modified from: Posamentier & James (1993)]*

(Posamentier & James 1993), but it must be borne in mind that in these regions, the coherence and lateral continuity of key boundaries and stratal patterns may be limited and, therefore, be of lesser or no predictive or correlative value at the regional scale.

11.6 CONCLUSION

Sequence stratigraphy has fundamentally changed the way in which we interpret the stratigraphical record. The main strengths of the model are its ability to explain the depositional patterns of sediments on a basinal scale with reference to changing sea level, subsidence and sediment supply, and to provide a chrono-stratigraphical template for correlation. It is also important for the predictive and contextual framework that it gives to facies sedimentology; most standard texts now incorporate sequence stratigraphical concepts into their sedimentological models, which are no longer considered in purely autocyclic terms (e.g. Walker & James 1992; Reading 1996). Sequence stratigraphy reinforces the view of the earth as a dynamic system and provides an allocyclic mechanism with an over-riding control on most sedimentary systems.

ACKNOWLEDGEMENTS

We are grateful to Matt Hart for drafting many of the figures and to John Howell for comments on the manuscript. S.J.V. would like to thank Trevor Elliott and members of the Liverpool STRAT group for their many discussions on sequence stratigraphy.

REFERENCES

Ahr, W.M. 1989. Sedimentary and tectonic controls on the development of an early Mississippian carbonate ramp, Sacramento Mountains area, New Mexico. In Crevello, P.D., Wilson, J.L., Sarg, J.F. & Read, J.F. (eds) *Controls on carbonate platform and basin development*. Society of Economic Paleontologists and Mineralogists Special Publication 44, 203–212.

Aitken, J.F. 1994. Coal in a sequence stratigraphic framework. *Geoscientist*, **4**, 9–12.

Aitken, J.F. & Flint, S.S. 1995. The application of high resolution sequence stratigraphy to fluvial systems: a case study from the Upper Carboniferous Breathitt Group, eastern Kentucky, U.S.A. *Sedimentology* , **42**, 3–30.

Aitken, J.F. & Flint, S.S. 1996. Variable expression of interfluvial sequence boundaries in the Breathitt Group (Pennsylvanian), eastern Kentucky, USA. In Howell, J.A. & Aitken, J.F. (eds) *High resolution sequence stratigraphy: innovations and applications*. Geological Society of London Special Publication 104, 193–206.

Allen, G.P. & Posamentier, H.W. 1993. Sequency stratigraphy and facies models of an incised valley fill: the Gironde estuary, France. *Journal of Sedimentary Petrology*, **63**, 378–391.

Allen, J.R.L. 1978. Studies in fluviatile sedimentation: an exploratory quantitative model for the architecture of avulsion-controlled alluvial suites. *Sedimentary Geology*, **21**, 129–147.

Autin, W.J., Burns, S.F., Miller, B.J., Saucier, R.T. & Snead, J.I. 1991. Quaternary geology of the lower Mississippi Valley. In Morrison, R.B. (ed.) *Quaternary nonglacial geology; conterminous U.S.* The Geology of North America, Geological Society of America K-2, 547–582.

Best, J.L. & Ashworth, P.J. 1997. Scour in large braided rivers and the recognition of sequence stratigraphic boundaries. *Nature*, **387**, 275–277.

Blum, M.D. 1990. Climatic and eustatic controls on Gulf coastal plain fluvial sedimentation: an example from the Late Quaternary of the Colorado River, Texas. In *Sequence stratigraphy as an exploration tool, concepts and practices in the Gulf Coast*. Gulf Coast Section of the Society of Economic Paleontologists and Minerologists Eleventh Annual Research Conference Program and Abstracts, 71–83.

Blum, M.D. 1994. Genesis and architecture of incised valley fill sequences: a Late Quaternary example from the Colorado River, Gulf Coastal Plain of Texas. In Weimer, P. & Posamentier, H.W. (eds) *Siliciclastic sequence stratigraphy – recent developments and applications*. American Association of Petroleum Geologists Memoir 58, 259–283.

Boreen, T.D. & James, N.P. 1993. Holocene sediment dynamics on a cool water carbonate shelf: Otway, southeastern Australia. *Journal of Sedimentary Petrology*, **63**, 574–588.

Boreen, T.D. & James, N.P. 1995. Stratigraphic sedimentology of Tertiary cool water limestones, SE Australia. *Journal of Sedimentary Research*, **B65**, 142–159.

Bridge, J.S. & Leeder, M.R. 1979. A simulation model of alluvial stratigraphy. *Sedimentology*, **26**, 617–644.

Bridges, P.H., Gutteridge, P. & Pickard, N.A.H. 1995. The environmental setting of Early Carboniferous mud-mounds. In Monty, C.L.V., Bosence, D.W.J., Bridges, P.H. & Pratt, B.R. (eds) *Carbonate mud mounds: their origin and evolution*. Special Publication of the International Association of Sedimentologists 23, 171–190.

Bryant, M., Falk, P. & Paola, C. 1995. Experimental study of avulsion frequency and rate of deposition. *Geology*, **23**, 365–368.

Burchette, T.P. & Wright, V.P. 1992. Carbonate ramp depositional systems. *Sedimentary Geology*, **79**, 3–57.

Carter, R.M., Abbott, S.T., Fulthorpe, C.S., Haywick, D.W. & Henderson, R.A. 1991. Application of global sea level and sequence-stratigraphic models in Southern Hemisphere Neogene strata from New Zealand. In Macdonald, D.I.M. (ed.) *Sedimentation, tectonics and eustasy: sea-level changes at active margins*. Special Publication of the International Association of Sedimentologists 12, 41–65.

Cartwright, J.A., Haddock, R.C. & Pinheiro, L.M. 1993. The lateral extent of sequence boundaries. In Williams, G.D. & Dobb, A. (eds) *Tectonics and seismic sequence stratigraphy.* Geological Society of London Special Publication 71, 15–34.

Cathles, L.M. & Hallam, A. 1991. Stress-induced changes in plate density, Vail sequences, epeirogeny and short-lived global sea-level fluctuations. *Tectonics,* **10**, 659–671.

Cecchi, M. 1992. Carbonate sequence stratigraphy: applications to the determination of play-models in the Upper Palaeozoic succession of the Barents Sea, offshore northern Norway. In Vorren, T.O., Bergsager, E., Dahl-Stamnes, Ø.A., Holter, E., Johansen, B., Lie, E. & Lund, T.B. (eds) *Arctic geology and petroleum potential.* Norwegian Petroleum Society (NPF) Special Publication 2. Elsevier, Amsterdam, 419–438.

Cloetingh, S. 1988. Intraplate stress: a tectonic cause for third-order cycles in apparent sea level? In Wilgus, C.K., Hastings, C.A., Kendall, C.G.St.C., Posamentier, H.W., Ross, C.A. & Van Wagoner, J.C. (eds) *Sea-level changes: an integrated approach.* Society of Economic Paleontologists and Mineralogists Special Publication 42, 19–29.

Cross, T.A. 1988. Controls on coal distribution in transgressive–regressive cycles, Upper Cretaceous, Western Interior, U.S.A. In Wilgus, C.K., Hastings, C.A., Kendall, C.G.St.C., Posamentier, H.W., Ross, C.A. & Van Wagoner, J.C. (eds) *Sea-level changes: an integrated approach.* Society of Economic Paleontologists and Mineralogists Special Publication 42, 371–380.

Dalrymple, R.W., Boyd, R. & Zaitlin, B.A. (eds) 1994. *Incised-valley systems: origin and sedimentary sequence.* Society of Economic Paleontologists and Mineralogists Special Publication 51.

Davies, J.R. 1983. *Stratigraphy, sedimentology and palaeontology of the Lower Carboniferous of Anglesey.* Unpublished PhD thesis, University of Keele, UK.

Eberli, G.P. & Ginsburg, R.N. 1989. Cenozoic progradation of northwestern Great Bahama Bank, a record of lateral platform growth and sea level fluctuations. In Crevello, P.D., Wilson, J.L., Sarg, J.F. & Read, J.F. (eds) *Controls on carbonate platform and basin development.* Society of Economic Paleontologists and Mineralogists Special Publication 44, 339–351.

Emery, D. & Myers, K.J. (eds) 1996. *Sequence stratigraphy.* Blackwell Science, Oxford.

Epting, M. 1989. Miocene carbonate buildups of central Luconia, offshore Sarawak. In Bally, A.W. (ed.) *Atlas of seismic stratigraphy, Vol. 3.* American Association of Petroleum Geologists Studies in Geology 27, 168–173.

Frazier, D.E. 1974. *Depositional episodes: their relationship to the Quaternary stratigraphic framework in the northwestern portion of the Gulf of Mexico Basin.* Bureau of Economic Geology GC74-1, University of Texas at Austin.

Galloway, W.E. 1989. Genetic stratigraphic sequences in basin analysis I: Architecture and genesis of flooding-surface bounded depositional units. *Bulletin of the American Association of Petroleum Geologists,* **73**, 125–142.

Gawthorpe, R.L. & Gutteridge, P. 1990. Geometry and evolution of platform margin bioclastic shoals, late Dinantian (Mississippian), Derbyshire, UK. In Tucker, M.E., Wilson, J.L., Crevello, P.D., Sarg, R.J. & Read, J.F. (eds) *Carbonate platforms: facies, sequences and evolution.* Special Publication of the International Association of Sedimentologists 9, 39–54.

Haq, B.U. 1991. Sequence stratigraphy, sea-level change, and significance for the deep sea. In Macdonald, D.I.M. (ed.) *Sedimentation, tectonics and eustasy: sea-level changes at active margins.* Special Publication of the International Association of Sedimentologists 12, 3–39.

Haq, B.U., Hardenbol, J. & Vail, P.R. 1987. Chronology of fluctuating sea levels since the Triassic. *Science,* **235**, 1153–1165.

Howell, J.A. & Aitken, J.F. (eds) 1996. *High resolution sequence stratigraphy: innovations and applications.* Geological Society of London Special Publication 104.

Jenkyns, H.C. 1974. Origin of red nodular limestone (Ammonitico Rosso, Knollenkalke) in the Mediterranean Jurassic: a diagenetic model. In Hsü, K.J. & Jenkyns, H.C. (eds) *Pelagic sediments: on land and under the sea.* Special Publication of the International Association of Sedimentologists 1, 249–271.

Jenkyns, H.C. 1991. Impact of Cretaceous sea level rise and anoxic events on the Mesozoic carbonate platform of Yugoslavia. *Bulletin of the American Association of Petroleum Geologists*, **75**, 1007–1017.

Jervey, M.T. 1988. Quantitative geological modeling of siliciclastic rock sequences and their seismic expressions. In Wilgus, C.K., Hastings, C.A., Kendall, C.G.St.C., Posamentier, H.W., Ross, C.A. & Van Wagoner, J.C. (eds) *Sea-level changes: an integrated approach*. Society of Economic Paleontologists and Mineralogists Special Publication 42, 47–69.

Kendall, A.C. 1988. Aspects of evaporite basin stratigraphy. In Schreiber, B.C. (ed.) *Evaporites and hydrocarbons*. Columbia University Press, New York, 11–65.

Kendall, A.C. 1992. Evaporites. In Walker, R.G. & James, N.P. (eds) *Facies models: response to sea-level change*. Geological Society of Canada, 375–409.

Kendall, A.C. & Harwood, G.M. 1989. Shallow water gypsum in the Castile Formation – significance and implications. In Harris, P.M. & Grover, G.A. (eds) *Subsurface and outcrop examination of the Capitan Shelf Margin, northern Delaware Basin*. Society of Economic Paleontologists and Mineralogists Core Workshop 13, 451–457.

Kocurek, G. & Havholm, K.G. 1994. Eolian event stratigraphy – a conceptual framework. In Weimer, P. & Posamentier, H.W. (eds) *Siliciclastic sequence stratigraphy – recent developments and applications*. American Association of Petroleum Geologists Memoir 58, 393–409.

Koss, J.E., Ethridge, F.G. & Schumm, S.A. 1994. An experimental study of the effects of base-level change on fluvial, coastal plain and shelf systems. *Journal of Sedimentary Research*, **B64**, 90–98.

Lane, E.W. 1955. The importance of fluvial morphology in hydraulic engineering. *American Society of Civil Engineers Proceedings*, **81**, 745.1–745.17.

Lawrence, D.T. 1994. Evaluation of eustasy, subsidence, and sediment input as controls on depositional sequence geometries and the synchroneity of sequence boundaries. In Weimer, P. & Posamentier, H.W. (eds) *Siliciclastic sequence stratigraphy – recent developments and applications*. American Association of Petroleum Geologists Memoir 58, 337–367.

Leeder, M.R. 1978. A quantative stratigraphic model for alluvium, with special reference to channel deposit density and interconnectedness. In Miall, A.D. (ed.) *Fluvial sedimentology*. Canadian Society of Petroleum Geologists Memoir 5, 587–596.

Leeder, M.R. & Gawthorpe, R.L. 1987. Sedimentary models for extensional tilt block/half-graben basins. In Coward, M.P., Dewey, J.F. & Hancock, P.L. (eds) *Continental extensional tectonics*. Geological Society of London Special Publication 28, 139–152.

Leeder, M.R. & Strudwick, A.E. 1987. Delta–marine interactions: a discussion of sedimentary models for Yoredale-type cyclicity in the Dinantian of Northern England. In Miller, J., Adams, A.E. & Wright, V.P. (eds) *European Dinantian environments*. John Wiley & Sons, Chichester, 115–130.

Leopold, L.B. & Bull, W.B. 1979. Base level, aggradation, and grade. *Proceedings of the American Philosophical Society*, **123**, 168–202.

Mackin, J. H. 1948. Concept of the graded river. *Bulletin of the Geological Society of America*, **59**, 463–512.

Martinez, J.I. & Hernandez, R. 1992. Evolution and drowning of the Late Cretaceous Venezuelan carbonate platform. *Journal of South American Earth Sciences*, **5**, 197–210.

McKinney, M.L. 1984. Suwannee Channel of the Paleogene coastal-plain – support for the carbonate suppression model of basin formation. *Geology*, **12**, 343–345.

Miall, A.D. 1991. Stratigraphic sequences and their chronostratigraphic correlation. *Journal of Sedimentary Petrology*, **61**, 497–505.

Mullins, H.T. & Neumann, A.C. 1979. Deep carbonate bank margin structure and sedimentation in the northern Bahamas. In Doyle, L.J. & Pilkey, O.H. (eds) *Geology of continental slopes*. Society of Economic Paleontologists and Mineralogists Special Publication 27, 165–192.

Nummedal, D., Riley, G.W. & Templet, P.L. 1993. High-resolution sequence architecture: a chronostratigraphic model based on equilibrium profile studies. In Posamentier, H.W.,

Summerhayers, C.P., Haq, B.U. & Allen, G.P. (eds) *Sequence stratigraphy and facies associations*. Special Publication of the International Association of Sedimentologists 18, 55–68.

Parkinson, N. & Summerhayes, C. 1985. Synchronous global sequence boundaries. *Bulletin of the American Association of Petroleum Geologists*, **69**, 685–687.

Payton, C.E. (ed.) 1977. *Seismic Stratigraphy – applications to hydrocarbon exploration*. American Association of Petroleum Geologists Memoir 26.

Pitman, W.C. 1978. Relationship between eustasy and stratigraphic sequences of passive margins. *Bulletin of the Geological Society of America*, **89**, 1389–1403.

Plint, A.G. 1996. Marine and non-marine systems tracts in fourth-order sequences in the Early–Middle Cenomanian, Dunvegan Alloformation, northeastern British Columbia, Canada. In Howell, J.A. & Aitken, J.F. (eds) *High resolution sequence stratigraphy: innovations and applications*. Geological Society of London Special Publication 104, 159–191.

Posamentier, H.W. 1988. Fluvial deposition in a sequence stratigraphic framework. In James, D.P. & Leckie, D.A. (eds) *Sequences, stratigraphy and sedimentology: surface and subsurface*. Canadian Society of Petroleum Geologists Memoir 15, 582–583.

Posamentier, H.W. & James, D.P. 1993. An overview of sequence-stratigraphic concepts: uses and abuses. In Posamentier, H.W., Summerhayers, C.P., Haq, B.U. & Allen, G.P. (eds) *Sequence stratigraphy and facies associations*. Special Publication of the International Association of Sedimentologists 18, 3–18.

Posamentier, H.W. & Vail, P.R. 1988. Eustatic controls on clastic deposition II – sequence and systems tract models. In Wilgus, C.K., Hastings, C.A., Kendall, C.G.St.C., Posamentier, H.W., Ross, C.A. & Van Wagoner, J.C. (eds) *Sea-level changes: an integrated approach*. Society of Economic Paleontologists and Mineralogists Special Publication 42, 125–154.

Posamentier, H.W., Jervey, M.T. & Vail, P.R. 1988. Eustatic controls on clastic deposition I – conceptual framework. In Wilgus, C.K., Hastings, C.A., Kendall, C.G.St.C., Posamentier, H.W., Ross, C.A. & Van Wagoner, J.C. (eds) *Sea-level changes: an integrated approach*. Society of Economic Paleontologists and Mineralogists Special Publication 42, 109–124.

Posamentier, H.W., Allen, G.P., James, D.P. & Tesson, M. 1992. Forced regressions in a sequence stratigraphic framework: concepts, examples, and exploration significance. *Bulletin of the American Association of Petroleum Geologists*, **76**, 1687–1709.

Quirk, D.G. 1996. 'Base level': a unifying concept in alluvial sequence stratigraphy. In Howell, J.A. & Aitken, J.F. (eds) *High resolution sequence stratigraphy: innovations and applications*. Geological Society of London Special Publication 104, 37–49.

Ramsbottom, W.H.C. 1973. Transgression and regression in the Dinantian: a new synthesis of British Dinantian stratigraphy. *Proceedings of the Yorkshire Geological Society*, **39**, 567–607.

Ramsbottom, W.H.C. 1978. Rates of transgression and regression in the Carboniferous of NW Europe. *Journal of the Geological Society of London*, **136**, 147–153.

Read, J.F. 1982. Carbonate platforms of passive (extensional) continental margins: types, characteristics and evolution. *Tectonophysics*, **81**, 195–212.

Read, J.F. 1985. Carbonate platform facies models. *Bulletin of the American Association of Petroleum Geologists*, **66**, 860–878.

Read, J.F. & Horbury, A.D. 1993. Eustatic and tectonic controls on porosity evolution beneath sequence-bounding unconformities and parasequence disconformities on carbonate platforms. In Horbury, A.D. & Robinson, A.G. (eds) *Diagenesis and basin development*. American Association of Petroleum Geologists Studies in Geology 36, 155–197.

Reading, H.R. (ed.) 1996. *Sedimentary environments: processes, facies and stratigraphy*, 3rd edition. Blackwell Science, Oxford.

Ross, W.C. 1989. Modeling base-level dynamics as a control on basin-fill geometries and facies distribution: a conceptual framework. In Cross, T.A. (ed.) *Quantitative dynamic stratigraphy*. Prentice Hall, Englewood Hills, New Jersey, 387–399.

Sarg, J.R. 1988. Carbonate sequence stratigraphy. In Wilgus, C.K., Hastings, C.A., Kendall, C.G.St.C., Posamentier, H.W., Ross, C.A. & Van Wagoner, J.C. (eds) *Sea-level changes: an integrated approach*. Society of Economic Paleontologists and Mineralogists Special Publication 42, 155–181.

Schlager, W. 1992. *Sedimentology and sequence stratigraphy of reefs and carbonate platforms.* American Association of Petroleum Geologists Continuing Education Course Note Series.

Schumm, S.A. 1977. *The fluvial system.* John Wiley & Sons, Chichester.

Schumm, S.A. 1993. River response to baselevel change: implications for sequence stratigraphy. *Journal of Geology*, **101**, 279–294.

Schumm, S.A. & Khan, H.R. 1972. Experimental study of channel patterns. *Bulletin of the Geological Society of America*, **83**, 1755–1770.

Shanley, K.W. & McCabe, P.J. 1991. Predicting alluvial architecture through sequence stratigraphy – an example from the Kaiparowits Plateau, Utah. *Geology*, **19**, 742–745.

Shanley, K.W. & McCabe, P.J. 1993. Alluvial architecture in a sequence stratigraphic framework: a case history from the Upper Cretaceous of southern Utah, USA. In Flint, S.S. & Bryant, I.D. (eds) *The geological modelling of hydrocarbon reservoirs and outcrop analogues.* Special Publication of the International Association of Sedimentologists 15, 21–56.

Shanley, K.W. & McCabe, P.J. 1994. Perspectives on the sequence stratigraphy of continental strata. *Bulletin of the American Association of Petroleum Geologists*, **78**, 544–568.

Shanley, K.W., McCabe, P.J. & Hettinger, R.D. 1992. Tidal influence in Cretaceous fluvial strata from Utah, USA: a key to sequence stratigraphic interpretation. *Sedimentology*, **39**, 905–930.

Sloss, L.L. 1963. Sequences in the cratonic interior of North America. *Bulletin of the Geological Society of America*, **64**, 93–113.

Smith, D.B. 1989. The late Permian palaeogeography of north-east England. *Proceedings of the Yorkshire Geological Society*, **47**, 285–312.

Southgate, P.N., Kennard, J.M., Jackson, M.J., O'Brien, P.E. & Sexton, M.J. 1994. Reciprocal lowstand clastic and highstand carbonate sedimentation, subsurface Devonian reef complex, Canning Basin, western Australia. In Weimer, P. & Posamentier, H.W. (eds) *Siliciclastic sequence stratigraphy – recent developments and applications.* American Association of Petroleum Geologists Memoir 58, 157–179.

Thomas, M.A. & Anderson, J.B. 1994. Sea-level controls on the facies architecture of the Trinity/Sabine incised-valley system, Texas continental shelf. In Dalrymple, R.W., Boyd, R. & Zaitlin, B.A. (eds) *Incised-valley systems: origin and sedimentary sequence.* Society of Economic Paleontologists and Mineralogists Special Publication 51, 63–82.

Tucker, M.E. 1985. Shallow marine carbonate facies and facies models. In Brenchley, P.J. & Williams, B.P.J. (eds) *Sedimentology: recent developments and applied aspects.* Geological Society of London Special Publication 18, 147–169.

Tucker, M.E. 1991. Sequence stratigraphy of carbonate–evaporite basins: models and application to the Upper Permian (Zechstein) of northeast England and adjoining North Sea. *Journal of the Geological Society of London*, **148**, 1019–1036.

Tucker, M.E. & Hollingworth, N.T.J. 1986. The Upper Permian reef complex (EZ1) of north east England: diagenesis in a marine evaporative setting. In Schroeder, J.H. & Purser, B.H. (eds) *Reef diagenesis.* Springer-Verlag, City, 270–290.

Underhill, J.R. 1991. Late Jurassic seismic sequences, Inner Moray Firth, UK: a critical test of a key segment of Exxon's original global cycle chart. *Basin Research*, **3**, 79–98.

Underhill, J.H. & Partington, M.A. 1994. Use of genetic sequence stratigraphy in defining and determining a regional tectonic control on the 'Mid Cimmerian Unconformity': implications for North Sea basin development and the global sea-level chart. In Weimer, P. & Posamentier, H.W. (eds) *Siliciclastic sequence stratigraphy – recent developments and applications.* American Association of Petroleum Geologists Memoir 58, 449–484.

Vail, P.R., Mitchum, R.M., Jr., Todd, R.G., Widmier, J.M., Thompson, S., III, Sangree, J.B., Bubb, J.N. & Hatlelid, W.G. 1977a. Seismic stratigraphy and global changes in sea level. In Payton, C.E. (ed.) *Seismic stratigraphy – applications to hydrocarbon exploration.* American Association of Petroleum Geologists Memoir 26, 49–212.

Vail, P.R., Mitchum, R.M., Jr. & Thompson, S., III. 1977b. Global cycles of relative changes in sea level. In Payton, C.E. (ed.) *Seismic stratigraphy – applications to hydrocarbon exploration.* American Association of Petroleum Geologists Memoir 26, 83–97.

Van Wagoner, J.C. 1995. Sequence stratigraphy and marine to nonmarine facies architecture of foreland basin strata, Book Cliffs, Utah, U.S.A. In Van Wagoner, J.C. & Bertram, G. (eds) *Sequence stratigraphy of foreland basin deposits: examples from the Cretaceous of North America.* American Association of Petroleum Geologists Memoir 64, 137–223.

Van Wagoner, J.C., Posamentier, H.W., Mitchum, R.M., Vail, P.R., Sarg, J.F., Louitt, T.S. & Hardenbol, J. 1988. An overview of the fundamentals of sequence stratigraphy and key definitions. In Wilgus, C.K., Hastings, C.A., Kendall, C.G.St.C., Posamentier, H.W., Ross, C.A. & Van Wagoner, J.C. (eds) *Sea-level changes: an integrated approach.* Society of Economic Paleontologists and Mineralogists Special Publication 42, 39–45.

Van Wagoner, J.C., Mitchum, R.M., Campion, K.M. & Rahmanian, V.D. 1990. *Siliclastic sequence stratigraphy in well logs, cores, and outcrops.* American Association of Petroleum Geologists Methods in Exploration Series, 7.

Walkden, G.M. & Davies, J.R. 1983. Polyphase erosion of subaerial omission surfaces in the Late Dinantian of Anglesey, North Wales. *Sedimentology*, **30**, 861–878.

Walker, R.G. & James, N.P. (eds) 1992. *Facies models: response to sea-level change.* Geological Society of Canada.

Westcott, W.A. 1993. Geomorphic thresholds and complex response of fluvial systems – some implications for sequence stratigraphy. *Bulletin of the American Association of Petroleum Geologists*, **77**, 1208–1218.

White, N. & McKenzie, D.P. 1988. Formation of the steer's head geometry of sedimentary basins by differential stretching of the crust and mantle. *Geology*, **16**, 250–253.

Wilgus, C.K., Hastings, C.A., Kendall, C.G.St.C., Posamentier, H.W., Ross, C.A. & Van Wagoner, J.C. (eds) 1988. *Sea-level changes: an integrated approach.* Society of Economic Paleontologists and Mineralogists Special Publication 42.

Williams, D.F. 1988. Evidence for and against sea-level changes from the stable isotopic record of the Cenozoic. In Wilgus, C.K., Hastings, C.A., Kendall, C.G.St.C., Posamentier, H.W., Ross, C.A. & Van Wagoner, J.C. (eds) *Sea-level changes: an integrated approach.* Society of Economic Paleontologists and Mineralogists Special Publication 42, 31–36.

Wood, L.J., Ethridge, F.G. & Schumm, S.A. 1993. The effects of rate of base-level fluctuation on coastal-plain, shelf and slope depositional systems: an experimental approach. In Posamentier, H.W., Summerhayers, C.P., Haq, B.U. & Allen, G.P. (eds) *Sequence stratigraphy and facies associations.* Special Publication of the International Association of Sedimentologists 18, 43–53.

Wright, V.P. 1994. Early Carboniferous carbonate systems; an alternative to the Cainozoic paradigm. *Sedimentary Geology*, **93**, 1–5.

Wright, V.P. & Burchette, T.P. 1996. Shallow-water carbonate environments. In: Reading, H.R. (ed.) *Sedimentary environments: processes, facies and stratigraphy*, 3rd Edition. Blackwell Science, Oxford, 325–394.

Wright, V.P. & Faulkner, T.J. 1990. Sediment dynamics of early Carboniferous ramps: a proposal. *Geological Journal*, **25**, 139–144.

Wright, V.P. & Marriott, S.B. 1993. The sequence stratigraphy of fluvial depositional systems: the role of floodplain sediment storage. *Sedimentary Geology*, **86**, 203–210.

12
Stratigraphical Applications of Radiogenic Isotope Geochemistry

Malcolm J. Hole

Radiogenic isotope geochemistry provides the most important method of assigning absolute ages to the stratigraphical record. Whereas most stratigraphical techniques provide information that enables the establishment of relative sequences of events in the geological record, none of these techniques can yield ages of rocks in years, or the dates of geological events. For example, although ammonite biozones can be used to correlate sedimentological successions over vast areas, they cannot be reliably used to determine how many millions of years are covered by the Jurassic period. Biostratigraphy is also inadequate in the consideration of palaeontologically barren rocks, such as Precambrian gneisses or some fluvial red-bed sequences. In these and other cases, radiogenic isotopes can be used to create an accurate sequence of events as well as an absolute age for their formation. Importantly, absolute dating of rocks and geological events allows for the calibration of the Chronostratigraphical Scale (see Chapter 13), and enables a global picture of geological events through earth history to be constructed. Knowing the absolute ages of rocks and dates of geological events is clearly crucial to our understanding of the evolution of the planet on which we live.

In general, isotopic techniques are applied to igneous and metamorphic rocks and/or their component minerals, although in some cases sedimentary rocks and their minerals can also be dated. For igneous rocks, and in particular granitoids, it may be possible to produce very precise dates of intrusion from an investigation of the isotopic properties of its component minerals. In metamorphic rocks, particularly if they are meta-igneous, absolute ages may be provided in the best cases for

Unlocking the Stratigraphical Record: Advances in Modern Stratigraphy. Edited by P. Doyle and M.R. Bennett.

not only the metamorphic events which created them, but also the initial crystallization of primary igneous rock. Isotopic dating of sedimentary rocks is not easy.
For example, the detrital material of clastic sedimentary rocks will provide only
information about the date of formation of the original source, and will not be
useful in the determination of the age of deposition. However, where authigenic
mineral growth during diagenesis has taken place, some dating is possible. Isotopic dating is also possible in geologically recent sediments using radiogenic
isotopes of carbon from included organic matter.

The purpose of this chapter is to summarize the commonly used isotopic methods
for generating absolutes ages for rocks, and to illustrate how these data can also be
used to constrain the ages of other geological events such as metamorphism and
diagenesis. It is not intended as a comprehensive account of dating using radiogenic
isotopes, and the review by Faure (1986) should be consulted for full details of all the
techniques described, and for the full derivation of all the relevant equations. In this
chapter, the principles of radioactive decay are first described and the methodology
behind deriving absolute ages will be discussed, using the rubidium–strontium
technique as an example. In the sections which follow, the most commonly used
methods in isotope geochronology are discussed in turn, with reference to examples
of their application taken from recent geological literature.

12.1 THE PRINCIPLES OF ISOTOPE GEOLOGY

An isotope is defined as one of two or more atoms that have the same atomic
number but which contain different numbers of neutrons. Simply put, this means
that an individual element, for example strontium (Sr, atomic number 38), always
has the same electronic configuration, for example the same number of protons
and electrons, and consequently behaves in a particular and unique manner in
geological systems. However, there are four isotopes of strontium, all of which
contain different numbers of neutrons. Because neutrons, unlike electrons, have a
discernible mass, the four isotopes of strontium all have different atomic masses.
By convention these are denoted ^{84}Sr, ^{86}Sr, ^{87}Sr and ^{88}Sr, the superscript giving the
atomic mass of the isotope in atomic mass units (amu). The critical point here is
that some isotopes, known as daughter isotopes, are produced by radioactive
decay of another isotope, the parent isotope, whilst others are totally stable and
their abundance does not change through geological time. There are a variety of
mechanisms of radioactive decay, and many radioactive isotopes have 'branched'
decay schemes where an individual parent isotope gives rise to a number of
different daughter isotopes. In addition, some decay series produce isotopes that
are intermediate between the parent and daughter isotopes of interest with respect
to geochronology. The example used here to illustrate the principles of isotope
geochronology is that of the decay of rubidium (Rb) to strontium (Sr).

For strontium, ^{87}Sr is produced from the decay of an isotope of another element,
$^{87}Rubidium$; ^{87}Rb is known as the parent isotope and ^{87}Sr the daughter isotope. The
other isotopes of strontium ^{84}Sr, ^{86}Sr and ^{88}Sr – are all stable. The rate at which ^{87}Rb
decays to ^{87}Sr is a physical constant, the decay constant (λ; Table 12.1). The decay

Table 12.1 *Commonly used half-lives and decay constants for isotope systems. Half-life (t$_{1/2}$) is related to decay constant by the expression; t$_{1/2}$ = ln 2/λ. Values as recommended by Steiger & Jäger (1977)*

Parent	Daughter	Decay constant (year^{-1})	Half-life
^{87}Rb	^{86}Sr	1.42×10^{-11}	4.88 Ga
^{40}K	^{40}Ar	5.81×10^{-9}	110 Ma
^{147}Sm	^{143}Nd	6.54×10^{-12}	108 Ga
^{176}Lu	^{176}Hf	1.96×10^{-11}	35.3 Ga
^{232}Th	^{208}Pb	4.947×10^{-11}	14 Ga
^{235}U	^{207}Pb	9.848×10^{-10}	704 Ma
^{238}U	^{206}Pb	1.551×10^{-10}	4.468 Ga

constant has been precisely determined in the laboratory, the internationally accepted value being 1.42×10^{-11} year^{-1} (Steiger & Jäger 1977), a value which is inversely proportional to the half-life ($t_{1/2}$) of $4.89 \pm 0.04 \times 10^{10}$ years. Importantly, the only way in which it is possible to produce ^{87}Sr is by radioactive decay of ^{87}Rb; that is, all ^{87}Sr currently present in the earth has been produced by radioactive decay. So how is this useful in determining the absolute ages of rocks?

12.1.1 Determining the Age of Rocks

Supposing we manufacture a piece of material that contains no strontium atoms at all, but contains a large concentration of rubidium atoms. We then leave that material for a few hundred million years, and ensure that no rubidium is lost, and no strontium gained during that period. If we were then to analyse it, we would find that it contained a number of atoms of ^{87}Sr which had been produced by radioactive decay of ^{87}Rb, such that the sample now has less atoms of rubidium and measurable amounts of strontium. Since all the ^{87}Sr has been produced by radioactive decay, and the rate at which decay occurs is well known, then by measuring the amount of ^{87}Sr in the sample using a mass spectrometer we can determine precisely how long decay has been taking place and therefore the absolute date in years on which the sample was manufactured. This is the underlying fundamental principle of isotope geochronology. Of course, natural rock systems are far more complex than this simple example, because all rocks will contain measurable amounts of rubidium and strontium at their time of formation. This example demonstrates the fundamental principles of isotope geochronology, which are essentially the same for all radiogenic isotope techniques. As the abundance of isotopes is measured using a mass spectrometer, it is worth considering some of the principles and practices of mass spectrometry.

12.1.2 Mass Spectrometry

Mass spectrometry relies on the basic principle that objects of different masses will follow different flight paths if accelerated with the same energy. The preparation and investigation of samples using mass spectrometry involves a set procedure which is described below using the Rb–Sr system as an example.

The sample for isotopic analysis is first dissolved in acids (normally hydrofluoric and nitric or perchloric acid) and then finally converted to a soluble chloride. The elements of interest are then chemically separated from the dissolved sample using ion exchange resins. In the Rb–Sr system, once the separation is complete there will be two solutions: one containing only rubidium, and one only strontium. The solutions are then evaporated down to leave a solid residue which represents the chloride of each element. This residue is then redissolved in a minuscule amount of ultra-pure water and put onto a metal filament (tungsten or rhenium). The water is evaporated off leaving a solid residue of the element of interest. For the analysis of the isotopes of strontium, once the sample has been placed in the mass spec-trometer, and an extremely high vacuum has been achieved, it is heated and thereby ionized. A strong accelerating voltage then hurls the individual ions down a curved flight tube which has a large and powerful electromagnet at its centre. Since the sample has been ionized and has a charge, its flight path can be controlled by altering the intensity of the magnetic field. So, because different isotopes have different masses, it is then possible by altering the magnetic field to accurately aim a beam of ions of a specific isotope at a target. The target also serves as a detector, or collector, and gives out an electrical signal that is proportional to the intensity of the ion beam of the isotope in question. In this way it is possible to accurately determine the intensity of the ion beams of all the isotopes of interest, and therefore their apparent concentrations. In practice, the concentrations of the radiogenic isotopes are normalized to a stable isotope, such that all data are reported as ratios. One advantage of this, is that it is not necessary to ensure 100% recovery of all the Sr in the sample during dissolution and chemical separation. This technique is known as solid source thermal ionization mass spectrometry.

The analysis of Rb/Sr ratios is in practice carried out by an elemental analytical method. Elemental analysis of rubidium and strontium is often achieved by using X-ray fluorescence spectrometry (XRF). This is ideal when a large amount of sample is available and the elements of interest are in high enough concentrations. The ^{87}Rb/^{86}Sr ratios of a sample can be calculated from the elemental Rb/Sr ratio if the ^{87}Sr/^{86}Sr ratio of the sample is known:

$$^{87}\text{Rb}/^{86}\text{Sr} = 2.69295 + (0.28304 \times {}^{87}\text{Sr}/^{86}\text{Sr}_\text{m}) \times \text{Rb}/\text{Sr} \tag{1}$$

where: ^{87}Sr/^{86}Sr$_\text{m}$ is the measured ^{87}Sr/^{86}Sr ratio; and Rb/Sr is the elemental ratio determined by XRF.

This analytical technique cannot be used for mineral separates because there will not be sufficient sample to allow an XRF pressed powder pellet, weighing some 15 g, to be made. In this case, the ^{87}Rb/^{86}Sr ratio is calculated by using a technique known as *isotope dilution*. This technique relies on 'spiking' a sample with a known concentration of a particular stable isotope of rubidium or strontium that has been artificially concentrated in the laboratory. If a known amount of the spike is added to a sample, then the difference between the concentration in the sample measured by mass spectrometry and that of the concentration added to the sample must be the unknown concentration in the sample. This technique is routinely applied to other isotope techniques such as the Sm–Nd and U–Pb methods because the absolute abundances of these elements are not great enough to analyse by other means.

Some dating techniques involve analysis of the isotopes of gasses (e.g. K–Ar; see below). In this case, instead of dissolving the sample and carrying out a chemical separation, the sample is fused in a vacuum using a radio-frequency induction furnace. This releases the gas in question, which is then introduced into the mass spectrometer. This technique is known as gas source analysis.

Clearly the above summary is very much a simplification, and modern mass spectrometers are technologically advanced (and costly) instruments. Nevertheless, the principles of analysis are important background to the following sections, which detail the individual techniques of isotopic dating available.

12.2 THE RUBIDIUM–STRONTIUM METHOD OF DATING

The Rb–Sr method of dating is one that has been used to great effect for many years, and it is particularly useful for the dating of granitic rocks that have undergone little or no alteration and have not been affected by metamorphism. Indeed the entire geochronological framework for the subduction-related igneous rocks of the Antarctic Peninsula, part of the proto-Pacific margin of Gondwana, is based on Rb–Sr geochronology (e.g. Pankhurst 1982; Pankhurst *et al.* 1988). In view of its importance, this technique is discussed in detail below.

12.2.1 Rb–Sr Whole-Rock Isochrons

Consider a magma of basic composition which has been emplaced, in a liquid state, into the continental crust above a subduction zone. This magma will begin to cool and consequently crystallization will proceed. Initially, the mineral phases dominating crystallization are hornblende and pyroxene. Neither of these mineral species contains any rubidium or strontium; that is, they have very low mineral/melt partition coefficients (k_D) such that as crystallization proceeds the volume of liquid remaining in the system decreases and the concentration of both rubidium and strontium increase (Figure 12.1). At some stage plagioclase begins to crystallize. Since plagioclase has an affinity to incorporate strontium ($k_D > 1$) into its crystal lattice but not rubidium, the remaining melt will have relatively more rubidium than strontium. If this process continues, then when crystallization of the granitoid is complete, there will be a large range of Rb/Sr ratios for samples within the same intrusion, which crystallized geologically *instantaneously*. Consequently, because some facies of the granitoid have more ^{87}Rb to decay to ^{87}Sr than others, over hundreds of millions of years the rate of generation of ^{87}Sr will be dependent on the amount of rubidium relative to strontium in different samples, as well as the amount of time that has elapsed since crystallization. Therefore, if a collection of whole-rock samples is made from a range of the facies in a granitoid

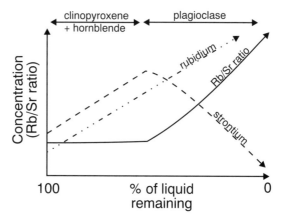

Figure 12.1 *Diagrammatic representation of the concentrations of Rb and Sr, and variations in the Rb/Sr ratio during the crystallization of a granitic magma. Note that for a crystallizing assemblage of clinopyroxene and amphibole, both Rb and Sr are incompatible and thus their concentrations increase with an increasing amount of crystallization. Once plagioclase is stable on the liquidus, because k_DSr_{plag} is > 1, the Sr concentration decreases, Rb continues to increase and thus the Rb/Sr ratio of the system increases. This gives a final suite of rocks with variable Rb/Sr ratios*

which all have different Rb/Sr ratios and which all crystallized at the same time, we can determine the age of crystallization.

For the Rb–Sr method, the stable isotope used for normalization is ^{86}Sr, and an equation can be derived:

$$^{87}Sr/^{86}Sr_m = {}^{87}Sr/^{86}Sr_i + {}^{87}Rb/^{86}Sr_m\,(e^{\lambda t} - 1) \qquad (2)$$

where $^{87}Sr/^{86}Sr_m$ is the $^{87}Sr/^{86}Sr$ ratio of a sample measured on a mass spectrometer today; $^{87}Sr/^{86}Sr_i$ is the $^{87}Sr/^{86}Sr$ ratio of the sample at the time of formation, known as the initial ratio or R0; $^{87}Rb/^{86}Sr_m$ is the $^{87}Rb/^{86}Sr$ ratio calculated from measurements made on a mass spectrometer or by other means today; λ is the decay constant, in this case 1.42×10^{-11} year^{-1}; and t is the time since the system crystallized.

Equation (2) has the general form of $y = mx + c$, the equation for a straight line. Thus a relatively simple binary plot of the two measured parameters $^{87}Rb/^{86}Sr_m$ versus $^{87}Sr/^{86}Sr_m$ will be a straight line for samples with the same age; this is known as an isochron diagram (*iso* = 'same'; *chron* = 'age') (Figure 12.2). An isochron diagram has an intercept which is $^{87}Sr/^{86}Sr_i$ and a slope equivalent to $e^{\lambda t} - 1$. The determination of the slope of an isochron diagram allows the age of the samples to be calculated:

$$t = 1/\lambda \times \ln\,(\text{slope} + 1) \qquad (3)$$

An isochron diagram for some real data for a granitoid, the Bildad Peak intrusion from the Antarctic Peninsula, is shown in Figure 12.3. An important feature of

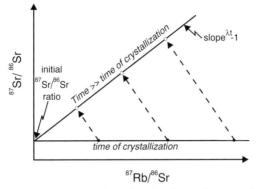

Figure 12.2 *Diagrammatic representation of an isochron diagram showing the evolution of $^{87}Sr/^{87}Sr$ and $^{87}Rb/^{86}Sr$ from the time of crystallization (solid circles) to a time much later than that of crystallization (open circles). Note that $^{87}Sr/^{86}Sr$ increases though time whereas $^{87}Rb/^{86}Sr$ decreases. The age of the intrusion (specifically, the time since crystallization) can be calculated from the slope of the isochron by applying equation (2). The intercept is the $^{87}Sr/^{86}Sr$ ratio of the system at the time of crystallization and has petrogenetic significance*

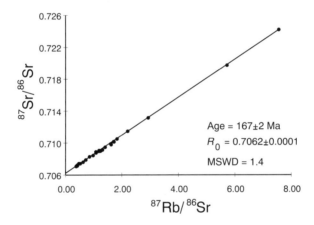

Figure 12.3 *Rb–Sr isochron for a zoned diorite–granodiorite–adamellite intrusion at Bildad Peak, Graham Land, Antarctic Peninsula. Note the broad spread of $^{87}Rb/^{86}Sr$ and $^{87}Sr/^{87}Sr$ ratios, which gives a statistically precise fit of the data to the isochron equation, and thus a low MSWD. [Based on data in: Pankhurst (1982)]*

the isochron diagram is the statistical 'goodness of fit' parameter which defines how well the data conform to the isochron equation. The most commonly used method of data manipulation is the York-Williams method which yields a parameter called the MSWD (mean standard weighted deviates), the value of which should be less than two to make an isochron statistically significant. The range of $^{87}Rb/^{86}Sr$ ratios is clearly an extremely important factor in allowing a statistically valid isochron to be produced; a limited range of $^{87}Rb/^{86}Sr$ ratios will not allow an accurate slope to be determined, largely due to the limits of accuracy and precision of the analytical technique. Consequently, there is little point in trying to date a suite of basalts by the Rb–Sr method, simply because all basalts have very low Rb/Sr ratios and the slope of a line cannot be accurately determined. This is an important limitation on the method. Additionally, the age of the samples is important in allowing an isochron diagram to be constructed; whatever the range in $^{87}Rb/^{86}Sr$ ratios, if the samples are too young then there will be little spread in $^{87}Sr/^{86}Sr$ ratios, and again a statistically significant age cannot be determined.

12.2.2 Determining the Age of Mineral Growth

So far the determination of ages based on whole-rock isochrons has been consider-ed. However, it is possible to date the growth of individual mineral grains in a sample as long as the mineral contains rubidium and strontium. Suitable minerals that are abundant in granitoids and are commonly dated include muscovite, biotite and K-feldspar. For example, if a hypothetical granitoid that was emplaced 300 million years ago was subjected to a thermal metamorphic event 150 million years ago at a temperature of *c.* 500 °C, this metamorphism would have caused changes to take place in the Rb–Sr system. This is because when minerals are heated to a certain temperature, they are no longer closed to diffusion of rubidium and strontium. This has the effect that individual mineral grains undergo isotopic homogenization, because rubidium and strontium are free to move in and out of the crystal lattice. If the minerals are then cooled, diffusion is stopped and the Rb–Sr system once more begins *in situ* radioactive decay (Figure 12.4). The tem-

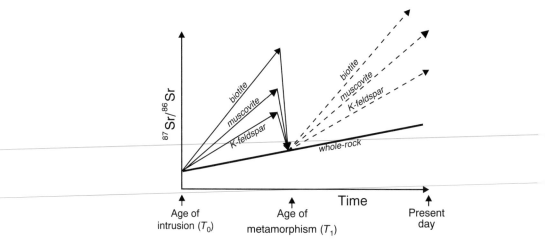

Figure 12.4 *Diagrammatic representation of a whole-rock isochron which has undergone a period of low-grade metamorphism during which the biotite, muscovite and K-feldspar have been taken above their blocking temperatures (Table 12.2) and thus have undergone isotopic homogenization. Post-metamorphism, once all the minerals have once again cooled below their respective blocking temperatures, in situ isotopic decay once again takes place. Note that the slope of the whole-rock isochron is undisturbed by the metamorphic event. A mineral isochron will yield the age when the minerals cooled below their blocking temperatures and the whole-rocks will yield the original age of crystallization*

perature at which elemental diffusion ceases is known as the *blocking temperature*, and above this temperature the isotopic system is homogenized. While the metamorphic event affects individual mineral grains, it does not have any effect on the scale of a whole-rock, so even if thermally metamorphosed, the whole-rock samples can still yield the age of crystallization (Figure 12.4).

In practice, it is possible to construct whole-rock–mineral and mineral–mineral isochrons. To do this, individual mineral species are physically separated from the whole-rock and their $^{87}Rb/^{86}Sr$ and $^{87}Sr/^{86}Sr$ ratios are measured. An isochron diagram is plotted for the minerals and whole rocks. If affected by thermal metamorphism, then the mineral Rb–Sr age will be younger than that provided by analysis of whole-rocks. The blocking temperatures for the Rb–Sr system for muscovite is *c.* 500 °C and for biotite 300 °C (Table 12.2). Therefore the mineral isochron records the time of metamorphism (time above 300 or 500 °C, depending on the mineral in question) and the time of intrusion of the whole rock body.

An example of the technique is provided by studies of the Carn Chuinneag granite of northern Scotland. Pidgeon & Johnson (1974) have determined that the whole-rocks yield an age of 548 ± 10 Ma, while a muscovite–biotite–K-feldspar isochron yields an age of 403 ± 5 Ma. The interpretation is that the granite was intruded 548 Ma ago and the granite was subsequently metamorphosed during the Caledonian Orogeny at 403 Ma. In this case both the age of the initial intrusion and the age of the metamorphism have been determined. It is even possible to refine this technique further by using more than one isotopic system on a range of minerals with different blocking temperatures, such that a whole sequence of cooling and heating events can be investigated, and in some cases it has even proved possible to determine cooling and uplift rates (e.g. Cliff 1985; Dempster 1985). This theme is explored later in this chapter.

Table 12.2 *Blocking (closure) temperatures for common minerals for different isotopic systems. FT = fission track method*

Mineral	Method	Closure temperature for mineral systems, $T(°C)$
Zircon	U–Pb	>800
Baddeleyite	U–Pb	>800
Monazite	U–Pb	700
Titanite	U–Pb	600
Garnet	U–Pb	>550
Garnet	Sm–Nd	>550
Hornblende	K–Ar	500
Muscovite	Rb–Sr	500
Muscovite	K–Ar	350
Apatite	U–Pb	350
Biotite	Rb–Sr	300
Biotite	K–Ar	280
K-Feldspar	K–Ar	200
Zircon	FT	200
Apatite	FT	120

12.2.3 Dating of Low-Grade Metamorphism and Diagenesis

During low-grade metamorphism and sometimes even during diagenesis, aqueous fluids and brines are an important agent of elemental mobility, particularly in mud rocks, but also for some igneous rocks. Under these conditions, rubidium and strontium are both mobile so the $^{87}Rb/^{86}Sr$ ratios measurable today are not the same as those originally imparted upon the samples at their time of formation. Fluid mobility is often pervasive, such that the whole-rocks may be seriously affected and their Sr-isotope characteristics homogenized, and therefore if an isochron can be produced from such samples it yields the age of a 'fluid event'.

This concept of re-setting of Rb–Sr isochrons during low-grade metamorphism has been shown to be geologically important by Evans (1991), who produced 22 whole-rock isochrons for volcanic and hypabyssal rocks from Wales which had clearly developed a hydrous secondary mineralogy. Combining all these isotopic analyses (some 200 samples in total) gives a mean age of 399 ± 9 Ma – an age that is synchronous with the Acadian deformation event in the area – leading Evans (1991) to conclude that all the isochrons were re-set during low-grade regional metamorphism.

12.3 THE SAMARIUM–NEODYMIUM METHOD OF DATING

Samarium (Sm; atomic number 62) and neodymium (Nd; atomic number 60) are both members of the group of elements known as the rare earth elements (REE). The REE are generally considered to be far more stable during metamorphism and diagenesis than Rb and Sr, although there are exceptions, particularly where the REE can be mobilized in fluids (e.g. Hole *et al.* 1992; Bouch *et al.* 1995). The isotopes of importance in the Sm–Nd system are the stable isotope of neodymium ^{144}Nd, to which all other isotopes are normalized, the parent isotope ^{147}Sm and its daughter isotope ^{143}Nd. The decay scheme is:

$$^{147}Sm \rightarrow {}^{143}Nd \text{ by } \alpha \text{ decay}$$

Decay is similar to the loss of a helium nucleus, such that the atomic masses of the parent and daughter isotopes have a difference of 4 amu. The form of the isochron equation for the Sm–Nd system is essentially the same as that for the Rb–Sr system:

$$^{143}Nd/^{144}Nd_m = {}^{143}Nd/^{144}Nd_i + {}^{147}Sm/^{144}Nd_m(e^{\lambda t} - 1) \qquad (4)$$

where $^{143}Nd/^{144}Nd_m$ is the $^{143}Nd/^{144}Nd$ measured today; $^{143}Nd/^{144}Nd_i$ is the initial $^{143}Nd/^{144}Nd$ ratio at the time of crystallization; $^{147}Sm/^{144}Nd_m$ is the measured $^{147}Sm/^{144}Nd$ ratio; λ is the decay constant, in this case 6.54×10^{-12} year^{-1}; and t is the age of the samples.

A plot of $^{147}Sm/^{144}Nd_m$ versus $^{143}Nd/^{144}Nd_m$ again has the form of a straight line and it is therefore possible to plot isochron diagrams. The decay constant for ^{147}Sm has been determined as 6.54×10^{-12} year^{-1}, which is nearly an order of magnitude longer than that for the Rb–Sr system. Consequently, the Sm–Nd isotope system takes considerably longer to generate radiogenic daughter isotopes than the Rb–Sr system, and this is one of the factors which makes it most applicable to studies of Palaeozoic and Precambrian rocks. However, since there is only a mass difference of 1 in 144 for the daughter and stable isotope, the physical process of analysis is much more difficult than that for the Rb–Sr system, where the mass difference is 1 in 86.

12.3.1 Sm–Nd Whole-Rock Isochrons and Their Applications

As both samarium and neodymium are members of the REE group, they tend to behave in a very similar manner in most geological systems. It is difficult to produce large variations in Sm/Nd ratios during normal magmatic processes; that is to say for most rock-forming minerals mineral/melt partition coefficients (k_D) are similar for both elements. Furthermore, granitoids are generally characterized by very low (*c.* <0.1) and relatively consistent $^{147}Sm/^{144}Nd$ ratios, rendering them difficult to date by this method. Nevertheless, in basic rocks which have flat REE patterns, and therefore moderate Sm/Nd ratios, there may be sufficient variations in $^{147}Sm/^{144}Nd$ ratios to generate measurable variations in $^{143}Nd/^{144}Nd$ ratios, as long as the samples are sufficiently old. Basalts and komatiites (very high magnesium basalts), which are common in Archaean and Proterozoic greenstone belts, lend themselves well to the Sm–Nd method of dating. Figure 12.5 shows Sm–Nd

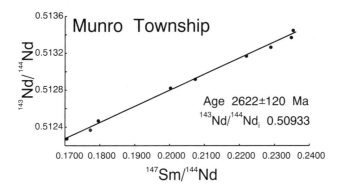

Figure 12.5 *Sm–Nd whole-rock isochron for Archaean komatiites and basalts from Canada. Note that the range of $^{147}Sm/^{144}Nd$ and $^{143}Nd/^{144}Nd$ ratios is much smaller than that for the Bildad Peak Rb–Sr isochron shown in Figure 12.3, and therefore the isochron is not as precise as can be seen from the errors on the ages. The initial $^{143}Nd/^{144}Nd$ ratios of these komatiites and basalts suggest they were generated from a predominantly mantle source region. [Based on data in: Zindler (1982)]*

isochrons for komatiites and tholeiites from Canada and it is clear that there is sufficient variability in Sm/Nd ratios to generate a range in $^{143}Nd/^{144}Nd$ ratios over the last two billion years. Sm–Nd systematics are, however, by no means restricted to mafic and ultramafic rocks as long as the samples in question are sufficiently old. For example, Hamilton *et al.* (1979) showed that it was possible to determine a precise age of 2920 ± 50 Ma for the Lewisian granulite and amphibolite facies gniesses from north-west Scotland; indeed the ages of some of the oldest rocks on earth, the Isua supracrustals of Greenland (3750 Ma), have been determined using Sm–Nd whole-rock isochrons.

12.3.2 Sm–Nd Whole-Rock–Mineral Isochrons

In common with the Rb–Sr method of dating, the Sm–Nd method of geochronological investigation is most important in certain mineral groups. The major requirement here is that the mineral species in question has sufficient concentrations of samarium and neodymium for analysis, and that it possesses a Sm/Nd ratio which is high enough to allow sufficient amounts of the daughter isotope be generated and to be measurable. One very common metamorphic mineral which fits these criteria beautifully is garnet. Garnet has a considerably greater affinity to incorporate heavy (H) REE into its lattice than the light (L) REE, and many garnets have LREE-depleted chondrite-normalized REE profiles. Since Nd is lighter than Sm, LREE-depleted minerals must have high Sm/Nd ratios. The blocking temperature for Sm–Nd systematics in garnets is $>550\,°C$ (Table 12.2) and therefore mineral or mineral–whole-rock isochrons involving garnet separates are ideal for the absolute dating of garnet amphibolite-grade metamorphism.

A fine example of the application of both Rb–Sr whole-rock and Sm–Nd garnet–whole-rock analysis to investigate the age of intrusion and subsequent amphibolite facies metamorphism of some orthogneisses is shown in Figure 12.6. Milne & Millar (1989) demonstrated that an Rb–Sr whole-rock isochron from the Target Hill metamorphic complex of the Antarctic Peninsula yielded an age of 410 ± 15 Ma, whereas a mixed Sm–Nd garnet–whole-rock isochron for the same complex yielded an age of 311 ± 8 Ma. Milne & Millar (1989) interpreted these data as indicative of a Silurian intrusive event which was followed by high-grade metamorphism during the late Carboniferous, resulting from subduction-related plutonism. These data were particularly significant stratigraphically because they demonstrated that much of the basement of the Antarctic Peninsula was of mid-Palaeozoic age, a hitherto unknown fact, and that there was possibly a widespread metamorphic event throughout the western margin of Gondwana during the Carboniferous.

12.3.3 Isochrons or Mixing Lines?

So far the use of isochrons to yield absolute ages of rocks and minerals has been discussed. In the preceding discussion, intrusions have been assumed to be a

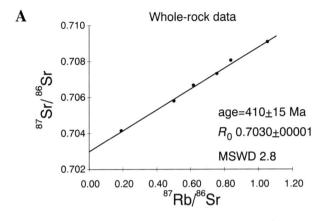

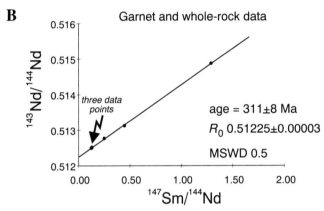

Figure 12.6 *Rb–Sr whole-rock (**A**) and Sm–Nd garnet–whole-rock (**B**) isochrons for ortho gniesses from Target Hill, Graham Land, Antarctic Peninsula. The age yielded by the Rb–Sr isochron has been interpreted as the age of the original intrusive precursor to the gneisses, whereas the Sm–Nd data have been interpreted as the age of garnet growth and thus metamorphism. The Target Hill gneisses are part of the country rock complex into which the Bildad Peak granitoid intruded, which has an age of 162 ± 2 Ma (Figure 12.3). Thus, the combination of the two isochrons for the Target Hill gneisses and that for the Bildad Peak granite gives three events in the area: intrusive magmatism at 410 ± 15 Ma, metamorphism at 311 ± 8 Ma and a further intrusive event at 162 ± 2 Ma. [Partially based on data in: Milne & Millar (1989; Target Hill)]*

single batch of magma from a single source region (e.g. the Bildad Peak intrusion in Antarctica). At active continental margins and in continental flood basalt provinces, prodigious volumes of magma are produced on a relatively short geological time-scale, often resulting in the evolution of complex magmatic plumbing systems. Consequently, there is frequently a great deal of opportunity for magmas to intimately intermingle, giving rise to hybrid intrusions and suites of volcanic rocks. Let us suppose that two batches of magma have different initial $^{143}Nd/^{144}Nd$ ratios and a range of $^{147}Sm/^{144}Nd$ ratios at their time of formation. If these two magmas are thoroughly mixed, then on a plot of $^{147}Sm/^{144}Nd$ versus

^{143}Nd/^{144}Nd ratios, mixing will produce a straight line *at the time of formation* of the mixed magma. Subsequent radioactive decay will increase the slope of that line and isotopic analysis of the samples at present day may produce an isochron which has apparent age significance. However, since there were two batches of magma with different initial ^{143}Nd/^{144}Nd ratios prior to crystallization, i.e. there was already a positive slope on an isochron diagram at the time of crystallization, then clearly the apparent isochron has no age significance, as it actually represents the effects of magma mixing and *in situ* radioactive decay. Interaction of a batch of magma with country rocks can also produce the same effect.

Arndt & Jenner (1986) showed that just such a situation had occurred during the genesis of the Kambalda komatiites and tholeiites of Australia (Figure 12.7). A whole-rock Sm–Nd isochron yielded an age of 3190 ± 85 Ma, but a Pb–Pb whole-rock isochron yielded an age of 2726 ± 34 Ma. Arndt & Jenner (1986) showed that the Sm–Nd systematics had been upset at the time of formation by contamination of the basalts by older granitic crust and metasedimentary rocks. This was in contrast to the Pb-isotope system, which was insensitive to the contamination because of the low abundances and unradiogenic nature of the lead in the crustal rocks. Use of the Sm–Nd method meant that the Kambalda volcanics had suddenly become younger by 700 Ma, the equivalent to the entire Phanerozoic, demonstrating the need for caution in such cases.

In another example, the Bildad Peak granitoid in Antarctica (Figure 12.3) yielded a precise Rb–Sr age of 167 ± 2 Ma which could also be interpreted as a mixing line. However, analysis of Nd-isotopes of a range of facies from the granitoid yielded precisely the same ^{143}Nd/^{144}Nd ratios at 162 Ma, implying that the entire

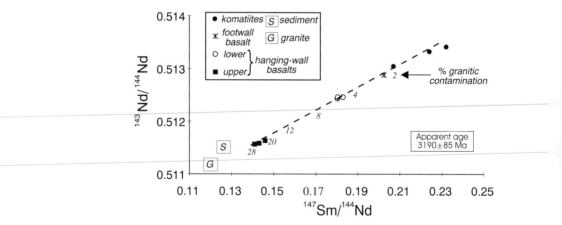

Figure 12.7 *Sm–Nd isochron diagram for the Archean Kambalda volcanic sequence of Australia. The apparent age of 3190 ± 85 Ma yielded by the Sm–Nd whole-rock technique has been shown by Arndt & Jenner (1986) to be a mixing line generated by variable interaction between a mafic magma and granitic and/or metasedimentary country rocks. The dashed line is the calculated mixing line, with the percentage contamination shown. A Pb–Pb whole-rock isochron yielded an age of 2726 ± 34 Ma, which is interpreted as the 'real' extrusive age of these rocks. [Based on data in: Chauvel et al. (1985)]*

intrusion was both in Sr- and Nd-isotopic equilibrium at the time of formation. It is practically impossible to see how this could happen if the isochron was a mixing line, given that rubidium, strontium, samarium and neodymium all behave in a different manner in geological systems. Therefore, by careful analysis of the data and using a combination of isotopic techniques, it is normally possible to distinguish between mixing lines and isochrons.

12.4 THE DATING OF ARCHEAN ROCKS USING HAFNIUM AND LUTETIUM ISOTOPES

The hafnium–lutetium method of dating is not commonly used in geochronological studies for reasons that will become apparent later in the chapter. However, the technique was pioneered by Patchett & Tatsumoto (1980) and it has been applied to the dating of Archaean gneisses and therefore is worth mentioning here. It is one of the most recently developed methods of absolute dating using radiogenic isotopes.

Lutetium (Lu, atomic number 71) is the heaviest of the REE (atomic mass 174.97) and its abundance is very low in all terrestrial rocks, being commonly in the range of 0.2 to *c.* 1.0 ppm. Hafnium (Hf, atomic number 72) exhibits a very similar chemical behaviour to zirconium (Zr, atomic number 40) and in many terrestrial rocks, particularly basalts, Zr/Hf ratios are relatively consistent at around 38–42. In intermediate and acid igneous rocks, the majority of the hafnium resides in zircons, and the fractionation of zircons during crystallization of acid intrusions is an important control on the generation of a range of Lu/Hf ratios in the wholerocks.

In terms of isotopes, the decay scheme of interest here is:

$$^{176}Lu \rightarrow {}^{176}Hf \text{ by } \beta \text{ decay} \tag{5}$$

For this system, the parent and daughter isotopes are normalized to the stable isotope ^{177}Hf, giving an isochron equation of

$$^{176}Hf/^{177}Hf_m = {}^{176}Hf/^{177}Hf_i + {}^{176}Lu/^{177}Hf \times (e^{\lambda t} - 1) \tag{6}$$

where $^{176}Hf/^{177}Hf_m$ is the measured $^{176}Hf/^{177}Hf$ ratio; $^{176}Hf/^{177}Hf_i$ is the initial ratio at the time of formation of the system; λ is the decay constant for $^{176}Lu = 1.94 \pm 0.07 \times 10^{-11}$ year^{-1}; and $^{176}Lu/^{177}Hf$ is the measured $^{176}Lu/^{177}Hf$ ratio at present day.

Again, this is the equation for a straight line with a slope of $e^{\lambda t} - 1$ and an intercept of $^{176}Hf/^{177}Hf_i$. In practice, this technique has not been used a great deal for three main reasons:

1. Terrestrial rocks have only limited variations in Lu/Hf ratios, and thus it is difficult to generate a spread of data on the *x*-axis of the isochron diagram and so the technique is only applicable to ancient rocks.

2. The low concentrations of Lu in most rocks make determination of Lu/Hf ratios less precise than other methods of dating.
3. The mass difference of 1 in 177 amu for the isotopes in question requires incredibly high-precision mass spectrometry.

Pettingill *et al.* (1981) successfully applied the Lu–Hf method to the dating of the Amitsôq gneisses of Greenland, some of the oldest rocks on earth (Figure 12.8). Using a combination of whole-rocks and zircons, a spread of ^{176}Lu/^{177}Hf ratios of only 0–0.025 was produced (compare to the spread of ^{87}Rb/^{86}Sr ratios for the Bildad Peak intrusion of 0.4–8.0), with a variation of ^{176}Hf/^{177}Hf$_m$ ratios of 0.2805 to *c.* 0.2820, illustrating one of the limitations of this technique. Nevertheless, an age of 3.59 ± 0.22 Ga was determined (Figure 12.8) which is concordant with a U–Pb zircon age of 3.65 Ga (Baadsgaard 1973) and, remarkably, also concordant with an Rb–Sr whole-rock isochron which yielded an age of 3.65 Ga (Moorbath *et al.* 1972). This is an interesting example of how robust the Rb–Sr whole-rock technique can be, even in ancient high-grade metamorphic rocks.

12.5 DATING ZIRCONS: URANIUM–LEAD GEOCHRONOLOGY

Zircons are a very common accessory mineral in intermediate, acid igneous and meta-igneous rocks. Zircon is incredibly resistant to mechanical breakdown and is chemically very stable. Its resistance to breakdown is demonstrated by its ubiquitous occurrence in clastic sedimentary rocks. Zircon contains appreciable amounts of uranium (U) and lead (Pb) and its closure temperature for uranium and lead is >800 °C. In consequence, absolute dating of zircons allows us to see through a myriad of post-zircon crystallization, metamorphic and structural events, allow-

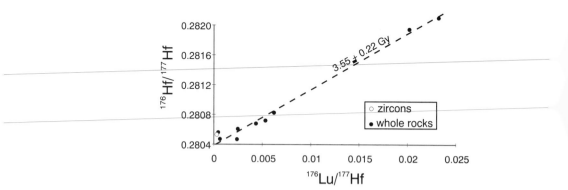

Figure 12.8 *Lu–Hf whole-rock–zircon isochron for the Amitsôq gneiss, West Greenland. Note the limited spread in ^{176}Lu/^{177}Hf and ^{176}Hf/^{177}Hf ratios, which is considerably less than that for the Sm–Nd isochron for the Target Hill gneisses (Figure 12.6). Nevertheless, this isochron yields a relatively precise age of 3.55 ± 0.22 Ga; an age that is concordant with U–Pb zircon and Rb–Sr whole-rock ages from the same locality. [Based on data in: Pettingill et al. (1981)]*

ing the determination of the time of zircon growth from a magma. In essence, if a rock contains zircons, then the absolute age of those mineral grains can be determined. Before dealing with zircon dating, it is necessary to consider briefly the theory of U–Pb geochronology.

12.5.1 The Principles of U–Pb Geochronology

The isotope geology of lead is extremely complicated, and only the most basic of concepts will be considered here. The reader is referred to Faure (1986) for full details. Lead is unusual isotopically in having three radiogenic parent isotopes, each of which decay to a different daughter isotope of lead:

$$^{235}U \rightarrow\ ^{207}Pb$$
$$^{238}U \rightarrow\ ^{206}Pb$$
$$^{232}Th \rightarrow\ ^{208}Pb$$

The stable isotope of lead which is used as the denominator in all equations is ^{204}Pb. The decay equations for each of these isotopic systems are given below:

$$^{206}Pb/^{204}Pb = {}^{206}Pb/^{204}Pb_i + {}^{238}U/^{204}Pb\ (e^{\lambda_1 t} - 1) \tag{7}$$

$$^{207}Pb/^{204}Pb = {}^{207}Pb/^{204}Pb_i + {}^{235}U/^{204}Pb\ (e^{\lambda_2 t} - 1) \tag{8}$$

$$^{208}Pb/^{204}Pb = {}^{208}Pb/^{204}Pb_i + {}^{232}Th/^{204}Pb\ (e^{\lambda_3 t} - 1) \tag{9}$$

where $^{206}Pb/^{204}Pb$, $^{207}Pb/^{204}Pb$ and $^{208}Pb/^{204}Pb$ are the isotope ratios at the time of analysis; $^{206}Pb/^{204}Pb_i$, $^{207}Pb/^{204}Pb_i$ and $^{208}Pb/^{204}Pb_i$ are the initial lead isotope ratios at the time of formation of the sample; $^{238}U/^{204}Pb$, $^{235}U/^{204}Pb$ and $^{232}Th/^{204}Pb$ are the isotope ratios at the time of analysis; λ_1, λ_2 and λ_3 are the decay constants for each system (Table 12.1); and t is the time elapsed since the closure of the mineral or rock.

Each of the parent isotopes has a different half-life (Table 12.1) and therefore decay constant. The isotopes of interest for U–Pb dating are ^{238}U and ^{235}U, the latter decaying some six times faster than the former. Unlike other radiogenic isotope systems, because there are two isotopes of uranium decaying to two isotopes of lead, the U–Pb system provides two independent geochronometers in any uranium- and lead-bearing mineral. Zircon is by far the commonest mineral phase used for U–Pb dating, but titanite (sphene) has also been used, particularly in the dating of basic intrusive rocks (e.g. Rogers & Dunning 1991). However, great care has to be taken to ensure that the titanites are not metasomatic in origin, thereby giving incorrect ages (Hole et al. 1992). Like all isotope geochronometers, it must be assumed that the zircon has been closed to uranium and lead since crystallization. This means that both the decay schemes in question should yield independent dates that are exactly the same, that is they are *concordant*.

Equation (7) can be written:

$$^{206}Pb*/^{238}U = e^{\lambda_1 t} - 1 \tag{10}$$

where $^{206}Pb*$ is the amount of radiogenic ^{206}Pb that has been produced in the sample, which from equation (7) is a function of the initial $^{206}Pb/^{204}Pb$ ratio and the $^{238}U/^{204}Pb$ ratio.

Similarly, equation (8) can be expressed as:

$$^{207}Pb*/^{235}U = e^{\lambda_2 t} - 1 \tag{11}$$

Using the above two equations it is possible to calculate $^{206}Pb*/^{238}U$ and $^{207}Pb*/^{235}U$ ratios for theoretical, specified values of t. This gives a set of data points on a plot of $^{207}Pb*/^{235}U$ versus $^{206}Pb*/^{238}U$ which define a curve and represent minerals that would have perfectly concordant ages. A concordant zircon would therefore plot on this curve at a position determined entirely by its age. This curve is known as the concordia diagram (Figure 12.9).

Unfortunately, it is rarely the case that minerals such as zircon have remained closed to uranium and lead. Since zircon contains appreciable amounts of the radioactive elements (which makes it useful for geochronology), a detrimental side-effect is that the grains can become radiation damaged or *metamict*, giving a brown colour in plane polarized light. In essence, the radioactive decay destroys the crystal structure, and uranium and lead may simply leak out in random proportions. This generally means that the ages obtained from the two geo-chronometers may not be the same and are said to be *discordant* (Figure 12.9).

The concordia diagram can be used to calculate the age of zircons even if they have experienced a period or periods of lead loss or uranium gain. The amount of

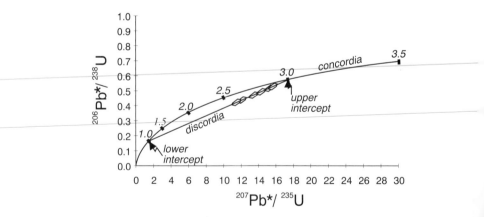

Figure 12.9   *Diagrammatic representation of the concordia plot. Data points that plot on the curve yield perfectly concordant ages in the $^{238}U-^{206}Pb$ and $^{235}U-^{207}Pb$ systems. If lead loss or uranium gain has taken place after crystallization, then data will form a straight line (collectively known as discordia) which will have two intercepts on concordia, both of which have age significance. The values on concordia are absolute ages in Ga*

lead loss varies between individual zircon grains as a function of size, concentration of uranium, and consequent amount of radiation damage, such that smaller grains and those with high uranium concentrations tend to lose more lead than larger grains or grains with low uranium concentrations. As a result, different zircon populations can be separated from a rock on the basis of size and optical properties. Since these zircon populations have all lost lead in differing quantities, they plot as a straight line, known as a *discordia*, on the concordia diagram. The discordia has two intercepts, a lower, younger intercept, and a higher, older intercept (Figure 12.9). The significance of these intercepts is discussed below.

12.5.2 Dating Zircons

Since zircons are so stable, there is always the possibility that a granitoid may contain zircons that have been derived from older crustal material during the rise of granitic magma through the continental crust. These derived zircons are known as inherited grains, and are problematical, as they do not yield the absolute age of the intrusion in which they are found. Since they are included within the granitic magma from older rocks, these allow merely a maximum age to be determined. It is particularly important to pay a great deal of attention to the size, shape and colour of zircons in a sample in order to establish their status. For example, euhedral, clear and uncracked grains are more likely to represent primary zircons which have crystallized directly from a magma than anhedral, metamict and cracked grains.

   In order to analyse zircons, it is first necessary to separate the different grain populations. This is done by the crushing and sieving of a large amount (tens of kilograms) of the whole-rock and hand picking the different populations under a binocular microscope, usually a rather tedious business. The different populations can either be analysed as bulk samples or, as is more common in recent studies, as single grains. For single-grain analysis of zircons that have a metamict rim and a euhedral core, it is often possible to remove some of the marginal material that has undergone lead loss by mechanical abrasion. Individual grains are simply placed in a stainless steel vessel and a low-pressure jet of compressed air is used to circulate the grains around the outer walls of the vessel. This has the effect of mechanically removing the metamict rims and can result in a considerable improvement in the concordance of individual grains.

   A good example of the use of U–Pb for dating a metamorphosed granitic intrusion is that of the Beinn Vurich intrusion in the Grampian Highlands of Scotland (see Chapter 4). Importantly, this intrusion has a well-defined relationship with the early tectonic activity of the Caledonian Orogeny, and as it has undergone moderate grade metamorphism, it contains garnet porphyroblasts. Initial attempts at dating the intrusion using the Rb–Sr whole-rock techniques yielded an age of 497 ± 37 Ma (Bell 1968). Since Rb–Sr resetting during metamorphism was likely, Pankhurst & Pidgeon (1976) separated bulk samples of zircons for U–Pb dating and were able to produce a discordia which had a lower intercept of c. 514 Ma, which they interpreted as the crystallization age. Subsequently,

Rogers *et al.* (1989) re-investigated the intrusion and carefully selected only the most euhedral uncracked grains, analysing them individually as single grains. From this analysis, Rogers *et al.* (1989) were able to produce a very precise age of 590 ± 2 Ma for the crystallization of the Beinn Vurich intrusion, and in addition they demonstrated that there was a population of inherited zircons derived from the crustal rocks with an age of *c.* 1448 Ma.

Taking this a step further, Pidgeon & Compston (1992) used an instrument called a SHRIMP (Super High Resolution Ion Micro-Probe) in dating the Beinn Vurich granite. This instrument has the capability to analyse the ages of individual zircons in a thin section without having to physically separate them from the whole-rock. This technique allows a much larger amount of data to be produced relatively rapidly, although the precision is not as good as that of mass spectrometric analysis of separated grains. The results of Pidgeon & Compston's (1992) study compare extremely well with those of Rogers *et al.* (1989), as they yielded two populations of zircons: one with ages of 597 ± 11 Ma for crystallization; the second with an age range of 1100–1200 Ma, representing the inherited component.

As a result of these U–Pb zircon analyses, the age of the intrusion Beinn Vurich granite is now considered to be about 100 Ma older than the original Rb–Sr age. Clearly, this has enormous significance in terms of the sequence of tectonic events during the Caledonian Orogeny, and also apparently requires the entire Dalradian metasedimentary succession to have been deposited during the Precambrian, and not at least partially during the Cambrian as originally thought (Rogers *et al.* 1989).

12.6 POTASSIUM–ARGON AND ARGON–ARGON DATING TECHNIQUES

Almost all terrestrial rocks contain measurable amounts of potassium (K), either in the form of potassium-bearing minerals, or as potassium in the glass of fine-grained igneous rocks. The naturally occurring radiogenic isotope ^{40}K decays to an isotope of argon, ^{40}Ar. Not all ^{40}K decays to ^{40}Ar (about 11% of it); the rest decays to ^{40}Ca. Therefore, three dating techniques are theoretically possible: K–Ar, Ar–Ar and K–Ca. However, ^{40}Ca is not very abundant, and there are practical difficulties in measuring its abundance by mass spectrometry.

As with most other radiogenic isotope dating techniques, if we can measure the amount of K and ^{40}Ar, and if the decay constant is known, then we can calculate the time since *in situ* radioactive decay began. It is important to remember, however, that this may not be the absolute age of the sample if it has been raised above its blocking temperature (Table 12.2) and the isotopic clock has been re-set. The K–Ar technique was first proven to be geologically useful in the late 1940s, and the Ar–Ar technique was first developed in the 1960s (Merrihue & Turner 1966).

The main difference in the K–Ar and Ar–Ar techniques of dating is related to the method of measurement of potassium. In the 'conventional' K–Ar method, potassium measurements are made by atomic absorption spectrometry on a portion of the sample, and a separate aliquot of the same sample is used to analyse the isotope ratio of interest, the $^{40}Ar/^{40}K$ ratio, the decay equation being

$$t = 1/\lambda \ln(^{40}Ar^*/^{40}K) \times (\lambda/\lambda e + 1) \tag{12}$$

where $^{40}Ar^*$ is the amount of radiogenically produced argon and the expression $\lambda/\lambda e$ represents the number of ^{40}K atoms that decay to ^{40}Ar. The half-life for the decay of ^{40}K is 1.25×10^9 years, giving a decay constant of 5.543×10^{-10} year^{-1}. Note that this decay constant is approximately one order of magnitude less than that for the Rb–Sr system, making the K–Ar method suitable for a broad age range of rocks and minerals. Because there are significant numbers of ^{40}Ar atoms in the atmosphere, a correction must be made so that the number of radiogenic atoms can be calculated. This is a routine calculation.

By contrast, the Ar–Ar technique does not require the measurement of potassium at all; it simply requires access to a very large and powerful nuclear reactor! The principle of this technique is relatively straightforward, but having radioactive samples in the laboratory is a slight drawback. Essentially, when bombarded with thermal neutrons, ^{39}K transmutes to ^{39}Ar. Since the ratio of ^{39}K to ^{40}K is constant, then the ratio of naturally produced radiogenic ^{40}Ar to the reactor produced ^{39}Ar is proportional to the age of the sample, such that

$$t = 1/\lambda \ln(1 + J \times {}^{40}Ar^*/^{39}Ar) \tag{13}$$

where $^{40}Ar^*$ is the amount of radiogenically produced ^{40}Ar; and J is a parameter related to the conditions in the nuclear reactor at the time of neutron bombardment.

Both the Ar–Ar and K–Ar methods can be used on whole-rocks and minerals and the usual closure temperature arguments apply. In practice, for both techniques, argon is released from the sample by fusing it in a radio-frequency furnace under a high vacuum. Any water is frozen out using liquid nitrogen, and the argon is introduced into a mass spectrometer and its isotopic composition measured.

12.6.1 The Pros and Cons of K–Ar and Ar–Ar Geochronology

Clearly, the actual physical laboratory procedures for K–Ar dating are rather more simple and less expensive than Ar–Ar dating. Nevertheless, basalts, particularly alkali basalts with relatively high K_2O, are ideally suited to the K–Ar and Ar–Ar methods of dating, and in essence there is no other technique that can be used to gain absolute ages of basalts. Like all other isotopic systems, determination of the age of crystallization of an alkali basalt is only possible if certain assumptions are satisfied in the natural system, the most important of which is that the mineral has remain closed to argon and potassium since crystallization.

A good example of one of the common problems that can be encountered in the dating of alkali basalts was documented by Hole *et al.* (1994) for some continental alkali basalts from the Jones Mountains, West Antarctica. K–Ar analysis of pillow basalts which unconformably overlie the glacially striated surface of a Cretaceous granitoid gave ages in the range 332 million to 6.1 million years, that is Early Carboniferous to Miocene – a highly improbable situation! The reasons for this are unclear, particularly given the samples were beautifully fresh, with no evidence of hydrothermal alteration. A possible explanation given for this 'ageing' effect is that the samples somehow incorporated excess argon during crystallization, such that there is more radiogenic argon than can be supported by the amount of potassium present, thereby giving an anomalously old age. Additionally, as both potassium and argon are extremely mobile during hydrothermal alteration or weathering (see Section 12.7.2), the assumption that the rock or mineral has remained closed to potassium and argon since crystallization is often not tenable.

One advantage of the Ar–Ar method in this connection is a technique known as *step heating*. Since the Ar–Ar technique only requires the analysis of the Ar isotopes once they have been irradiated, the rate at which argon is released from a mineral grain or whole-rock sample can be monitored and a number of age measurements can be made for a range of temperatures. During alteration or low-grade metamorphism, less strongly bonded argon within the crystal lattice is preferentially removed over more strongly bonded argon. In this way, the loosely bonded argon is re-set more easily than the strongly bonded argon. When the sample is step-heated in the laboratory, the loosely bonded argon is released at lower temperatures and will yield an apparent age which represents a younger metamorphic event. The apparent ages of the argon released at high temperatures are therefore more likely to reflect the age of mineral growth than the age of the metamorphism.

In practice, there is often a plateau in ages calculated by step-heating. Statistically, this is likely to be the actual crystallization age of a mineral or whole rock sample, and younger ages therefore reflect post-crystallization events. Consequently, as in the Rb–Sr whole-rock mineral method, useful geological data other than simply the age of crystallization can be gained by this method. An even more innovative refinement of the Ar–Ar technique is that of laser ablation, where the argon is released into the mass spectrometer by bombardment with a laser beam. This allows incredibly good spatial resolution of sampling and enables very small samples to be analysed.

12.6.2 Specific Stratigraphical Applications of the Ar–Ar Method

The Ar–Ar technique is very useful in dating 'young' potassium-bearing minerals, such as biotite, muscovite, sanidine and potassium-rich volcanic glass. This is important in the dating of bedded tuffs, ignimbrites and lava flows which may be interbedded and concordant with sedimentary units. Absolute dating of these volcanic and volcaniclastic rocks can therefore be used to provide absolute age ranges of intercalated sedimentary units, marker horizons or biozones. The potential of this technique for correlation, and for the construction of stratigraphical frameworks within sedimentary basins, particularly where the basin-fill is faunally barren, is immense. In some cases, the resolution can be extremely fine-scale; so much so that there are even archaeological applications of the Ar–Ar method. For example, Deino & Potts (1992) dated tephra deposits from Olorgesailie in southern Kenya, which is known for its early hominid remains, and showed that by careful analysis of pumice samples it was possible to resolve the stratigraphy over a range of 1051 ± 14 Ka to 49 ± 6 Ka.

12.7 THE APPLICATION OF ISOTOPIC TECHNIQUES TO ECONOMIC GEOLOGY

In addition to the absolute dating of the crystallization of magmas, and of metamorphic events, there are a number of radiogenic isotope techniques useful in the determination of the stratigraphy of faunally barren hydrocarbon reservoirs, and in the absolute geochronology of ore deposits. The purpose of this section is to highlight some examples of these techniques.

12.7.1 Hydrocarbon Reservoir Correlation

Faunally barren continental red-bed sequences are common hydrocarbon reservoirs, particularly within the North Sea. The subsurface correlation of individual stratigraphical horizons within such sequences is a non-trivial problem, as poor correlation of reservoir rocks between wells can result in 'missing' the reservoir interval when drilling new exploration wells. Clearly, this can mean the loss of millions of barrels of oil and millions of dollars. Consequently a number of new methods of correlation of such sequences have been developed. Two of these techniques are described below: the use of Sm–Nd 'model ages' and SHRIMP U–Pb dating of zircons.

Neodymium isotope model ages and hydrocarbon reservoir correlation

One of the fundamental features of the distribution of samarium and neodymium in the earth today, is that the mantle has a considerably higher $^{147}Sm/^{144}Nd$ ratio (*c.* 0.222) than 'average' continental crust (*c.* 0.1). Analyses of upper-mantle samples, i.e. modern mid-ocean ridge basalts (MORB), exhibit a range of $^{147}Sm/^{144}Nd$ ratios, with an average value of about 0.22, and these samples have present-day $^{143}Nd/^{144}Nd$ ratios in the range 0.5129–0.5132. A large database of REE analyses of sedimentary rocks, and in particular sandstones, is available in the literature (e.g. Mearns *et al.* 1989) and these rocks have typical $^{147}Sm/^{144}Nd$ ratios in the region 0.09–0.11 (considerably lower than that for the upper mantle), and highly variable $^{143}Nd/^{144}Nd$ ratios. This is because lighter elements, in this case neodymium, are more readily extracted from the mantle than heavier elements (i.e. samarium), leading to crust that is rich in neodymium relative to samarium, and leaving a mantle residue that is poor in neodymium relative to samarium. The so-called 'depleted' mantle source generates far more of the daughter isotope

^{143}Nd by the decay of ^{147}Sm than the crustal reservoir, simply because the mantle has a higher Sm/Nd ratio than the crust. A consensus has been reached as to numerical values for 'average' upper-mantle Nd-isotopic compositions, known as DMUR (depleted mantle uniform reservoir), which has present-day ^{147}Sm/^{143}Nd and ^{143}Nd/^{144}Nd ratios of 0.22 and 0.51303 respectively (Mearns 1988). By applying the age equation (4) above, the neodymium isotopic evolution of DMUR can therefore be calculated through time (Figure 12.10).

Let us say that at around 250 Ma, close to the Permian–Triassic boundary, a major fluvial system existed in a given basin. In this hypothetical basin, rivers were flowing from the north, eroding Proterozoic mantle-derived granites, and depositing them as fluvial sandstones. Other rivers were flowing from the south, where the crust was considerably younger, composed of Cambrian mantle-derived granitoids. The neodymium isotopic composition of the two sandstones will therefore be controlled by the neodymium isotopic composition of the two granitoids of differing ages, because it is not possible to significantly fractionate samarium from neodymium by sedimentary depositional processes. Now, since both granitoids, regardless of their age, will have approximately the same ^{147}Sm/^{144}Nd ratios, their ^{143}Nd/^{144}Nd ratios will be dependent on the isotopic composition of the mantle at the time of their formation. This is illustrated in Figure 12.10, where the evolution of the ^{143}Nd/^{144}Nd ratio is plotted as a function of time, such that the slope of the lines represents the ^{147}Sm/^{144}Nd ratio. The slope of the evolution line for DMUR is steep and negative because it has a high ^{147}Sm/^{144}Nd ratio. The sandstones, on the other hand, have shallower slopes because they have low ^{147}Sm/^{144}Nd ratios. The provenance of sandstone 'a' was a mantle-derived granite of Cambrian age. If we measure the ^{147}Sm/^{144}Nd and ^{143}Nd/^{144}Nd ratios of sandstone 'a' today, we can calculate how the neodymium-isotope system would have evolved through time.

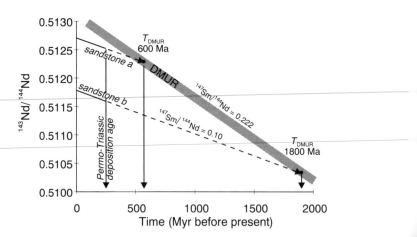

Figure 12.10 *The growth of ^{143}Nd/^{144}Nd as a function of time for a uniform mantle reservoir (DMUR) and two Permo-Triassic sandstones each formed from mantle-derived granites of different ages. Sandstone 'a' extrapolates back to the DMUR growth line at 600 Ma, and sandstone 'b' at 1800 Ma. Therefore, while the two sandstones have precisely the same stratigraphical ages, their model ages (T_{DMUR}) reflect differences in provenance*

Figure 12.10 shows that the neodymium evolution line for sandstone 'a' intersects DMUR at 600 Ma. This is the Cambrian age of the granite from which it was derived. Similarly, sandstone 'b' extrapolates back to DMUR at 1800 Ma, reflecting the Proterozoic age of the granite from which it was derived. However, we know that both sandstones are actually Permo-Triassic in age, so the 600 Ma and 1800 Ma ages are model ages (by convention T_{DMUR}), reflecting provenance, and can also be described as crustal residence ages as model ages reflect the last time the sample was in neodymium-isotopic equilibrium with DMUR.

The use of model ages is fraught with problems, because in reality the provenance of sandstones is mixed, and because obviously not all granites are entirely mantle derived. In essence, variability in model ages gives some kind of indication of mean crustal residence time and mean age of all the different rock materials contributing to the sedimentary system. This technique has, however, been used to great effect in correlating hydrocarbon reservoirs. A good example is the study of Mearns *et al.* (1989) of the late Triassic to early Jurassic Statfjord Formation of the Snorre oil field in the North Sea. In their study, Mearns *et al.* (1989) identified specific horizons where model ages changed from 2100 Ma to about 1700 Ma, reflecting a change from a dominantly Archaean (Lewisian) source to an average Proterozoic source. Changes in model ages such as these are potentially correlatable over considerable distances. Whilst Sm–Nd isotopic data are not routinely acquired, their use is growing rapidly within the oil industry and this technique adds another tool to the armoury of hydrocarbon exploration techniques.

SHRIMP zircon ages and hydrocarbon reservoir correlation

Detrital heavy minerals such as rutile, zircon, garnet, apatite and monazite are common in most sandstones. Their relative proportions are controlled by the provenance of the sediment, and the ratios of these minerals are often diagnostic of particular sedimentary horizons, thereby enabling detailed stratigraphical correlation schemes to be constructed. Unfortunately, however, during diagenesis some of the minerals tend to be dissolved, particularly in paleosols which have acid pore waters at the time of deposition. As a result, the heavy mineral populations can be modified at the time of deposition such that they do not correctly reflect provenance, and so correlation becomes impossible.

Morton *et al.* (1996) approached this problem by obtaining SHRIMP dates for zircons in the Statfjord Formation of the North Sea. Previous correlation in this formation relied on a dramatic change in the amount of garnet relative to zircon in the heavy mineral assemblage at a particular depth (Morton & Berge 1995), and this was considered to represent a change in provenance which could be correlated across the entire field. More than 300 individual SHRIMP U–Pb analyses showed that there was no difference in the age distribution of zircons with depth, and importantly, the change in provenance which was thought to be correlatable on the basis of heavy minerals was not evident. A large population of zircons yielded ages between 400 and 500 Ma, implying derivation from Caledonian intrusions, and another cluster of ages in the range 2700–3000 Ma implied an ultimate

Archaean (Lewisian) source. The important conclusion reached from this study was that the decrease in garnet abundance at a particular depth was due not to a change in provenance but to dissolution of garnet soon after, or during, transport and deposition.

In this case, zircon dating demonstrated that the original stratigraphical correlations based on heavy mineral populations were erroneous. However, obtaining SHRIMP data in the quantities used by Morton *et al.* (1996) is an incredibly costly and time-consuming business, so it is unlikely that it will become a routine tool in hydrocarbon reservoir studies. Nevertheless, in certain specific cases SHRIMP zircon ages could be an important tool in stratigraphical correlation.

12.7.2 The Dating of Hydrothermal Ore Deposits

Hydrothermal fluids cause dramatic alteration of original mineral assemblages of rocks. Most geochronologists go to great lengths to avoid the analysis of hydrothermally altered samples simply because the ages yielded will not be related to the time of intrusion of an igneous rock, the very information that is normally being sought. Nevertheless, of great importance to economic aspects of ore deposits is knowledge of the time of ore formation and its relationship to other geological events, so that regional predictions of the pattern of mineralization can be made. Since ore deposition takes place over a wide range of temperatures (magmatic to weathering at ambient temperatures), this is one field where the accuracy and application of absolute geochronology is tested to extremes, and great care must be taken in choosing a suitable technique for a particular ore body.

Many hydrothermal deposits contain secondary minerals that are common as primary minerals in igneous rocks such as micas (sericite) and K-feldspars, the dating of which has already been discussed. Galena, which is uranium-free, is suitable for dating using the Pb–Pb method, but the ages yielded are model-dependent (Faure 1986). Of particular interest of late has been the dating of supergene gold, silver and tin-bearing deposits. Such deposits frequently contain potassium-bearing phases such as those from the alunite–jarosite solid solution series ($KAl_3(OH)_6(SO_4)_2$-$KFe_3(OH)_6(SO4)_2$) which are particularly suitable for K–Ar and therefore Ar–Ar analysis. This subject will not be treated in detail here, although the recent review by Sillitoe & McKee (1996) provides much detail of the technique.

12.8 GEOCHRONOLOGICAL TOOLS FOR DATING 'YOUNG' EVENTS

The above sections summarize the commonly used isotope systems for dating geological events from the Precambrian to the late Tertiary. However, they are not the only tools available, as improvements in the techniques of modern mass spectrometry in particular have led to the development of a large number of geochronological tools of use in the recent geological and archaeological record. Here, a brief appraisal of these techniques will be given.

12.8.1 Fission Track Dating

Mineral grains which contain significant concentrations of radioactive isotopes are physically damaged by *in situ* decay, the degree of radiation damage being proportional to both the age of the sample and the concentration of the radioactive element present. This radiation damage is considered to be largely the result of the spontaneous fission of ^{238}U. If a sample of a radioactive isotope-bearing mineral is etched with acid, the radiation damage is seen under the microscope as a series of dark lines reflecting trails of radiation damage resulting from the transfer of energy from the decay particles to the atoms in the structure of the crystal. The density (i.e. the number per unit volume) of these tracks is proportional to the amount of radiation damage and therefore the age of the sample. If a sample is heated to a significant degree, the fission tracks anneal, rather like the blocking temperature for radiogenic isotope systems. Annealing temperatures for fission tracks are at the very low temperature end of the blocking temperature spectrum (Table 12.2).

Zircon and apatite (blocking temperatures 200 °C and 120 °C, respectively) are particularly suited to fission track dating because both minerals contain significant concentrations of uranium and both are very common as accessories in a variety of rock types. Due to these low blocking temperatures, the response of isotopic systems to uplift and cooling are expanded to considerably lower temperatures than that of most radiogenic isotope systems. Therefore, the use of a combination of radiogenic isotopes on whole-rocks and minerals, and fission track dating, gives an overall view of events that have affected, for example, a granitic intrusion over the temperature range of >800 to *c*. 120 °C.

Holliday (1993) and Cope (1994) have used fission track dating of apatite to reconstruct the pattern uplift and erosion in the British Isles during the Mesozoic, and have estimated that up to 2 km of Mesozoic 'cover' was removed from the current structural highs of the English Lake District and the Northern Peninnes. Indeed, Cope (1994) provides detailed arguments that the current pattern of outcrop of the Mesozoic strata of south-east England is the result of domal uplift during the late Cretaceous or early Tertiary, possibly as a result of the thermal uplift caused by a mantle plume. This is the subject of significant debate amongst igneous petrologists working on the plume-related British Tertiary Igneous Province, but nevertheless this example shows that radiogenic isotopes can provide important information other than just absolute ages for the crystallization of igneous rocks.

12.8.2 Cosmogenic Radionuclides

Technological improvements in mass spectrometry in the last decade have led to the development of a whole new generation of techniques that require very high precision, low-level isotope analysis. Cosmogenic radionuclides are produced by the interaction of cosmic rays with atoms at the earth's surface; that is, the production of these radionuclides yields data concerning the period of exposure of a sample at the surface of the earth. Radionuclides of interest are ^{10}Be, ^{26}Al, ^{32}Si and ^{36}Cl. All of these radionuclides have short half-lives in the region of 1 to 2 million years and therefore they are only of use for dating relatively recent events. Importantly, these radionuclides can be used to determine:

1. the residence times of meteorites at the earth's surface – a common application of ^{10}Be to determine the age of meteorite impacts in the Antarctic ice sheets;
2. the age of recent marine sediments, continental sediment and manganese nodules (^{10}Be and ^{26}Al);
3. the age of biogenic silica (^{32}Si); and
4. the dating of continental ice sheets (^{10}Be, ^{26}Al and ^{36}Cl).

12.8.3 Carbon-14 Dating

Probably the most well-known and familiar concept in the radiogenic dating of materials is that of carbon-14 dating. Carbon-14 is produced in the atmosphere by a number of different nuclear reactions involving cosmic-ray-produced neutrons interacting with stable isotopes of oxygen, nitrogen and carbon. The most important of these for dating purposes is the interaction of neutrons with ^{14}N. In living plants, ^{14}CO$_2$ enters tissues by photosynthesis of absorption through the roots. Since ^{14}C is continually being produced in the atmosphere, but is also continuously decaying, the amount of ^{14}C present reaches an equilibrium state known as secular equilibrium. This equilibrium state is destroyed once the tissue dies because no more ^{14}C can enter the tissue, but it continues *in situ* decay. It is the rate of *in situ* decay in dead tissue that can be used to generate carbon-14 dates. The half-life of ^{14}C is accepted as 5568 ± 30 years and therefore it has a large decay constant (1.209 ± 10^{-4} year^{-1}). Consequently, much of the ^{14}C in a sample decays away to below detection limits in a matter of tens of thousands of years. However, carbon-14 dating is used for dating Recent or late Quaternary deposits, often in conjunction with other techniques for young rocks.

12.9 SUMMARY

In this chapter it has been shown that there are a diversity of applications of radiogenic isotope geochemistry to basic stratigraphical problems, not all of them requiring the generation of absolute age dates. Radiogenic isotopes can not only be used to determine the crystallization ages of igneous rocks, but can also provide fundamental information on later metamorphic and tectonic uplift affecting the same rocks. A particularly good example of the application of a range of isotope techniques to a stratigraphical problem is that of Dempster (1985) who used a range of Rb–Sr and K–Ar mineral dates to produce very detailed absolute ages for a number of metamorphic and uplift events that affected the Dalradian rocks of the central highlands of Scotland over a one million year period. Since high-grade metamorphism affected the Dalradian, and because there are no useful fossils within these rocks, radiogenic isotope geochronology represents the only method of unravelling the complexity of the geology of the area. In combination with the U–Pb zircon ages produced by Rogers *et al.* (1989) and Pidgeon & Compston (1992) for the metamorphosed Ben Vurich granite (see Sections 12.6.2 and 4.2.7) which intrudes the Dalradian, these studies have allowed a very detailed assessment of the absolute time-scale for metamorphism and intrusion during the Caledonian orogeny to be developed. However, it must be remembered that before any isotopic studies could be carried out, a detailed structural map was produced to facilitate the production of an accurate lithostratigraphy.

In addition, the studies of Pankhurst (1982), Milne & Millar (1989), Hole *et al.* (1991) and Hole & Larter (1993) demonstrate how Rb–Sr whole-rock, Sm–Nd garnet–whole-rock and K–Ar whole-rock dating can be used to develop a stratigraphy of events in the structurally and magmatically complex, faunally barren geological setting of an active continental margin. This work demonstrates that the basic stratigraphical framework for the igneous and metamorphic rocks of the Antarctic Peninsula is almost entirely based on radiometric dating techniques, covering an age range of *c.* 400 Ma to the Early Quaternary (Figures 12.3 and 12.6).

12.9.1 Which Technique for Which Rocks?

This chapter has highlighted the fact that there are a large number of naturally occurring radiogenic isotopes, all of which may have applications for geochronology. Indeed, not all the available techniques have been described; merely the ones that are commonly applied to stratigraphical problems and are frequently published in the geological literature. However, a common question facing geologists is: 'Which isotopic technique should I apply, and what will the date mean ?' Clearly, some techniques, such as Lu–Hf dating are very specific to Archaean rocks, whereas other techniques, such as Rb–Sr and Ar–Ar, are not really restricted in terms of the age of the samples in question, but these techniques are only applicable, or are best suited, to certain rock types. Figure 12.11 gives some guidelines on the choice of a radiogenic isotope technique for different rock types throughout the stratigraphical column, and gives an indication of what that technique will yield in terms of a geological event. It must be emphasized, however, that this figure is by no means comprehensive.

Recent developments in mass spectrometry and general improvements in computer and electronic technology have caused an explosion in new techniques with specific stratigraphical applications. Radiogenic isotope geochemistry has even invaded the sedimentological strong-hold of reservoir geology, and there is a growing need in hydrocarbon exploration and production for high-precision radiogenic isotope analysis. One fact, however, must remain foremost in the minds of geologists; that there is no substitute for the acquisition of basic,

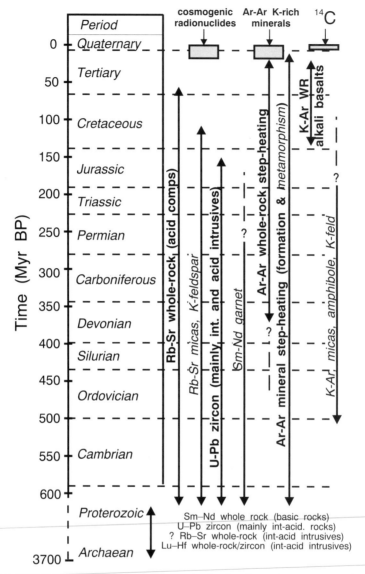

Figure 12.11   *Application of different methods of radiogenic isotope geochronology to different rock types of differing ages. Methods in italics will yield dates that are likely to reflect predominantly metamorphic/uplift events, whereas methods in normal text are more likely to
reflect magmatic crystallization*

descriptive information about rocks. Simple cross-cutting relationships for igneous rocks can often provide important stratigraphical information without
recourse to expensive 'high-tech' isotope analysis. Similarly, in sedimentary successions, there will never be a substitute for biostratigraphy, which has the ability to define stratigraphy to a far higher resolution than isotope geochronology

could ever do. Radiogenic isotope geochemistry is a vitally important toolbox for solving stratigraphical problems. However, like any other toolbox, it is critical that the correct tool for the job is selected and used in the correct manner.

ACKNOWLEDGEMENT

Bob Pankhurst helped materially with the production of this chapter.

REFERENCES

Arndt, N.T. & Jenner, G.A. 1986. Crustally contaminated komatiites and basalts from Kambalda, Western Australia. *Chemical Geology*, **56**, 229–255.

Baadsgaard, H. 1973. U–Th–Pb dates on zircons from the early Precambrian Amitsôq gneisses, Godthaab District, West Greenland. *Earth and Planetary Science Letters*, **19**, 22–28.

Bell, K. 1968. Age relations and provenance of the Dalradian series of Scotland. *Bulletin of the Geological Society of America*, **79**, 1167–1194.

Bouch J.E.E., Hole, M.J., Trewin, N.H. & Morton, A.C. 1995. Low temperature aqueous mobility of the rare earth elements during sandstone diagenesis. *Journal of the Geological Society of London*, **152**, 895–898.

Chauvel, C., Dupre, B. & Jenner, G.A. 1985. The Sm–Nd age of the Kambalda volcanics is 500 Ma too old! *Earth and Planetary Science Letters*, **74**, 315–423.

Cliff, R.A. 1985. Isotopic dating in metamorphic belts. *Journal of the Geological Society of London*, **142**, 97–110.

Cope, J.C.W. 1994. A latest Cretaceous hotspot and the southeasterly tilt of Britain. *Journal of the Geological Society*, **151**, 905–908.

Deino, A. & Potts, R. 1992. Age-probability spectra for examination of single-crystal ^{40}Ar/^{39}Ar dating results: examples from Olorgesailie, southern Kenya Rift. *Quaternary International*, **13/14**, 47–53.

Dempster, T.J. 1985. Uplift patterns and orogenic evolution in the Scottish Dalradian. *Journal of the Geological Society of London*, **142**, 111–128.

Evans, J.A. 1991. Resetting of Rb–Sr whole-rock ages during Acadian low-grade metamorphism in N Wales. *Journal of the Geological Society*, **148**, 703–710.

Faure, G. 1986. Principles of Isotope Geology. Wiley, New York.

Hamilton, P.J., Evensen, N.M. & O'Nions, R.K. 1979. Sm–Nd systematics of Lewisian gneisses: implications for the origin of granulites. *Nature*, **277**, 25–27.

Hole, M.J. & Larter, R.D. 1993. Trench proximal volcanism following ridge crest–trench collision along the Antarctic Peninsula. *Tectonics*, **12**, 897–910.

Hole, M.J., Pankhurst, R.J. & Saunders, A.D. 1991. The geochemical evolution of the Antarctic Peninsula magmatic arc: the importance of mantle–crust interaction during granitoid genesis. In Thomson, M.R.A., Crame, J.A. and Thomson, J.W. (eds) *The geological evolution of Antarctica*. Cambridge University Press, Cambridge, 369–374.

Hole, M.J., Trewin, N.H. & Still, J. 1992. The mobility of high field strength, rare earth elements and yttrium during late diagenesis. *Journal of the Geological Society of London*, **149**, 689–692.

Hole, M.J., Storey, B.C. & LeMasurier, W.E. 1994. Tectonic setting and geochemistry of Late Cenozoic alkalic basalts from the Jones Mountains, West Antarctica. *Antarctic Science*, **6**, 85–92.

Holliday, D.W. 1993. Mesozoic cover over northern England: interpretation of apatite fission track data. *Journal of the Geological Society of London*, **150**, 657–660.

Mearns, E.W. 1988. A samarium–neodymium isotopic survey of modern river sediments from northern Britain. *Chemical Geology, Isotope Geoscience Section*, **73**, 1–13.

Mearns, E.W., Knarud, R., Raestad, N., Stanley, K.O. & Stockbridge, C.P. 1989. Samarium–neodymium isotope stratigraphy of the Lunde and Statfjord Formations of Snorre Oil Field, northern North Sea. *Journal of the Geological Society of London*, **146**, 217–228.

Merrihue, C.M. & Turner, G. 1966. Potassium–argon dating by activation with fast neutrons. *Journal of Geophysical Research*, **71**, 2852–2857.

Milne, A.J. & Millar, I.L. 1989. The significance of mid-Paleozoic basement in Graham Land, Antarctica. *Journal of the Geological Society*, **146**, 207–210.

Moorbath, S., O'Nions, R.K., Pankhurst, R.J., Gale, N.H. & McGregor, V. 1972. Further Rb–Sr age determinations on the very early Precambrian rocks of the Godthaarb District, West Greenland. *Nature*, **240**, 78–82

Morton, A.C. & Berge, C. 1995. Heavy mineral suites in the Statfjord and Nansen Formations of the Brent Field, North Sea: a new tool for reservoir correlation and subdivision. *Petroleum Geoscience*, **1**, 355–364.

Morton, A.C., Claoué-Long, J.C. & Berge, C. 1996. SHRIMP constraints on sediment provenance and transport history in the Mesozoic Statfjord Formation, North Sea. *Journal of the Geological Society of London*, **153**, 912–929.

Pankhurst, R.J. 1982. Rb–Sr geochronology of Graham Land, Antarctica. *Journal of the Geological Society of London*, **139**, 701–711.

Pankhurst, R.J. & Pidgeon R.T. 1976. Inherited isotope systems and the source region pre-history of early Caledonian granites in the Dalradian series of Scotland. *Earth and Planetary Science Letters*, **31**, 55–68.

Pankhurst, R.J., Hole, M.J. & Brook, M. 1988. Isotope evidence for the origin of Andean granitoids. *Philosophical Transactions of the Royal Society of Edinburgh, Earth Sciences*, **79**, 123–133.

Patchett, P.J. & Tatsumoto, M. 1980. A routine high-precision method for Lu–Hf isotope geochemistry and geochronology. *Contributions to Mineralogy and Petrology*, **75**, 263–268.

Pettingill, H.S.P., Patchett, P.J., Tatsumoto, M. & Moorbath, S. 1981. Lu–Hf total-rock age for the Amitsôq Gneiss, West Greenland. *Earth and Planetary Science Letters*, **55**, 150–156.

Pidgeon, R.T. & Compston, W.A. 1992. A SHRIMP ion microprobe study of inherited and magmatic zircons from four Scottish Caledonian granites. In Brown, P.E. & Chappell B.W. (eds) Second Hutton Symposium on the Origin of Granites and Related Rocks. *Transactions of the Royal Society of Edinburgh: Earth Sciences*, **83**, 473–483.

Pidgeon, R.T. & Johnson, M.R.W. 1974. A comparison of zircon U–Pb and whole-rock Rb–Sr systems in three phases of the Carn Chuinneag granite, northern Scotland. *Earth and Planetary Science Letters*, **24**, 105–112.

Rogers, G. & Dunning, G.R. 1991. Geochronology of appinitic and related granitic magmatism in the W Highlands of Scotland: constraints on the timing of transcurrent fault movement. *Journal of the Geological Society of London*, **148**, 17–27.

Rogers, G., Dempster, T.J., Bluck, B.J. & Tanner, P.W.G. 1989. A high-precision U–Pb age for the Ben Vuirich granite: implications for evolution of the Scottish Dalradian Supergroup. *Journal of the Geological Society of London*, **146**, 789–798.

Sillitoe, R.H. & McKee, E.H. 1996. Age of supergene oxidation and enrichment in the Chilean porphyry copper province. *Economic Geology*, **91**, 164–179.

Steiger, R.H. & Jäger, E. 1977. Subcomission on geochronology: convention on the use of decay constants in geo- and cosmo-chronology. *Earth and Planetary Science Letters*, **36**, 359–362.

Zindler, A. 1982. Nd and Sr isotopic studies of komatiites and related rocks. In Arndt, N.T. & Nisbet, E.G. (eds) *Komatiites*. George Allen & Unwin, London, 399–420.

13
Chronostratigraphy (Global Standard Stratigraphy): A Personal Perspective

Charles H. Holland

Stratigraphy is the study of successions of rocks and the interpretation of these as sequences of events in the history of the earth. It is a central discipline of the geological sciences. In stratigraphy, correlation is the heart of the matter. This requires a rigorous standard, which is provided by chronostratigraphy. Historically, chronostratigraphical divisions are the descendants of time-rock divisions, originally conceived as divisions of rock representing specific divisions of time. The process of standardization of chronostratigraphy, or better, of a global standard stratigraphy, continues actively by international effort. Its completion is of some urgency. Once the global standard is agreed, that is when stratigraphical boundaries between the various divisions of the global standard hierarchy are defined at specific points in boundary stratotype sections, correlation with it of rocks elsewhere in the world (Figure 13.1) can be achieved by all the possible methods described elsewhere in this book. In most cases, in the Phanerozoic at least, this will be through biostratigraphy (see Chapter 5). In this way, an agreed stratigraphical framework, tied to time, will be provided within which all the interesting possibilities of basin analysis, structural and magmatic processes, plate tectonics, palaeogeography, palaeoclimatology, and organic evolution can be pursued.

The important divisions in the chronostratigraphical scale are the *system*, *series* and *stage*. This hierarchy has evolved since the nineteenth century until the names

Unlocking the Stratigraphical Record: Advances in Modern Stratigraphy. Edited by P. Doyle and M.R. Bennett.
© 1998 John Wiley & Sons Ltd.

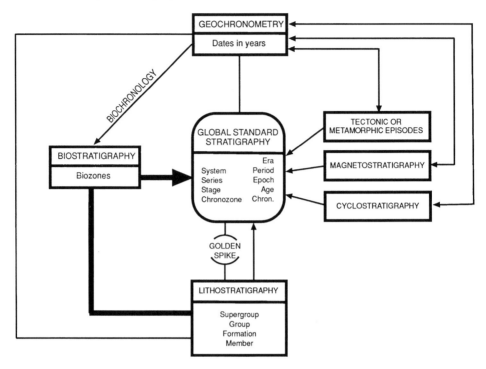

Figure 13.1   *Stratigraphical procedures and classification. [Modified from: Holland & Bassett (1989)]*

of the systems, at least, are in general use. As originally defined, these systems were geographically based and really lithostratigraphical in nature, taking this to include broad palaeontological as well as lithological criteria. For example, in the summer of 1831 Murchison and Sedgwick set off separately into what the former had called 'the interminable greywacke'. Their work resulted in the introduction, in 1835, of the names Cambrian and Silurian, the former by Sedgwick after the old name for Wales, and the latter by Murchison, after the *Silures*, an ancient tribe in the Welsh Borderland. For further examples, higher in the stratigraphical column, the Permian System was named by Murchison from what he had seen in a more exotic territory, the province of Perm, east of the Ural mountains; the Triassic system related to the threefold expression of those rocks in Germany; the Jurassic to the Jura Mountains in France and Switzerland, where those strata are so splendidly displayed. The various divisions of the Cenozoic Erathem have a more strictly palaeontological origin, beginning with Charles Lyell's comparison between their faunas and those of the present day.

A modern Phanerozoic (post-Precambrian) list of the systems and their groupings is shown in Figure 13.2. Corresponding terms such as 'Jurassic Period' are required only for use in language; thus one cannot say that the dinosaurs lived in the Jurassic System. The wider divisional term *erathem* is seldom used, the names often standing (if somewhat improperly) alone, or being treated as eras.

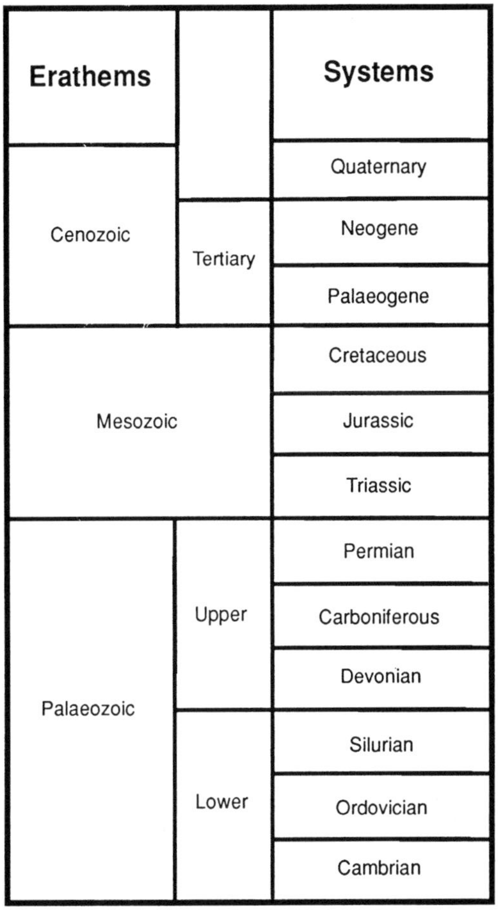

Erathems		Systems
Cenozoic		Quaternary
	Tertiary	Neogene
		Palaeogene
Mesozoic		Cretaceous
		Jurassic
		Triassic
Palaeozoic	Upper	Permian
		Carboniferous
		Devonian
	Lower	Silurian
		Ordovician
		Cambrian

Figure 13.2 *Phanerozoic systems and their groupings*

Some series names also relate to nineteenth century work; others such as the Přídolí Series (Silurian) are of recent origin. The stages are a different matter. They originated in France in the 1840s when d'Orbigny tired of the multiplication of local lithological terms. He used geographical names for his stages, but, in defining them, observed that 'Geologists in their classifications allow themselves to be influenced by the lithology of the beds, while I take for my starting point . . . the annihilation of an assemblage of life-forms and its replacement by another'. He felt 'bound to adopt them for the double reason that there is nothing arbitrary about them and that they are, on the contrary, the expression of the divisions which nature has delineated with bold strokes across the whole earth'.

So, already in retrospect, we see a confusion with biostratigraphy. Hancock (1977), in a very useful review of the whole matter, saw Oppel's use of zones in 1856, for the first time in a modern way, as its birth. Oppel was to begin to use zones to build stages. The matter is a complex one, which Hancock summarized by saying that 'stages were introduced more rapidly than zones, but were more often misunderstood and met with more opposition than zones'. Resistance to the use of stages came strongly in Britain and Germany, where stratigraphical divisions were already well established, and in the United States, where the necessary palaeontology had yet to be done. At the Bologna International Geological Congress in 1880, the stages did fall into place in the stratigraphical hierarchy below series. In spite of this, Mesozoic stratigraphers, with their splendid ammonite succession, for a long time showed limited interest in this concept, being content with an increasingly precise biostratigraphy.

Arkell (1933), in his introductory chapter to *The Jurassic System in Great Britain*, not only clarified historical matters concerning stage and zone, but also attempted to eliminate confusion between time and stratal terms. He explained a whole family of zonal terms (I would now call them biozonal) and gave their chronological equivalents. Trueman's epibole (Holland 1989) fell into place for the strata deposited during Buckman's hemera, itself regarded by Arkell as an acme zone.

The birth of chronostratigraphy itself can be put at 1941. North American stratigraphers, as was the case with others undertaking geological exploration in virgin territories, needed mappable units regardless of their biostratigraphy. Such were the groups and formations of lithostratigraphy. It fell to Schenck & Muller (1941), however, to recognize that there was a third kind of stratigraphy, represented by the systems and their subdivisions, which they defined as time-rock units. By the time of publication of the *American Code of Stratigraphic Nomenclature* in 1970, the relevant heading read 'Time Stratigraphic (Chronostratigraphic Units)'.

Some confusion between time and rocks continued. *The International Stratigraphic Guide* (Hedberg 1976: 67) defined a chronostratigraphical unit as 'a body of rock strata that is unified by being the rocks formed during a specific interval of geologic time'. Chronstratigraphical Units were said to be 'bounded by isochronous surfaces'.

In the meantime, matters were becoming clarified in a separate way. The Silurian–Devonian Boundary Committee of the Commission on Stratigraphy of the International Union of Geological Sciences, constituted during the 21st International Geological Congress at Copenhagen in 1960, presented a final report at Montreal in 1972. This may seem to be a long time for such decision-making, but the committee had to contend with a 'lost series' (the Přídolí Series); lost that is to say in previous correlations considerably in error because the belief had persisted that the latest graptoloids must be Silurian and not Devonian in age.

The choice of horizon therefore ultimately depended upon a compromise, raising the traditional British boundary and lowering that in use in Central Europe. The horizon having been agreed as at the base of the *Monograptus uniformis* Biozone, submissions were made and field visits undertaken. Morocco, Nevada, Podolia (Ukraine) and Czechoslovakia became the final short list. Barrande's classical ground to the south-west of Prague won the day. There the locality of Klonk was chosen, rather than the tectonically more complex Karlstein. The late Anders Martinsson and I were in a minority who would have placed the boundary at the base of Bed 20, which is only a few centimetres thick, but the majority favoured the point where *Monograptus uniformis* was first seen in the bed. It has since emerged that Bed 20 is a turbidite.

One important consequence of this work was the establishment of a set of criteria which the Committee had felt to be important in exercising their choice (McLaren *in* Martinsson 1977). They included faunal and floral development, stratigraphical considerations, structural considerations, facies and biodiversity, geographical accessibility, and the possibility of preservation of sections. To these should now be added the potential for magnetostratigraphy and radiometric dating.

An embryonic Sub-Commission on Silurian Stratigraphy met in Montreal in 1972. This was also the time at which Academician Vladimir Vasilievich Menner, as Chairman of the Commission on Stratigraphy, finally brought together his scheme for a full set of Sub-Commissions, Committees and Working Groups to undertake the immense task of rationalizing and standardizing the whole of stratigraphy. In 1974 in Birmingham, the Silurian and Devonian Sub-Commissions were formally constituted, as was an Ordoviocian–Silurian Boundary Working Group. All benefited from the spirit of international co-operation which was a feature of their parent body, the Silurian–Devonian Boundary Committee.

The Working Group operated as had its predecessor (Cocks & Rickards 1988) and eventually had a short list comprising Charles Lapworth's classic ground at Dob's Linn in the Southern Uplands of Scotland, the Canadian island of Anticost and sections near Oslo. The British section was eventually chosen on grounds of historical priority and graptolite biostratigraphy, though the others have very high merits. The decision was ratified by the International Union of Geological Sciences in 1985. The agreed horizon at Dob's Linn was at the base of the *Parakidograptus acuminatus* Biozone, rather than that of the *Glyptograptus persculptus* Biozone which formerly had been used in Britain.

The Silurian Sub-Commission undertook an eight-year programme of work, which, by 1985, had resulted in a full treatment of chronostratigraphy (Figures 13.3 and 13.4) (Holland & Bassett 1989). In this respect, it led the way and so adds to our examples. The Devonian Sub-Commission has almost reached a similar state. Some chronostratigraphical boundaries are being resolved much more slowly than are others; but there is an urgency about such work, held up as it is by nationalism, by reluctance to make decisions, by lack of resources and because many systems do not have the cosmopolitan faunas of the Silurian. It is also important to realize that there are no ideal sections; it is unreasonable to expect the fulfilment in one place of all those criteria referred to above. We must make do with the best presently available. I once quoted as follows from a context far from geological: 'The point could also be made that many new intellectual departures have become possible only after the luxuriant complexities accumulated before them have once more been reduced to surveyable simplicity' (Berger 1971: 117). It is surprising how quickly things settle down and use of the standardized divisions, perhaps once unpopular with some, become uniformly used.

The particular contribution to chronostratigraphy of the Geological Society of London came through its succession of stratigraphical committees which began work in 1965 under the chairmanship of Neville George. These committees emphasized the importance of defining each chronostratigraphical division at an

GLOBAL STANDARD STRATIGRAPHY				LOCATION OF BASAL BOUNDARY STRATOTYPE
SILURIAN SYSTEM	UPPER SILURIAN		PŘÍDOLÍ SERIES	(division into stages to await necessity)	BARRANDIAN PRAGUE BASIN (Pozary Section)
		LUDLOW SERIES	LUDFORDIAN STAGE	LUDLOW DISTRICT (Sunnyhill Quarry)
			GORSTIAN STAGE	LUDLOW DISTRICT (Pitch Coppice)
	LOWER SILURIAN	WENLOCK SERIES	HOMERIAN STAGE	WENLOCK DISTRICT (Whitwell Coppice)
			SHEINWOODIAN STAGE	WENLOCK DISTRICT (Hughley Brook)
		LLANDOVERY SERIES	TELYCHIAN STAGE	LLANDOVERY DISTRICT (Cefn Cerig section)
			AERONIAN STAGE	LLANDOVERY DISTRICT (Cefn Coed - Aeron Farm)
			RHUDDANIAN STAGE	SOUTHERN UPLANDS OF SCOTLAND (Dob's Linn)

Figure 13.3 *Chronostratigraphical or Global Standard Stratigraphical classification of the Silurian System. [Reproduced with permission from the National Museum of Wales, from: Holland & Bassett (1989)]*

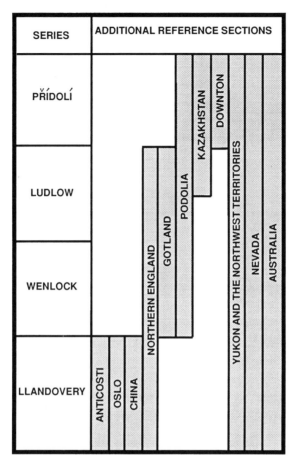

Figure 13.4 *Additional reference sections for the global standard stratigraphy of the Silurian System, showing the ranges for which they are particularly relevant. They support, but do not replace, the authority of the boundary stratotype with its golden spike. [Modified from: Holland & Bassett (1989)]*

internationally agreed point in the boundary stratotype section. 'The purists saw the point as geometrical, in the sense of infinitely small dimension, but practicality dictated the notion of hammering a spike (actually or symbolically) into the rock. I remain unclear as to how we began to use the term "golden spike", though a comparison has been made with the procedure of the early American railway builders' (Holland 1986: 7). Deliberations were often vigorous, some of us taking the not unreasonable view that one cannot hammer spikes into time.

The activities of the various subsidiary bodies of the International Commission on Stratigraphy have resulted in immensely useful work. As to the value of the golden spike itself, some continue to think that what matters is the choice of horizon for the boundary in question, which alone might solve the boundary

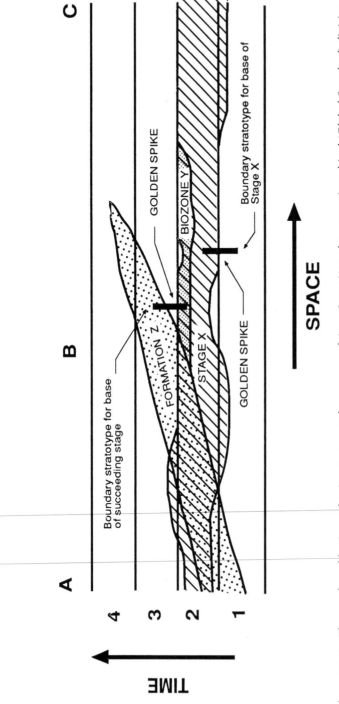

Figure 13.5 *The numbered horizontal strips represent the passage of time. Stage X, a chronostratigraphical (Global Standard) division, in-dicated by oblique ornament, is extended through space (A, B and C) by correlation involving the total available palaeontological and physical evidence. Its boundaries approach time planes, but can be known to coincide with these only at the boundary stratotypes for the base of the stage and for that of the stage above (which defines the top of Stage X). Biozone Y is a biostratigraphical division used in extending the upper boundary of the stage, but only in the area where its diagnostic fauna (or flora) is known to occur; it mostly provides a close approximation to a time boundary. A flagrantly diachronous lithostratigraphical division is added for completeness*

problems that have beset stratigraphy as the world has become geologically better known. It seems to me that there is a gain in that this point in the boundary stratotype section provides a rigorous standard analogous to the holotype in palaeontology. In addition, there is a philosophical gain. Here and here alone we know that a defined point in rock represents precisely a defined point in time (Figures 13.1 and 13.5). Elsewhere, we use biostratigraphy and, in certain situations and perhaps increasingly, other methods of correlation to allow a close approximation to the representation in rock of this plane in time; but we do not, and may never, know how closely we achieve it.

Some have questioned the importance of chronostratigraphy itself as distinct from biostratigraphy. Such an inclination would be more prevalent with Mesozoic stratigraphers. This is partly a matter of history but also because of the special value of ammonites in very detailed correlation (Callomon 1995). One of the troubles has been the supposition mentioned earlier that chronostratigraphical divisions have boundaries which are defined as, and actually known to be, time planes. Such hypothetical definitions may lack appeal. The very term chronostratigraphy has this unfortunate connotation. The use of *global standard stratigraphy* is better.

At the bottom of the global standard hierarchy comes the *chronozone*. Two such were defined in the Wenlock Series of the Silurian System as an example of procedure and because they are good cases (Bassett *et al.* 1975). However, divisions of this smallest scale are not yet the concern of the Commission on Stratigraphy. Palaeozoic stratigraphers are beginning to emulate the work of their Mesozoic colleagues in moving towards a more strictly standard biozonation. It is to be hoped that the latter will gradually lose any remaining antipathy to a global standard stratigraphy with its geographically named divisions and boundaries defined by golden spikes. It seems that future work should move along different pathways towards this desirable end, which surely is a well understood, internationally agreed standard, under the umbrella of which all kinds of other interesting work can continue. There may be a remote future when we can talk of a detailed stratigraphy simply in years; presently it is but a dream.

Finally, it has to be said that a different situation exists in the search for a global standard stratigraphy of the Precambrian. Procedures as outlined above can at present move only into the very youngest Proterozoic rocks (Holland 1986). For rocks older than this, standard internationally accepted stratigraphical classification can presently be achieved only by fitting rock successions, and their tectonic and metamorphic history, into an agreed geochronometric framework.

REFERENCES

Arkell, W.J. 1933. *The Jurassic System in Great Britain*. Clarendon Press, Oxford.
Bassett, M.G., Cocks, L.R.M., Holland, C.H., Rickards, R.B. & Warren, P.T. 1975. *The type Wenlock Series*. Report of the Institute of Geological Sciences, No. 75/13.
Berger, P.L. 1971. *A rumour of angels*. Penguin Books, Harmondsworth.

Callomon, J.H. 1995. Time from fossils: S.S. Buckman and Jurassic high resolution geo-chronology. In Le Bas, M.J. (ed.) *Milestones in geology*. Geological Society of London, Memoir No. 16, 127–150.

Cocks, L.R.M. & Rickards, R.B. 1988. A global analysis of the Ordovician–Silurian boundary. *Bulletin of the British Museum Natural History (Geology)*, **43**, 1–394.

Hancock, J.M. 1977. The historic development of concepts of biostratigraphic correlation. In Kauffman, E.G. & Hazel, J.E. (eds) *Concepts and methods of biostratigraphy*. Dowden, Hutchison & Ross, Stroudsburg.

Hedberg, H.D. (ed.) 1976. *International stratigraphic guide: a guide to stratigraphic classification, terminology & procedure*. John Wiley & Sons, New York.

Holland, C.H. 1986. Does the golden spike still glitter? *Journal of the Geological Society of London*, **143**, 3–21.

Holland, C.H. 1989. Trueman's epibole. *Proceedings of the Geologists' Association*, **100**, 457–460.

Holland, C.H. & Bassett, M.G. (eds) 1989. *A Global Standard for the Silurian System*. National Museum of Wales, Geological Series No. 10.

Martinsson, A. 1977. *The Silurian–Devonian boundary*. International Union of Geological Sciences, Series A, No. 5. Schweizerbart'sche Verlagsbuchhandlung, Stuttgart.

Schenck, H.G. & Muller, S.W. 1941. Stratigraphic terminology. *Bulletin of the Geological Society of America*, **52**, 1419–1426.

Part II
INTERPRETING THE RECORD

14
Interpreting the Record: Facies Analysis

Duncan Pirrie

The stratigraphical column provides a signature of changing environmental conditions at the earth's surface through geological time. Facies analysis is one of the
most fundamental tools in the understanding of that record. The earth's surface
today is a mosaic of different environments, each characterized by a range of
physical, chemical and biological processes. If we assume that ancient environments were also characterized by a similar range of discrete physical, chemical or
biological processes, then detailed interpretation of the geological record can be
achieved through an understanding of how the combined processes operated.
Once we can recognize discrete depositional environments then we can start to
investigate the causes of environmental change during earth history. Modern-day
environments are dynamic and change with time either as a result of natural
sedimentary processes, so-called autocyclic controls (e.g. the lateral migration of
an active fluvial channel), or due to external forcing mechanisms, or allocyclic
controls (e.g. the expansion of desert conditions as a result of changing climate).
We can also speculate as to whether environmental changes recorded in the
geological record are driven by either autocyclic or allocyclic controls. Such work
not only allows the understanding of the geological evolution of the earth and the
interplay between processes operating at the earth's surface, but also has direct
practical application in the extraction and utilization of earths natural resources.

This chapter aims to introduce the principles and application of facies analysis
in the interpretation of depositional environments in the rock record. However,
within a single chapter it is not possible to summarize the key facies which

Unlocking the Stratigraphical Record: Advances in Modern Stratigraphy. Edited by P. Doyle and M.R. Bennett.
© 1998 John Wiley & Sons Ltd.

characterize each environment at the earth's surface. For this, the reader should refer to the detailed textbooks of Reading (1996) and Walker (1984), and for facies analysis in a sequence stratigraphical framework, Walker & James (1992), Emery & Myers (1996) and also Reading (1996). Several case studies are given here to illustrate the application of facies analysis in both clastic and carbonate sedimentology, and for additional references see the collection of papers in Plint (1995).

Successful facies analysis requires accurate and detailed observation (either in the field, core store, or laboratory) and a thorough understanding of sedimentary processes. A grounding in these principles is given in the texts by Collinson & Thompson (1982), Leeder (1982), Allen (1985), Graham (1988), Tucker & Wright (1990) and Tucker (1991, 1995). Anderton (1985) provides a good historical review of facies analysis. In the first part of this chapter, a number of key concepts in facies analysis are introduced, before methods of facies analysis are reviewed in the last part.

14.1 KEY CONCEPTS IN FACIES ANALYSIS

Facies analysis has at its heart four fundamental concepts: (1) the identification of facies at all scales of subdivision; (2) the definition of facies associations; (3) the development of facies models; and (4) the interpretation of the environmental significance of the facies model. Each of these key concepts is considered in turn below.

14.1.1 Facies, Subfacies and Microfacies

Facies

The term facies has been used with a variety of different meanings (Reading 1978). Gressly (1838) introduced the term, and used it to mean the combined lithological and palaeontological aspects of a stratigraphical unit (Walker 1992). Subsequently it has been used in three discrete ways (Figure 14.1). First, it has been used in a purely descriptive sense for a body of rock with specified characteristics. In a sedimentary rock these characteristics may include lithology; composition; colour; geometry; physical, chemical or biological sedimentary structures; and micro- and macrofossil content. A stratigraphical sequence may therefore be subdivided into a number of distinct descriptive facies based on the physical, biological and/or chemical characteristics (e.g. normally graded sandstone facies). It is important to note that as each facies is defined on the basis of a discrete set of characteristics, then it is reasonable to assume that it represents an individual depositional process although that process may occur in several different environments. Secondly, the term is often used in a genetic sense, based on the interpretation of the depositional process. For example, a normally graded sandstone with a massive bed base, passing up into parallel and ripple cross lamination may be referred to as a 'turbidite facies' based on the interpretation that the bed was deposited from a

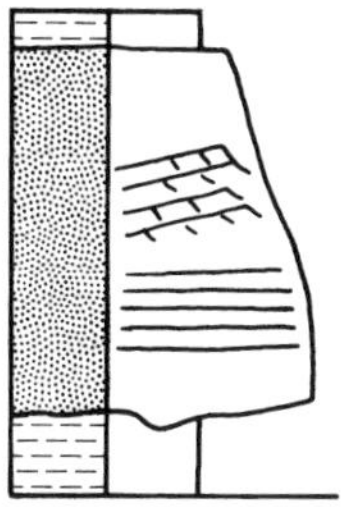

	Descriptive	Interpretative	
		Process (genetic)	Environmental
	Normally graded very coarse to very fine-grained sandstone	Bouma T_{abc} turbidite	Submarine fan lobe
	Purely descriptive classification scheme	Interpretation of the process of sediment transport and deposition	Interpretation of the environment of deposition

Figure 14.1 *Uses of the term facies. In facies analysis, the use of the term in a purely descriptive sense is preferred*

waning low-concentration turbidity current. Thirdly, the term may be used for an interpreted depositional environment, for example a submarine fan lobe facies. The different uses of the term are justifiable, as long as the sense in which the term is being used is clearly defined. The term *lithofacies* has been used where the facies characteristics are based primarily on lithological information, and *biofacies* where the focus is on palaeontological data. In addition, the term *palynofacies* is used by some authors where emphasis has been placed on the palynomorph/palynodebris content of a sediment (e.g. Whitaker *et al.* 1992). Ideally, if we are going to achieve a detailed environmental reconstruction based on facies analysis, we should combine together *all* of the available data, be it palaeontological or lithological, into our facies description and interpretation.

It is, however, important to recognize that the genetic and environmental usage of the term facies is *interpretative* and therefore subjective, and will depend upon the experience and training of the individual, along with the scientific framework within which the work is carried out. For example, the sedimentary structure 'hummocky cross stratification' was only described and interpreted as being deposited by storm processes during the 1970s and early 1980s (Cheel & Leckie 1993). The identification of this structure and its environmental significance led to the reinterpretation of previously described sandstone–mudstone sequences in terms of sedimentation on a storm-dominated shelf (e.g. Brenchley *et al.* 1979). Similarly the Jackfork Group of the Ouachita Mountains of North America, has classically been considered to represent deposition in a turbidite-dominated submarine fan (Shanmugam & Moiola 1995). Reinterpretation of the process sedimentology of the facies present, led Shanmugam & Moiola (1995) to suggest instead that the sequence is dominated by debris flow and slump deposits, with very few true turbidity current deposits (cf. Statt *et al.* 1997). The change in interpretation is not purely of academic interest, as the depositional processes control the likely sediment body geometry and, therefore, hydrocarbon reservoir potential (e.g. Shanmugam *et al.* 1995). Consequently, in rigorous facies analysis, the use of the term in a descriptive sense is recommended, as it can be defined objectively. Facies descriptions should therefore be clearly recognizable to other workers.

Subfacies and microfacies

Individual facies may be further subdivided into subfacies and/or microfacies. Subfacies are useful where there are minor variations in sediment characteristics which do not warrant the definition of new facies, but which may be significant in the detailed understanding of depositional processes or sub-environments (Figure 14.2). These subfacies would be defined in the field (see Section 14.2.3). Microfacies are defined by Flügel (1982: 1) as 'the total of all the palaeontological and sedimentological criteria which can be classified in thin sections, peels and polished slabs'. Although useful in all types of sedimentary rocks, microfacies have most commonly been used in carbonate facies analysis, where the component grain types are commonly more environmentally specific than in clastic sedimentology. In addition, microfacies analysis has recently been usefully employed in the understand-

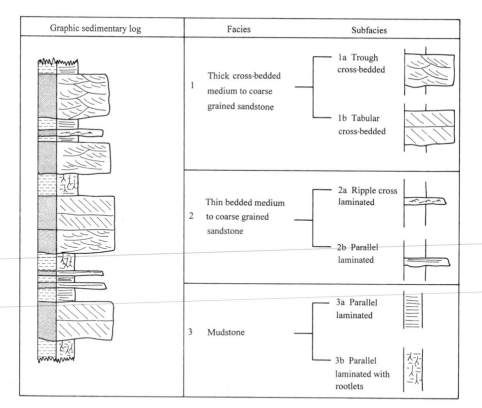

Figure 14.2 *Graphic sedimentary logs measured through sedimentary successions can be subdivided using a hierarchical scheme into initially facies and then subfacies. Subfacies are useful where minor variations in sediment characteristics occur, but are not sufficient to warrant the definition of new facies*

ing of mudstone-dominated sequences, where field-based descriptions are integrated with thin section, mineralogical and scanning electron microscope studies (e.g. Schieber 1989; Macquaker & Gawthorpe 1993; Macquaker 1994). In microfacies analysis, facies defined in the field may be further subdivided on the basis of laboratory thin section analysis of the component grain types and microtextures to help refine the understanding of depositional processes and environments (see Section 14.2). Microfacies are still defined on the basis of objective observations, but the correct recognition of their spatial and stratigraphical distribution will depend in part on the sampling density.

14.1.2 Facies Associations and Facies Successions

Individual facies (which will normally relate to an individual depositional process) can be grouped together to form facies associations. *Facies associations* are defined as 'groups of facies that occur together and are considered to be genetically or environmentally related' (Reading 1978: 5; Figure 14.3). Therefore, in clastic facies analysis, whilst individual facies are process related and are usually not environmentally specific (e.g. migrating unidirectional current ripples), facies associations are environment specific (e.g. a braided river). Only rarely are individual clastic facies environment specific. In contrast, in carbonate facies analysis, individual facies may, in some cases, be environmentally specific because, as stated by James & Kendall (1992: 265), 'the sediments are born, not made'. In other words, sediments that are biochemically or directly chemically precipitated in place are much more likely to be environmentally specific than those which have undergone physical transport and deposition. Consequently, whilst in clastic facies analysis, individual processes (facies) need to be combined together into facies associations to define environments (see Section 14.2.4), in carbonate sedimentology, individual facies may be interpreted in terms of environment. The definition of facies associations is usually rather subjective (see Section 14.2.4). The term *facies sequence* or *facies succession* is used where 'a series of facies which pass gradually from one into the other' (Reading 1978: 5); for example, in a coarsening-upwards succession characteristic of the progradation of a delta front, or a fining-upwards succession, which may represent the gradual infilling and abandonment of a fluvial channel. Walker (1992) advocates the use of the term 'facies succession' rather than 'facies sequence', due to the widespread use of the term 'sequence' in sequence stratigraphy, although Reading & Levell (1996) argue that this is unnecessary. Effectively, a facies succession emphasizes gradualistic changes between facies associations which can be interpreted in terms of the lateral migration of originally adjacent depositional environments (Figure 14.3). This was first recognized by Walther's Law of the Correlation of Facies, in which he stated 'it is a basic statement of far reaching significance that only those facies and facies areas can be superimposed primarily which can be observed beside each other at the present time' (Walther 1894 *in* Reading 1978: 4), that is a conformable vertical succession of facies (or facies associations) was deposited in originally laterally adjacent environments. The most widely used example to

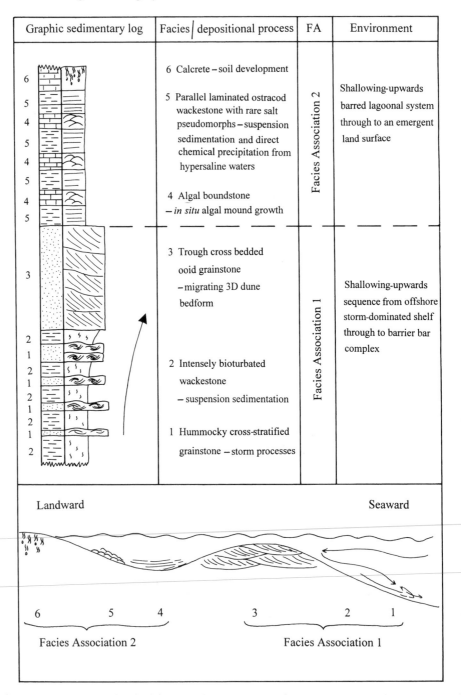

Figure 14.3 *Once individual facies, which represent discrete depositional processes, are defined, they can be combined together into facies associations which can then be interpreted in terms of depositional environment. Whilst individual facies are rarely environment-specific, combinations of facies (facies associations) usually are*

explain this principle is the progradation of a delta, with a prodelta facies associ-ation gradually passing into delta front and delta top facies associations (Figure 14.4). However, in some cases, Walther's Law does not hold true; for example, in some carbonate sequences where the vertical facies changes may be due to con-tinued growth of skeletal carbonates in response to changing sea-level, without any lateral migration of depositional environments. Effectively Walther's Law applies where autocyclic controls on sedimentation occur, such as the lateral switching of a delta lobe, which will result in a vertical facies sequence represen-tative of laterally adjacent environments. In contrast, if the controls on sedimenta-tion are allocyclic, such as abrupt changes in eustatic sea-level, then different environments can be juxtaposed together in a vertical sequence. The most signifi-cant part of Walther's Law, as noted by Reading (1978) and Walker (1984), is therefore the nature of the boundaries between the facies associations. Gradual boundaries imply that the different facies associations and therefore environ-ments must have been laterally equivalent. However, if the boundary is abrupt then it is possible that environments which were not originally adjacent to one another have been superimposed. For example, within the Miocene Waitemata Basin of North Island, New Zealand, shallow marine sediments deposited along a rocky shoreline are abruptly overlain by a range of facies interpreted to represent bathyal (1000–2000 m) water depths, with the transition occurring in possibly as little as 1 Ma (Ricketts *et al.* 1989). With the advent of sequence stratigraphy, the significance of abrupt contacts between stratigraphical units has been more fully recognized, and commonly can be used to aid regional correlation.

14.1.3 Facies Models

As discussed above, once a stratigraphical unit has been subdivided into separate descriptive facies (which are interpreted in terms of depositional process), they may be combined together as facies associations, that is a range of processes inferred to have occurred within a single environment. To interpret the deposi-tional environment, the unknown facies association has to be compared with a previously defined summary, or model, of the possible depositional environ-ments. This facies model is defined by Walker (1992: 6) as 'a general summary of a given depositional system'. Facies models are normally based on a combination of both present-day examples and previously interpreted ancient equivalents, al-though in some cases this is not possible, as comparable examples may not exist today. The facies model is developed by combining all of the data from a number of examples (both modern and ancient) and then removing the local variability, so that the resultant model characterizes the general depositional system. Conse-quently, a facies model is not fixed, and as new studies are added, it should evolve to take into account the greater data set.

Walker (1992) argues that there are a limited number of facies models as there are a limited number of possible depositional systems. This view has led some authors to define a limited range of facies and facies associations for discrete depositional systems, with the adoption of letter codes to signify individual facies.

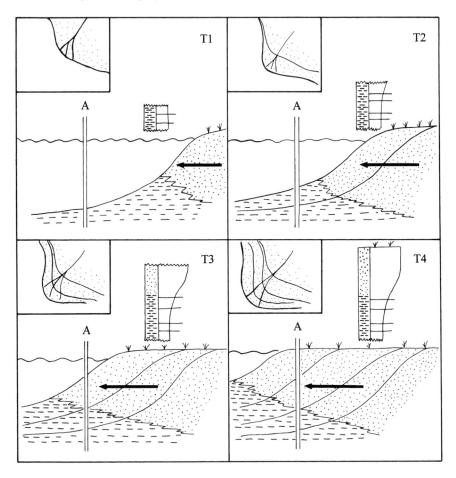

Figure 14.4 *Walther's Law of the Correlation of Facies suggests that a conformable vertical
succession of facies (or facies associations) was deposited in laterally adjacent environments.
This can be clearly illustrated if we consider the progradation of a delta through time. In the
diagram above we can see what happens as we go through four time slices, T1 to T4. With
time, the shoreline progrades from right to left, so that at a single location depicting a vertical
sequence (A), we see the gradual transition from a mudstone-dominated to a sandstone-
dominated succession (i.e. a coarsening-up cycle), which represents the progradation of
laterally adjacent depositional environments*

For example, Miall (1992) developed a widely adopted facies classification scheme
for fluvial systems, as he considered there to be a limited number of possible facies.
In contrast, some authors argue that there is an infinite number of facies or
combinations of facies, and therefore possible facies models, and that an over-rigid
classification with a limited numbers of facies and facies models will obscure
significant detail (see Anderton 1985; Walker 1992). Consequently, each ancient or
modern example of a specific environment is a unique set of data and the facies
model will purely reflect this unique data set. The significance of this depends in

part upon the aims of the individual study. If the aim is to achieve a broad environmental interpretation, then direct application of a pre-existing facies model may provide an adequate answer. However, if the aim is to provide a detailed environmental interpretation and understand the complex interplay of controls, then a much more rigorous and critical appraisal of the pre-defined facies models is required.

In some cases the field or laboratory data will not 'fit' a pre-existing model; this is not because the data are incorrect, but because the model they are being compared with may have been based on a limited number of examples, or biased towards specific case studies, or controlling factors within the depositional environment. For example, alluvial fan facies models for semi-arid climates were originally based largely on the Trollheim fan in California. However, this fan has subsequently been considered to be rather atypical of this depositional environment (Blair & McPherson 1992, 1993; Hooke 1993). In this case, the facies model must be adapted to take into account the newly available data and will therefore continue to evolve. There is a risk with the very formally defined facies models typified by Miall (1992) that new data sets may be 'forced' to fit the pre-existing model, rather than allowing the model to adapt.

Walker (1992) has argued that a facies model has four main uses:

1. it must act as a norm for the purposes of comparison;
2. it must act as a framework and guide for future observations;
3. it must act as a predictor in new geological situations; and
4. it must act as an integrated basis for the interpretation of the depositional system that it represents.

As a norm the facies model gives us something we can compare our unknown facies association against. This allows us to identify not only the similarities but also the differences between the unknown and the norm, and if it is different, to consider why. The facies model summarizes all of the key features of a depositional setting, hence it can act as a useful framework or guide when examining a new example (i.e. it helps to identify what key features we need to look out for). However, although this framework helps to ensure that the necessary data needed for comparison with the pre-defined facies model are collected, it is important to note that this should not limit our observations to those features that we seek; instead we should fully describe all aspects of the stratigraphical unit.

The third use of facies models as a predictive tool is of considerable economic significance. Once we have compared our unknown against a facies model, and are confident that our interpretation fits all of the available data, then we can use the facies model as a predictive tool. For example, in a hydrocarbon exploration programme, where the interpretation is based on limited data, the facies model may allow the distinction between a fluvially dominated and a wave-dominated delta, which may lead to a prediction of the likely geometry of the sandstone reservoir unit. The accuracy of any prediction will depend not only on the suitability of the facies model, but also on the amount of available data. For example, in the northern North Sea, the reservoir sandstones forming the Rannoch, Etive and Ness formations of the Middle Jurassic Brent Group in the North Sea, have long been interpreted as a wave-dominated delta prograding from south to north, but an increase in well data and improved facies models have allowed a more detailed facies analysis, and resultant palaeogeographical interpretations have been critical for hydrocarbon reservoir appraisal (Richards 1992; Figure 14.5). The fourth use for facies models is to act as an integrated basis for interpretation, where, by combining data from numerous examples, the local variability has been removed, allowing a clearer interpretation (Walker 1992).

Anderton (1985) argues that whilst a facies model can be used as a framework for future observations and as a predictive tool, it cannot be used as a norm for comparison or as a basis for interpretation. This follows on from the previous discussion as to whether there are a limited or an infinite number of facies, models

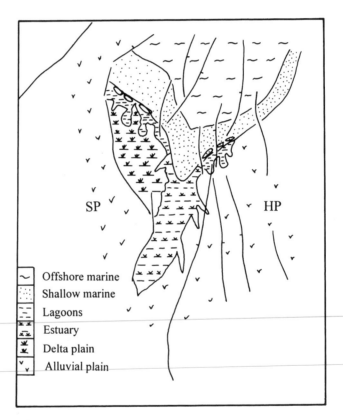

Figure 14.5   *Within the northern North Sea, increasingly detailed well coverage, combined with improved facies models, has led to the development of complex palaeographical interpretations for individual time slices. For example, the Middle Jurassic Brent Group, long interpreted as a deltaic unit prograding from south to north, can now be interpreted in terms of a complex mosaic of offshore marine, shallow marine, lagoonal, estuarine, delta plain and alluvial plain environments. Detailed facies analysis can therefore aid our understanding of hydrocarbon reservoir potential. SP, Shetland Platform; HP, Horda Platform. [Modified from Richards (1992)]*

and environments. Anderton (1985: 33) states that 'if, like me, you have a more nihilistic view of life, the universe and everything, then you have to admit an infinite number of environments, facies and models. Every descriptive sedimentary facies is then unique and has a unique interpretation in terms of a facies model.' Consequently, he does not consider it possible for a facies model to be used as a norm for comparison nor a more powerful basis for interpretation, because, by definition, every study will result in a discrete facies model.

Most facies models incorporate information on modern environments. Therefore, before embarking on facies analysis we need to consider two further points:

1. Is the axiom 'the present is the key to the past' justifiable in terms of facies analysis?, and
2. What is the significance of rare 'catastrophic' events versus gradualism or actualism in the geological record?

Are modern environments comparable with examples from the geological record, when the biosphere and the atmosphere were different? For clastic sediments, many ancient environments do appear equivalent, at least during the Phanerozoic, although there are ancient environments, such as the Cretaceous shallow epeiric seas, for which we do not have a modern analogue. In carbonate facies analysis, the situation is more complex. Not only are carbonate-secreting organisms controlled spatially by changing environmental parameters, but also temporally as a result of evolution (Figure 14.6). In addition, ocean chemistry and therefore primary sedi-

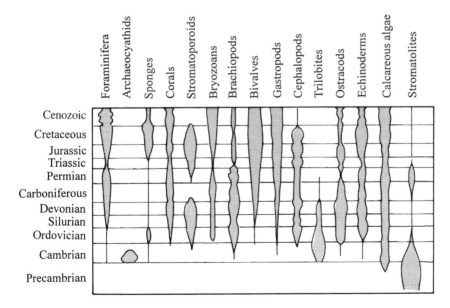

Figure 14.6 *Carbonate facies may be more difficult to interpret environmentally than clastic sediments because the relative contribution of different carbonate-secreting organisms has changed through earth history. [Modified from: Tucker (1991)]*

ment mineralogy has also changed through earth history. Consequently, the direct comparison of a modern and ancient clastic shelf is significantly easier, as at least the grain types are the same, whereas in carbonate sedimentology the grain types themselves have changed throughout earth history, and we know little about the ecology of some extant species. Facies models for ancient evaporites are also restricted as the few Holocene analogues that do exist are all very small; there are no modern examples of the 'saline giants' of the past (James & Kendall 1992; Kendall & Harwood 1996).

The second problem facing the use of modern environments in facies analysis is the relationship between gradualism/actualism and catastrophism in the geological record. Ager (1993) provides a thought-provoking discussion in *The New Catastrophism*, in which he argues that rare episodic events are significant in the geological record. On the modern continental shelf, the daily gradual processes may lead to relatively little deposition in comparison with the effects of a once in every 100 or 1000 year high-intensity storm event. Our understanding of the potential magnitude of natural processes in modern environments is limited by the length of time that we have been observing them. This is probably most acutely understood in the field of volcanology, where recent eruptions such as Mount St Helens have been on a much smaller scale than even historically documented eruptions such as Krakatau in 1883 (Francis 1993). High-magnitude events, depositing a larger sediment thickness, have a greater preservation potential and will therefore be well represented in the ancient rock record, but our understanding of such processes based on our observation of modern environments is limited.

14.1.4 Facies Analysis

Having considered the meaning of the terms facies, facies association and facies model, we must now consider what we mean by the term facies analysis. Anderton (1985: 36) has provided a clear definition of the term as 'the description and classification of any body of sediment followed by the interpretation of its process and environments of deposition, usually in terms of a facies model'. Many workers would expand this definition to incorporate volcaniclastic rocks (including epiclastic, hyaloclastic, autoclastic and pyroclastic volcanic rocks), which can also be interpreted in terms of process and environment (e.g. Suthren 1985; Cas & Wright 1987; Lajoie & Stix 1992; Orton 1996). Facies analysis has five main stages:

1. detailed description of the stratigraphical unit;
2. its subdivision into a number of descriptive facies, and the quantification of the relationships (if any) between the facies;
3. the interpretation of each facies in terms of depositional process;
4. the compilation of the individual facies into a number of facies associations; and
5. the comparison of the facies associations with pre-existing facies models to allow an environmental interpretation (Figure 14.7).

This methodology is described in more detail in the next section.

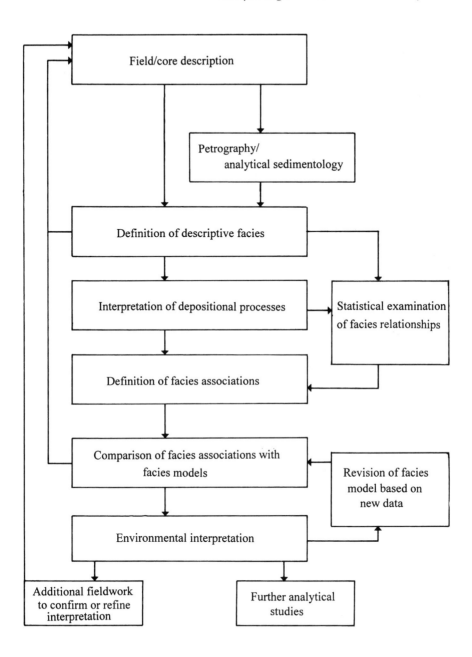

Figure 14.7 *Flow chart showing the methodology adopted in facies analysis*

14.2 METHODOLOGY OF FACIES ANALYSIS

Facies analysis comprises five phases, and is reliant on the accuracy of the data collected and the interpretational models employed. Five phases are described in detail below: data collection; petrography and provenance; facies classification and interpretation; the definition of facies associations; and the development of facies models.

14.2.1 Data Collection and Presentation

The first stage in facies analysis is the collection of the primary data, whether in the field at outcrop or by the examination of cores. The construction of graphic logs has been discussed by Tucker (1995), Graham (1988) and Anderton (1985), amongst others. They should form the basis for facies analysis, but will also be the frame-work not only for the lithostratigraphy, but also for biostratigraphy. High-resol-ution stratigraphical control is critical in the interpretation of three-dimensional sediment body geometry and lateral (i.e. spatial) distribution of environments. Stratigraphical control in facies analysis was traditionally based on biostrati-graphical techniques, whilst sequence stratigraphy today provides a valuable correlation tool. Ideally, in a field-based project, the measurement of graphic logs should go hand-in-hand with geological mapping, since together they form the basis of the regional lithostratigraphy (e.g. Pirrie *et al.* 1991).

The measurement of graphic logs

Graphic logs can be measured from essentially one-dimensional sections through a stratigraphical unit (be it an exploration core, or a stream section), from a two-dimensional cliff or quarry face or from a three-dimensional series of out-crops, cores or quarry faces. If dealing with a two-dimensional face it is important that the geometry of individual units is fully described and that the graphic logs reflect the true variability present. This is best achieved by initially sketching or photographing the entire face and identifying the main units present, and then choosing a series of lines of section which fully characterize the succession. In some cases this may be accomplished with a single graphic log, whilst in other cases a number of overlapping cross-correlated logs may be required. In expo-sures where there is marked lateral variation in bed geometry, the whole face should be photographed from a fixed distance with each photographic frame overlapping, so that a photographic montage can be constructed. This can then be used to aid the recognition of the bounding surfaces between individual beds or units and to understand the lateral variations in bed geometry. It is important to ensure that the position of the graphic logs is clearly recorded to allow cross-reference between the logs and the photographic montage (Figure 14.8). In areas where there is effectively three-dimensional exposure, then numerous sections can be measured and cross-correlated to allow a clear understanding of the

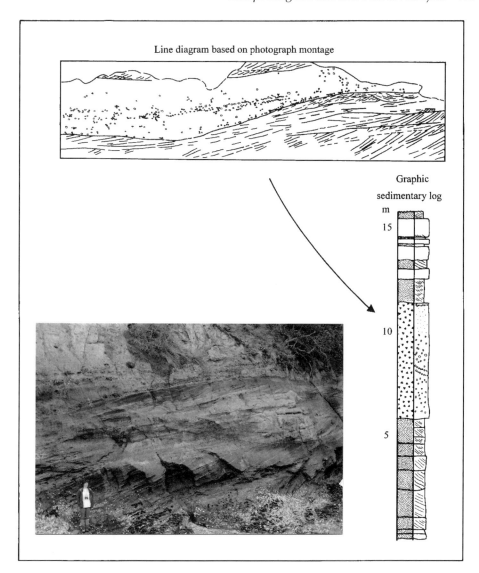

Figure 14.8 *Typical data that may be collected during fieldwork. A photomontage was taken of the cliff section and then used to provide a line diagram, highlighting the key bounding surfaces through a complex tidal sandstone sequence. Additional, more detailed, field photographs can also be cross-referenced to the photomontage. The sequence shown is from the Puketi Formation, Kaipara, North Island, New Zealand, where shallow tidal sand bars are interbedded with primary and reworked pyroclastic flow deposits*

spatial facies variations. This type of information can usefully be presented with the use of fence diagrams.

The graphic logs measured should be as detailed as possible in the field time available, which will largely depend upon the scope and aim of the project. Detailed bed-by-bed logging at a centimetre-scale is slow work, with possibly only 10 m logged per day if the sequence is quite complex; in lithologically monotonous successions, a thicker interval may be logged. Graphic logs should normally be measured from the base of the section to the top. Individual bed thicknesses should be recorded along with the nature of the bed bounding surfaces. The lithology should be fully described and any sedimentary structures should be recorded through the processes of sketching, photographing and measuring. In general, field notes need to be as quantitative as possible. For example, it is all too easy to record a unit as parallel laminated without noting the spacing of the laminae nor what is defining the lamination. It is essential to record as many palaeocurrent data as possible, always noting the type and scale of the sedimentary structure the measurement is from. In sequences where bedding is dipping at greater than 10°, both the palaeocurrent and the bedding orientation should be noted, so that the palaeocurrent data can subsequently be corrected for tectonic tilt (Tucker 1995). In coarser-grained clastic sediments, the field description should include quantified clast composition data, to assist in the subsequent interpretation of the sediment provenance. Palaeontological data, with detailed descriptions of not only the presence or absence of any flora/fauna or trace fossils, but also a full description of relevant taphonomic details, should also be given. The positions of samples collected, whether for petrographic, geochemical or micropalaeontological analysis, should be clearly annotated on to the logs, as should the position of any field photographs taken. Detailed sampling will be required if a microfacies study is required.

During the data collection it is common to have one or more facies models in mind, either based on previous work or on preliminary field interpretations. This can be useful as its acts as a framework for your observations, ensuring that key features, important in terms of the interpretation of depositional process, are not overlooked; it also acts as a prompt for the search of key data in poorly exposed sections. In addition, subdivision of the sequence into descriptive facies is commonly attempted in the field, but it is important to ensure that each bed or unit is still fully described. Repeatedly noting the repetition of a previously recognized facies should be avoided because subtle variations, which may be significant in terms of either process or environment, may often be missed.

Presentation and interpretation of graphic logs

Once collected, the field data can then be presented as graphic sedimentary logs. Logs can either be drawn up 'free hand' as realistic representations of the field data, or may be produced in a more stylized form. Usually a set of standardized symbols are used to depict the lithologies and structures observed. Many companies have 'in-house' standardized symbols and styles for the presentation of graphic logs. Although this aids communication between different workers, the use of very stylized logs may lead to the loss of information that may have aided interpretation (Anderton 1985). Over the last 10 years, computer-based graphics packages have become widely available, leading to the production of some car- tographically very poor logs. With modern scanning technology, a sensible com- promise is to draw-up graphic logs by hand whilst in the field and then, if necessary, scan them into the computer, and edit them using a suitable graphics package.

14.2.2 Petrography, Provenance and Microfacies

Petrographic studies should ideally be carried out as a routine part of any facies analysis, providing additional descriptive data concerning sediment texture and mineralogy. In many cases, these data will only provide supporting information to aid your field-based facies scheme. For example, in clastic sediments, provenance studies based either on point counting of thin sections, heavy mineral analysis or clast composition data should be carefully integrated with both the facies scheme and the palaeocurrent data. In coarser-grained carbonate sediments, facies analy- sis can be based primarily on field data (e.g. Mutti *et al.* 1996); however, microfacies analysis has been widely used in carbonate sedimentology, and increasingly in the understanding of mudstone-dominated sequences. In carbonates, the field identi- fication of the component grain types can often be difficult due to either grain size or diagenetic alteration. Consequently, thin sections, peels or polished slabs are commonly needed to allow the identification of both skeletal and non-skeletal grain types and rock textures. This information can be either qualitative thin section description leading to limestone classification, or it may be quantitative, based on point counting the individual grain types present.

The biogenic and direct chemical origin of carbonate grain types means that unless they have been subsequently reworked, they are much more diagnostic of the original depositional environment than clastic grains. This has led some authors to try to erect standardized microfacies types for carbonates, each of which has a restricted environmental interpretation (Tucker & Wright 1990). Whilst useful for comparison, these standardized microfacies have not kept pace with the more detailed modern understanding of carbonate depositional environments, hence their value is limited. Diagenetic information can also be usefully incorpor- ated into carbonate microfacies analysis, as early diagenesis related to the deposi- tional pore-water conditions is very common in many limestones, and can be interpreted from both petrographic and geochemical data, if there has not been significant subsequent diagenetic over-printing.

Flügel (1982) provides a thorough review of microfacies analysis of lime- stones. A good recent example of the combination of both field data and micro- facies analysis is provided by Gawthorpe & Gutteridge (1990) for Late Dinantian (Mississippian) bioclastic shoals in Derbyshire. In this study, detailed sedimen- tary logs and sketches of the main quarry faces were combined with petro- graphic studies, resulting in the recognition of four distinct microfacies, defined on the basis of the skeletal grain types present, and subsequently interpreted in terms of both depositional processes and environment. Similarly, Arnaud-Vanneau & Arnaud (1990) combined field data and microfacies analysis in the interpretation of carbonate shelf sedimentation in the Cretaceous of France and Switzerland.

Recently, a number of authors have effectively used microfacies analysis in the description and interpretation of mudstone-dominated sequences. Schieber (1989) described six distinct shale facies from the mid-Proterozoic Newland Formation, Montana, USA, based on the petrographic examination of 500 thin sections combined with an examination of mineralogy on the basis of X-ray diffraction. Each facies defined was then interpreted in terms of discrete depositional environments. Similarly, the facies present in organic-rich Jurassic mudstones of both the Kimmeridge Clay Formation (Macquaker & Gawthorpe 1993) and the Oxford Clay Formation (Macquaker 1994) of the UK have been described on the basis of combined petrographic, scanning electron microscope and organic geochemical data, leading to an interpretation of the facies present in terms of depositional environment.

14.2.3 Facies Classification and Interpretation

Once the field data have been collected, graphic logs prepared and any laboratory analytical or petrographic data incorporated, then the sequence can be subdivided into descriptive sedimentary facies. If beds were not grouped together into descriptive facies, then each individual bed in a succession would have to be interpreted separately in terms of depositional process; for example, Ineson (1989) was able to summarize 1800 m of detailed graphic logs into nine main descriptive facies. The definition of facies is probably the most difficult stage of facies analysis as it is essentially an arbitrary process, which aims to subdivide the complete succession into a manageable number of descriptive facies. The number of facies defined will depend upon the variability of the section studied, the thickness of strata measured and the individual worker's own approach as either a 'lumper' with a very limited number of facies, or a 'splitter' with a greater number of facies. Ideally the facies scheme has to be manageable, whilst conveying the actual variation seen, and Anderton (1985) has suggested that most sequences can be divided into between two and ten facies. In some facies models, a considerably larger number of facies have been recognized (e.g. in deep-sea clastics; Pickering *et al.* 1986); hence any facies scheme developed will need to reflect this complexity.

Facies schemes which are both manageable and reflect the true variability of a succession can be achieved by dividing up the measured section using a hierarchical classification scheme. For example, visual examination of a graphic log may allow the immediate distinction of three major facies on the basis of lithology (mudstone, sandstone, conglomerate) (Figure 14.9). The sandstones could then, for example, be further subdivided on the basis of bed thickness, geometry or grain size, or on the basis of internal sedimentary structures. The sandstones shown in the graphic log in Figure 14.9 can be divided into:

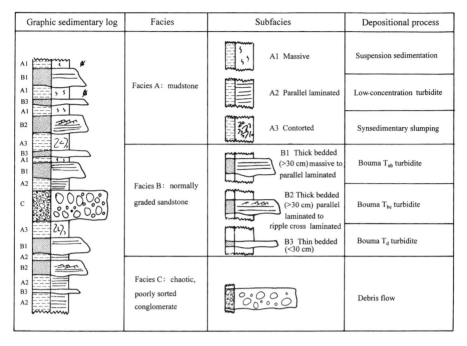

Graphic sedimentary log	Facies	Subfacies	Depositional process
		A1 Massive	Suspension sedimentation
	Facies A: mudstone	A2 Parallel laminated	Low-concentration turbidite
		A3 Contorted	Synsedimentary slumping
	Facies B: normally graded sandstone	B1 Thick bedded (>30 cm) massive to parallel laminated	Bouma T$_{ab}$ turbidite
		B2 Thick bedded (>30 cm) parallel laminated to ripple cross laminated	Bouma T$_{bc}$ turbidite
		B3 Thin bedded (<30 cm)	Bouma T$_{d}$ turbidite
	Facies C: chaotic, poorly sorted conglomerate		Debris flow

Figure 14.9 *Diagram illustrating a hierarchical classification scheme used in facies analysis. The graphical sedimentary log can be subdivided on the basis of lithology into three facies – mudstone, sandstone, conglomerate – which can then be further subdivided on the basis of the presence or absence of specific sedimentary structures. Each subfacies can then be interpreted in terms of depositional process. Thus whilst all of the normally graded sandstones represent deposition from a waning sandy turbidity current, the discrete subfacies can be interpreted in terms of different intervals of the Bouma turbidite sequence, allowing a more detailed interpretation*

1. thickly bedded (>30 cm), coarse to fine grained, normally graded sandstones (Facies B1 and B2); and
2. thinly bedded (<30 cm), fine to very fine grained, normally graded sandstones (Facies B3).

Commonly, for clarity of writing, facies are assigned a numeric or letter code (e.g. Facies A, Facies B, etc.) as shown above. In the initial stages of interpretation, over-subdivision, emphasizing the true variability, is beneficial; the facies can always be combined together at a later stage if the variability is not significant in terms of either depositional process or environment. Hence, using our example above, Facies B may be further subdivided into subfacies B1, B2 and so on, based upon the organization of internal sedimentary structures (e.g. subfacies B1, massive sandstone overlain by parallel lamination, and subfacies B2 parallel lamination overlain by ripple cross lamination). In this case all of the subfacies included in Facies B represent deposition from a waning low-density turbidity current, but a more detailed process interpretation can be achieved by identifying the discrete

Bouma intervals present, so that Subfacies B1 may be interpreted as a T_{ab} turbidite deposit, whilst Subfacies B2 represents a T_{bc} turbidite deposit. By following this hierarchical classification scheme, the measured graphic log can be divided into a relatively limited number of descriptive facies, and by the end of the interpretation, all beds should have been assigned to one individual facies.

It is quite probable that the number of beds assigned to different facies will vary, with some facies based on only a very limited number of beds. This is to be expected as these may represent the rare or episodic events within the depositional environment. The key difficulty in facies classification is that the individual worker has to define the hierarchical scheme that they are using (i.e. assign a particular significance level to each field observation), and consequently different workers may end up with differing descriptive facies for the same succession.

Based on the summary of the characteristics of each descriptive facies, the facies can then be interpreted in terms of depositional process by comparison with previously published facies schemes, processes observed in modern environments and/or experimental studies on sediment transport. Some facies will be easier to interpret in terms of depositional process and/or have greater environmental significance than others. For example, a tabular cross-bedded sandstone in isolation can only be interpreted in terms of the lateral migration of a straight crested dune bedform, whilst a mudstone with either rootlets or calcrete can be interpreted as representing a combination of suspension sedimentation and either colonization by terrestrial plants or the precipitation of carbonate minerals from groundwaters as part of a pedogenic profile. Similarly in carbonate sediments, a cross-bedded grainstone will have less direct environmental significance than a fenestral limestone. In some cases (e.g. where primary sedimentary structures are lacking), it may be impossible to interpret the depositional process for an individual facies. This is particularly true in shallow marine sequences where intense bioturbation may homogenize the primary sedimentary record, and in carbonate sequences that have an intensive diagenetic overprint. In such cases an environmental interpretation can only be made by looking at the surrounding facies. Once the facies interpretation is complete, a range of different depositional processes will have been identified, which, when combined together, will allow an environmental interpretation.

14.2.4 Facies Relationships and Facies Associations

Once the individual depositional facies have been interpreted, the relationships between facies can be investigated, leading to the definition of facies associations. Early attempts at understanding facies relationships led to the construction of facies relationship diagrams, which depicted whether the transition between facies was sharp or gradual, and the frequency of each type of transition. There may be no relationship between facies transitions if the section measured represents a range of processes occurring at random within the original depositional environment. Alternatively, the facies transitions may be interrelated in environments where the individual processes are genetically linked; for example, siltstones interpreted as deposited from suspension on a floodplain during overbank flooding, overlain by coals formed during the accumulation of organic matter in a waterlogged area.

Subsequently, facies relationships can be examined by counting the number of times there is an observed transition from one facies to another, and comparing this observed value with the calculated probability that this facies transition occurred at random (Reading 1978; Walker 1979). This should be carried out by (1) tabulating the observed numbers of transitions recognized between each facies; and (2) converting these observed values to relative frequencies based on the total number of facies transitions observed, when compared to row totals. A predicted table of facies transitions is then calculated based on the null hypothesis that facies transitions occur at random and are purely a function of the relative abundance of the facies (and therefore the potential number of facies transitions within the succession). The table of random probabilities can then be subtracted from the table of observed transitions so that any significant variations from a random distribution are highlighted (Walker 1979). The values obtained can be tested statistically using, for example, a χ^2 (chi squared) test.

Harper (1984) provides an alternative approach to testing the significance of the observed minus expected facies transition values obtained. The use of traditional facies transition matrices, is however, problematic, as it is not always possible to recognize a transition from one facies to itself, as it depends upon how prominent the bounding surface between individual depositional units is. For example, whilst in some cases it is possible to identify amalgamation surfaces between turbiditic sandstones, many more are probably not easily identified, hence the true number of discrete depositional events cannot be identified; within a hemipelagic mudstone, for example, the recognition of such boundaries is even more difficult. Consequently, the facies transition matrix is not statistically correct, as in a random distribution it should be possible to have transitions from one facies to itself (Walker 1984). This problem was overcome by the use of a statistical test known as markov chain analysis (Carr 1982; Powers & Easterling 1982). In markov chain analysis, the matrices are 'embedded matrices' because they contain structural zeros as a result of the methodology rather than the actual data (e.g. Facies A to Facies A transitions). The statistical method that has to be adopted to deal with embedded matrices is summarized by Graham (1988).

Either based upon a rigorous statistical assessment of facies relationships as discussed above, or more commonly based on a more subjective examination of the graphic sedimentary logs, it is possible to define facies associations. Facies associations are defined as 'groups of facies that occur together and are considered to be genetically or environmentally related' (Reading 1978: 5), and are the key to environmental interpretation. Effectively, the facies association represents the range of processes operating within the depositional environment. A good example is provided by Ineson (1989) for a submarine fan and slope apron sequence in Antarctica. In conjunction with regional mapping, nearly 1800 m of detailed graphic sedimentary logs were measured through the Cretaceous Kotick Point and Whisky Bay formations, and were divided into nine descriptive sedimentary facies (each of which was in turn subdivided into subfacies). Following the interpretation of the individual depositional processes for each facies, the facies were combined together with the definition of three facies associations: a mudstone association, sandstone–breccia association and a conglomerate association. Each separate facies association was then interpreted in terms of depositional environment, based on the range of processes operating and by comparison with previously published facies schemes. This led Ineson (1989) to interpret the mudstone association as representing deposition at the base of a slope, the sandstone–breccia association as representing base of slope deposition at the mouths of slope gullies, and the conglomerate association as the inner channelized region of a coarse gravelly submarine fan.

14.2.5 Facies Models and Depositional Environments

Once the facies association has been defined it can be compared with previously defined facies models. The textbooks by Walker (1984), Walker & James (1992) and Reading (1996) provide summaries of facies models for the main depositional environments and are a valuable introduction to the relevant literature. The environmental interpretation of the facies association will be based upon the range of processes recognized for each of the component facies. In many cases, some facies will have much greater environmental significance than others, but it is important to note that the interpretation has to account for the complete range of interpreted processes. The facies association can be compared with the facies model, and the controls on sedimentation assessed. The vertical sequence of facies associations can be interpreted in terms of changing environments of deposition through time. Similarly, laterally equivalent facies associations for the same bio- or chronostratigraphical time interval can be compared, allowing an interpretation of the spatial distribution of environments for a single time slice. This interpretation can be enhanced by considering the facies analysis within a sequence stratigraphical framework. Once the spatial variation in environments has been identified for discrete intervals of earth history, then palaeogeographical maps or three-dimensional block diagrams depicting the environmental interpretation can be constructed. It is at this stage that the facies interpretation and facies model can be used as a predictive tool to extend the study laterally into areas with less complete outcrop or well control.

Once an environmental interpretation has been reached, additional related work can be considered. The broad understanding of depositional environment may lead on to additional fieldwork to refine the previous study and provide a more detailed interpretation. In clastic sedimentology, provenance studies can usefully be integrated with the understanding of depositional processes and environments, as they may influence sediment mineralogy due to selective grain sorting (e.g. Kairo *et al.* 1993). These data may also then feed back into the original depositional model. Facies-related diagenesis has been clearly recognized in both clastic and carbonate sediments. For example, Spark & Trewin (1986) clearly show facies-related diagenetic differences in the Triassic–Jurassic Claymore Oilfield of the North Sea. Sandstone reservoir characteristics are widely recognized to be facies controlled both as a result of primary depositional variation in grain size and sorting, and also due to facies-related diagenesis. Examples include the Leman Sandstone Formation of the Rotliegend Group, Ravenspurn North Field in the southern North Sea (Ketter 1991), and the Tirrawarra Oil Field of South Australia (Hamlin *et al.* 1996). Detailed facies analysis can commonly aid hydrocarbon exploration and production. For example, Jobe & Saller (1995) examined the depositional environments of the Upper Harrodsburg Limestone in the Illinois Basin. They recognized three main facies: (1) bryozoan boundstones; (2) bryozoan grainstones; and (3) fossiliferous wackestones. The bryozoan boundstones were deposited as bryozoan mounds which were fringed by the bryozoan grainstones (formed as a result of fragmentation of the mounds), passing laterally and vertically into the deeper water fossiliferous wackestones. Within the Illinois Basin, only the bryozoan grainstones have reservoir potential, and were effectively sealed by the fossiliferous wackestones. The distribution of the grainstones, was however, directly linked to the location of the bryozoan mounds, which could potentially be identified using three-dimensional seismics. In this case a thorough understanding of the depositional environment controls the oil exploration programme.

14.3 SUMMARY

Facies analysis provides a powerful tool with which to unlock the environmental record contained within the stratigraphical column. Detailed field or core-store observations lead to the definition of descriptive sedimentary facies which are then interpreted in terms of process sedimentology. These descriptive facies, when combined into facies associations and compared with previously defined facies models, allow detailed environmental interpretation. However, facies analysis should not be carried out in isolation, and should be fully integrated with the macro- and microfossil record, taphonomic and palaeoecological studies, and analytical mineralogical or geochemical data, to provide a combined environmental interpretation consistent with all of the available data.

Facies analysis has become a routine technique in the description of sedimentary and volcaniclastic rocks. Our current understanding of processes, resultant facies and depositional environments will continue to evolve, as long as we continue to make detailed observations and interpretations during laboratory simulation of natural processes, field-based studies on modern and ancient environments, or laboratory examination of rock samples. Such work, carried out within a high-resolution stratigraphical framework, can be combined with detailed analytical sedimentology and micro- and macropalaeontology to provide a detailed interpretation of changing environmental conditions at the earth's surface.

ACKNOWLEDGEMENTS

I am very grateful to H. G. Reading, J. D. Marshall, A. E. Mather and M. R. F. Shaikh for commenting on draft versions of this chapter.

REFERENCES

Ager, D. 1993. *The new catastrophism.* Cambridge University Press, Cambridge.

Allen, J.R.L. 1985. *Principles of physical sedimentology.* George Allen & Unwin, London.

Anderton, R. 1985. Clastic facies models and facies analysis. In Brenchley, P.J. & Williams, B.P.J. (eds) *Sedimentology, recent developments and applied aspects.* The Geological Society of London, 31–47.

Arnaud-Vanneau, A. & Arnaud, H. 1990. Hauterivian to Lower Aptian carbonate shelf sedimentation and sequence stratigraphy in the Jura and northern Subalpine chains (southeastern France and Swiss Jura). *Special Publication of the International Association of Sedimentologists*, **9**, 203–233.

Blair, T.C. & McPherson, J.G. 1992. The Trollheim alluvial fan and facies model revisited. *Bulletin of the Geological Society of America*, **104**, 762–769.

Blair, T.C. & McPherson, J.G. 1993. The Trollheim alluvial fan and facies model revisited: discussion and reply. *Bulletin of the Geological Society of America*, **105**, 563–567.

Brenchley, P.J., Newall, G. & Stanistreet, I.G. 1979. A storm surge origin for sandstone beds in an epicontinental platform sequence, Ordovician, Norway. *Sedimentary Geology*, **22**, 185–217.

Carr, T.R. 1982. Log-linear models, Markov chains and cyclic sedimentation. *Journal of Sedimentary Petrology*, **52**, 905–912.

Cas, R.A.F. & Wright, J.V. 1987. *Volcanic successions, modern and ancient.* Allen & Unwin, London.

Cheel, R.J. & Leckie, D.A. 1993. Hummocky cross-stratification. In Wright, V.P. (ed.) *Sedimentology review 1.* Blackwell Scientific Publications, Oxford, 103–122.

Collinson, J.D. & Thompson, D.B. 1982. *Sedimentary structures.* George Allen & Unwin, London.

Emery, D. & Myers, K.J. 1996. *Sequence stratigraphy.* Blackwell Scientific Publications, Oxford.

Flügel, E. 1982. *Microfacies analysis of limestones.* Springer-Verlag, Berlin.

Francis, P. 1993. *Volcanoes, a planetary perspective.* Clarendon Press, Oxford.

Gawthorpe, R.L. & Gutteridge, P. 1990. Geometry and evolution of platform-margin bio-clastic shoals, late Dinantian (Mississippian), Derbyshire, UK. *Special Publication of the International Association of Sedimentologists*, **9**, 39–54.

Graham, J. 1988. Collection and analysis of field data. In Tucker, M.E. (ed.) *Techniques in sedimentology.* Blackwell Scientific Publications, Oxford, 5–62.

Gressly, A. 1838. Observations géologiques dur le Jura Soleurois. *Neve Denkschriften der Allgemeinen Schweizerischen Gesellschaft für die Gesammten Naturwissenschaften*, **2**, 1–112.

Hamlin, H.S., Dutton, S.P., Seggie, R.J. & Tyler, N. 1996. Depositional controls on reservoir properties in a braid-delta sandstone, Tirrawarra Oil Field, South Australia. *Bulletin of the American Association of Petroleum Geologists*, **80**, 139–156.

Harper, C.W. 1984. Improved methods of facies sequence analysis. In Walker, R.G. (ed.) *Facies models*, 2nd edition. Geoscience Canada, Reprint Series 1, 11–13.

Hooke, R. LeB. 1993. The Trollheim alluvial fan and facies model revisited: discussion and reply. *Bulletin of the Geological Society of America*, **105**, 563–567.

Ineson, J.R. 1989. Coarse-grained submarine fan and slope apron deposits in a Cretaceous back-arc basin, Antarctica. *Sedimentology*, **36**, 793–819.

James, N.P. & Kendall, A.C. 1992. Introduction to carbonate and evaporite facies models. In Walker, R.G. & James, N.P. (eds) *Facies models, response to sea level change.* Geological Association of Canada, 265–275.

Jobe, H. & Saller, A. 1995. Oil reservoirs in grainstone aprons around bryozoan mounds, Upper Harrodsburg Limestone, Mississippian, Illinois Basin. *Bulletin of the American Association of Petroleum Geologists*, **79**, 783–800.

Kairo, S., Suttner, L.J. & Dutta, P.K. 1993. Variability in sandstone composition as a function of depositional environment in coarse-grained delta systems. In Johnsson, M.J. & Basu,

A. (eds) *Processes controlling the composition of clastic sediments.* Geological Association of America, Special Paper 284, 263–283.

Kendall, A.C. & Harwood, G.M. 1996. Marine evaporites: arid shorelines and basins. In Reading, H.G. (ed.) *Sedimentary environments: processes, facies and stratigraphy.* Blackwell Scientific Publications, Oxford, 281–324.

Ketter, F.J. 1991. The Ravenspurn North Field, Blocks 42/30, 43/26a, UK North Sea. In Abbotts, I.L. (ed.) *United Kingdom oil and gas fields, 25 years commemorative volume.* The Geological Society of London, Memoir 14, 459–467.

Lajoie, J. & Stix, J. 1992. Volcaniclastic Rocks. In Walker, R.G. & James, N.P. (eds.) *Facies models, response to sea level change.* Geological Association of Canada, 101–118.

Leeder, M.R. 1982. *Sedimentology, process and product.* George Allen & Unwin, London.

Macquaker, J.H.S. 1994. A lithofacies study of the Peterborough Member, Oxford Clay Formation (Jurassic), UK: an example of sediment bypass in a mudstone succession. *Journal of the Geological Society of London,* **151,** 161–172.

Macquaker, J.H.S. & Gawthorpe, R.L. 1993. Mudstone lithofacies in the Kimmeridge Clay Formation, Wessex Basin, southern England: implications for the origin and controls of the distribution of mudstones. *Journal of Sedimentary Petrology,* **63,** 1129–1143.

Miall, A.D. 1992. Alluvial deposits. In Walker, R.G. & James, N.P. (eds.) *Facies models, response to sea level change.* Geological Association of Canada, 119–142.

Mutti, M., Bernoulli, D., Eberli, G.P. & Vecsei, A. 1996. Depositional geometries and facies associations in an Upper Cretaceous prograding carbonate platform margin (Orfento Supersequence, Maiella, Italy). *Journal of Sedimentary Research, Section B,* **66,** 749–765.

Orton, G.J. 1996. Volcanic Environments. In Reading, H.G. (ed.) *Sedimentary environments: processes, facies and stratigraphy.* Blackwell Scientific Publications, Oxford, 485–567.

Pickering, K.T., Stow, D.A.V., Watson, M. & Hiscott, R.N. 1986. Deep-water facies, processes and models: a review and classification scheme for modern and ancient sediments. *Earth Science Reviews,* **22,** 75–174.

Pirrie, D., Crame, J.A. & Riding, J.B. 1991. Late Cretaceous stratigraphy and sedimentology of Cape Lamb, Vega Island, Antarctica. *Cretaceous Research,* **12,** 227–258.

Plint, A.G. 1995. *Sedimentary facies analysis; a tribute to the research and teaching of Harold G. Reading.* Special Publication of the International Association of Sedimentologists, No. 22.

Powers, D.W. & Easterling, R.G. 1982. Improved methodology for using embedded markov chains to describe cyclical sediments. *Journal of Sedimentary Petrology,* **52,** 913–923.

Reading, H.G. 1978. Facies. In Reading, H.G. (ed.) *Sedimentary environments and facies.* 1st edition. Blackwell Scientific Publications, Oxford, 4–14.

Reading, H.G. 1996. *Sedimentary environments: processes, facies and stratigraphy,* 3rd edition. Blackwell Scientific Publications, Oxford.

Reading, H.G. & Levell, B.K. 1996. Controls on the sedimentary rock record. In Reading, H.G. (ed.) *Sedimentary environments: processes, facies and stratigraphy.* Blackwell Scientific Publications, Oxford, 5–36.

Richards, P.C. 1992. An introduction to the Brent Group: a literature review. In Morton, A.C., Haszeldine, R.S., Giles, M.R. & Brown, S. (eds) *Geology of the Brent Group.* Special Publication of the Geological Society of London, No. 61, 15–26.

Ricketts, B.D., Ballance, P.F., Hayward, B.W. & Meyer, W. 1989. Basal Waitemata Group lithofacies: rapid subsidence in an early Miocene interarc basin, New Zealand. *Sedimentology,* **36,** 559–580.

Schieber, J. 1989. Facies and origin of shales from the mid-Proterozoic Newland Formation, Belt Basin, Montana, USA. *Sedimentology,* **36,** 203–219.

Shanmugam, G. & Moiola, R.J. 1995. Reinterpretation of depositional processes in a classic flysch sequence (Pennsylvanian Jackfork Group), Ouachita Mountains, Arkansas and Oklahoma. *Bulletin of the American Association of Petroleum Geologists,* **79,** 672–695.

Shanmugam, G., Bloch, R.B., Mitchell, S.M., Beamish, G.W.J., Hodgkinson, R.J., Damuth, J.E., Straume, T., Syvertsen, S.E. & Shields, K.E. 1995. Basin-floor fans in the North Sea: sequence stratigraphic models vs. sedimentary facies. *Bulletin of the American Association of Petroleum Geologists,* **79,** 477–512.

Slatt, R.M., Weiner, P. & Stone, C.G. 1997. Reinterpretation of depositional processes in a classic flysch sequence (Pennsylvanian Jackfork Group), Ovachita Mountains, Arkansar and Oklahoma: discussion. Bulletin of the American Association of the Petroleum Geologists, **81**, 449–459.

Spark, I.S.C. & Trewin, N.H. 1986. Facies-related diagenesis in the main Claymore oilfield sandstones. *Clay Minerals*, **21**, 479–496.

Suthren, R.J. 1985. Facies analysis of volcaniclastic sediments: a review. In Brenchley, P.J. & Williams, B.P.J. (eds) *Sedimentology, recent developments and applied aspects*. The Geological Society of London, 123–146.

Tucker, M.E. 1991. *Sedimentary petrology*, 2nd edition. Blackwell Scientific Publications, Oxford.

Tucker, M.E. 1995. *Sedimentary rocks in the field*, 2nd edition. John Wiley and Sons, Chichester.

Tucker, M.E. & Wright, V.P. 1990. *Carbonate sedimentology*. Blackwell Scientific Publications, Oxford.

Walker, R.G. 1979. Facies and facies models. 1. General introduction. In Walker, R.G. (ed.) *Facies models*, 1st edition. Geoscience Canada Reprint Series 1, 1–7.

Walker, R.G. 1984. General introduction: facies, facies sequences and facies models. In Walker, R.G. (ed.) *Facies models*, 2nd edition. Geoscience Canada Reprint Series 1, 1–9.

Walker, R.G. 1992. Facies, facies models and modern stratigraphic concepts. In Walker, R.G. & James, N.P. (eds) *Facies models, response to sea level change*. Geological Association of Canada, 1–14.

Walker, R.G. & James, N.P. (eds) 1992. *Facies models, response to sea level change*. Geological Association of Canada.

Whitaker, M.F., Giles, M.R. & Cannon, J.C. 1992. Palynological review of the Brent Group, UK sector, North Sea. In Morton, A.C., Haszeldine, R.S., Giles, M.R. & Brown, S. (eds) *Geology of the Brent Group*. Special Publication of the Geological Society of London, No. 61, 169–202.

15
Interpreting Sea Level

Anthony Hallam

Near the end of the last century the Austrian geologist Eduard Suess published his great synthesis *Das Antlitz der Erde* (1888), which achieved a wider readership once it had been translated as *The Face of the Earth* (Suess 1906). He was the first person to take up the idea of global sea-level changes as a general working theory, and introduced the term 'eustatic' for such changes (Hallam 1992a,b). Suess proposed that past sea-level changes could be studied by using three different, though not unrelated, methods. First, the position above present sea level of ancient strandlines, such as the raised beaches widely recognizable along present coastlines, could be determined and their ages fixed. Secondly, the areal spread of marine sediments of a given age and the inferred extent of epicontinental sea could be analysed. Thirdly, vertical changes in the stratigraphical succession could be studied with a view to determining alternations of marine and non-marine sediments or changes in the depth of deposition of marine sediments. The first method is of necessity restricted to the Late Quaternary, but the other two can be applied to the whole Phanerozoic.

The notion of eustasy was quickly accepted for the Quaternary, with a clear association being inferred respectively between falls and rises of sea level and the growth and retreat of ice caps. For much of this century the concept of eustasy for the rest of the Phanerozoic has, however, proved more controversial, despite the strong advocacy of such leading stratigraphers as Grabau (1936) and Umbgrove (1939). In recent decades the subject has been revitalized as a result of two principal developments. First, improvements in biostratigraphy have allowed increasingly refined correlation across the world, thereby providing the all-important opportunity of distinguishing global from local and regional events.

Unlocking the Stratigraphical Record: Advances in Modern Stratigraphy. Edited by P. Doyle and M.R. Bennett.
© 1998 John Wiley & Sons Ltd.

The second development concerns the adoption by Exxon geologists successively of seismic and sequence stratigraphy as tools for regional and global correlation.

There are alternative approaches to the analysis of sea-level change, which inevitably show some overlap; they involve facies analysis and sequence stratigraphy, and will be considered below. Before this is done it is important to make a distinction between *relative* and *global* or *eustatic* change. Recognizing relative sea-level change within a given region such as a marine sedimentary basin is generally not difficult, but inferring eustatic control depends crucially upon good biostratigraphy that allows for global correlation. This is especially pertinent for presumed eustatic cycles of short duration, which may be beyond the limit of biostratigraphical resolution. Relative depth of sea for a given depositional succession clearly depends on three factors: rate of subsidence, rate of sedimentation, and rate of sea-level change. Normally the first two factors, controlled essentially by regional tectonics, are the most important. Disentangling the third can be difficult, and without adequate biostratigraphical control it is well nigh impossible.

An attempt will now be made to outline the facies analysis and sequence stratigraphical approaches to analysing sea-level change, and to compare results obtained from these two techniques for the Jurassic. There follows an account of how hypsometric curves may be used to attempt estimates of the amount of sea-level change, and the chapter ends with a discussion of minor cyclicity and possible underlying mechanisms.

15.1 THE FACIES ANALYSIS APPROACH

Consider the facies succession diagram of Figure 15.1, signifying deposition in a shallow epicontinental marine basin. The left-hand column signifies a shallowing-upwards marly and calcareous succession, as determined from a combination of lithofacies and biofacies data; the right-hand column represents a shallowing-upward siliciclastic succession. The bases of the sedimentary cycles are characterized by condensed beds, shown as a concentration of eroded pebbles and phosphatic nodules on the left and a shell concentration on the right; these are interchangeable or may be mixed. Similarly the laminated, organic-rich shale near the base of the right-hand cycle succession could also be portrayed in an equivalent position in the left-hand succession. As the amount of accommodation space decreases with falling relative sea level, coarse siliclastics tend to prograde towards the basin centre, as indicated in the diagram. Within a regime of calcareous sedimentation, the response may be one of sediment starvation, with the development of one or more hardgrounds by submarine and/or subaerial cementation.

Taken on its own, the right-hand succession could represent a normal delta progradation, involving phenomena intrinsic to a particular sedimentation regime, but if it can be correlated by good biostratigraphy to a shallowing-upwards regime elsewhere in the basin in non-siliciclastic sediments, as in the left-hand succession, some explanation involving relative sea-level fall is called for. In a

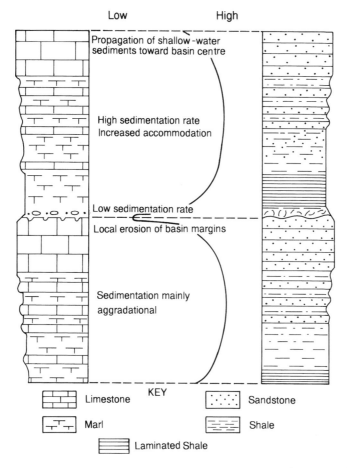

Figure 15.1 *Correlation of shallowing-upward successions in carbonate and siliciclastic strata, and their relationship to relative sea-level change. [Reproduced with permission from the Society for Sedimentary Geology, from: Hallam (1988)]*

siliciclastic context this has been called *forced regression* (Plint 1991; Posamentier *et al.* 1992), although that term has been used, strictly speaking, for events involving erosive contacts between shelf mudstones and sandstones. The relative sea-level curve corresponding to the two shallowing-upwards successions, also portrayed in Figure 15.1, is cuspate in character. On the face of it this seems to imply relatively rapid sea-level fall and relatively low sea-level rise. However, since the initial sea-level rise phase is characterized by relatively low sedimentation rate, a characteristic phenomenon recognizable throughout the Phanerozoic (Hallam 1992a), a truer picture of sea-level change would be characterized by a more symmetrical sinusoidal curve.

Figure 15.2 illustrates how relative sea-level fluctuations can affect different sedimentary regimes within a basin and adjacent basin margin and carbonate platform. Two episodes of shallowing give rise to spreads of sand into the left-hand

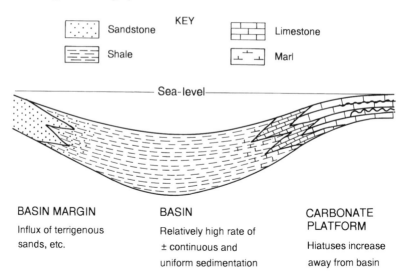

KEY

Sandstone Limestone

Shale Marl

Sea-level

BASIN MARGIN	BASIN	CARBONATE PLATFORM
Influx of terrigenous sands, etc.	Relatively high rate of ± continuous and uniform sedimentation	Hiatuses increase away from basin

Figure 15.2 *Illustration of how sea-level fluctuations affect different sedimentary regimes. Two episodes of shallowing give rise to spreads of sand into the left-hand basin margin and corresponding spreads of shallower and deeper neritic limestone facies into marls and shales of the right-hand basin margin. These episodes of sea-level fall are represented by hiatuses in the carbonate platform succession on the right and lack of any representation in terms of facies changes in the basin centre. [Reproduced with permission from the* Annual Review of Earth and Planetary Science, *from: Hallam (1984)]*

basin margin and corresponding spreads of shallower and deeper neritic limestone facies into marls and shales of the right-hand basin margin. These episodes of sea-level fall are represented by hiatuses in the carbonate platform succession and lack any representation in terms of facies change in the basin centre. Clearly the basin margins, with their pronounced changes of facies up the sedimentary succession, are the best location for analysing relative sea-level change. Towards the basin centre the local rate of subsidence might have been sufficient to swamp any more extensive regional effects, and the hiatuses in the carbonate platform succession could easily be missed, or difficult to date precisely because of a lack of good biostratigraphical indicators in the shallow neritic limestones.

A good example to illustrate how a eustatic sea-level curve can be derived from regional data involves Jurassic marine strata, which are favoured by the abundance of ammonites. Ammonites are among the best of all biostratigraphical indicators, with temporal precision in favourable cases being as little as a few hundred thousand years (Hallam 1992a). Figure 15.3 shows a number of relative sea-level curves derived from facies analysis across the Middle–Upper Jurassic (Callovian–Oxfordian) boundary in western Europe, extending from north-west Scotland (Staffin and Elgol) through England and France to Spain, Switzerland and southern Germany (Norris & Hallam 1995). Correlation to the precision of ammonite subzones is possible across this whole region, which embraces an area exceeding a million square kilometres, far more extensive than the many component basins and intervening swells. Yet there is a consistent pattern of relative sea-level rise in the Late Callovian, stabilizing or going slightly into reverse across

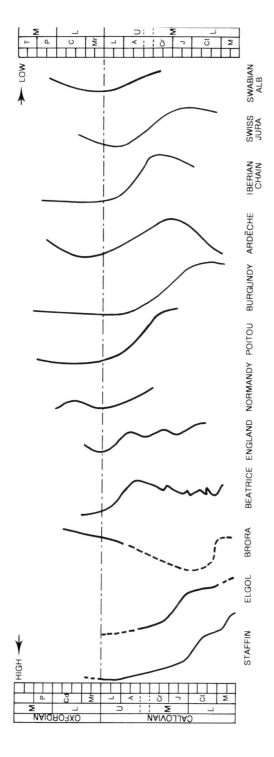

Figure 15.3 *Bathymetric curves for Callovian–Oxfordian boundary sections in western Europe. [Reprinted from Palaeogeography, palaeoclimatology, palaeoecology, 116, Norriss & Hallam, Facies variations across the Middle-Upper Jurrasic boundary in western Europe and the relationship to sea-level changes, 33–42, © 1995, with kind permission of Elsevier Science – NL, Sara BurgerharStaat 25, 1055 KV Amsterdam, The Netherlands]*

the Callovian–Oxfordian boundary. The only exception is the Brora section, but this is readily explicable by local sedimentation conditions giving rise to a thick, coarse siliciclastic succession. Most of the sections do not contain coarse siliciclastics and the southern European ones are mainly in condensed limestone–marl successions.

The clear implication of this pattern of change is that there is an overprint of sea-level change which is independent of a series of basins and swells related to regional tectonics, and hence is likely to be eustatic in origin. This is confirmed by a less intensive global survey using the best available ammonite data (Figure 15.4).

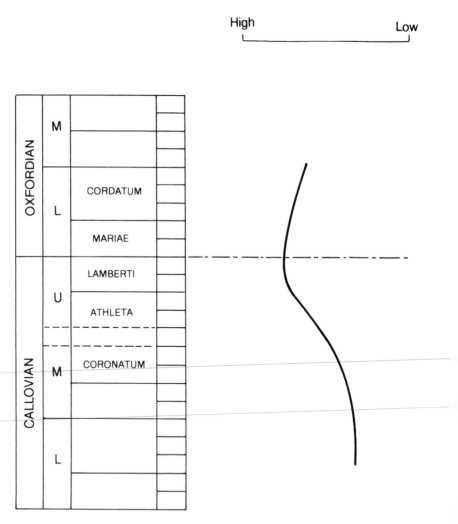

Figure 15.4 *Proposed eustatic curve for the Callovian–Lower Oxfordian interval [Reprinted from* palaeogeography, palaeoclimatology, palaeoecology, 116, *Norriss & Hallam, Facies variations accross the Middle-Upper Jurrasic boundary in western Europe and the relationship to sea-level changes, 33–42, © 1995, with kind permission of Elsevier Science – NL, Sara Burgerharstaat 25, 1055 KV Amsterdam, The Netherlands]*

15.2 THE SEQUENCE STRATIGRAPHICAL APPROACH

This approach to the study of sea-level change derives directly from the earlier attempt by Exxon geologists to establish a eustatic curve from seismic stratigraphical data on continental margins (Vail *et al.* 1977). Relative sea-level rise for a given region was established by coastal onlap of unconformity-bound sedimentary successions termed sequences, and sea-level fall by the downward shift of coastal onlap. Correlation across the world suggested eustatic control. The result was a saw-tooth eustatic curve extending back to the Triassic, with apparently gradual rises being punctuated by geologically instantaneous falls. This much-cited work was subjected to a number of criticisms, as reviewed by Hallam (1992a), and Vail and his colleagues soon acknowledged that their saw-tooth curve reflected coastal onlap rather than sea level, whose fluctuation would have been temporally more symmetrical.

Notable criticisms were put forward by Miall (1986) and Christie-Blick *et al.* (1990). Miall argued that seismic reflections overlapping unconformities may often be submarine fans and not coastal sediments. If this is the case, the magnitude of the downward shift bears no relation to the amount by which sea level has fallen. Christie-Blick and his colleagues maintained that the amplitude of eustatic fluctuations cannot be inferred from seismic stratigraphical data alone because coastal aggradation is primarily the result of basin subsidence, not sea-level rise, and downward shifts in onlap reflect only the rate of sea-level fall relative to the rate of basin subsidence. Furthermore, it is not clear to them how the measurement of a single seismic section can be objectively regarded as most representative for the coastal overlap chart of a basin, because aggradation varies within the basin according to the local rate of subsidence. This is reinforced by the detailed seismic sequence analysis undertaken for the Canterbury Basin, New Zealand, by Fulthorpe & Carter (1989), who draw attention to the ambiguities and subjectivity involved in seismic data interpretation.

Partly in response to such criticisms, the Exxon geologists subsequently endeavoured to incorporate the best available biostratigraphical and chronostratigraphical data obtainable from both onshore and offshore sections, laying more emphasis on data more widely available in the public domain, namely stratigraphical sequences recognizable in outcrop. The result was their famous eustatic sea-level curve for the Mesozoic and Cenozoic (Haq *et al.* 1987). This revised analysis has in turn been subjected to some strong criticism. For example, Carter *et al.* (1991) pointed out that sequence stratigraphy embraces two distinct conceptual models: (1) sea-level behaviour through time; and (2) the stratigraphical record produced during a single sea-level cycle. Though these two models are interrelated they are logically distinct and therefore it is necessary to test them separately. Carter *et al.* suspect that the Exxon purported global sea-level curve in general represents a patchwork through time of many different local sea-level curves.

The criticisms of Miall (1994) are particularly trenchant and need to be considered at some length. The critical test of the Exxon chart is to demonstrate that successions of cycles of precisely similar age do indeed exist in many tectonically independent basins around the world. Many sequence studies have begun by using the chart as a template for stratigraphical correlation; a tactic that implies circular reasoning. Miall argues that current dating techniques do not permit the level of accuracy and precision in sequence correlation indicated in the chart, whose density of stratigraphical events is such that there is literally 'an event for every occasion' (Miall 1992). Dickinson (1993) demonstrated that the average duration of so-called third-order cycles in the Exxon chart increases with age, and suggested that this reflects a decrease in the quality of sequence data in older sections. In other words, the event spacing is at least in part an artefact of the data quality.

In his work, Miall emphasizes the incompleteness of the stratigraphical record: although time is continuous the stratigraphical record is not. Sequences deposited during the time spans between breaks may not actually correlate in time at all; they could, in principle, be markedly diachronous. Kidwell (1988) showed that the major break in sedimentation on an open shelf occurs as a result of erosion during lowstand and sedimentary bypass or starvation during transgression, whereas in marginal-marine environments the major break occurs during regression. The sequence boundary unconformities are therefore offset by as much as a half cycle between locations at the basin margin and basin centre. Problems involved in the identification of sequence boundaries include: (1) confusion of these boundaries with ravinement surfaces produced by wave erosion during transgression; and (2) apparent sequence boundaries formed by submarine erosion. An unconformity represents a finite timespan that may represent the amalgamation of more than one event, regional tectonic and/or eustatic. Cycles of tectonic origin could have regional or even continental extent.

15.3 COMPARISON OF THE FACIES AND SEQUENCE STRATIGRAPHICAL APPROACHES

Perhaps the most useful way of evaluating the respective merits of the facies analysis and sequence stratigraphical approaches to eustasy is to compare sea-level curves established by use of these alternative methods. Points of agreement can reinforce confidence in interpretation, while points of difference can focus attention on areas that require more thorough evaluation. Two examples will be considered. The first deals with what in the Vail *et al.* (1977) terminology would be termed third-order cycles (duration 1–10 million years) in the Jurassic; the second with first- and second-order cycles, for the whole Phanerozoic.

Figure 15.5 illustrates alternative Jurassic sea-level curves derived respectively from facies analysis in Europe, the global spread of marine sediments, and stratigraphical location of major transgressions and regressions (Hallam 1988), and sequence stratigraphy, mainly in western Europe (Haq *et al.* 1987). There is an agreement about the more or less progressive rise in sea level through the course of the period, and in particular details, such as important episodes of sea-level rise in the Toarcian, Bajocian and Callovian. There are also notable points of difference. The Exxon curve fails to recognize important sea-level falls at the end of the

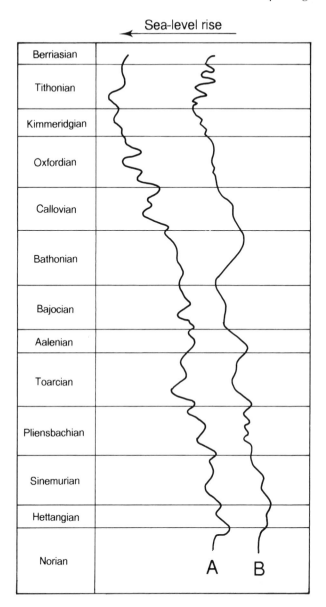

Figure 15.5 *Alternative Jurassic sea-level curves.* **A.** *Hallam (1988).* **B.** *Haq et al. (1987).*
[Modified from: Hallam (1992a)]

Triassic (Norian) and Callovian. The revised version of the Callovian–Oxfordian sea-level curve referred to earlier (Norris & Hallam 1995) is indeed closer to that of Haq *et al.*, the reason being that Hallam (1988) was misled by the widespread occurrence of Late Callovian hiatuses in Europe and elsewhere. Such hiatuses, and associated hardgrounds in calcareous strata, have frequently in the past been associated with regressions, recent examples from the Upper Cretaceous Chalk of Europe being cited by Hancock (1990). This is because hiatuses and the sub-contemporaneous cementation required to produce hardgrounds have generally been thought to form preferentially in shallower-water regimes. It now appears more likely, however, that these hiatuses in an open shelf setting are associated with sea-level rise (cf. Kidwell (1988) for the Miocene). On the other hand, Haq *et al.* (1987) have missed clear evidence of an end-Triassic regression (Hallam 1990).

More striking differences concern the Exxon curve indicating a markedly oscillatory pattern in the Tithonian, and pronounced sea-level falls in the Aalenian and Bathonian. As pointed out by Hallam (1988), the curve of Haq *et al.* (1987) is very similar to one produced by Vail & Todd (1981) on the basis solely of an analysis of seismic lines in the Moray Firth, off north-eastern Scotland. Vail and Todd's curve was claimed to be eustatic, which carries a clear implication that the Exxon geologists believe that eustatic overprint would override regional tectonics. However, Hallam (1988) pointed out that the Moray Firth, and indeed the North Sea in general, was subjected in the Late Jurassic to significant tensional tectonics involving the formation of half-grabens, which would have overridden any eustatic signal. With regard to the pronounced Middle Jurassic lowstands in the Exxon curve, this was interpreted by Hallam & Sellwood (1976) as the consequence of regression induced by thermal doming centred on the North Sea region. Both these interpretations involving regional tectonics rather than eustasy have been supported by the thorough and detailed analyses of Underhill (1991) and Underhill & Partington (1993).

The Phanerozoic picture is portrayed in a similar way in Figure 15.6. The left-hand curve of Hallam (1984) is based partly on facies analysis and partly on the areal spread of epicontinental seas through time, as indicated by palaeogeographical maps. The right-hand curve of Vail *et al.* (1977) is based on seismic stratigraphy down to the Triassic, and sequence stratigraphy on the North American continent for the Palaeozoic. The saw-tooth character of their curve signifies that it was drawn at a time when it was believed by the Exxon geologists that coastal overlap could be directly equated with sea level. Both curves show a broad agreement in recognizing what Fischer (1984) has termed two Phanerozoic supercycles. There was a more or less progressive rise of sea level through the Early Palaeozoic to an Ordovician–Silurian peak followed by a fall towards the end of the era. Thereafter there was a further major rise to a Late Cretaceous peak followed by a further fall. A number of the second-order cycles are comparable but one of the most notable differences concerns the Oligocene sea-level fall, which is shown to be far more pronounced in the Vail curve. Further research by Cenozoic workers suggests that this was a considerable overestimate (Hallam 1992a).

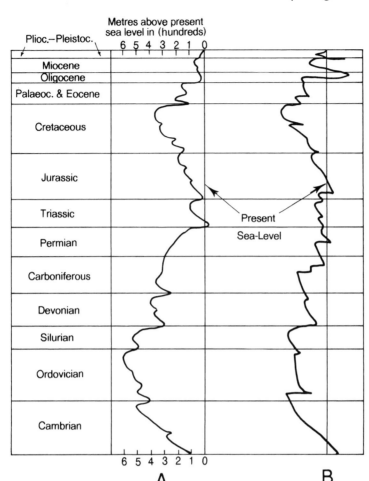

Figure 15.6 *Eustatic curves for the Phanerozoic.* **A.** *After Hallam (1984).* **B.** *After Vail et al. (1977). [Modified from: Hallam (1992a)]*

15.4 THE USE OF HYPSOMETRIC CURVES FOR ATTEMPTING ESTIMATES OF THE AMOUNT OF SEA-LEVEL CHANGE

The hypsometry of a topographical surface is its cumulative areal frequency with respect to elevation. Continental area-elevation distributions form sigmoidal curves that are convex up at low elevations, concave up at high elevations, and have an inflexion point close to sea level (Figure 15.7). The rate of change of elevation with respect to cumulative area is termed 'hypsometric slope' and is a measure of the rate of potential flooding of a landmass within a given elevation range (Algeo & Seslavinsky 1995). This inflexion point is the elevation at which the slope of the curve is gentlest and, consequently, at which a landmass is potentially the most rapidly floodable. The sigmoidal shape of continental hypsometric curves is probably controlled by a balance between net continent-interior erosion and net continent-margin deposition. For epeirogenically stable landmasses, the inflexion point coincides approximately with the boundary between erosional and depositional regimes, usually at the up-dip margin of coastal-plain wedges at an elevation of several tens of metres above sea level.

Hypsometric curves can be used to convert marine flooding data to sea-level elevations, and secular changes in sea-level elevation can be reconstructed from a series of equal area palaeogeographical maps (Figure 15.7). Although the hypsometric method does not yield as high a temporal resolution as seismic or sequence stratigraphical analysis owing to the use of relatively coarse time slices, it is a more reliable way of estimating absolute eustatic elevations than that put

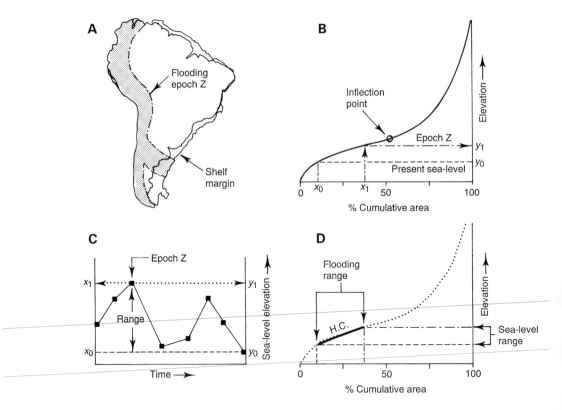

Figure 15.7   *Example of calculation of sea-level elevations using hypsometry and continental flooding data.* **A.** *The area flooded for a given epoch is measured from an equal-area palaeogeographic map relative to a specified reference level, such as the shelf margin.* **B.** *Fractional area flooded* (x_1) *is converted to sea-level elevation* (y_1) *using a landmass-specific hypsometric curve.* **C.** *A secular sea-level elevation curve is constructed from a series of palaeogeographic maps.* **D.** *Given time-invariant continental coastline hypsometry, a special flooding range yields a corresponding range of sea-level elevations. A 'hypsometric chord' (H.C.) is a linear representation of part of a continental area-elevation distribution.* [Reproduced with permission from the American Journal of Science, from: Algeo & Seslavinsky (1995)]

forward by Vail *et al.* (1977) on the basis of seismic stratigraphy. It can, furthermore, be applied to the whole Phanerozoic.

An attempt was made by Hallam (1984) to estimate changing eustatic elevations through the Phanerozoic by utilizing palaeogeographical maps for the United States and the former Soviet Union on the assumption that, collectively embracing a third of the world's land area today, they would give a fair reflection of the global picture. It was also assumed that continental hypsometry had not changed through time. The results as shown in Figure 15.6 indicate a peak elevation in the Late Ordovician of about 600 m above present sea level.

A more recent and comprehensive analysis of Palaeozoic data by Algeo & Seslavinsky (1995) has made it clear that this figure should be revised downward. It turns out that continental hypsometry is closely linked to landmass area, so that corrections for variable landmass area are necessary in all hypsometric analyses of flooding records. A large supercontinent such as Pangaea would have stood correspondingly higher above the ocean than the present array of smaller continents. For the Early Palaeozoic it is now generally acknowledged that there were even more continents, of correspondingly smaller size, and therefore greater propensity for flooding. Algeo and Seslavinsky analysed the flooding records of 13 Palaeozoic landmasses and established that those with the best records for eustasy are Baltica, North China/United China, Kazakchstania and Siberia. Laurentia and Laurussia, although having a high-quality flooding record, exhibit almost no eustatic component to variance, making their use as a global standard questionable. Their analysis suggests that the mid-Ordovician maximum was between about 100 and 225 m above present sea level.

15.5 MINOR CYCLICITY IN CARBONATE SUCCESSIONS

Whereas traditional stratigraphical units, such as groups, formations and members, may show a characteristic range of dimensions, none need exhibit a distinct modal thickness. Each category comprises only part of a continuous series of decreasingly thick stratal entities. In contrast, following the pioneer treatment of Vail *et al.* (1977), a general perception has arisen of rank or caste ordering of sequence stratigraphical entities, with each possessing a characteristic duration. Supposed thickness hierarchies within many cratonic carbonate successions have been interpreted as prime evidence for multiple periodicities of sea-level change and orbitally forced origins of cyclic peritidal carbonates (e.g. Crevello 1991; Osleger & Read 1991). Furthermore, on the basis of amplitudes and durations of relative sea-level change for the rhythmic generation of units of differing thickness, Goldhammer *et al.* (1993) relate each size group to accumulation during either predominantly tectonic or glacioeustatic change.

However, if the stratal elements in question comprise a continuum of thickness and durations, the basis for inferring periodic accumulation is cast in doubt, as is the basis for inferring a hierarchical scheme of sediment accumulation as readily tangible manifestations of stratigraphical cyclicity. Drummond & Wilkinson (1996) derived size–frequency relations for a variety of stratal units and observed lognormal or exponential distributions. It is thought likely that the lognormal distributions are a consequence of the deficiency of the smallest units and that exponentional distributions reflect the truer situation, similar to many scalar parameters in natural systems. Drummond and Wilkinson therefore interpret the lognormal distributions of measured sizes as due to incomplete sampling of naturally occurring exponential decreases in thickness frequency. This also applies to the numerous sea-level cycles of Haq *et al.* (1987).

Another problem arises about inferring true rhythmicity. Goldhammer *et al.* (1990) point out that non-rhythmic patterns of stratigraphical occurrence do not necessarily mean that the intensity of depositional forces varied in an arhythmic manner. The predominance of exponential size distributions in sediment accumu- lation does not unequivocally preclude interpretations of periodic accumulation. Computational simulation of cyclic sediment accumulation (e.g. Franseen *et al.* 1991; Osleger & Read 1993) demonstrate that recurrent sea-level change does not necessarily result in corresponding synthetic cycles. Coincidence of long-term emergence with short-term eustatic highstands may result in non-depositional 'missed beats', whereas continuous submergence and maintenance of finite ac- commodation space during short-term sea-level falls may result in continuous accumulation or cycle amalgamation (Goldhammer *et al.* 1990). Goldhammer *et al.* (1991) suggest that the Carboniferous carbonates of the Paradox Basin of the western United States display three superimposed orders of cyclicity, with 29 Ka 'fifth-order' cycles grouped into 257 Ka 'fourth-order' sequences grouped into 2–3 Ma 'third-order' sequences. If sedimentary units do indeed occur in such distinct modal groups then a genetic link between magnitude and temporal recurrence of extra-basinal process and hierarchy of sedimentary response seems entirely reasonable. The work of Drummond & Wilkinson (1996) casts serious doubt on such an inference.

These authors demonstrate that many groupings of sedimentary units exhibit lognormal frequency distributions wherein most of the population has an ex- ponentially increasing frequency of occurrence with linearly decreasing class size. Such thickness distributions typify a wide range of sedimentary entities, including individual lithofacies, formally named stratigraphical units and cyclic peritidal lithofacies associations, as well as durations of unconformity-bounded strati- graphical sequences and the magnitude and durations of presumed changes in global sea level. These distributions indicate that most natural populations of sedimentary units comprise a non-modal series of increasing frequency with decreasing size.

Another major difficulty derives from biostratigraphy which, as indicated ear- lier, provides the key evidence for eustatic control. Miller *et al.* (1987) point out that Cenozoic biostratigraphy cannot provide resolution any more precise than for third-order cycles in the Exxon scheme, and this is equally true for the Mesozoic and Palaeozoic (Hallam 1992a). Consequently, claims that the so-called fourth- and fifth-order cycles characteristic of many carbonate platform successions are necessarily controlled by eustatic changes should be viewed with scepticism.

As Satterley (1996a) points out, on carbonate platforms only a limited range of facies can occur in a cycle and there is no easy way to distinguish eustatic and tectonic cycles and autocycles. As regards the widespread invocation of orbitally forced 'Milankovitch' cycles, a useful record is only preserved in successions if post-depositional modification such as erosion, bioturbation and diagenesis is relatively unimportant, and indeed the presence of a useful Milankovitch climatic record in shallow-water carbonate environments is highly equivocal. Algeo & Wilkinson (1988) maintain that such environments are an inappropriate place to look for a Milankovitch record because of the degree of unpredictability involved in sedimentation on carbonate platforms. Indeed, carbonate sedimentation in Florida Bay has generated no fewer than three shallowing-up cycles during the last sea-level rise (Drummond & Wilkinson 1993).

As regards the Alpine Triassic, Goldhammer *et al.* (1990, 1991) have argued that climatic oscillations control fourth- and fifth-order eustatic sea-level oscillations via fluctuations in volume of polar ice caps and Alpine glaciers or, supposedly, by thermal contraction and expansion of seawater. The Milankovitch theory of cycle origins assumes the following:

1. a relatively complete stratigraphical record;
2. periodic (non-random) cycle deposition;
3. a 'lag time' before deposition can restart after a sea-level rise; and
4. that one cycle is deposited per sea-level rise.

According to Satterley's (1996a) review, all of these assumptions are unreliable. In another paper Satterley (1996b) provides evidence for autocyclicity in the Upper Triassic Dachstein Limestone of the Northern Calcareous Alps, with the deposition of laterally discontinuous cycles within a complex palaeogeography of tidal islands and coexistent subtidal areas and channels. The islands were in a more or less constant state of change. Changes in external factors like tectonics or sea level are likely to have been met with a complex response, and the resultant stratal succession need not be an accurate reflection of changes in the forcing factors. The limited lateral continuity of the so-called Lofer cycles and lateral facies interfingering suggest that the eustatic model on its own is insufficient to explain the cyclicity. For these cycles and others in the Southern Alps, Satterley (1996a) favours the operation of a non-linear and complex dynamic system, as described by Nicolis & Nicolis (1991). In such systems there is no periodicity and hence predictability. Random fluctuations characterize complex physical systems and can give rise to repetitious patterns. At some critical times the environmental system can rapidly switch from one stable regime to another, with the potential to generate large shifts in depositional style in response to small environmental perturbations.

15.6 UNDERLYING MECHANISMS

The most obvious mechanism for causing eustatic sea-level changes, as long recognized, is the melting and freezing of ice caps, so-called *glacioeustasy*. While this has clearly been important in the Quaternary, for most of Phanerozoic history there is no evidence for such ice caps, or the strong latitudinal climatic gradients associated with their presence. The only times prior to the Oligocene for which one can infer their presence with confidence are the latest Ordovician and the Late Carboniferous and Early Permian (Frakes *et al.* 1992). Consequently the claim, for example, by Osleger & Read (1991) for Late Cambrian glacioeustatic sea-level oscillations of ~20 m, to account for minor carbonate cyclicity, is highly implausible. It is possible that the alternating thermal expansion and contraction of seawater related to climatic change could cause eustatic oscillations but these would be unlikely to be more than a few metres at most (Hallam 1992a). In view of the uncertainties indicated in the previous section about the intepretation of minor carbonate cyclicity, there appear to be no good grounds for invoking eustasy as a general explanation.

Longer-term changes where biostratigraphical correlation on a global scale can be invoked are a different matter, and there are many well-supported examples of eustasy throughout the Phanerozoic, independent of the work of Exxon geologists (Hallam 1992a). The causal mechanism for most of these must involve tectonics in some way. There are a number of possibilities for *tectonoeustasy*. The most obvious one to produce long-term sea-level changes, such as the first- and second-order cycles in the Exxon scheme, are variations in the volume of oceanic ridges related to sea-floor spreading, involving either variation in spreading rate or the length of ridges related to oceanic opening, or both (Hallam 1992a). In addition, it has been argued that the mid-Cretaceous sea-level rise could have been induced by a mantle superplume beneath the Pacific, but Hardebeck & Anderson's (1996) recent analysis suggests that the breakup of Pangaea is a more viable explanation.

It is quite likely that some of the extensive marine inundations of the continents that have manifestly taken place at different times in the Phanerozoic are not strictly eustatic but neither are they confined to the limited areas of tens of thousands of square kilometres involved in normal regional tectonics. Consequently Gurnis (1993) has put forward an interesting spherical model of mantle flow constrained by locations of trenches, which can be used to predict the dynamic topography of the earth's surface, and hence the marine inundation of continents. For past periods of high sea level the predicted pattern of flooding correlates well with the geological record. The high spatial correlation may result from increased plate velocities at these times, leading to increased rates of subduction, subsidence and inundation of convergent margins.

A problem persists with the interpretation of Exxon-type third-order cycles at times when it is not plausible to invoke glacioeustasy, because the rate and frequency of probable sea-level change, admittedly very uncertain quantities, were likely to have been too high to be satisfactorily accounted for by the tectonoeustatic mechanisms mentioned above (Hallam 1992a). Variations in within-plate lithospheric stress have been invoked by Cloetingh *et al.* (1985) to account for such cycles, but their model relates only to particular sedimentary basins, whereas many third-order cycles can be traced far more extensively. Such phenomena in non-glacial times may in fact be explicable by stress-induced changes in plate density (Cathles & Hallam 1991). These changes imply extensive regional epeirogenic movements rather than true eustasy and as such may account for many events traceable over areas much larger than sedimentary basins but not recognizable across the whole world. On rarer occasions, however, an elastic snap-back mechanism associated with the rapid formation of a new rift could perhaps explain global regressions and subsequent transgressions, such as often associated with mass extinctions of marine faunas (Hallam 1989). The rate of sea-level change implied is much more rapid than that associated with tectonoeustasy and compares more with glacioeustatic phenomena.

15.7 CONCLUSIONS

Having been largely ignored for much of this century, the study of pre-Quaternary sea-level changes is now a flourishing field of study, with an ever burgeoning literature. Relationships with other important phenomena, such as organic radiations and extinctions, have been demonstrated (Hallam 1992a). Much remains to be done, however. There is a paramount need for better stratigraphical correlation, in order to test rigorously various claims of global, rather than regional, tectonic control, and especially for estimates of the amount and rate of sea-level change. Difficult as it is to determine reliably depths of deposition of ancient marine sediments, as well as precise ages, even approximate, order-of-magnitude estimates can be helpful for distinguishing the underlying mechanisms (Hallam 1997).

REFERENCES

Algeo, T.J. & Seslavinsky, K.B. 1995. The Paleozoic world: continental flooding, hypsometry, and sea level. *American Journal of Science,* **295**, 787–822.
Algeo, T.J. & Wilkinson, B.H. 1988. Periodicity of mesoscale Phanerozoic sedimentary cycles and the role of Milankovitch orbital modulation. *Journal of Geology,* **96**, 313–322.
Carter, R.M., Abbott, S.T., Fulthorpe, C.S., Haywick, D.W. & Henderson, R.A. 1991. Application of global sea level and sequence-stratigraphical models in Southern Hemisphere Neogene strata from New Zealand. In Macdonald, D.I.M. (ed.) *Sedimentation, tectonics and eustasy.* International Association of Sedimentologists, Special Publication 12, 41–65.
Cathles, L.M. & Hallam, A. 1991. Stress-induced changes in plate density, Vail sequences, epeirogeny and short-lived global sea-level fluctuations. *Tectonics,* **10**, 659–671.
Christie-Blick, N., Mountain, G.S. & Miller, K.G. 1990. Seismic stratigraphical record of sea-level change. In Revelle, R.R. (ed.) *Sea-level change.* National Academy Press, Washington, 116–140.
Cloetingh, S., McQueen, H. & Lambeck, K. 1985. On a tectonic mechanism for regional sea level variations. *Earth and Planetary Science Letters,* **75**, 157–166.
Crevello, P.D. 1991. High-frequency carbonate cycles and stacking patterns: interplay of orbital forcing and subsidence on Lower Jurassic rift platforms, High Atlas, Morocco. *Kansas Geological Survey Bulletin,* **233**, 207–230.
Dickinson, W.R. 1993. The Exxon global cycle chart: an event for every occasion? Discussion. *Geology,* **21**, 282–283.
Drummond, C.N. & Wilkinson, B.H. 1993. Carbonate cycle stacking patterns and hierarchies of orbitally forced eustatic sea level change. *Journal of Sedimentary Petrology,* **63**, 369–377.
Drummond, C.N. & Wilkinson, B.H. 1996. Stratal thickness frequencies and the prevalence of orderedness in stratigraphical sequences. *Journal of Geology,* **104**, 1–18.

Fischer, A.G. 1984. The two Phanerozoic supercycles. In Berggren, W.A. & Van Couvering, J.A. (eds) *Catastrophes and earth history*. Princeton University Press, Princeton, 351–376.

Frakes, L.A., Francis, J.E. & Syktus, J.I. 1992. *Climate modes of the Phanerozoic*. Cambridge University Press, Cambridge.

Franseen, E.K., Watney, W.L., Kendall, C.G.St.C. & Ross, W. (eds) 1991. *Sedimentary modelling: computer simulations and methods for improved parameter definition*. Kansas Geological Survey Bulletin, 233.

Fulthorpe, C.S. & Carter, R.M. 1989. Test of seismic sequence methodology on a southern hemisphere passive margin: the Canterbury Basin, New Zealand. *Marine and Petroleum Geology*, **6**, 348–359.

Goldhammer, R.K., Dunn, P.A. & Hardie, L.A. 1990. Depositional cycles, composite sea level changes, cycle stacking patterns, and the hierarchy of stratigraphical forcing: examples from Alpine Triassic platform carbonates. *Bulletin of the Geological Society of America*, **102**, 535–562.

Goldhammer, R.K., Oswald, E.J. & Dunn, P.A. 1991. Hierarchy of stratigraphical forcing: example from Middle Pennsylvanian shelf carbonates of the Paradox Basin. *Kansas Geological Survey Bulletin*, **233**, 361–413.

Goldhammer, R.K., Lehmann, P.J. & Dunn, P.A. 1993. The origin of high-frequency platform carbonate cycles and third-order sequences (Lower Ordovician El Paso Group, West Texas): constraints from outcrop data and stratigraphical modelling. *Journal of Sedimentary Petrology*, **63**, 318–359.

Grabau, A.W. 1936. Oscillation or pulsation? *Report of the 16th International Geological Congress*, **1**, 539–552.

Gurnis, M. 1993. Phanerozoic marine inundation of continents driven by dynamic topography above subducting slabs. *Nature*, **364**, 589–593.

Hallam, A. 1984. Pre-Quaternary changes of sea level. *Annual Review of Earth and Planetary Sciences*, **12**, 205–243.

Hallam, A. 1988. A re-evaluation of Jurassic eustasy in the light of new data and the revised Exxon curve. In Wilgus, C.K., Hastings, B.S., Kendall, C.G.St.C., Posamentier, H.W., Ross, C.A. & Van Wagoner, J.C. (eds) *Sea-level change: an integrated approach*. Society of Economic Paleontologists and Mineralogists Special Publication 42, 261–273.

Hallam, A. 1989. The case for sea-level change as a dominant causal factor in mass extinction or marine invertebrates. *Philosophical Transactions of the Royal Society of London*, **B325**, 437–455.

Hallam, A. 1990. The end-Triassic mass extinction event. *Geological Society of America Special Paper*, **247**, 577–583.

Hallam, A. 1992a. *Phanerozoic sea-level changes*. Columbia University Press, New York.

Hallam, A. 1992b. Eduard Suess and European thought on Phanerozoic eustasy. *Geological Society of America Memoir*, **180**, 25–29.

Hallam, A. 1997. Estimates of the amount and rate of sea-level change across the Rhaetian–Heltangian and Pliensbachian–Toarcian boundaries (latest Triassic to early Jurassic). *Journal of the Geological Society of London*, **154**, 773–780.

Hallam, A. & Sellwood, B.W. 1976. Middle Mesozoic sedimentation in relation to tectonics in the British area. *Journal of Geology*, **84**, 302–321.

Hancock, J.M. 1990. Sea-level changes in the British region during the late Cretaceous. *Proceedings of the Geologists' Association*, **100**, 565–594.

Haq, B.U., Hardenbol, J. & Vail, P.R. 1987. Chronology of fluctuating sea levels since the Triassic. *Science*, **235**, 1156–1167.

Hardebeck, J. & Anderson, D.L. 1996. Eustasy as a test of a Cretaceous superplume hypothesis. *Earth and Planetary Science Letters*, **137**, 101–108.

Kidwell, S.M. 1988. Reciprocal sedimentation and non correlative hiatuses in marine-paralic siliciclastics: Miocene outcrop evidence. *Geology*, **16**, 609–612.

Miall, A.D. 1986. Eustatic sea level changes interpreted from seismic stratigraphy: a critique of the methodology with particular reference to the North Sea Jurassic record. *Bulletin of the American Association of Petroleum Geologists*, **70**, 131–137.

Miall, A.D. 1992. Exxon global cycle chart: an event for every occasion? *Geology*, **20**, 787–790.

Miall, A.D. 1994. Sequence stratigraphy and chronostratigraphy: problems of definition and precision in correlation, and their implications for global eustasy. *Geoscience Canada*, **21**, 1–26.

Miller, K.G., Fairbanks, R.G. & Mountain, G.S. 1987. Tertiary oxygen isotope synthesis, sea level history, and continental margin erosion. *Paleoceanography*, **2**, 1–19.

Nicolis, G. & Nicolis, C. 1991. Nonlinear dynamic systems in the geosciences. *Kansas Geological Survey Bulletin*, **233**, 33–42.

Norris, M. & Hallam, A. 1995. Facies variations across the Middle–Upper Jurassic boundary in Western Europe and the relationship to sea-level changes. *Palaeogeography, Palaeoclimatology, Palaeoecology*, **116**, 189–245.

Osleger, D. & Read, J.R. 1991. Relation of eustasy to stacking patterns of meter-scale carbonate cycles, Late Cambrian, U.S.A. *Journal of Sedimentary Petrology*, **61**, 1225–1252.

Osleger, D. & Read, J.F. 1993. Comparative analysis of methods used to define eustatic variations in outcrop: Late Cambrian interbasinal sequence development. *American Journal of Science*, **293**, 157–216.

Plint, A.G. 1991. High-frequency relative sea level oscillations in Upper Cretaceous shelf clastics of the Alberta foreland basin: possible evidence for glacio-eustatic control? In Macdonald, D.I.M. (ed.) *Sedimentation, tectonics and eustasy*. International Association of Sedimentologists, Special Publication 12, 409–428.

Posamentier, H.W., Allen, G.P., James, D.P. & Tesson, M. 1992. Forced regressions in a sequence stratigraphical framework: concepts, examples, and exploration significance. *Bulletin of the American Association of Petroleum Geologists*, **76**, 1687–1709.

Satterley, A.K. 1996a. The interpretation of cyclic successions of the Middle and Upper Triassic of the Northern and Southern Alps. *Earth Science Reviews*, **40**, 181–207.

Satterley, A.K. 1996b. Cyclic carbonate sedimentation in the Dachstein Limestone, Austria: the role of patterns of sediment supply and tectonics in a platform–reef–basin system. *Journal of Sedimentary Research*, **66**, 307–322.

Suess, E. 1888. *Das Antlitz der Erde*, vol. 2. Tempsky-Freytag, Prague, Vienna, Leipzig.

Suess, E. 1906. *The Face of the Earth*, vol. 2, translated by W.J. Sollas. Clarendon Press, Oxford.

Umbgrove, J.H.F. 1939. On rhythms in the history of the Earth. *Geological Magazine*, **76**, 116–129.

Underhill, J.R. 1991. Late Jurassic seismic sequences, Inner Moray Firth, UK: a critical test of a key segment of Exxon's original global cycle chart. *Basin Research*, **3**, 79–98.

Underhill, J.R. & Partington, M.A. 1993. Jurassic thermal doming and deflation in the North Sea: implications of the sequence stratigraphical evidence. In Parker, J.R. (ed.) *Petroleum geology of northwest Europe: proceedings of the 4th conference*. Geological Society of London, 337–345.

Vail, P.R. & Todd, R.G. 1981. Northern North Sea Jurassic unconformities, chronostratigraphy and sea-level changes from seismic stratigraphy. In Illing, L.V. & Hobson, G. (eds) *Petroleum geology of the continental shelf of north-west Europe*. Heyden, London, 216–235.

Vail, P.R., Mitchum, R.M., Todd, R.G., Widmier, J.M., Thompson, S., Songree, J.B., Bubb, J.N. & Hatlelid, W.G. 1977. Seismic stratigraphy and global changes of sea level. *American Association of Petroleum Geologists Memoir*, **26**, 49–212.

16
Interpreting Palaeoenvironments from Fossils

Peter Doyle and Matthew R. Bennett

Fossils provide some of the most potent tools in the determination of ancient environments. Both body fossils (the physical remains of animals and plants) and trace fossils (their tracks, trails burrows and excreta) provide information which is of great value in interpreting the sedimentary record, through the application of palaeoenvironmental analysis (e.g. Ager 1963; Frey 1975; Boucot 1981; Dodd & Stanton 1990; Goldring 1991; Bosence & Allison 1995).

Palaeoenvironmental analysis is the determination of the nature of the ancient environment using the sediments, their geochemistry and their palaeoecology. Effectively palaeoenvironmental analysis entails piecing together, fragment by fragment, the nature of an ancient environment from all the available evidence (Figure 16.1). The technique has a long history, dating back to the 1840s and Edward Forbes (Craig 1966; Dodd & Stanton 1990): 'By carefully observing the mineral character of the former bottom, by noticing the associations of species and their relative depth . . . our conclusions [about palaeoenvironments] will doubt-less approximate to the truth' (Forbes 1844: 172).

As discussed in Chapter 14 the sedimentary rocks themselves provide important environmental evidence in determining the nature of an ancient environment. The original environment and conditions of deposition of each sedimentary rock body may be deduced through the concept of sedimentary facies: the sum total of all the characteristics of a rock unit. Geochemical analyses, particularly carbon and oxygen stable isotope studies from skeletal and whole-rock carbonates, can give important information on ancient climates (see Chapter 17) and biological productivities.

Unlocking the Stratigraphical Record: Advances in Modern Stratigraphy. Edited by P. Doyle and M.R. Bennett.
© 1998 John Wiley & Sons Ltd.

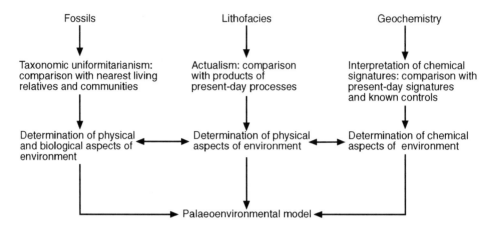

Figure 16.1   *The palaeoenvironmental model: determination of palaeoenvironments from fossils, lithofacies and geochemistry*

Fossils are an integral part of sedimentary facies, and palaeoecology – the study of the ecological relationships of ancient organisms – is arguably the most important component in palaeoenvironmental analysis (e.g. Dodd & Stanton 1990; Bosence & Allison 1995). This is simply because palaeoecology involves the identification of fossil assemblages as once-living communities of organisms bounded in space by aspects of their environment, the so-called limiting factors. It is these limiting factors which control the diversity and abundance of organisms, and through comparison with their living relatives, fossils can provide a detailed picture of the nature of their ancient environments. Fundamentally, fossils can provide information about sea-levels, climate, salinity, oxygen, temperature, and a whole range of environmental parameters that no other element of facies analysis can achieve.

This chapter is intended to provide a basis for understanding the importance of palaeoecology in the determination of palaeoenvironments derived from the stratigraphical record. It is not intended to be comprehensive, but simply to introduce the importance of fossils in the interpretation of ancient environments. The chapter first introduces the most fundamental concept in palaeoenvironmental analysis, taxonomic uniformitarianism, before going on to discuss the principles of palaeoecology. The applications of palaeoenvironmental analysis are demonstrated in two case studies which examine how environmental parameters may be derived from the fossil record.

16.1 TAXONOMIC UNIFORMITARIANISM

Uniformitarianism is a fundamental geological concept commonly expressed in the phrase: 'the present is the key to the past'. This principle is usually applied in stratigraphy in order to provide the basis for the interpretation of rock bodies as the products of specific ancient environments. This process entails comparison of the products of present-day environments with those of the geological past, and is the most fundamental aspect of facies analysis (Figure 16.1; also see Chapter 14). From this comparison, geologists can make the assumption that the environmental conditions which operate today also operated in the past to create the rock record. A similar comparison is made in geochemical analysis, where relative concentrations of stable isotopes known to be sensitive to environmental parameters are compared with their Recent counterparts (Figure 16.1).

Taxonomic uniformitarianism uses the same basic principle applied to the biological components of an environment (Dodd & Stanton 1990). It assumes that the study of present-day organisms holds the key to interpreting the life history and ecological tolerances of the organisms of the geological past. Taxonomic uniformitarianism also assumes that the animals now preserved as fossil remains once functioned in a similar manner to, and under similar environmental constraints as, their nearest living relatives. This principle is demonstrated by Kanazawa (1992), who studied the detailed burrowing behaviour of living echinoids in order to determine the mode of life of fossil relatives (Figure 16.2), and Linsley *et al.* (1978), who interpreted the function of the marginal frills of the fossil

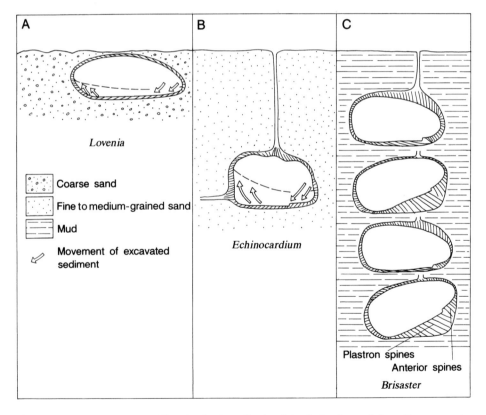

Figure 16.2 *Taxonomic uniformitarianism. Study of the behaviour of these Recent spatangoid echinoids provides an interpretative model for three shapes of burrowing echinoids in the fossil record (**A–C**). [Modified from: Kanazawa (1992)]*

gastropod *Euomphalopterus* with reference to the living genus *Xenophora* (Figure 16.3).

Taxonomic uniformitarianism is the fundamental tool for the interpretation of fossil assemblages as ancient ecologies, and is the basis for palaeoecology and therefore palaeoenvironmental analysis from fossils. As in the determination of the life histories of individual taxa, research in this area involves an interaction between palaeontology and ecology, particularly through the detailed analysis of present-day environments by palaeontologists ('actuo-palaeontology'). Importantly, it assumes that the ecological associations of living organisms provide the key to understanding and interpreting past ecological relationships. This approach is well illustrated by the work of Schäfer (1972) in the North Sea, whose studies of the interaction of living organisms with reference to their interpretation in the fossil record are very much in the spirit of Edward Forbes' pioneering work.

The uniformitarian approach is also directly applicable to the interpretation of trace fossils (Frey 1975; Frey & Seilacher 1980; Bromley 1990; Pemberton *et al.* 1990). Trace fossils may be classified according to ethology, that is the study of the behaviour of the organisms that created them. This is interpreted directly from the traces preserved in the fossil record (Figure 16.4). Some are obvious, such as the footprints of dinosaurs and birds, but others are less so, and require comparison with modern equivalents (Bromley 1990). The problems encountered with changes in ecological tolerances of body fossils through geological time are in most cases not encountered with trace fossils, as the morphology and environment of trace fossils appears to have remained invariate. In general, broadly similar groups of organisms, producing broadly similar sets of tracks and trails, can be encountered throughout the geological record of life on earth.

However, taxonomic uniformitarianism has some drawbacks. In particular, the accuracy of the approach is constrained by geological time and evolution, as well as by the vagaries of preservation of the fossil record (Dodd & Stanton 1990; Bottjer *et al.* 1995). Effectively, the more recent the fossil, the greater the chance that there is a closely related organism living today. With increasing age, comparisons become more difficult, and in many cases, especially with the most ancient rocks, interpretation is more a matter of inspired guesswork. In some cases, strict uniformitarian models may be inappropriate where equivalent taxa or phenomena are unknown or relatively unknown in present-day environments.

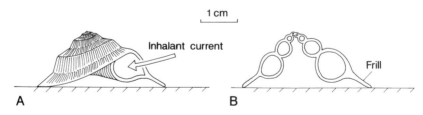

Figure 16.3 *Taxonomic uniformitarianism. The interpretation of frills on the whorls of the Palaeozoic gastropod* Euomphalopterus *as props was achieved through comparison with the living gastropod* Xenophora. *[Modified from: Linsley et al. (1978)]*

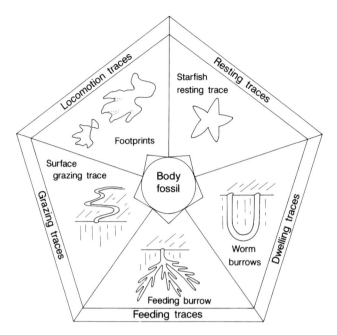

Figure 16.4 *The ethological classification of trace fossils. [Reproduced with permission from: John Wiley & Sons, from: Doyle (1996)]*

Good examples of this are the Mesozoic–Cenozoic carbonate mound buildups now considered to be the result of deep-water cold-methane seeps, promoting the development of chemosynthetic rather than photosynthetic communities (e.g. Campbell *et al.* 1993). This is at odds with former palaeoenvironmental models for these mounds which were constrained by uniformitarian comparisons with living, shallow-water, hermatypic reef corals dependent on photosynthetic zooxanthellae algae (Bottjer *et al.* 1995).

Equally subtle is the possibility that certain organisms have evolved and changed their environmental preferences through geological time, and live today in very different environments than they did in the geological past without significant morphological change (Dodd & Stanton 1990). This can be seen in the progressive onshore–offshore displacement of a great many organisms during geological time. This point is illustrated by the producers of traces such as *Palaeodictyon*, which migrated from a shallow to a deep water setting during the Palaeozoic (Crimes & Anderson 1985). Finally, and perhaps most importantly, the taxonomic uniformitarian interpretation of fossils and communities is constrained by preservation potential, particularly with soft-bodied and delicate organisms which are easily destroyed, modified by diagenesis or disarticulated. This has played an important part in the re-interpretation of Cambrian organisms and ecosystems from the Burgess Shale of British Columbia, particularly in the reconstruction of organisms such as *Anomalocaris* and *Hallucigenia* (e.g. Bengston 1991; Figure 16.5).

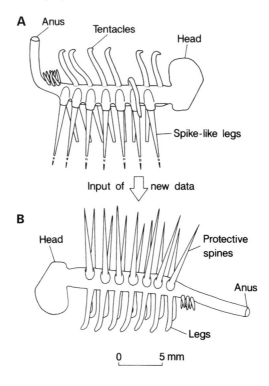

Figure 16.5   *The initial interpretation of the mode of life of the Cambrian organism Hal-lucigenia was hampered by poor preservation. New discoveries in China in which two pairs of 'tentacles' were preserved led to a reappraisal of its mode of life. [Reproduced with permission from John Wiley & Sons, from: Doyle (1996)]*

16.2 PALAEOECOLOGY

Palaeoecology can be broadly defined as the study of the interaction of fossil organisms with each other, and with the environment in which they lived in the geological past. Like ecology, palaeoecology is an extremely important tool in the determination of the nature of the biosphere and in recording changes in environment. Taxonomic uniformitarianism is again the key to this concept. Palaeoecology can be viewed from many perspectives, but traditionally considered, it has two branches (Ager 1963; Craig 1966), namely:

1. *palaeosynecology*, or the relationship of groups of organisms to each other and to their environment; and
2. *palaeoautecology*, or the relationship of individual or small groups of organisms to their environment.

16.2.1 Palaeosynecology

Palaeosynecology is concerned with the relationship of groups of organisms with each other and their environment. In biology, synecology forms much of the detailed subject matter of ecology, as it concentrates on the interactions of organisms, such as symbiosis or predation, or in the study of recurrent communities of organisms (Ager 1963; Craig 1966; Dodd & Stanton 1990).

It is in the study of ancient communities that palaeosynecology has the greatest role to play in palaeoenvironmental analysis. Definitions of the term 'community' vary, but in essence it is a recurrent group of organisms from one or more species that occupy the same, spatially limited, habitat in life (e.g. Dodd & Stanton 1990). The spatial limitation of communities is determined by physical, chemical and biological aspects of the environment. Successful initial colonization and evolution of communities is associated with the stability of environment through time, and with the effective partitioning of resources, referred to as the trophic structure (Dodd & Stanton 1990). Present-day communities are clearly limited in this way, and the well-known, depth-related zonation of organisms in intertidal and subtidal settings is a good example (Petersen 1915). The identification of fossil intertidal and subtidal communities has enabled the determination of palaeoenvironmental factors such as changes in sea-level and relative onlap of transgressive sequences (e.g. Ziegler 1965; Scott 1974; Lockley 1983; Doyle *et al.* 1997; Figure 16.6).

In palaeoecology, however, the identification of communities and the interpretation of their limitations and trophic structure is much more difficult (Ager 1963; Scott 1974, 1978; Boucot 1981, 1990; Dodd & Stanton 1990). For one thing, it is important to remember that all fossil assemblages are death assemblages, as only in exceptional circumstances do they come anywhere near the nature of the living community – the life assemblage, or biocoenosis. Relatively few fossils are found in their life position, and most have been transported, some of which may be derived from a completely different habitat. In a great many cases, physical transport also leads to the displacement of fossil shells and other skeletal parts far from their point of life, and death. Swimming or floating organisms, for example, may drift a great distance after their death, before they finally sink and become incorporated into the sedimentary record. Even bottom-dwelling organisms such as gastropods may float from their point of death due to the buildup of gasses during the decay process. Such displaced fossils are said to be allochthonous, and this has a great bearing on the interpretation of assemblages of fossils as once-living communities of animals and plants. In some cases, allochthonous fossils may be identified by the physical damage left by the action of transport, or by the accumulation or particular orientation of skeletal remains according to prevailing current or wind direction. As physical transport is common before burial, recognition of in-place, or autochthonous, fossils is certain only in cases where the fossil is found in life position, such as within burrows, or growing in life attitude, as with corals and tree stumps. This is relatively rare, and in most cases, fossils can be described only as semi-autochthonous, that is that they have undergone at least some transport, often exhibiting a stable position relative to the surrounding sediment. Figure 16.7 illustrates the hypothetical case. To the left of the diagram,

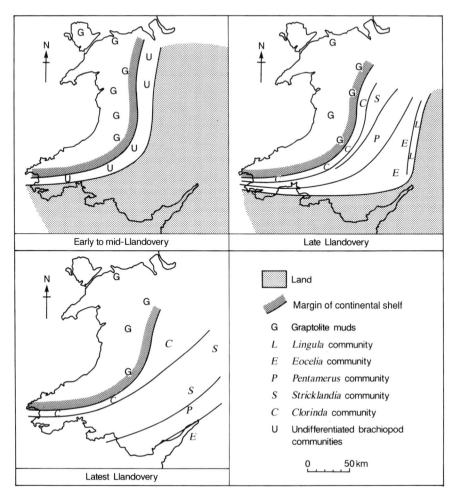

Figure 16.6   *Depth and substrate-related brachiopod assemblages from the Ordovician of Wales. Plotting the position of these assemblages through time provided a key to the progress of a transgressive episode. [Modified from: Ziegler (1965)]*

the organism progresses from life through death and burial, leaving the fossil (A) close to life position. This is truly autochthonous. Fossil B has been transported after death but before burial; fossil C represents an organism that died away from its normal habitat, but was not transported after death. Both fossils B and C are semi-autochthonous in position. Fossil D represents an organism which died away from its normal habitat and was transported after death. This fossil is truly allochthonous. Fossils A–D can all be eroded from their sediments and be reworked as allochthonous fossils.

The implications of transport and reworking are that for most fossil assemblages, there will not only be allochthonous fossils present that were not native to the original community, or even reworked fossils from a different time-frame, but

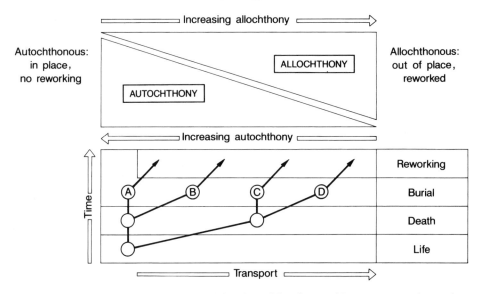

Figure 16.7 *Transport and reworking of fossils and fossil assemblages. See text for explanation. [Reproduced with permission from John Wiley & Sons, from: Doyle (1996)]*

also in most cases the soft-bodied organisms will be absent. In reality, only exceptionally preserved conservation *lagerstätten* such as the Cambrian Burgess Shale of British Columbia can come remotely close to representing the living assemblage as it originally existed, but even then allochthonous fossils may be present, and other organisms absent.

Clearly, the only semi-reliable tools of community analysis are: (1) the recurrence of the association, increasing the probability that it is ecologically relevant; and (2) authchthony, which demonstrates whether an organism has been significantly transported or not (Ager 1963; Scott 1974; Dodd & Stanton 1990). It is therefore preferable to use the term *assemblage* to describe a grouping of fossils preserved in a given rock unit, as it may not necessarily reflect the nature of the original community. Recurring assemblages suggest a greater ecological significance, and these are referred to as *associations* (Boucot 1981, 1990; Dodd & Stanton 1990). There has been a great deal of literature that has attempted to discuss the relevance of community analysis (e.g. Gray *et al.* 1981), and the recognition of what were once actual living communities is a challenging task for any palaeoecologist (e.g. McKerrow 1978).

Perhaps the only reliable indicators of pre-existing communities are trace fossils (Pemberton *et al.* 1990). Firstly, trace fossils are conservative: the same types of traces are intimately associated with certain sedimentary environments throughout geological time. Secondly, trace fossils are usually found *in situ*, and are therefore more reliable indicators of the actual living communities than body fossils, which are often reworked or transported. Thirdly, trace fossils indicate the former presence of soft-bodied organisms in an environment and, in some cases, may be the only fossil record within a given sedimentary environment. Finally, trace fossils provide important information about organism–sediment interactions which is rarely gained from body fossils alone.

16.2.2 Palaeoautecology

Palaeoautecology is concerned with the relationship of individual or small groups of organisms to their environment. It is primarily concerned with the tolerance levels of individual species, and the ability of communities of organisms to thrive in specific sets of environmental conditions, that is the limiting factors (Ager 1963; Craig 1966; Ziegler 1983; Dodd & Stanton 1990). These limiting factors control directly the diversity of species and abundance of individuals in living communities, and are therefore of great value in the interpretation of the nature of ancient ecologies and in palaeoenvironmental analysis. The basis for this interpretation is in taxonomic uniformitarianism, as the response of living organisms to limiting factors today forms the basis for the interpretation of their close relatives in the geological past. As before, the time factor is of paramount importance, and with increasing age, the confidence levels for the interpretation of response to limiting factors decrease.

Limiting factors

Some of the most significant limiting factors are light, climate, oxygen, salinity, water depth, substrate, water turbulence and food supply. In most cases these factors are closely interrelated, and both the diversity and distribution of organisms may be controlled by several such factors. The principal limiting factors are discussed below.

1. *Light.* Light promotes photosynthesis and is therefore important to the success of plants and the food chains that they support. Light is most intense at the equator, and the angle of incidence of the sunlight reaching the earth decreases towards the poles. This has bearing on the success of photosynthesizing plants, for example, and the densest vegetation is situated at the equator. The discovery of vent faunas along mid-ocean ridges has indicated that not all organisms are dependent on light, however, the community being sustained by the process of chemosymbiosis (e.g. Cavanaugh 1985).

2. *Climate.* Climate has varied appreciably through geological time, and it is possible to observe changes in the geographical distribution of organisms through time (Brenchley 1984). This is particularly true of land plants, and the interpretation of the climatic significance of fossil land plants is of paramount importance to the understanding of ancient climates through the application of taxonomic uniformitarianism (Spicer 1990). Some aquatic organisms may also be restricted to certain temperature regimes, such as the warm-water corals (Figure 16.8), and we know from their consistently restricted distribution patterns that certain extinct organisms, such as the Cretaceous rudist bivalves for example, were also limited in this way. Other organisms may be pandemic and

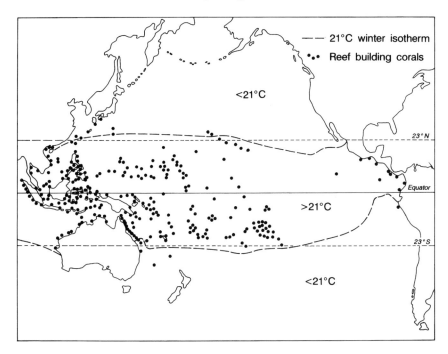

Figure 16.8 *Latitudinal and temperature restriction on the distribution of living hermatypic corals. [Modified from: Ziegler (1983)]*

of limited value in interpreting the relative temperature of ancient water masses. Isotopic methods of palaeotemperature determination, using the oxygen and carbon isotopes of molluscs such as oysters and belemnites, of brachiopods, and of foraminifera also provide valuable insights into ancient climatic conditions.

3. *Oxygen*. Oxygen is an important limiter. For example, in aquatic environments with reduced oxygen, such as in stagnant, anaerobic, water bodies, faunal diversity is often dramatically reduced, following a set pattern of reduction in diversity and shell thickness with reducing oxygen levels (Rhoads & Morse 1971; Figure 16.9). This diversity decline is observed in both shelly and soft-bodied benthonic organisms and their traces, although nektonic organisms may be unaffected. The interpretation of benthonic oxygen levels from the diversity of fossil assemblages is now a well-used and valuable interpretative tool below (e.g. Savrda & Bottjer 1986, 1987), and is further discussed in Section 16.4.2.

4. *Salinity*. Salinity is of significance to aquatic organisms. Although some organisms can survive in fresh and marine water conditions, and others can manage in brackish conditions, normal salinity is essential to most marine organisms. In most cases, there is a distinct diversity decline with increasing or decreasing salinity (Figure 16.10). In addition, it is possible to recognize euryhaline organisms, such as bivalves, gastropods and ostracods, which are tolerant to salinity

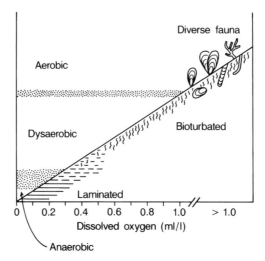

Figure 16.9 *The relationship of benthonic organisms to dissolved oxygen levels. With decreasing oxygen levels, shelly fauna decreases in diversity, followed by soft-bodied infauna. This relationship defines the three main oxygen-related biofacies: anaerobic, dysaerobic and aerobic. [Modified from: Rhoads & Morse (1971)]*

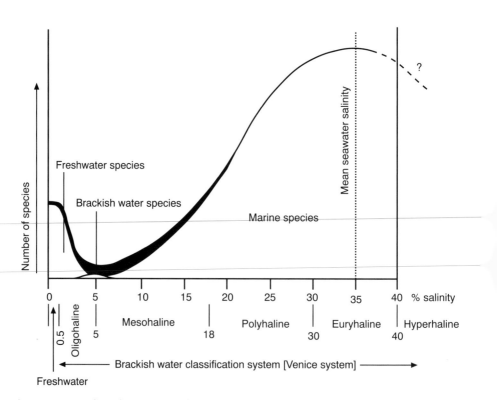

Figure 16.10 *The relationship of faunal diversity, expressed as the number of species present, with salinity. [Modified from: Hudson (1990)]*

fluctuations, and stenohaline organisms, such as corals, bryozoa, articulate brachiopods and cephalopods, which can only flourish in normal marine salinity levels. These can be valuable palaeoenvironmental tools (e.g. Hudson 1990; Fürsich 1994), and are further discussed in Section 16.4.1.

5. *Water depth*. Aquatic organisms are limited by water depth, largely because the deepest waters have the least amount of light penetration, a factor particularly important in the development of reefs (Perrin *et al.* 1995), although free-swimming organisms such as ammonites may also have been limited by increasing water pressure (Westermann 1972; Figure 16.11). Commonly, depth and substrate are closely related, as there is often an onshore–offshore gradient of firm to soft substrata (e.g. Ziegler 1965; Lockley 1983; Doyle *et al.* 1997). This is illustrated by the concept of ichnofacies, that is assemblages of trace fossils which seem to have a bathymetric relationship (Seilacher 1967; Figure 16.12).

6. *Substrate*. Of particular importance to bottom-dwelling organisms is the nature of the substrate, which determines whether they are able to construct burrows, or live on a firm surface without sinking, for example (Goldring 1995). As already noted, substrate is interlinked with water depth, and this is of great

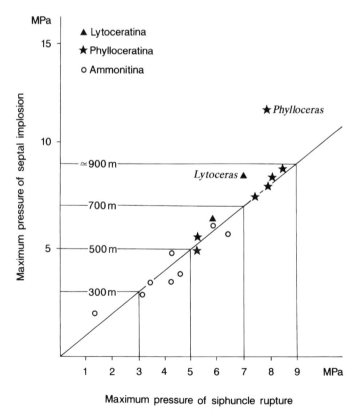

Figure 16.11   The relationship of ammonite suborders to water depth, expressed as the structural strength of their shells. [Modified from: Westermann (1972)]

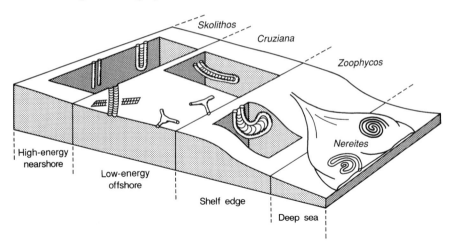

Figure 16.12   *The ichnofacies concept. The depth and substrate relationship of recurrent trace fossil assemblages. [Modified from: Seilacher (1967)]*

importance to the ichnofacies and other biofacies, such as those recognized by Ziegler (1965), which may actually be controlled by overall substrate type (Figures 16.6 and 16.12). However, substrate type also has a relationship with turbulence, which is important to some suspension feeders who require continuous agitation of suspended food particles.

7. *Food supply.* Food and nutrient supplies are of great importance to living organisms, as they ultimately determine success or failure in a given environment. The greatest diversity of marine organisms live on the continental shelf, within the zone known as the photic zone. This is the depth of maximum light penetration (approximately 200 m) and is the area of greatest food resources for many marine organisms.

8. *Biotic factors.* Such factors as competition for habitat space, and the relative distribution of prey and predators, for example, may also limit organisms. This is particularly important in the delimitation of intertidal communities, for example. However, the extent of the impact and influence of such factors is often difficult to detect in the geological record.

Limiting factors directly influence the distribution, diversity and abundance (density) of living organisms. Trace producers are also subject to the same range of limiting factors as any other organisms, and in turn these can determine the distribution, abundance and diversity of trace fossils. Determination of community structure is therefore an important tool in palaeoenvironmental analysis, and it is further discussed in Section 16.3.3. Limiting factors may have a great influence on the success of the individual or community, while at a much larger scale they have a direct impact on the distribution of plants and animals across the globe, the subject of biogeography and its geological counterpart, palaeobiogeography (see Chapter 18).

Biofacies and ichnofacies

Specific recurrent fossil assemblages can often be associated with certain sedimentary facies suggesting overall control by particular limiting factors, such as salinity, depth or oxygenation. These assemblages are usually referred to as biofacies, as the interpretation of the sum total of their characteristics is important in the determination of the limiting factors of the palaeoenvironments that produced them. The term is most commonly used to express the lateral or stratigraphical variation in groups of organisms in relation to changes in their environment, and where the assemblage is composed entirely of trace fossils, the term *ichnofacies* may be used (Brenchley 1990; Bromley 1990; Figure 16.12).

The concept of biofacies (and ichnofacies) is an extremely important one as it reflects change within an environment without the formality of communities. Biofacies are equally as recurrent and spatially limited as communities but here it is the broad scale attributes of the fauna that are important, not the individual components. The recognition of biofacies, however, need not be imprecise. For example, the oxygen-related biofacies of Rhoads & Morse (1971), based on uniformitarian examination of recent environments, provides a yardstick for semiquantitative estimates of oxygen-levels in the geological past (Figure 16.9). Similar examples exist for salinity levels (Hudson 1990; Fürsich 1994), and water depth (Ziegler 1965; Seilacher 1967; Pemberton *et al.* 1990; Perrin *et al.* 1995; Figures 16.6 and 16.12). In some cases, taphonomic 'biofacies' capable of environmental interpretation may be recognized. Examples include the increasing comminution of brachiopod valves with increasing depth in the Cretaceous (Middlemiss 1962), or a taphonomic range of barnacle shells from intact and *in situ*, to highly comminuted, again interpreted as an increasingly offshore trend (Doyle *et al.* 1997; Figure 16.13).

16.3 PRACTICAL ASPECTS OF PALAEOENVIRONMENTAL ANALYSIS

In seeking to determine the nature of ancient ecologies, and interpret the palaeoenvironmental factors that controlled them, five elements are of importance: (1) taxonomic uniformitarianism; (2) determination of autochthony; (3) recurrence of association; (4) population structure; and (5) comparison with other data sources. The basis for taxonomic uniformitarianism has already been discussed, and the remaining elements are detailed below. Further information is contained in the texts by Ager (1963), Boucot (1981), Briggs & Crowther (1990), Dodd & Stanton (1990), Goldring (1991) and Doyle (1996).

16.3.1 Determination of Autochthony

As discussed in Section 16.2.1, fossil assemblages are entities created by virtue of an interplay of two factors: the nature of the original community and the nature of the sedimentary environment. The original community structure, the life

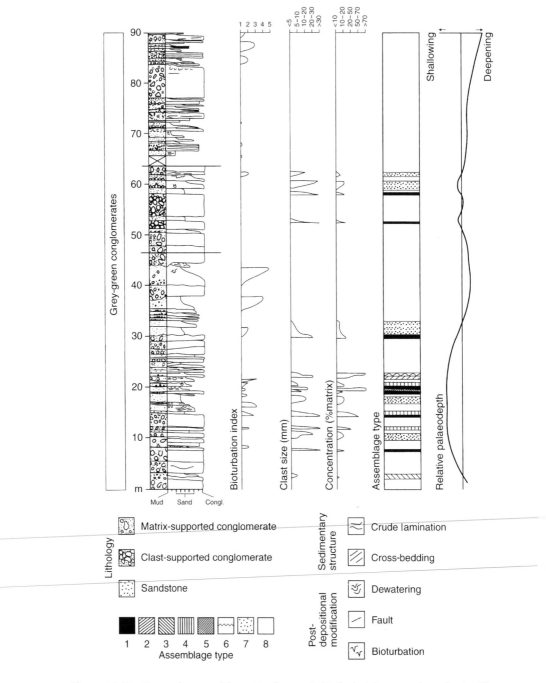

Figure 16.13 *Barnacle assemblages (taphonomic biofacies) from southern Spain. The assemblages vary from well-preserved in situ barnacle assemblages (type 1) to comminuted barnacle fragments (type 5). This is interpreted as a primary depth relationship. [Modified from: Doyle et al. (1997)]*

assemblage, may be complex and composed of a number of biological entities, with both hard- and soft-bodied organisms. In the fossil record, soft-bodied organisms are rarely preserved, except under exceptional circumstances of low oxygen and high sedimentation rate. As the soft-bodied components are usually lost to diagenesis and biochemical breakdown, most fossil assemblages do not closely resemble the living community. This is because the assemblage will be represented mostly by the shelly components or other hard body parts of a range of organisms, which upon their death are available for incorporation into the sedimentary record, either directly after death, or eventually after a period of transport (Figure 16.7). To approximate to the living community, assemblages should consist of largely autochthonous species. In general, autochthonous fossils are those which, through direct observation, can be proven to be in life position. This is most applicable to cementing or otherwise sessile organisms on the sediment surface, or to burrowing organisms clearly preserved in a life position in the sediment. Most other fossils, such as mobile benthonic, nektonic or planktonic organisms, or sessile benthonic organisms since removed from their point of attachment or washed from their burrows, are open to transport by a variety of wave and current activities, especially in the shallow marine environment. Determining the extent of such transport is critical in any consideration of the ecological significance of the assemblage.

There are several criteria which may be used to determine the extent of transport and reworking. These mainly reflect the interpretation of the original life position of the organism, and the extent of sorting and/or damage to the skeletal materials, and as such they are necessarily an imperfect measure, but nevertheless serve as a valuable guide (Goldring 1991).

16.3.2 Recurrence of Association

Assemblages may be found to recur at intervals through the geological record, and as such may be considered as: (1) accumulations of specific skeletal components, usually by physical sedimentological processes; or (2) approximations to the original community structure, with a recurrent set of organisms that reflect the prevailing environmental parameters.

Skeletal accumulations or 'shell beds' are primarily a function of sedimentological processes such as accumulation or winnowing by prevailing currents and reduced sedimentation rates. Such shell beds are composed of highly autochthonous fossils and although they can provide valuable information about palaeocurrent directions, for example, they can give little direct evidence about community structure and limiting factors prevailing at the time of death (e.g. Kidwell 1991). By contrast, fossil assemblages which approximate to the original community structure are clearly valuable in determining limiting factors. Typically, such assemblages have a recurrent association of trace and body fossil taxa, and they may also have a consistent association with particular lithofacies. In these cases, it is the frequency of their reappearance in the fossil record which is illustrative of their value in determining the environmental parameters that limited them.

16.3.3 Determination of Population Structure

Limiting factors commonly control the density (abundance) and diversity of particular assemblages. The interpretation of relative abundance and diversity in fossil assemblages and associations can help assess the nature of the palaeoenvironment.

Diversity is a valuable indicator of the nature and hostility of an environment, and together with the density or abundance of individuals of a particular species, it is an important palaeoecological tool (Dodd & Stanton 1990; Figure 16.14). Generally, environments of a relatively 'hostile' nature, such as with high or low salinities, or low oxygen, are characterized by an assemblage that is low in diversity, often being monospecific. Conversely, the abundance of individuals may be high. In these stressed environments, most taxa are excluded, including most predators, so that the available resources are used by a limited number of successful species. This leads to the development of the low diversity but high density of the assemblage. Often, the colonization of such habitats may be rapid, the species concerned being opportunists (Levinton 1969; Figure 16.14).

In less limited environments, such as normal marine salinities and fully oxygenated conditions for example, the diversity of species may be high, but the abundance of individuals may be correspondingly lower. This reflects the partitioning of the same resources available to the low-diversity assemblage between a greater number of species, with a corresponding increase in predators. In such environments, although there are many species producing young, density is kept low through the competition for resources, and by the action of the predators. Species in these environments are known as equilibrium species, maintained in equilibrium by the quantity of the resource (Levinton 1969; Figure 16.14).

Populations can also be analysed from the point of view of age and size frequency. Some assemblages may represent mass mortalities with young, juveniles and adults present in the same assemblage, as it would in a normal, living population. In such cases, it is possible to get a clearer idea about the nature of the

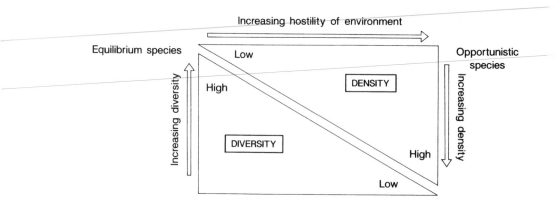

Figure 16.14 *The relationship of diversity and density with increasing environmental hostility. [Reproduced with permission from John Wiley & Sons, from: Doyle (1996)]*

population and its success in its environment through analyses of the age and size distributions of individuals in the population. This can also be achieved without mass mortality in trilobite and ostracod assemblages because of the presence of the moult stages (Dodd & Stanton 1990; Goldring 1991).

16.3.4 Comparison with Facies and Other Data Sources

Fossils are just part of the story, as the sediments which contain them provide as many clues to the nature of the environment as the fossils themselves (Figure 16.1). It is therefore good practice to consider fossils and sediments together in any palaeoenvironmental study (Chapter 14). Clearly, this is most appropriate with benthonic organisms associated with particular substrates, and it should be expected that fossil communities would be associated with specific substrates. A good example is the association of rock-boring bivalves, cementing bivalves and barnacles with the hard substrate rocky shoreline; a repetitive association for much of the Mesozoic and Cenozoic (Doyle *et al.* 1997; Figure 16.13).

In addition, isotope analyses of shell chemistry are appropriate in determining the ratio of stable isotopes of carbon, and oxygen in the skeletal carbonates (Figure 16.1). As these will have been precipitated in equilibrium with the water column, it is possible to determine enrichments in particular isotopes which are clear indicators of temperature and salinity variations (e.g. Tan & Hudson 1974; Fürsich 1994).

16.4 PALAEOENVIRONMENTAL CASE STUDIES

Two case studies, based for the most part on specific papers, are presented in this section in order to demonstrate the palaeoenvironmental potential of the fossil record. Both case studies demonstrate the ability of the taxonomic uniformitarian approach in the determination of semi-quantitative estimates of environmental conditions, particularly salinity and oxygen levels.

16.4.1 Interpreting Ancient Salinity Levels

Most marine and freshwater organisms are limited in their abundance and distribution by salinity. Measured in parts per thousand (‰), normal marine salinities are around 35‰ while freshwaters are between 0 and 0.5‰ (Figure 16.10). Surprisingly, although environments in which levels of salinity fluctuated in the geological past must have been common, relatively little is known about the nature of their fauna.

Salinity levels are prone to fluctuation, and sedimentary facies associations suggest that there have been at least two settings where this has been especially the case in the geological past:

1. in largely land-locked marine basins separated from the main ocean circulation through tectonic or other events; and
2. in the boundary between marine and terrestrial environments, at the coast.

Although recognition of the former is possible through the association of sedimentary bodies such as evaporites, the more subtle salinity fluctuations of the coastal region are difficult to assess directly from sedimentary facies, and are based largely upon the interpretation of the ecological tolerances of the contained fossils. The recognition of salinity-controlled benthonic assemblages is therefore of great value in the interpretation of the nature of the ancient sedimentary environment. An excellent review of the palaeoecology and evolution of salinity-controlled fossil assemblages is given in Fürsich (1994), but one of the most important preliminary studies in this field remains that of Hudson (1963).

The Middle Jurassic Great Estuarine Group

In 1963 John Hudson completed a study of the palaeoecology and sedimentology of the Great Estuarine Group, a succession of Middle Jurassic limestones, sandstones and shales which crop out in the Isle of Skye in north-west Scotland. As illustrated by its name, the Great Estuarine Group had long been considered to have been deposited under non-marine or brackish water conditions. The limestone and shale formations that characterize the group have lithologies which are laterally persistent, but which vary rapidly from bed to bed up the succession; their contained fauna demonstrates the same characteristics. These formations also show evidence of shallow-water conditions, with mud-cracks providing evidence for periodic desiccation of the sediments, and the most likely interpretation of the environment is that of a series of lagoons partially separated by a bar from the open sea. Sandstones represent the periodic growth of deltas into the lagoons. Hudson (1963) compared the Middle Jurassic environment of the Inner Hebrides with the present-day Texas coast, where a series of lagoons with brackish water conditions have developed.

General characteristics of the fauna

The Great Estuarine Group is dominated largely by molluscan fossils. For the most part, Hudson found that these fossils occur in assemblages of semi-autochthonous shells. They were rarely found in life position, and the valves were mostly disarticulated and a little sorted by wave action. However, Hudson considered that these shells were not fully allochthonous, and that they are broadly representative of the local environment, because: (1) many were found in recurring associations; (2) there was little evidence of strong current activity in the surrounding shale beds; and (3) many of the assemblages were completely different even though separated by a few centimetres of sediment.

Characteristics of recent brackish and Great Estuarine Group faunas

The most striking characteristics of modern brackish faunas, such as those of the Texas lagoons, are that their diversity is reduced, although individuals may occur in great numbers (i.e. with a high density), and that they are composed mostly of marine species. A series of criteria for the recognition of salinity-controlled assemblages, based upon the observation of modern brackish water faunas, was reviewed by Hudson (1963, 1980, 1990), and more recently by Fürsich (1994). Hudson recognized that salinity-controlled assemblages would have the following characteristics:

1. low-diversity faunas, often forming dense shell beds;
2. lacking in stenohaline organisms which have a narrow salinity tolerance range, such as corals, cephalopods, echinoderms and most bryozoans;
3. dominated by euryhaline organisms with a broad salinity range, such as bivalves, gastropods and ostracods; and
4. associated with marine and freshwater sediments demonstrating the facies relationships.

Hudson (1963) found that the Great Estuarine Group fauna fitted these criteria (Figure 16.15):

1. the total number of fossil species, of all types, was 50. This was extremely poor in comparison with contemporaneous faunas from fully marine limestones in England, which have over 400 molluscan species alone;
2. corals and cephalopods are unknown, and other stenohaline forms (echinoid and bryozoan fragments) are limited to one small part of the Great Estuarine Group;
3. bivalves, gastropods and ostracods dominate the fauna; and
4. marine beds occur above and below the Great Estuarine Group; one or two freshwater beds occur within it, and drifted plant remains are common. No evaporites, indicative of hypersaline conditions, are recorded.

Hudson (1963) was able to recognize a range of salinity-controlled assemblages within the Great Estuarine Group, based on the known tolerance levels of living organisms in similar environments, such as in the coastal lagoons of Texas. These assemblages are composed of molluscan groups, as follows (Figure 16.15):

1. *Unio, Viviparis, Neomiodon* and *Euestheria*. *Unio, Viviparis* and *Euestheria* live in freshwaters today, although the first two can tolerate low salinities in modern estuarine settings. *Neomiodon* is extinct, and may have been euryhaline, as it occurs with both freshwater and marine fossils. Considered to be typical of fresh to brackish-freshwater (oligiohaline) conditions in the range of 0–5‰;
2. *Liostraea* and *Mytilus* commonly occur in monotypic shell beds, resembling living oyster and mussel beds. These were considered by Hudson to be representative of brackish-marine, mesohaline faunas, with a salinity range of 9–16.5‰;

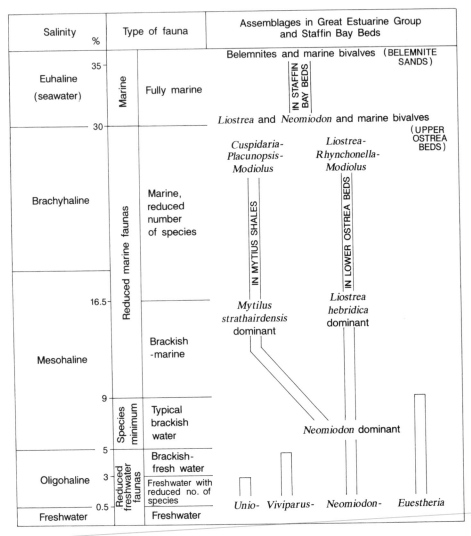

Figure 16.15   *Salinity-controlled assemblages from the Great Estuarine Group, Scotland.*
[Modified from: Hudson (1963)]

3. *Liostrea* and *Mytilus* also occur with fully marine organisms such as rhynchonel-
   lid brachiopods. Hudson considered this assemblage to be representative of
   reduced-marine salinities, brachyhaline faunas, with a salinity range of
   16.5–30‰;
4. cephalopods occur in the sandstones above the Great Estuarine Group, and are
   demonstrative of fully marine, euhaline conditions in the range of 30–35‰.

These results were later substantiated through carbon and oxygen stable isotope
analyses of the skeletal carbonates (Tan & Hudson 1974).

Conclusion

This example demonstrates the principles of palaeoenvironmental analysis. The assemblages initially recognized by Hudson were considered in terms of their autochthony and occurrence. The environmental parameters limiting the diversity and abundance of the molluscs were deduced through taxonomic uniformitarian comparison with living relatives. Sedimentary associations and isotopic data provided corroborating evidence. This kind of approach is now taken to be an important method for the determination of salinity levels in ancient successions, and has been successfully used by other authors (e.g. Fürsich 1981; Fürsich & Werner 1984, 1986) as well as in later studies by Hudson and his co-workers (Hudson *et al.* 1995).

16.4.2 Palaeo-oxygenation Studies

Black shales are relatively common in the geological record, and consist of finely laminated shales commonly containing organic carbon of up to 1–20% by weight (Wignall 1994). As such, these shales are economically important, and it has been estimated that around 70% of the world's mineral oil resources are derived from Mesozoic black shales. Black shales develop in oxygen-deficient basins, where the water column becomes stratified, commonly with surface waters having near-normal oxygen saturation levels of 6–8.5 ml O_2/l (millilitres of oxygen per litre of water), and bottom waters which may be oxygen deficient, with levels ranging from 1 to 0 ml/l of dissolved oxygen. It is generally agreed that stratification occurs because of differences in temperature or salinity between the lower and upper levels of the water column, generally producing high-density bottom waters and low-density surface waters. Mixing may occur periodically.

Understanding the levels of oxygenation in ancient sedimentary basins is important for many reasons. First, the economic importance of black shales is such that a clearer understanding of the conditions that formed them is important. Secondly, we now know that the early development of life on earth was in low oxygen conditions, and therefore a study of the interaction of organisms with their environment through geological time is clearly of the greatest value to our appreciation of the evolution of life.

Palaeo-oxygenation models

One of the most important contributions to our understanding of the level of palaeo-oxygenation in ancient sedimentary basins was made by Donald Rhoads and John Morse in 1971 (Figure 16.9). Rhoads & Morse (1971) used taxonomic uniformitarian models to interpret the level of oxygen saturation in bottom waters of the early Cambrian, and determine its bearing on the early diversification of the fossil metazoans. Rhoads and Morse set out to study the relationship of organism diversity with oxygen levels in two basins known to have low levels of dissolved oxygen: the Black Sea, and the continental borderland basins of southern California. Both basins are enclosed, with little connection with well-oxygenated oceanic waters. A third basin, the Gulf of California, was also examined; here, although there is open connection with oxygenated surface waters, bottom waters were found to have low oxygen levels. In all three examples, a clear diversity decline was noted with increasing depth and decreasing levels of dissolved oxygen. The greatest diversities were found to be associated with oxygen levels in excess of 1.0 ml O_2/l, and water depths of less than 150 m. In general, organisms with heavily calcified shells were not found in abundance in oxygen levels less than 1.0 ml O_2/l, with poorly calcified organisms and soft bodied taxa occurring at much lower oxygen levels (Figure 16.9).

From their studies in these Recent environments, Rhoads & Morse (1971) concluded that:

1. diversity declines sharply as oxygen levels fall below 1.0–0.5 ml O_2/l;
2. soft-bodied organisms are the commonest benthonic organisms below 1.0 ml O_2/l; and
3. heavily calcified taxa are largely restricted to oxygen levels of greater than 1.0 ml O_2/l.

They recognized three biofacies, that is broad assemblages of taxa which can be used to interpret the nature of the ancient environment (Figure 16.9):

1. the aerobic biofacies, comprising a diverse, heavily calcified fauna in well-bioturbated sediments;
2. the dysaerobic biofacies, comprising soft-bodied, burrowing species in bioturbated sediments; and
3. the anaerobic biofacies, comprising well-laminated sediments apparently devoid of life.

The disappearance of heavily calcified taxa in bottom waters with less than 1.0 ml O_2/l of dissolved oxygen was interpreted as a result of the stress in the production and retention of a shell in such low-oxygen, high CO_2 and lowered pH conditions. Rhoads & Morse (1971) were able to demonstrate that the three biofacies recognizable in modern environments are commonly represented in the fossil record.

Trace fossils and palaeo-oxygenation

The Rhoads and Morse model relies upon the recognition of the disappearance of shelly fossils and well-bioturbated sediments (aerobic to dysaerobic facies), and the transition from bioturbated to laminated sediments (dysaerobic to anaerobic facies). Work by Savrda & Bottjer (1986, 1987) confirmed these broad conclusions, but they were able to take the level of precision in the recognition of oxygen-controlled biofacies still further through the study of the diversity of trace fossil assemblages in black shale successions.

Savrda & Bottjer (1986) used three indices of oxygenation based upon trace fossils (Figure 16.16): (1) diversity of the trace assemblage; (2) maximum burrow

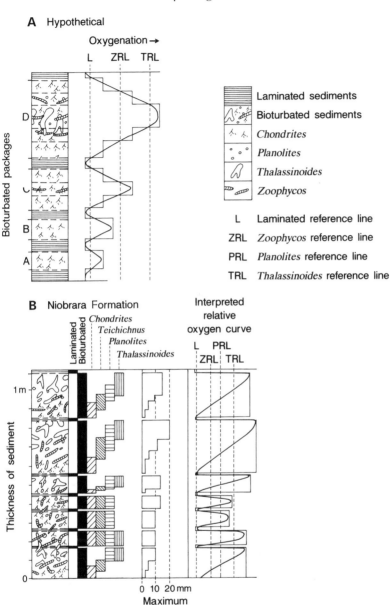

Figure 16.16 *Relative oxygenation curves derived from trace fossil data (diversity and burrow diameter).* **A.** *Hypothetical model showing the construction of the oxygenation curve.* **B.** *Example from the Cretaceous Niobrara Formation, California.* [*Modified from: Savrda & Bottjer (1986)*]

diameter; and (3) tiering of the trace fauna, that is the relationship of trace fossils with the sedimentary succession and each other. In general, low-oxygen conditions are associated with low-diversity trace fossil assemblages, and it has been demonstrated that the feeding burrow *Chondrites* is usually the first to colonize laminated black shale sequences (Bromley & Ekdale 1984). Burrow diameter is taken as an indicator of oxygen levels primarily because, as recognized from recent successions, as the concentration of dissolved oxygen in bottom water decreases, so does the relative size of benthonic organisms. Clearly, large burrows are indicative of larger organisms and therefore increased oxygen levels. Tiering – the vertical stacking of traces within the sediment – reflects the successive colonization of a substratum by a group of organisms. It is represented by the cross-cutting relationships of traces as organisms burrow deeper in to the sediment in search of food or dwelling space. Put simply, in low-oxygen conditions, there will be few tiers, or levels of colonization, while in increased oxygen conditions, the number of tiers will increase, with organisms mining successively lower levels, and cross cutting the traces already created. In general, as the oxygen level decreases, so ichnogenera feeding at successively deeper levels in the sediment progressively disappear: shallow, large burrow diameter *Planolites* or *Thalassinoides* go first followed by complex feeding burrows such as *Zoophycos* and, finally, by the deeper burrowing, feeding trace *Chondrites*.

Taken together, these lines of evidence can be used to interpret not only the ancient levels of oxygen, but also the colonization history of the basin with respect to increased or decreased oxygen levels, in the creation of relative oxygenation curves.

Relative oxygenation curves

Using trace fossil assemblages, Savrda & Bottjer (1986) were able to reconstruct the pattern of oxygenation in two basins: the Miocene Monterey Formation in California, and the Cretaceous Niobrara Formation in Colorado (Figure 16.16). Successive levels of oxygenation were demonstrated graphically using first appearances of three ichnotaxa: *Planolites*, *Thalassinoides* and *Zoophycos*, and the presence or absence of laminated facies. *Chondrites* was found to occur in all facies except the laminated beds. These first appearances, plotted as lines on the graph, represent thresholds of oxygen above which each of these traces could be produced within the fabric of the sediment. This was correlated with burrow diameter, and a curve of oxygenation created to demonstrate the oxygenation history of the basins (Figure 16.16).

Conclusion

This case study demonstrates the application of the taxonomic uniformitarian principle in palaeoenvironmental studies, and the particular value of trace fossils. Trace fossils are rarely anything other than autochthonous. Rhoads & Morse' (1971) original study identified the critical threshold of oxygen with biofacies, or assemblages directly related to levels of oxygen. Savrda & Bottjer (1986) have emonstrated that these assemblages are recurrent throughout the geological cord. Detailed study of modern analogues has shown that the number of stacked ers, burrow diameter and diversity are directly related with oxygen levels, and erefore has led to the acceptance that it is possible to determine oxygenation ents based upon the nature of the colonization history of the successive traces. ater work by Savrda & Bottjer (1987, 1991) and Bottjer & Savrda (1990) has ecognized the existence of other oxygen-related biofacies, and the technique first etermined by them using trace fossils has been used to advantage by a number of ithors in the determination of ancient oxygen levels using both body and trace ssils (e.g. Doyle & Whitham 1991).

16.5 SUMMARY

alaeoenvironmental analysis is the interpretation of ancient environments rough the study of ancient sediments and their fossils. Sedimentary facies covides some of the information, but alone it can only go so far. Fossils, as nce-living animals and plants, had a wide range of ecological tolerances, and an nderstanding of the palaeoecology of these ancient organisms can give a valuable sight into the nature of ancient environments.

Palaeoenvironmental analysis from fossils is based on the principle of tax-nomic uniformitarianism, which assumes that the study of present-day organ-ms holds the key to interpreting the life history and ecological tolerances of xtinct species. It is a principle which holds for the recent past and which decreases validity with advancing age. Taxonomic uniformitarianism is the basis for alaeoecology, the study of the ecological relationships of fossil organisms, both rith each other and with their environment. There are two basic components to alaeoecology: (1) palaeosynecology, the interaction of groups of organisms with ach other and with their environment; and (2) palaeoautecology, the relationship f individual organisms with their environment. Both have the potential to pro-ide information, through taxonomic uniformitarianism, which will enhance the nesse of our interpretation of the nature of ancient environments.

REFERENCES

ger, D.J. 1963. *Principles of paleoecology*. McGraw-Hill, New York.

engston, S. 1991. Odd-balls from the Cambrian start to get even. *Nature,* **351**, 184–185.

osence, D.W.J. & Allison, P.A. (eds) 1995. *Marine palaeoenvironmental analysis from fossils.* Geological Society Special Publication No. 83.

ottjer, D.J. & Savrda, C.E. 1990. Oxygen levels from biofacies and trace fossils. In Briggs, D.E.G. & Crowther, P.R. (eds) *Palaeobiology – a synthesis.* Blackwell Scientific Publications, Oxford, 408–410.

ottjer, D.J., Campbell, K.A., Schubert, J.K. & Droser, M.L. 1995. Palaeoecological models, non–uniformitarianism, and tracking the changing ecology of the past. In Bosence, D.W.J. & Allison, P.A. (eds) *Marine palaeoenvironmental analysis from fossils.* Geological Society Special Publication No. 83, 7–26.

oucott, A.J. 1981. *Principles of benthic marine paleoecology.* Academic Press, New York.

Boucot, A.J. 1990. Evolution of communities. In Briggs, D.E.G. & Crowther, P.R. (eds) *Palaeobiology – a synthesis*. Blackwell Scientific Publications, Oxford, 391–394.

Brenchley, P.J. (ed.) 1984. *Fossils and climate*. John Wiley, Chichester.

Brenchley, P.J. 1990. Biofacies. In Briggs, D.E.G. & Crowther, P.R. (eds) *Palaeobiology – synthesis*. Blackwell Scientific Publications, Oxford, 395–400.

Briggs, D.E.G. & Crowther, P.R. (eds) *Palaeobiology – a synthesis*. Blackwell Scientific Publications, Oxford.

Bromley, R.G. 1990. *Trace fossils, biology and taphonomy*. Unwin Hyman, London.

Bromley, R.G. & Ekdale, A.A. 1984. *Chondrites*: a trace fossil indicator of anoxia in sediments. *Science*, **224**, 872–874.

Campbell, K.A., Bottjer, D.J. & Carlson, C. 1993. Fossil cold seep limestones and associated chemosymbiotic macroinvertebrate faunas, Jurassic–Cretaceous Great Valley Group California. In Graham, S.A. & Lowe, D.R. (eds) *Advances in the sedimentary geology of the Great Valley Group, Sacramento Valley*. Pacific Section SEPM, 73, 37–50.

Cavanaugh, C.M. 1985. Symbiosis of chemoautotrophic bacteria and marine invertebrate from hydrothermal vents and reducing sediments. *Bulletin of the Biological Society of Washington*, **6**, 373–388.

Craig, G.Y. 1966. Concepts in palaeoecology. *Earth Science Reviews*, **2**, 127–155.

Crimes, T.P. & Anderson, M.M. 1985. Trace fossils from Late Precambrian–Early Cambrian strata of southeastern Newfoundland (Canada): temporal and environmental considerations. *Journal of Paleontology*, **59**, 310–343.

Dodd, J.R. & Stanton, R.J. 1990. *Paleoecology, concepts and applications*, 2nd edition. John Wiley, New York.

Doyle, P. 1996. *Understanding fossils. An introduction to invertebrate palaeontology*. John Wiley Chichester.

Doyle, P. & Whitham, A.G. 1991. Palaeoenvironments of the Nordenskjöld Formation: an Antarctic Late Jurassic–Early Cretaceous black shale–tuff sequence. In Tyson, R.V. & Pearson, T.H. (eds) *Modern and ancient continental shelf anoxia*. Geological Society Special Publication No. 58, 397–414.

Doyle, P., Mather, A.E., Bennett, M.R. & Bussell, M.A. 1997. Barnacle assemblages from the Miocene of southern Spain and their palaeoenvironmental significance. *Lethaia*, **29** 267–274.

Forbes, E. 1844. Report on the Mollusca and Radiata of the Aegean Sea, and on their distribution, considered as bearing on geology. *Reports of the British Association for the Advancement of Science*, **1843**, 130–193.

Frey, R.W. (ed.) 1975. *The study of trace fossils. A synthesis of principles, problems and procedures in ichnology*. Springer-Verlag, Berlin.

Frey, R.W. & Seilacher, A. 1980. Uniformity in marine invertebrate ichnology. *Lethaia*, **13** 183–207.

Fürsich, F.T. 1981. Salinity-controlled benthic associations from the Upper Jurassic of Portugal. *Lethaia*, **14**, 203–223.

Fürsich, F.T. 1994. Palaeoecology and evolution of Mesozoic salinity-controlled benthic macroinvertebrate associations. *Lethaia*, **26**, 327–346.

Fürsich, F.T. & Werner, W. 1984. Salinity zonation of benthic associations in the Upper Jurassic of the Lusitanian Basin (Portugal). *Geobios, Mémoire Spéciale*, **8**, 85–92.

Fürsich, F.T. & Werner, W. 1986. Benthic associations and their environmental significance in the Lusitanian Basin (Upper jurassic, Portugal). *Neues Jahrbuch für Geologie und Paläontologie, Abhandlungen*, **172**, 271–329.

Goldring, R. 1991. *Fossils in the field. Information potential and analysis*. Longman, Harlow.

Goldring, R. 1995. Organisms and the substrate: response and effect. In Bosence, D.W.J. & Allison, P.A. (eds) *Marine palaeoenvironmental analysis from fossils*. Geological Society Special Publication No. 83, 151–180.

Gray, J., Boucot, A.J. & Berry, W.B.N. (eds) 1981. *Communities of the past*. Hutchinson Ross Stroudsberg.

Hudson, J.D. 1963. The recognition of salinity-controlled mollusc assemblages in the Great

Estuarine Series (Middle Jurassic) of the Inner Hebrides. *Palaeontology*, **6**, 318–326.

Hudson, J.D. 1980. Aspects of brackish-water facies and faunas from the Jurassic of north–west Scotland. *Proceedings of the Geologists' Association*, **91**, 99–105.

Hudson, J.D. 1990. Salinity from faunal analysis and geochemistry. In Briggs, D.E.G. & Crowther, P.R. (eds) *Palaeobiology – a synthesis*. Blackwell Scientific Publications, Oxford, 406–408.

Hudson, J.D., Clements, R.G., Riding, J.B., Wakefield, M.I. & Walton, W. 1995. Jurassic paleosalinities and brackish-water communities – a case study. *Palaios*, **10**, 392–407.

Kanazawa, K. 1992. Adaptation of test shape for burrowing and locomotion in spatangoid echinoids. *Palaeontology*, **35**, 733–750.

Kidwell, S.M. 1991. Taphonomic feedback (live/dead interactions) in the genesis of bioclastic beds: keys to reconstructing sedimentary dynamics. In Einsele, G., Ricken, W. & Seilacher, A. (eds) *Cycles and events in stratigraphy*. Springer, Berlin, 268–282.

Levinton, J.S. 1969. The paleoecological significance of opportunistic species. *Lethaia*, **3**, 69–78.

Linsley, R.M., Yochelson, E.L. & Rohr, D.M. 1978. A reinterpretation of the mode of life of some Palaeozoic frilled gastropods. *Lethaia*, **11**, 105–112.

Lockley, M.G. 1983. A review of brachiopod dominated palaeocommunities from the Type Ordovician. *Palaeontology*, **26**, 111–145.

McKerrow, W.S. (ed.) 1978. *The ecology of fossils: an illustrated guide*. Duckworth, London.

Middlemiss, F.A. 1962. Brachiopod ecology and Lower Greensand palaeogeography. *Palaeontology*, **5**, 253–267.

Pemberton, S.G., Frey, R.W. & Saunders, T.D.A. 1990. Trace fossils. In Briggs, D.E.G. & Crowther, P.R. (eds) *Palaeobiology – a synthesis*. Blackwell Scientific Publications, Oxford, 355–362.

Perrin, C., Bosence, D. & Rosen, B. 1995. Quantitative approaches to palaeozonation and palaeobathymetry of corals and coralline algae in Cenozoic reefs. In Bosence, D.W.J. & Allison, P.A. (eds) *Marine palaeoenvironmental analysis from fossils*. Geological Society Special Publication No. 83, 181–229.

Petersen, C.G.J. 1915. On the animal communities of the sea bottom in the Skagerrak, the Christiana Fjord and the Danish waters. *Report of the Danish Biological Station*, **23**, 3–28.

Rhoads, D.C. & Morse, J.W. 1971. Evolutionary and ecologic significance of oxygen-deficient marine basins. *Lethaia*, **4**, 413–428.

Savrda, C.E. & Bottjer, D.J. 1986. Trace fossil model for the reconstruction of paleoxygenation in bottom waters. *Geology*, **14**, 1–6.

Savrda, C.E. & Bottjer, D.J. 1987. The exaerobic zone, a new oxygen-deficient marine biofacies. *Nature*, **327**, 54–56.

Savrda, C.E. & Bottjer, D.J. 1991. Oxygen-related biofacies in marine strata: an overview and update. In Tyson, R.V. & Pearson, T.H. (eds) *Modern and ancient continental shelf anoxia*. Geological Society Special Publication No. 58, 397–414.

Schäfer, W. 1972. *Ecology and palaeoecology of marine environments*. Oliver & Boyd, Edinburgh.

Scott, R.W. 1974. Bay and shoreface benthic commuities in the Lower Cretaceous. *Lethaia*, **7**, 315–330.

Scott, R.W. 1978. Approaches to trophic analysis of paleocommunities. *Lethaia*, **11**, 1–14.

Seilacher, A. 1967. Bathymetry of trace fossils. *Marine Geology*, **5**, 413–428.

Spicer, R.A. 1990. Climate from plants. In Briggs, D.E.G. & Crowther, P.R. (eds) *Palaeobiology – a synthesis*. Blackwell Scientific Publications, Oxford, 401–403.

Tan, F.C. & Hudson, J.D. 1974. Isotopic studies on the palaeoecology and diagenesis of the Great Estuarine Series (Jurassic) of Scotland. *Scottish Journal of Geology*, **10**, 91–128.

Westermann, G.E.G. 1972. Strength of concave septa and depth limits of fossil cephalopods. *Lethaia*, **6**, 383–403.

Wignall, P.B. 1994. *Black Shales*. Clarendon Press, Oxford.

Ziegler, A.M. 1965. Silurian marine communities and their environmental significance. *Nature*, **207**, 270–272.

Ziegler. B. 1983. *Introduction to palaeobiology. General palaeontology*. Ellis Horwood, Chichester.

17
Interpreting Palaeoclimates

Jane E. Francis

Palaeoclimatology, the study of climates during the geological past, is one of the most topical areas of research in the geosciences at present. The threat of future climate change caused by higher levels of greenhouse gases, which would drastically alter many aspects of our environment, has prompted much research to try to understand how our complex climate system works. Only by understanding how climate has evolved over million of years can we identify important climate cycles with a frequency in excess of the short climate records we possess. These climate cycles have the potential to have a profound effect on our environment.

Understanding our climate history in the geological past is also important for climatologists trying to construct accurate numerical computer models of our present climate system to use for predicting future climate change. It is obviously not possible to check the accuracy of models that are predicting the future so climatologists must turn to the past to see if their models can accurately simulate ancient climates. It is therefore the role of the geoscientist to collect as many data as possible from the rock record to provide the hard evidence with which to test the models and set the limits (the boundary conditions) within which the modeller must work. Once this information is collated it has great value as a tool for the prediction of the distribution of economically important resources whose formation is controlled by climate, such as oil, bauxites and coal, and may help explain dramatic events in the past, such as mass extinctions.

The techniques of palaeoclimatic interpretation from the stratigraphical record are varied. This chapter aims to introduce the principal sources of information on past climates and gives some examples of how they have been used to determine ancient climate conditions. It is important to understand the techniques of palaeo-

Unlocking the Stratigraphical Record: Advances in Modern Stratigraphy. Edited by P. Doyle and M.R. Bennett.
© 1998 John Wiley & Sons Ltd.

climate interpretation in order to evaluate the validity of the data obtained from them. One single source of palaeoclimate data, although interesting in itself, is not sufficient for reconstructing robust palaeoclimate scenarios, so for every case study as many sources of data as possible should be assembled.

17.1 INTERPRETING PALAEOCLIMATES

Basic information about past climates comes from understanding how climate influences certain sedimentary systems, floras and faunas on earth today and extrapolating this information back to interpret geological evidence. A uniformitarian approach is used in most cases, that is assuming environments in the past were influenced by the same parameters as at present. However, it is becoming increasingly apparent that some ancient environments cannot be found on earth today, such as the presence of warmth-loving vegetation and animals living near the poles. In these cases it is vital to carefully interpret all potential sources of environmental information to reconstruct these unique situations from primary data.

Several books and papers deal with palaeoclimatology in detail (e.g. Briden & Irving 1964; Frakes 1979; Parrish *et al.* 1982; Crowley & North 1991; Frakes *et al.* 1992) so only a brief resume is presented here.

17.1.1 Climatically Sensitive Sediments

The formation of some rock types is directly influenced by aspects of climate. Some of the most useful are coals, evaporites, glacial deposits and carbonates.

Coal

The presence of coal, initially formed from the accumulation of plant material as peat, is generally taken to indicate warm wet humid climates ideal for lush plant growth, and where the rainfall is higher than the rate of evaporation, such as in equatorial regions. However, rainfall is a more important factor than temperature, as are high water-tables and waterlogged swamps (mires) which are required to preserve the peat. Large volumes of peat also form today in high latitudes, such as the Canadian and Siberian tundra regions, because although the cool climates restrict the rate of plant growth, the cool temperatures slow the decay processes, allowing peat to accumulate. In addition, basin topography and the tectonic setting have to be suitable (McCabe 1984; McCabe & Parrish 1992).

In the past, the most abundant coal deposits were formed during the Carboniferous when large subsiding continental areas were situated in low latitudes and experienced hot and humid climates. The great Carboniferous forests were composed of the pithy-stemmed clubmosses and lycopods, such as *Lepidodendron*, *Sigillaria* and *Calamites*, which grew to giant sizes in the hot wet conditions and formed thick layers of peat as they collapsed into waterlogged swamps. The disappearance or decrease in size of these water-loving plants at the end of the

Carboniferous marked the onset of much drier conditions in low-latitude regions during the Permian. However, in much higher southern latitudes of Gondwana during the Permian the climate remained cool and wet as the major Palaeozoic ice sheets melted, and a different suite of plants thrived in cool swampy conditions. Extensive forests dominated by glossopterid plants lived on all southern continents, even within the southern polar region (Francis *et al.* 1994) and their remains form extensive and some economically important coal deposits today (Figure 17.1).

Carbonates

In the marine realm, carbonate sediments are often used as indicators of warm

Figure 17.1 *Coal seam of Permian age from the Weller Coal Measures, Allan Hills, Transantarctic Mountains. The coal is composed mostly of glossopterid leaves and woody stems. The seam is approximately 3 m thick. Permian coal in Antarctica is not of economic importance*

ocean waters. Carbonate sediments of Bahamian type (including reef-building hermatypic corals, some algae and ooids) are important indicators of warm marine seas. At present, such deposits are limited to tropical and subtropical latitudes up to 30 °N and 30 °S. Favourable light levels, a function of latitude, are an important control, as is temperature (Ziegler *et al.* 1984). Reefs are also used as indicators of warm seas, the modern Australian Barrier Reef often being used as a classic comparison. However, many fossil reefs are composed of different communities of organisms which may have had different ecological tolerances in the past. For example, in the Cretaceous, reefs were dominated by rudist bivalves, whereas in Permian and Triassic times various types of algae and sponges formed reefs (Flügel 1994; Sellwood & Price 1994).

A different suite of carbonates also form today in cool temperate waters. These are composed of benthic foraminifera, red algae, barnacles, molluscs and bryozoa. These carbonate deposits form in much higher latitudes under cooler conditions. Identification of the carbonate constituents is therefore important to distinguish between cool-water and warm-water carbonates for palaeoclimate interpretation (James & Bone 1991).

Evaporites

Evaporites, such as anhydrite, gypsum and halite, are used as a guide to aridity in the past. Their formation requires evaporation rates to exceed precipitation, at least seasonally, and to exceed water inflow into the evaporating basin (Frakes 1979). At present, marine coastal evaporites are found in latitudes 8–45°N and 5–27°S and non-marine evaporites incontinental interiors up to 50° latitude (Parrish *et al.* 1982). However, it is apparent from palaeoclimatic studies that the distribution of evaporites (and coals) has not been strictly zonal (parallel to latitudes) in the past but related more closely to the location of atmospheric pressure cells in relation to ancient palaeogeographies.

For example, the Triassic was shown by Gordon (1975) to have been the time of greatest formation of evaporites. At this time all landmasses were joined together into the supercontinent of Pangea and the size and position of this huge landmass profoundly influenced global climates. The seasonal formation of intense high and low pressure cells over mid-latitude land areas (monsoonal circulation) would have resulted in seasonal air flows that would have simply blown over dry land and crossed the equator without collecting moisture from the ocean. This would have left continental interiors very dry. Seasonal rainfall would have fallen only along the coasts of Tethys and in western high mid latitudes (Parrish 1993). Many Triassic evaporite deposits were formed on the shores of Tethys in shallow coastal lagoons that provided the ideal setting for evaporite formation (Parrish *et al.* 1982).

Glacial deposits

Evidence for glaciation and the presence of thick ice sheets can be obtained from a variety of sources. The most convincing are striated pavements, that is surfaces of bedrock with grooves scratched by debris frozen into the base of moving ice glaciers (Figure 17.2). The orientation of ice movement and therefore in some cases the position of glacial centres can also be determined. For example, studies of Carboniferous and Permian glacial deposits in the Southern Hemisphere enabled Crowell & Frakes (1975) and Visser (1993) to reconstruct the location of ice centres over Gondwanan continents and to determine the direction of movement of ice lobes from these centres. They showed that, for example, glacial deposits in South America were transported by glacial tongues that originated in Africa, and they were able to link glacial rocks as distant as India with ice centres in the rest of Gondwana. They also showed that the location of the main ice masses coincided with high-altitude cooling on mountain tops, created by active tectonic belts.

Glacial tillites can provide information about ice passage but, in the absence of other glacial features, tillites can sometimes be hard to distinguish from other diamictites, such as debris flow deposits, which may have formed under totally different conditions. Ice-rafted dropstones and varves indicate that ice formed, at least seasonally, and produced dumps of ice-carried debris or seasonal lake sediments (see Bennett *et al.* 1996). In addition, glendonite nodules (stellate aggregates of crystal pseudomorphs of the mineral ikaite which is stable only at low temperatures) have also been used as evidence for cold climates, particularly for the Permian from which sequences they were originally described (Brandley & Krause 1994; Sellwood & Price 1994).

Figure 17.2 *Glacial striations of Permian age on Precambrian bedrock, Hallett Cove, South Australia. These striations were formed as ice sheets moved across this region of Australia during the later stages of the great Gondwanan glaciation during the Late Palaeozoic*

Aeolian sediments and red beds

The distribution of red beds and aeolian sediments can also provide some indication of controlling climatic parameters. Interestingly, these two sediment types do not tend to plot parallel to latitudinal zones for much of the Phanerozoic (Briden & Irving 1964) but instead related more to palaeogeography and related climatic zones, much like coals and evaporites. Aeolian deposits can provide important information on prevailing wind directions. For example, the Upper Triassic Chinle Formation in south-western United States includes some aeolian sand sheets deposited on the Colorado Plateau during the Late Triassic. Climate models predict that summer low-pressure cells over this part of the vast Pangean continent would have been so intense that the resulting westerly winds would have been strong enough to have drawn moisture from adjacent western oceans into this region, completely reversing the normal easterly equatorial air flow. Studies of the aeolian deposits do indeed show that the wind direction was westerly to north-westerly at that time, supporting model predictions for such unusual conditions (Parrish & Peterson 1988; Dubiel *et al.* 1991).

Red beds were once considered as classic indicators of desert conditions. However, it is now believed that the main factor governing their formation (particularly reddened fluvial and alluvial sediments) is not solely aridity but the seasonal nature of rainfall. Alternating wet and dry periods govern the mobilization and precipitation of iron minerals. Therefore reddening can occur in a range of environments, from those which are generally arid with a short season of rainfall to those which are seasonally very wet (Van Houten 1982; Kraus & Bown 1986; Parrish 1993; Dubiel & Smoot 1994).

There are several other climatically sensitive sediments that have been used to determine climate. For example, the chemistry and structure of fossil soils can also yield information about ancient temperatures and aspects of precipitation and evaporation (Figure 17.3; Wright 1986). Certain clay minerals tend to form under specific climate settings (Chamley 1989). Bauxites are limited to tropical and subtropical settings with high rainfalls (Frakes 1979; Bardossy 1994).

17.1.2 Palaeontological Evidence

Fossil plants are the principal source of information about terrestrial climates. They can be used in several ways to determine climate (e.g. Creber & Chaloner 1984; Upchurch & Wolfe 1987; Chaloner & Creber 1990; Spicer 1990).

Fossil plant assemblages can be compared to similar living floras and the climate under which the modern floras live then extrapolated back into the past (the 'nearest living relative' approach; see Chapter 16). This is the most common method used to interpret climate from floras as it can give broad indications of climate regimes. However, it must be used with caution as it is often difficult to match fossil plants with direct living descendants. This method also assumes that floras lived under the same environmental conditions in the past as they do today, yet it is apparent that some fossil floras represent unique ecosystems that are not

Figure 17.3 *Fossil soil of Late Jurassic age, Lulworth Formation, Purbeck Group, southern England. This paleosol, called the Great Dirt Bed, is similar to a rendzina, a soil developed on carbonate bedrock. It consists of a carbonaceous upper A horizon on top of a C horizon of weathered limestone. The white clasts within the soil are fragments of calcrete, a secondary deposit formed by precipitation of dissolved carbonate influenced by a strongly seasonal climate (Francis 1984). Black lens cap for scale*

represented on earth at present; for example, there is no direct living equivalent of the Cretaceous, Tertiary or Permian forests that lived in high latitudes under the stressful polar light regime (Creber & Chaloner 1985; Francis 1990).

However, fossil plants can provide information about past growing conditions and climate that is independent of taxa but related more to physical properties of the plant and its response to aspects of climate (physiognomic analysis). For example, leaf size can be related to light levels, temperature and water availability, with large leaves dominating the understorey of humid forests and the smallest leaves occurring in dry or cold conditions. In addition, leaf shape and leaf margin analysis of angiosperms (flowering plants) can be used to determine temperatures, water/humidity and aridity (Wolfe 1985, 1993; Wolfe & Upchurch 1986, 1987; Spicer 1990). The abundance of smooth (entire) margined leaves compared to toothed margined leaves in a flora is strongly correlated to mean annual temperature in modern climates (leaves with smooth margins are more common in warm climates and toothed margined leaves are more prevalent in cool climates). This is a feature that can be relatively easily deduced from fossil angiosperm assemblages.

Physiognomic attributes of fossil leaves from a number of plant assemblages in North America were used by Wolfe & Upchurch (1987) to deduce mean annual temperatures for low to mid palaeolatitudes during the Cretaceous and Tertiary (Figure 17.4). Their measurements of leaf size variation and the percentage of entire-margined species showed that temperatures reached their warmest (about 25°C mean annual temperature) for the Cretaceous during the Coniacian–Santonian, cooled slightly during the latest Cretaceous, but then warmed again near the Cretaceous/Tertiary boundary. Leaf size in the Early Tertiary (Palaeocene) floras was much larger than that in the Cretaceous and may reflect increased precipitation at that time.

While this method is very valuable for providing estimates of several climate parameters (although preservational bias has to be taken into account), it is not applicable to conifer or other non-angiosperm plants, so this method is not suitable for floras older than the Cretaceous before which angiosperms had not evolved. In a more recent development using statistical techniques (Wolfe 1993), characteristics of fossil land plants can be analysed and directly related to environmental parameters in a more quantitative way (Kovach & Spicer 1996). In this method the statistical correspondence of certain modern features of vegetation to climate features are assessed and then similar significant features of fossil plant assemblages can be interpreted in terms of climate in some detail. By studying features of plant assemblages and leaf morphology and using methods mentioned above, Spicer & Parrish (1990) were able to obtain information about the palaeoclimates of northern Alaska during the Late Cretaceous. The floras consisted mainly of stream-side angiosperms with conifers, ferns and cycadophytes; most plants were probably deciduous, which would have helped them survive the long dark winter months at these high latitudes. Floral analysis provided estimates of mean annual temperatures for the Cenomanian of about 10°C at 75–85 °N, rising to 13°C in the Coniacian, and dropping to 5°C in the Maastrichtian (Spicer & Parrish 1990).

Other features of fossil leaves that can be related to climate include the nature of the leaf cuticle; for example, thick waxy cuticles are common in plants that tolerate

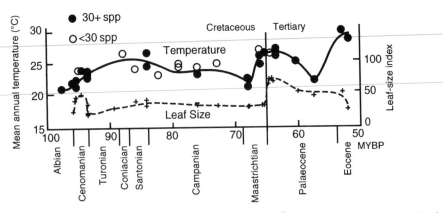

Figure 17.4 *Late Cretaceous and Tertiary palaeotemperatures determined from average leaf size and percentage of entire-margined species in fossil plant assemblages from low to mid palaeolatitudes of North America. [Modified from: Upchurch & Wolfe (1987)]*

arid hot climates and thin leaves are typical of deciduous, cool climates. The density of stomata on fossil leaves can now also be used as a tool to assess climate, especially atmospheric carbon dioxide concentrations in the geological past (McElwain & Chaloner 1996).

The anatomical structure of fossil wood is also an important store of palaeoclimate data and can yield detailed information about annual climates and even about conditions during specific parts of a year. Growth rings in fossil wood record patterns of annual growth because the formation of cells that make up a tree ring is directly influenced by water availability and temperature. The rings can be measured and statistically analysed to reveal annual growth rates and measurement of the variability of the growth from year to year. Within an annual ring, cell size variation can be used to determine growth patterns within a season. These data can then be interpreted in terms of the influence exerted by certain climatic factors, such as temperature changes, water availability, seasonality, light levels, productivity of the forests and traumatic events such as insect attack or fire (Creber & Chaloner 1984; Francis 1986).

For example, growth rings in fossil wood of Late Cretaceous age from Antarctica are uniformly wide, indicating that Cretaceous polar climates were warm even at latitudes as high as about 70°S (Jefferson 1982; Francis 1986; Figure 17.5). Likewise in Permian wood from Antarctica, growing at even higher latitudes of 80°S, growth rings are as wide as 14 mm and consist of a huge number of cells (up to 194) (Figure 17.6). This shows that, despite the high latitudes and low light levels in winter, the trees were well adapted to such conditions and grew well during warmer spring growing seasons when sunlight was available for 24 hours a day (Francis *et al.* 1994). In a different setting, trees of Cupressus type grew in semi-arid climates on the southern shores of England during the Jurassic. The growth rings in these trees are very narrow (<1 mm) and irregular in width, reflecting intermittent growth under a seasonally arid climate where water supply was limited and only periodically available. Partial or false rings show that drought often affected tree growth during the growing season (Figure 17.7). Evidence for palaeoclimate from adjacent sediments, fossil soils and faunas also supports the idea that these mid-latitude Jurassic climates were strongly seasonal, with winter rainfall and summer droughts, similar to modern climates in Mediterranean regions (Francis 1983, 1984).

The uniformitarian approach can also be applied to fossil faunas for climate evidence, particularly for Tertiary faunas for which living relatives can be linked with some confidence. For example, certain types of foraminifera inhabit specific temperature zones in the ocean today. Shifts of the palaeolatitudinal range of comparable fossil foraminifera through geological time can be interpreted as changing temperature zones in the ocean (Haq 1982; McGowran 1990). Shells of marine molluscs also yield palaeotemperature data, even to seasonal level (Jones & Quitmyer 1996), through isotope analysis. For earlier geological periods the global distribution has been the only palaeontological evidence for global conditions. For example, during the Early Palaeozoic the distribution of marine faunas such as trilobites and graptolites shows that some were equatorial warm-water types and some cold-water polar types (Berry & Wilde 1990; Cocks & Fortey 1990), although

Figure 17.5   *Growth rings in Late Cretaceous fossil angiosperm wood from the Antarctic Peninsula. The average ring width here is 3.5 mm, indicating this tree grew well in a warm temperate climate with good water supply. This wood is of Nothofagus type (southern beech) and shows clusters of large vessels (used for water transport) concentrated along the early part of each growth ring*

this is based on a rather circular argument that polar zones were naturally cooler than tropical regions.

Chemical signals

One of the most useful sources of palaeotemperature data comes from the analysis of oxygen isotope ratios (^{18}O:^{16}O) in shells of marine fossils and sediments (Hudson & Anderson 1989; Marshall 1992). The difference in isotopic ratios between calcium-carbonate shell material and the water from which it precipitates

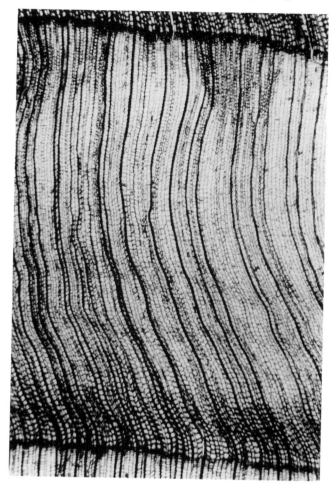

Figure 17.6 *Growth ring in fossil wood of Permian age from Allan Hills, Transantarctic Mountains. This ring shows the larger cells that form at the start of the growing season (the earlywood) followed by cells of decreasing diameter as the growing season progressed. The end of the growing season is marked by a narrow band of thin, dark cells (the latewood). The large number of cells within the ring indicates that the climate was very favourable for tree growth and that it was well adapted to growth at high latitudes with continuous summer sunlight and long dark winters. Ring width is 3 mm*

is a function of temperature (Anderson 1990). The precision of this method allows palaeotemperatures of ancient waters to be determined to ±0.5 °C. Complications include the fact that some shells fractionate calcium carbonate in slightly different ratios, and that the isotopic ratio of seawater is enriched in ^{18}O during glaciations (Anderson 1990). In rocks older than the Cretaceous, changing ocean chemistry and diagenetic alteration of shelly material also makes isotopic analysis of older material more dubious. However, once potential diagenetic problems have been eliminated by careful study of the source material, isotope

Figure 17.7   *Growth rings in fossil wood of Late Jurassic age from the Lulworth Formation, Purbeck Group, southern England. These rings are not only narrow in width (average 1 mm) but are also very irregular, in response to intermittent water availability. Some rings are also false rings, formed in response to a period of drought during the growing season. Evidence from the wood and associated sediments and fossils shows that this tree grew under a seasonal Mediterranean type climate. Thin section. Transverse (vertical) dimension of view represents 13 mm*

studies can provide vital information about marine temperatures and the nature of ocean circulation.

In particular, oxygen isotope data for Cretaceous and younger deposits are very detailed and extremely valuable since cores of well-preserved marine sediments have been provided by deep-sea drilling. This has produced a detailed picture of changing ocean temperatures for the Cretaceous and Tertiary from northern mid-latitude oceans (Figure 17.8). This curve illustrates that during the Late Cretaceous to Eocene both surface and bottom waters were warm, reflecting times of peak greenhouse warming. The temperature gradient from top to bottom of the ocean was small because there was no cold deep water in the oceans. However, dramatic drops in the isotopic ratio at certain points during the Tertiary (including the Eocene/Oligocene boundary and during the Miocene) reflect strong cooling of ocean waters as ice built up on Antarctica and global climates cooled towards the present ice age. The data show that oceanic circulation changed at the Eocene/Oligocene boundary so that, while surface waters remained relatively warm, deep benthic waters became sourced from a separate, much colder body of water generated from the ice caps that had accumulated on Antarctica.

Carbon isotope analyses cannot directly provide numerical data for climate

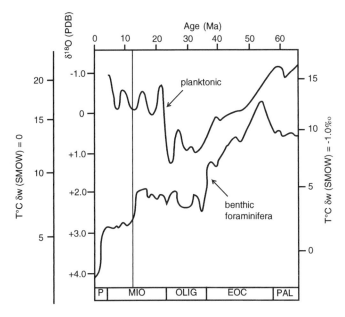

Figure 17.8 *Generalized curves for oxygen isotope data for Cenozoic benthic and planktonic foraminifera. Two temperature scales are given for ocean states before 15 Ma before major ice presence on Antarctica, and after 15 Ma when an extensive ice cap was present. Bottom water temperatures serve as a proxy for high-latitude marine surface temperatures and illustrate the onset of cold bottom water as the Antarctic ice cap built up in the Miocene. [Modified from: Hudson & Anderson (1989)]*

reconstructions but they are used to determine the state of carbon cycling in the oceans and on land, which itself is interlinked to the atmospheric carbon cycle and climate. Carbon isotope data provide information about productivity of organic carbon by plants on land or in the sea and the nature of marine and atmospheric systems during its subsequent oxidation or burial as coal or organic-rich marine shales (Arthur 1982; Post *et al.* 1990). This may reveal details concerning, for example, the oxygen state of the oceans or the carbon dioxide levels in the atmosphere. Modelling by Berner (1992) and Worsley *et al.* (1994) have predicted past global levels of carbon dioxide that have been used for important comparisons with geological data. Their models show that carbon dioxide levels were high during past greenhouse phases, such as during the Cretaceous, and low during times of cold climates, such as during the Permian–Carboniferous glaciations, although the Ordovician–Silurian glaciation is the exception in that a glaciation seems to have occurred at times of high carbon dioxide levels.

Climate modelling

Climate modelling has been one of the major advances in palaeoclimatology since the early 1980s, particularly for the Cretaceous (Barron & Washington 1984, 1985). The production of global maps predicting specific climate zones for times in the past has concentrated the minds of geologists and spurred them to use their geological data to test and verify the model simulations. While the often colourful and detailed maps produced by climate modellers provide a wealth of information that looks reassuringly realistic, it must be remembered that they are still only predictions, not factual climate records, and it is crucial that geological evidence for climate from rocks and fossils is used to challenge or verify the models. There are three types of models frequently used in palaeoclimate research.

1. *Conceptual models.* These are based on understanding the fundamentals of atmospheric and oceanic circulation of present climate systems; they are the most simple and are useful for understanding how continental positions would have influenced climate patterns (Parrish 1993). In essence, the patterns of atmospheric pressure and rainfall systems of the present world are applied to different palaeogeographies in the past. These models are relatively simple and predict only the seasonal positions of atmospheric pressure cells, wind directions, and areas of high and low rainfall. However, the results have been successfully correlated with the geographical occurrences of certain sediments (such as coal and evaporites), floras and faunas to provide some indication of the controlling factors on their distribution (e.g. Parrish *et al.* 1982).

2. *Numerical models.* Numerical models are a powerful tool for predicting palaeoclimates. Energy balance models (EBMs) are based on the laws of energy conservation and therefore consider only radiation flux and temperatures. The models compute simple formulas for various flux terms (Crowley & North 1991). By slightly altering parameters in the calculations, the sensitivity and stability of the climate response can be assessed. Seasonal EBMs have been useful for palaeoclimate studies because they can test how factors such as landmass distribution and orbital cycles have influenced climate change (Crowley & North 1991).

   Crowley *et al.* (1987) used EBMs to model Late Ordovician climates in order to understand the factors which caused glaciation at this time. Their model results showed that if a pole is located over the centre of a large continent the seasonal extremes experienced there do not allow an ice cap to form; any ice that forms during the cold winters then melts during hot summers (approximately 20 °C predicted for Ordovician summers). However, if the pole is positioned at the edge of a continent, as it was over Gondwana during the Late Ordovician, the moderating effect of the ocean keeps summer temperatures low (0 °C predicted), allowing ice to build up (Figure 17.9). In experiments with EBMs, during which they increased global warmth to mimic the high carbon dioxide levels predicted by Berner (1992) for the Ordovician, they discovered that the seasonal cycle was not affected so much as to preclude the formation of ice.

   Crowley *et al.* (1987) also modelled climates for the Devonian. By this time the South Pole was situated closer to the continental interior (Scotese & McKerrow 1990), and Crowley *et al.* predicted high summer temperatures that would have melted winter ice and prohibited the build-up of an ice sheet. This may have been the reason for the lack of an extensive phase of glaciation in the Devonian.

3. *Three-dimensional global models.* These climate models, otherwise known as general circulation models or GCMs, have been developed for simulating

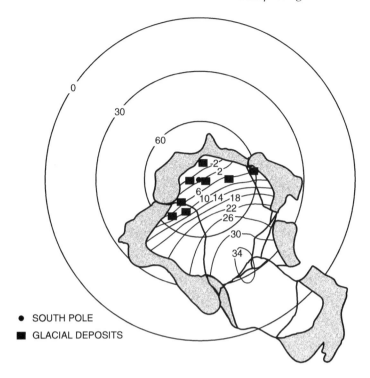

Figure 17.9 *Realistic geography simulation for Gondwana during the Late Ordovician. The contours on the continents represent predicted palaeotemperatures for January (summer) in the southern polar region and show that coastal regions would experience freezing temperatures. The location of known glacial deposits (rectangles) are shown (from Caputo & Crowell 1985). The shaded areas represent flooded continental margins with shallow seas. [Modified from: Crowley et al. (1987)]*

present-day climates and predicting future changes (Crowley & North 1991). However, they can only be thoroughly tested by using them to simulate climates in the geological past and then comparing their predictions with palaeoclimate information from geological sources. GCMs use mathematical equations governing many physical climate processes of both atmosphere and oceans to simulate climate for snapshots of geological time. Boundary conditions used for these models include continental position, elevation, atmospheric composition, solar forcing, orbital parameters and sea surface temperatures (Valdes 1994). Although they are continually being improved and are a powerful tool that has revolutionized palaeoclimate research, the models still have many problems, particularly their present inability to accurately simulate the complex interactions between the ocean and atmosphere.

More recently, many other types of models have been used to assess changing geological conditions that may have affected past climates in less direct ways. One interesting modelling experiment was performed by Moore & Worsley (1994) to see if tectonic uplift and mountain-building could cause climate cooling. For example, the formation of the Himalayas in the Tertiary and the resulting increase in weathering of the uplifted rocks on the Tibetan Plateau is proposed to have caused a decrease in atmospheric carbon dioxide levels and a climate cooling that intensified the Cenozoic Ice Age, according to Raymo & Ruddiman (1992). Similarly, the uplift of mountain belts along the margins of Gondwana during the Late Palaeozoic may have prompted the onset of the Gondwanan ice age. The modelling of Moore & Worsley (1994) assumes that uplifted land would weather more rapidly, the resulting breakdown of silicates in the rock allowing liberated calcite to react with carbon dioxide to form carbonates. This would have caused a decrease in atmospheric carbon dioxide levels and cooled the climate. Their results (Figure 17.10) not only show that times of mountain-building correlate reasonably well with known ice ages in the geological past but that for most of the Phanerozoic the earth may have been cool enough for glaciations to occur if other conditions were suitable. They showed that mountain ranges situated in polar regions may not necessarily have had a strong cooling effect since weathering rates under polar climates would have been relatively low. However, if mountains were formed in low latitudes where weathering rates would have been high due to warm wet tropical climates, the resulting large carbon dioxide drawdown would have cooled

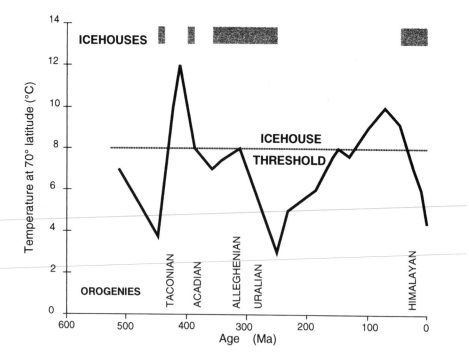

Figure 17.10   *Calculated palaeotemperatures for the Phanerozoic, from modelling by Moore & Worsley (1994) and Worsley et al. (1994). The borderline temperature for ice-cap formation is 8 ± 2 °C. Temperatures above this are considered too warm for ice formation but temperatures below have a high probability of ice-cap formation. The timing of major orogenies and glaciations are also shown. [Modified from: Moore & Worsley (1994)]*

climates strongly enough for ice caps to form on low land at high latitudes. As Moore & Worsley (1994) admit, the effect of weathering may be subtle and may have simply intensified already cold climates rather than starting ice ages, but their models highlight the important climatic effects of mountains and should prompt geologists to estimate more accurately the location and height of past mountain ranges.

17.2 SUMMARY

Palaeoclimate studies focus on both determining the climate states of the earth during the geological past and understanding how the climate system worked to produce those ancient environments. The careful observation, collection and interpretation of geological evidence from as many independent sources as possible is the most reliable way of determining details of ancient climate states. For many geological periods, palaeoclimatic data are still new, and more rigorous methods for interpreting the data that we already have are needed. The most exciting aspect of this geological research is that we can study unique environments that existed on earth in the past that are no longer present in our modern world, such as forests and dinosaurs living in warm climates in the polar regions and the extreme environments in continental interiors on supercontinental landmasses such as Pangaea.

A collaboration of both geological investigations and palaeoclimate modelling has revolutionized palaeoclimatology and is helping us to understand how the complex climate system works both now and in the past, and understand which features of ancient environments were the strongest forcing factors for climate change. The climate models show that there are many factors which we have yet to take full account of, including the effect and location of mountain belts, the influence of orbital cycles on climates during greenhouse climates, the complexities of the carbon cycle (especially in the oceans), the location of major expanses of water within continental regions, and how weathering, volcanism and vegetation change have affected climate. While we now know much about specific climate states in the past, what we have yet to understand properly is what caused the climate to change from one state to another, both on short time scales of a million years or less, to the major changes from greenhouse to icehouse world on time scales of hundreds of millions of years.

REFERENCES

Anderson, G.Y. 1990. Temperature from oxygen isotype ratios. In Briggs, D.E.G. & Crowther, P.R. (eds) *Palaeobiology; a synthesis*. Blackwell Scientific Publications, Oxford, 403–406.

Arthur, M.A. 1982. The carbon cycle: controls on atmospheric CO_2 and climate in the geologic past. In Berger, W.H. & Crowell, J.C. (eds) *Climate in earth history*. National Academy Press, Washington, 55–67.

Bardossy, G. 1994. Carboniferous to Jurassic bauxite deposits and paleoclimatic and paleogeographic indicators. In Embry, A.F., Beauchamp, B. & Glass, D.J. (eds) *Pangea:*

global environments and resources. Canadian Society of Petroleum Geologists Memoir 17, 283–293.

Barron, E.J. & Washington, W.M. 1984. The role of geographic variables in explaining palaeoclimates: results from Cretaceous climate model sensitivity studies. *Journal of Geophysical Research*, **89**, 1267–1279.

Barron, E.J. & Washington, W.M. 1985. Warm Cretaceous climates: high atmospheric CO_2 as a plausible mechanism. In Sundquist, E.T. & Broecker, W.S. (eds) *The carbon cycle and atmospheric CO_2: natural variations Archean to present*. American Geophysical Union Geophysical Monograph No. 32, 546–553.

Bennett, M.R., Doyle, P. & Mather, A.E. 1996. Dropstones: their origin and significance. *Palaeogeography, Palaeoclimatology, Palaeoecology*, **121**, 331–339.

Berner, R.A. 1992. Palaeo-CO_2 and climate. *Nature*, **358**, 114.

Berry, W.B.N. & Wilde, P. 1990. Graptolite biogeography: implications for palaeogeography and palaeoceanography. In McKerrow, W.S. & Scotese, C.R. (eds) *Palaeozoic Palaeogeography and Biogeography*. Geological Society Memoir No. 12, 129–137.

Brandley, R.T. & Krause, F.F. 1994. Thinolite-type pseudomorphs after Ikaite: indicators of cold water on the subequatorial western margin of Lower Carboniferous North America. In Embry, A.F., Beauchamp, B. & Glass, D.J. (eds) *Pangea: global environments and resources*. Canadian Society of Petroleum Geologists Memoir 17, 333–344.

Briden, J.C. & Irving, E. 1964. Paleolatitude spectra of sedimentary paleoclimate indicators. In Nairn, A.E.M. (ed.) *Problems in paleoclimatology*. John Wiley, New York, 199–224.

Caputo, M.V. & Crowell, J.C. 1985. Migration of glacial centres across Gondwana during the Paleozoic Era. *Bulletin of the Geological Society of America*, **96**, 1020–1036.

Chaloner, W.G. & Creber, G.T. 1990. Do fossil plants give a climatic signal? *Journal of the Geological Society of London*, **147**, 343–350.

Chamley, H. 1989. *Clay sedimentology*. Springer-Verlag, Berlin.

Cocks, L.R.M. & Fortey, R.A. 1990. Biogeography of Ordovician and Silurian faunas. In McKerrow, W.S. & Scotese, C.R. (eds) *Palaeozoic Palaeogeography and Palaeobiogeography*. Geological Society Memoir No. 12, 97–104.

Creber, G.T. & Chaloner, W.G. 1984. Influence of environmental factors on the wood structure of living and fossil trees. *Botanical Review*, **50**, 357–448.

Creber, G.T. & Chaloner, W.G. 1985. Tree growth in the Mesozoic and early Tertiary and the reconstruction of palaeoclimates. *Palaeogeography, Palaeoclimatology, Palaeoecology*, **52**, 35–60.

Crowell, J.C. & Frakes, L.A. 1975. The Late Palaeozoic glaciation. In Campbell, K.S. (ed.) *Gondwana geology*. Australian National University Press, Canberra, 313–331.

Crowley, T.J. & North, G.R. 1991. *Paleoclimatology*. Oxford University Press and Clarendon Press, Oxford.

Crowley, T.J., Mengel, J.G. & Short, D.A. 1987. Gondwanaland's seasonal cycle. *Nature*, **329**, 803–807.

Dubiel, R.F. & Smoot, J.P. 1994. Criteria for interpreting paleoclimate from red beds – a tool for Pangean reconstructions. In Embry, A.F., Beauchamp, B. & Glass, D.J. (eds) *Pangea: global environments and resources*. Canadian Society of Petroleum Geologists Memoir 17, 295–310.

Dubiel, R.F., Parrish, J.T., Parrish, J.M. & Good, S.C. 1991. The Pangaean megamonsoon: evidence from the Upper Triassic Chinle Formation, Colorado Plateau. *Palaios*, **6**, 347–370.

Flügel, E. 1994. Pangean shelf carbonates: controls and paleoclimatic significance of Permian and Triassic reefs. *Geological Society of America Special Paper*, **288**, 247–266.

Frakes, L.A. 1979. *Climates throughout geologic time*. Elsevier, Amsterdam.

Frakes, L.A., Francis, J.E. & Syktus, J.I. 1992. *Climate modes of the Phanerozoic*. Cambridge University Press, Cambridge.

Francis, J.E. 1983. The dominant conifer of the Jurassic Purbeck Formation. *Palaeontology*, **26**, 277–294.

Francis, J.E. 1984. The seasonal environment of the Purbeck (Upper Jurassic) fossil forests. *Palaeogeography, Palaeoclimatology, Palaeoecology*, **48**, 285–307.

Francis, J.E. 1986. Growth rings in Cretaceous and Tertiary wood from Antarctica and its palaeoclimatic implications. *Palaeontology*, **48**, 285–307.

Francis, J.E. 1990. Polar fossil forests. *Geology Today*, **6**, 92–95.

Francis, J.E., Wolfe, K.J., Arnott, M.J. & Barrett, P.J. 1994. Permian climates of the southern margins of Pangea: evidence from fossil wood in Antarctica. In Embry, A.F., Beauchamp, B. & Glass, D.J. (eds) *Pangea: global environments and resources*. Canadian Society of Petroleum Geologists Memoir 17, 275–282.

Gordon, W.A. 1975. Distribution by latitude of Phanerozoic evaporite deposits. *Journal of Geology*, **83**, 671–684.

Haq, B.U. 1982. Climatic acme events in the sea and on the land. In Berger, W.H. & Crowell, J.C. (eds) *Climates in earth history*. National Academy Press, Washington, 126–132.

Hudson, J.D. & Anderson, T.F. 1989. Ocean temperatures and isotopic compositions through time. *Transactions of the Royal Society of Edinburgh*, **80**, 183–192.

James, N. & Bone, Y. 1991. Origin of a cool water Oligocene–Miocene deep shelf limestone, Eucla Platform, southern Australia. *Sedimentology*, **38**, 323–342.

Jefferson, T.H. 1982. Fossil forests from the Lower Cretaceous of Alexander Island, Antarctica. *Palaeontology*, **25**, 681–708.

Jones, D.S. & Quitmyer, I.R. 1996. Marking time with bivalve shells: oxygen isotopes and season of annual increment formation. *Palaios*, **11**, 340–346.

Kovach, W.L. & Spicer, R.A. 1996. Canonical correspondence analysis of leaf physiognomy: a contribution to the development of a new palaeoclimatological tool. *Palaeoclimates*, **2**, 125–138.

Kraus, M.J. & Bown, T.M. 1986. Paleosols and time resolution in alluvial stratigraphy. In Wright, V.P. (ed.) *Paleosols – their recognition and interpretation*. Blackwell Scientific, Oxford, 180–207.

Marshall, J.D. 1992. Climatic and oceanographic isotopic signals from the carbonate rock record. *Geological Magazine*, **129**, 143–160.

McCabe, P.J. 1984. Depositional environments of coal and coal-bearing strata. *Special Publications of the International Association of Sedimentologists*, **7**, 13–42.

McCabe, P.J. & Parrish, J.T. 1992. Tectonic and climatic controls on the distribution and quality of Cretaceous coals. *Geological Society of America Special Publication*, **267**, 1–15.

McElwain, J.C. & Chaloner, W.G. 1996. The fossil cuticle as a skeletal record of environmental change. *Palaios*, **11**, 376–388.

McGowran, B. 1990. Fifty million years ago. *American Scientist*, **78**, 30–39.

Moore, T.L. & Worsley, T.R. 1994. Orogenic enhancement of weathering and continental ice-sheet initiation. *Geological Society of America Special Paper*, **288**, 75–89.

Parrish, J.T. 1993. Climate of the supercontinent Pangea. *Journal of Geology*, **101**, 215–233.

Parrish, J.T. & Peterson, F. 1988. Wind directions predicted from global circulation models and wind directions determined from eolian sandstones of the western United States – a comparison. *Sedimentary Geology*, **56**, 261–282.

Parrish, J.T., Ziegler, A.M. & Scotese, C.R. 1982. Rainfall patterns and the distribution of coals and evaporites in the Mesozoic and Cenozoic. *Palaeogeography, Palaeoclimatology, Palaeoecology*, **40**, 67–101.

Post, W.M., Peng, T.-H., Emanuel, W.R., King, A.W., Dale, V.H. & DeAngelis, D.L. 1990. The global carbon cycle. *American Scientist*, **78**, 310–326.

Raymo, M.E. & Ruddiman, W.F. 1992. Tectonic forcing of late Cenozoic climate. *Nature*, **359**, 117–122.

Scotese, C.R. & McKerrow, W.S. 1990. Revised world maps and introduction. In McKerrow, W.S. & Scotese, C.R. (eds) *Palaeozoic Palaeogeography and Biogeography*. Geological Society Memoir No. 12, 1–21.

Sellwood, B.W. & Price, G.D. 1994. Sedimentary facies as indicators of palaeoclimate. In Allen, J.R.L., Hoskins, B.J., Sellwood, B.W., Spicer, R.A. & Valdes, P.J. (eds) *Palaeoclimates and their modelling*. The Royal Society & Chapman & Hall, London, 17–25.

Spicer, R.A. 1990. Climate from plants. In Briggs, D.E. & Crowther, P.R. (eds) *Paleobiology: a synthesis*. Blackwell Scientific Publications, Oxford, 401–403.

Spicer, R.A. & Parrish, J.T. 1990. Late Cretaceous–early Tertiary palaeoclimates of the northern high latitudes: a quantitative view. *Journal of the Geological Society of London*, **147**, 329–341.

Upchurch, G.R. & Wolfe, J.A. 1987. Mid Cretaceous to Early Tertiary vegetation and climate: evidence from fossil leaves and woods. In Friis, E.M., Chaloner, W.G. & Crane, P.R. (eds) *The origins of angiosperms and their biological consequences*. Cambridge University Press, Cambridge, 75–105.

Valdes, P. 1994. Atmospheric general circulation models of the Jurassic. In Allen, J.R.L., Hoskins, B.J., Sellwood, B.W., Spicer, R.A. & Valdes, P.J. (eds) *Palaeoclimates and their modelling*. Chapman and Hall, London, 109–118.

Van Houten, F.B. 1982. Red beds. *McGraw-Hill Encyclopedia of Science and Technology*, 5th edition. McGraw-Hill, New York, 441–442.

Visser, J. 1993. A reconstruction of the late Palaeozoic ice sheet on southwestern Gondwana. In Findlay, R.H., Unrug, R., Banks, M.R. & Veevers, J.J. (eds) *Gondwana Eight. Assembly, evolution and dispersal*. Hobart, Tasmania. A.A. Balkema, Rotterdam, 449–458.

Wolfe, J.A. 1985. Distribution of major vegetational types during the Tertiary. In Sundquist, E.T. & Broecker, W.S. (eds) *The carbon cycle and atmospheric CO_2: natural variations Archean to present*. American Geophysical Union Geophysical Monograph No. 32, 357–375.

Wolfe, J.A. 1993. A method of obtaining climatic parameters from leaf assemblages. *US Geological Survey Bulletin*, **2040**, 1–71.

Wolfe, J.A. & Upchurch, G.R. 1986. Vegetation, climatic and floral changes across the Cretaceous–Tertiary boundary. *Nature*, **324**, 148–152.

Wolfe, J.A. & Upchurch, G.R. 1987. North American non-marine climates and vegetation during the Late Cretaceous. *Palaeogeography, Palaeoclimatology, Palaeoecology*, **61**, 33–77.

Worsley, T.R., Moore, T.L., Fraticelli, C.M. & Scotese, C.R. 1994. Phanerozoic CO_2 levels and global temperatures inferred from changing paleogeography. *Geological Society of America Special Paper*, **288**, 57–73.

Wright, V.P. 1986. *Paleosols, their recognition and interpretation*. Blackwell Scientific, Oxford.

Ziegler, A.M., Hulver, M.L., Lottes, A.L. & Schmactenberg, W.F. 1984. Uniformitarianism and palaeoclimates: inferences from the distribution of carbonate rocks. In Brenchley, P.J. (ed.) *Fossils and climate*. John Wiley and Sons, Chichester, 3–25.

18
Interpreting Orogenic Belts: Principles and Examples

David A. T. Harper

Mountains have been the primary focus of geological research throughout most of the history of the earth sciences. These complex belts contain the critical evidence for large-scale vertical and horizontal movements of crust, and demonstrate the former existence of oceans and palaeotectonic terranes. They provide a framework within which to organize geological events through, for example, the recognition of unconformities developed during terrane collision, uplift and orogenic collapse. Moreover, the range of basins associated with the life span of an orogenic zone contain a detailed record of the provenance and type of crust involved in an orogeny. The study and interpretation of mountain belts is not easy; a huge variety of techniques have been applied to their investigation (see Chapter 4). Nevertheless, stratigraphical techniques involving superposition (see Chapter 2) and biostratigraphy (see Chapter 4) together with biogeography have proved fundamental in developing a framework for the formation of orogenic belts. The stratigraphy of mountains leads directly into the demonstration and understanding of global tectonic processes and therefore to the core of geological research.

The importance of fossil data and stratigraphy in the study of these complex tectonic zones of the crust was in the vanguard of investigations during the nineteenth century (Lyell 1850). The stratigraphical analysis of mountain belts (Trümpy 1971), however, has not been superseded as a core activity and remains central to the elucidation of orogenic zones and orogenic processes. The transfer of classical stratigraphical principles from the relatively undeformed platform successions into the highly deformed rocks of mountain belts has required some

Unlocking the Stratigraphical Record: Advances in Modern Stratigraphy. Edited by P. Doyle and M.R. Bennett.

modification of existing procedures and the development of a variety of new techniques, principally involving the extraction and analysis of fossils from meta-morphosed and tectonized rocks (Bruton & Harper 1992).

The tectonostratigraphy, or the main tectonic divisions, of the European mountain belts has been assembled through the detailed stratigraphical analysis of the Caledonides, the Variscides and the Alpides based on modifications of the classic principles of superposition and biostratigraphy. Moreover, the age of magmatic and orogenic events within orogenic zones is commonly constrained within fos-siliferous successions and both the origins and history of individual parts of many mountain belts can be determined on the basis of faunal and floral provinciality. The emplacement of thrust sheets may be tracked within stratigraphical frame-works, emphasized by flysch development and shadowed in the deposition of sediment in adjacent molasse troughs. Stratigraphy has therefore provided the essential key to understanding the evolution of Phanerozoic mountain belts.

This chapter first reviews the historical development of terrane analysis, empha-sizing the role played by palaeobiogeography. This is followed by a discussion of some of the basic principles of modern terrane analysis, including a review of statistical methods of applied palaeobiogeography. In the second part of the chapter a series of examples of orogenic belts and their component terranes is presented, which emphasizes the role of palaeobiogeography in the analysis of complex orogenic belts.

18.1 HISTORICAL PERSPECTIVES

During the nineteenth century, stratigraphical principles and the foundation of many of the geological systems were established in the relatively undeformed strata of the plains of Europe although some, such as the Cambrian and Silurian, were based on deformed strata in Wales. Superposition, correlation and develop-ment of global chronostratigraphy are associated with fossiliferous, 'layer-cake' successions of horizontal bedded units. One of the earliest and most fundamental questions to be answered was whether the principles of classical stratigraphy could be routinely applied to the complex and deformed rocks of mountain belts. In fact, the complex structure and history of mountain belts can only be eluci-dated with reference to traditional stratigraphical constraints and frameworks. For example, attempts during the early nineteenth century to unravel the struc-ture of the Alpine Belt were unsuccessful in the absence of a stratigraphical framework, because the relative sequences of events were not established; realis-tic reconstructions of the Alps were possible only later that century, when the stratigraphy of adjacent, undeformed platform and basin successions had been defined and related to the deformed rocks of the orogen (Trümpy 1971). Rudolf Trümpy's review reinforced Bailey's (1935) tabulation of the main units within the Helvetics and the Prealps which strongly emphasized both the palaeontologi-cal and stratigraphical controls on structural data in the Alpine belt. These studies serve to demonstrate that basic stratigraphical concepts and principles have been of fundamental importance in the development of our understanding of orogeny and orogenic processes. In addition, the biogeographical signatures of fossil assemblages have proved invaluable in tracking the origins and movements of individual parts of orogenic zones prior to collision. Both biostratigraphy and biogeography are now used with increasing sophistication in the analysis of mountain belts. The initial observations on fossils from mountains were often simplistic but with incredible implications.

Darwin's discovery of shelly fossils at the summit of the Peuquenes, over 4.5 km above sea level on the western seaboard of South America, provided graphic evidence of large-scale vertical movements in mountain belts. Darwin (1839) noted the occurrence of the distinctive Lower Jurassic oyster *Gryphaea* together with an ammonite and forms resembling the 'turritellæ and terebratulæ', concluding that the formation correlated with the middle parts of the secondary series (the Mesozoic rocks) of Europe. Clearly, since these Jurassic rocks were deposited in marine environments, their fossils demonstrated that the Andes had been elevated up to 5 km above sea level.

The first indications that large-scale horizontal movements could occur in the earth's crust were suggested by Escher von der Linth's work in the Swiss Alps. Von der Linth recognized, during the early 1840s, Jurassic marine strata with characteristic ammonites which were tectonically overlain by Palaeozoic basement including the Permian Verrucano conglomerates in the Glarus region of Switzerland. Later discoveries in the north-west Highlands of Scotland, by Calloway and Lapworth in the early 1880s, demonstrated that sheets of metamorphics, the Moine succession, were thrust westwards over fossiliferous Cambrian and Ordovician platform strata. Torneböhm established similar evidence for nappe translation, but from west to east, in the Scandinavian Caledonides during the mid 1880s. Nevertheless, the large-scale horizontal movements of thrust sheets common in the assembly of both the Caledonides and the Alps was not demonstrated in the Variscides until relatively recently. These studies clearly showed that sheets of rock had moved horizontally – but could entire continents move?

Wegener (1929) emphasized that a number of late Palaeozoic and Mesozoic fossils such as the Permian reptile *Mesosaurus* and the Permo-Carboniferous *Glossopteris* flora were common to parts of the now disparate southern continents. Rather than invoke trans-oceanic landbridges, Wegener argued that these areas once formed a coherent landmass during the late Palaeozoic and had simply drifted apart during the Mesozoic. Large-scale continental movements could be demonstrated on the basis of the contrasting biogeographical signatures of biotic assemblages evolving in different provinces. Nevertheless, throughout most of the first half of the twentieth century, ideas about the evolution of mountain belts were rooted in geosynclinal theory; tectonic deformation was associated with thick accumulations of sediment in static troughs usually developing and deforming adjacent to continental margins and platforms.

During the 1960s the discovery of seafloor spreading prompted a rethink of planetary processes. The unifying concept of plate tectonics, developed by geophysicists, at last satisfied the large compendium of palaeontological data explicable only in terms of continental drift. The planet was covered with mobile plates interacting at spreading ridges, subduction complexes and transcurrent fault zones. Large-scale translation of lithospheric plates could be modelled and the existence of ancient continents and oceans together with their interactions could be predicted and discovered, fossilized, in mountain belts. This provided the theoretical framework within which to interpret the information provided by fossil and stratigraphical data.

This was further enhanced by the work of J. Tuzo Wilson who predicted the prior existence of a proto-Atlantic Ocean, separating Europe from North America during the Early Palaeozoic (Wilson 1966). Brachiopods, trilobites and graptolites provided essential age constraints on the Lower Palaeozoic sections on the platforms of Laurentia, Baltica and Avalonia. However, it was clear that these fossil faunas had contrasting biogeographical signals. During the Ordovician brachiopods were part of discrete European and North American provinces, trilobites developed in the separate Asaphid and the Bathyurid provinces, whereas the graptolites could be assigned to either Atlantic or Pacific provinces. A faunal suture could be recognized along the length of the Appalachian–Caledonide belt implying an ancient ocean, pre-dating the present Atlantic Ocean, dividing North America from Europe during the Early Palaeozoic.

The first plate tectonic models for the Caledonides (Dewey 1969), and other orogenic belts, invoked simple orthogonal opening and closure of oceanic systems. The reality was to prove much more complex. The discovery of strike-slip movements and rotations of crustal blocks prohibited the continued application of simple two-dimensional models for ocean systems. Both continental and oceanic plates were clearly capable of hitherto unpredictable movements with respect to current models. A new range of techniques was necessary to track their movements and determine their history.

Terrane analysis (Coney *et al.* 1980) has provided a modern, synthetic method of analysing the evolution of mountain belts. Each orogenic belt is considered as a collage of discrete terranes docking as a consequence of orogeny. Each terrane has a characteristic stratigraphy, i.e. magmatic, sedimentary and tectonic history together with specific faunal and floral signatures. Terranes may be further distinguished on the basis of metallogenic and isotopic signatures. Terranes developed seaward of continental margins are not easily constrained geographically and are usually described as suspect. Orogenic zones, however, are sequentially constructed by the amalgamation or docking of terranes; biostratigraphical and biogeographical controls have proved essential components of this type of analysis.

18.2 BASIC PRINCIPLES

The rocks of orogenic belts, such as marbles, slates and schists, yield fossils that are usually rare, poorly preserved and commonly deformed. Nevertheless, this fossil material has helped determine the finite strain and thermal maturity of strata (Bruton & Harper 1992) but, more importantly, palaeontological data have provided biostratigraphical and biogeographical constraints on the development of orogenic zones.

Classical stratigraphical techniques, using both superposition and biostratigraphy, have been used to establish the age of tectonic events by constraining deformed strata and associated unconformities (Fortey & Cocks 1990). The deformation, uplift and collapse of strata can be constrained between the age of the tectonized basement rocks themselves, and that of the overlying cover sequence. For example, in the Caledonides of western Norway marine sequences with deformed mid-Silurian brachiopods are unconformably overlain by Old Red Sandstone-type molasse successions, constraining a late Silurian, Scandian tectonic event affecting the entire Scandinavian Caledonides. Phases of magmatism signalled by volcanic and volcaniclastic strata commonly interdigitate with fossiliferous units that also provide time constraints on their development. Within the Caledonides of western Ireland, a range of volcanic and volcaniclastic rocks punctuate fossiliferous marine Ordovician and Silurian strata (Ryan & Dewey 1991; Williams & Harper 1991). The ages of the Ordovician arc-related successions have recently been refined by graptolite (Williams & Harper 1994) and shelly (Williams & Curry 1985) data, whereas in the Silurian overlap successions, mass mortality horizons are associated with volcaniclastic surges (Harper *et al.* 1995) overlain by bentonites and volcanogenic turbidites; these later magmatic events are bracketed by abundant shelly faunas and rarer graptolite assemblages within a late Llandovery to early Wenlock window. Two main volcanic phases, during the early Ordovician and the early to mid-Silurian are constrained by palaeontological data.

Simple structural patterns may be established on the basis of facing directions and way-up criteria developed from palaeontological data. In the Southern Uplands of Scotland, the basic pattern of inverted sequences of graptolitic shales and turbidites within larger packages of strata, separated by major tectonic structures, themselves dipping southwards but not inverted, was established over a century ago through the detailed mapping of graptolite biozones (Lapworth 1878). The recognition of sequential assemblages of graptolite faunas allowed strata to be placed in order through zones of considerable structural complexity. Although the definition and recognition of sequential graptolite assemblages allow the overall structure of the Ordovician and Silurian rocks of southern Scotland to be determined, suggesting an accretionary prism, controversy still surrounds the origin and development of this part of the Caledonides (Armstrong *et al.* 1996).

The complex orogenic zones of the European mountain belts, dominated by nappe tectonics, have been reconstructed on the basis of stratigraphical principles commonly with reference to palaeontological data. First, the age and way-up of strata within each package or individual thrust sheet may be described by biostratigraphical data. Secondly, their original location and environmental settings are suggested by biogeographical and sedimentological data. Thirdly, the timing of deformation can be established by the recognition of syn-orogenic flysch and unconformities within the package. Finally, the translation of the thrust sheets may be monitored by the changing character and provenance of sediments in adjacent molasse troughs fed by the deforming and rising thrust sheets. Within the Alpine chain such stratigraphical studies have formed the basis of the descriptions of the dynamics and structure of the belt (Trümpy 1971).

The stratigraphical control of sea-level changes within orogenic belts provides another opportunity to track tectonic events. For example, in the Ordovician successions of Wales and the Anglo-Welsh area, marked divergences from global sea level or palaeobathymetric curves, manifest in regional unconformities, imply tectonism during the Arenig and late Caradoc (Fortey & Cocks 1986). These local events influenced sea levels more significantly than global processes, such as the fluctutations of ice sheets, operating with greater effect elsewhere. The description of palaeobathymetric change has wide application in the studies of mountain belts and tectonic processes.

Terrane analysis has, more recently, provided a more sophisticated range of techniques for the reconstruction of ancient and modern mountain belts (Howell 1989). Orogenic zones contain a wide variety of crustal material, including blocks of continental crust and obducted flakes of oceanic crust associated with a variety of microplates including fragments of continental margins, volcanic arcs, sedimentary basins and seamounts. Many of these tectonic units have been entrained within nappes or thrust sheets. Mountain belts consist of diverse collages of crustal material. Terrane analysis has firstly sought to define, organize and classify the constituent parts of these belts. Each terrane is fault bounded and contrasts in its biotic, magmatic, metamorphic, metallogenic and sedimentary history with that of adjacent blocks. In fact, juxtaposed terranes have quite different stratigraphies, suggesting contrasting origins and geological histories. The second element of terrane analysis concerns the time of terrane docking. Techniques such as sedimentary linkage, sedimentary overlap, plutonic stitching and metamorphic welding can, with varying degrees of precision, help date docking events along with tectonic features such as melanges, thrusts and strike-slip faults (Figure 18.1).

Terranes that developed seaward to continental margin sequences are essentially unconstrained and are termed suspect. These terranes, more than any others, have been the subject of intense study to first establish their origins, secondly monitor their subsequent movements, and thirdly constrain their arrival and time of docking. Many suspect terranes are associated with oceanic crust and commonly carry exotic biotas. The first detailed terrane analysis tracked the movement of the large Wrangellia Terrane, now stretching from Alaska to Oregon, in the Western Cordillera of North America (Jones *et al.* 1977); the changing latitude of the giant terrane was monitored with reference to the biogeographic patterns of fusilinoid foraminifera in the orogen (Monger & Ross 1971).

The western seaboard of North America is currently a region of active tectonism; the periodic movement of the San Andreas Fault allows the translation of the Pacific plate past the North American continent. A history of terrane movement can be traced back through the Cenozoic and Mesozoic as the Western Cordillera was gradually constructed from a mosaic of disparate terranes (Coney *et al.* 1980). At least nine major allochthonous terranes are now recognized within the cordillera (Jones *et al.* 1983), although there may be as many as 50, including terranes of Tethyan origin and a variety of island complexes (Newton 1987). A relatively simple plot of the current positions in the cordillera of low-latitude Tethyan faunas from time slices in the Permian, Triassic and Jurassic relative to the migrating boundaries between the Tethyan and Boreal provinces on the North American craton, demonstrates clearly the northward displacement, along the North American margin, of these terranes through time (Hallam 1986).

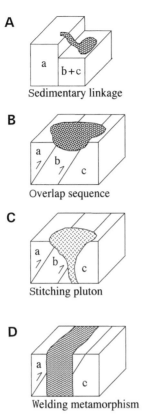

A

a | b+c
Sedimentary linkage

B

a / b / c
Overlap sequence

C

a / b / c
Stitching pluton

D

a / c
Welding metamorphism

Figure 18.1 *Techniques for terrane analysis. **A.** Sedimentary linkage. **B.** Overlap sequence. **C.** Stitching pluton. **D.** Welding metamorphism*

Palaeontological data have further applications within regional and global tectonics; fossils have contributed to strain analysis (Ramsay 1967) and the understanding of distortion, together with palaeobarometry and palaeothermometry (Bruton & Harper 1992). Nevertheless, the temporal and spatial properties of fossils, biostratigraphy and biogeography, remain the core palaeontological techniques for the study of mountain belts. The construction of stratigraphical successions within mountain belts, where biostratigraphical data constrain magmatic and tectonic events, is the basis for the case studies of the European mountain belts that follow.

18.3 STATISTICAL ANALYSIS IN BIOGEOGRAPHY

Computer-based algorithms have provided a more scientific approach to both biostratigraphical and biogeographical studies. In the majority of mountain belts biostratigraphical data are too sparse for formal graphical and statistical analysis. On the other hand, information is adequate to statistically frame and test biogeographical hypotheses. The wide range of investigative techniques available ar polarized into two main groups, based mainly on either dispersal or vicarianc models. To date, phenetic methods have been most widely used in biogeographi cal analysis (e.g. Harper 1992), commonly in support of dispersal models, where as cladistic methodologies have been used to support vicariance models (e.g Fortey & Mellish 1992). In fact, both groups of methods are equally useful in mos situations independent of any theoretical considerations since both organiz structured data and allow hypotheses to be tested regarding biological distribu tions.

Most biogeographical information is based on binary data; an organism i either present or absent. Clearly, rare taxa are overemphasized and common tax underplayed. Irrespective of the rarity or abundance of an organism, its presenc is a spike in an overall taxon range. These raw data may be converted into a wid variety of binary coefficients for further analysis. Unfortunately in some data set the biogeographical signals may be masked by other factors such as palaeoen vironmental controls and local community structure. Therefore the resultin, palaeobiogeographical patterns may be a complex of provincial, environmenta and biological fingerprints. To minimize the controls of environment, many bio geographical analyses compare similar environments or groups of communitie Statistical analysis proceeds with the preparation of a matrix of presence an absence of all taxa across all the sites investigated. Phenetic methods develop thi by the computation of a square similarity or distance coefficient matrix.

Two main groups of coefficients have been used to assess affinities an palaeogeographic distributions of faunas and floras. Distance coefficients effec tively measure the distance apart of sites, whereas similarity coefficients comput the closeness of assemblages. In most cases the distance coefficient (D) is given b $D = 1 - S$, where S is the similarity coefficient. Four of the more common coeffi cients are as follows:

1. the Dice coefficient $= 2A/2A + B + C$;
2. the Jaccard coefficient $= A/A + B + C$;
3. the simple matching coefficient $= A + D/A + B + C + D$; and
4. the Simpson coefficient $= A/A + E$.

In all these equations, A is the number of taxa common to any two samples, B i the number in the first sample, C is the number in the second sample, D is th number of taxa absent from both samples, and E is the smaller of the two sample (B and C) compared.

These analyses will produce symmetrical matrices of distance or similarit coefficients. However, two complementary types of matrix can be generated intersite comparisons or Q-mode analysis provides a matrix of distances or simi larities between localities or sites based on the distribution of the variates or taxa R-mode analysis generates a matrix of distances or similarities between tax occurring across localities or sites which now act as the variates. The most effi cient and informative analyses use a cross plot of the Q- and R-mode cluster against the original, ordered data matrix.

Smith (1990) has simulated the values for a range of similarity coefficients using
a Monte Carlo method in order to test the efficacy of each measure. The method
generates a large number of coefficient values from randomly generated data sets;
each value may then be plotted against sample size. The Dice coefficient operates
best on sparse data sets whereas both the Dice and Simpson coefficients function
well with large data matrices; the Jaccard coefficient was found to be less reliable
(Smith 1990). More recently, Shi (1993) has compared the efficiency of 39 binary
similarity coefficients. Each of these measures was evaluated against a series of
criteria. Not surprisingly, none of the coefficients tested met all the criteria; nevertheless, 7 of the 39 techniques gave adequate results.

Cluster analysis generates dendrograms through a choice of linkage methods.
This approach assumes the existence of data point clusters within some form of
hierarchical structure. Ordination techniques such as principle component analysis (PCA) and correspondence analysis (CA) reduce the dimensionality of the
measurement space: they allow the main directions of variation in the sample to
be isolated and interpreted. In contrast to implicit assumptions of hierarchical
structure in cluster analysis, ordination techniques do not require *a priori* groups
or discontinuities within the raw data set. Seriation is an ordering technique that
will rearrange a square matrix to maximize gradational structure within the data
set. In theory at least, biogeographical data, largely controlled by climatic gradients, are ideal for seriation procedures (Parkes *et al.* 1990).

Cladistic methods have been applied specifically to problems of vicariance
biogeography although the technique works equally well, as a method of or-
ganizing data sets, without any such *a priori* assumptions (Newton 1990). Taxonomic cladistics assumes that any pair of organisms, at some stage in the past,
shared a common ancestor. New taxa are recognized on the basis of derived
features or apomorphies; these taxa also have shared features in common with
their ancestors. When translated into biogeographical terms, and assuming a
vicariance model, a now fragmented province shared a common origin with
other provinces. Taxa that have appeared since fragmentation are analogous to
apomorphic characters, and define a new province. Shared taxa, however, can
group together similar provinces that presumably were derived from a common
biotic realm.

An example of the application of these techniques is in the manipulation of the
large database available for the distribution of early Ordovician brachiopods in the
North Atlantic region. Phenetic and cladistic analyses of the database (Figure 18.2)
have helped fingerprint a variety of terranes that developed within or marginal to
the early Palaeozoic Iapetus Ocean (Harper *et al.* 1996). In addition, assuming a
faunal gradient occurred across the early Ordovician Iapetus Ocean, a seriation of
presence and absence data will help resolve spatial trends. Data from sites around
the Iapetus Ocean are far from homogeneous: some samples are large and diverse
while others are represented by only a few taxa. Nevertheless cross-latitudinal
trends from high (Celtic) to low (Toquima-Table Head) are apparent with broadly
similar patterns to those established by other phenetic and cladistic methods
(Parkes *et al.* 1990).

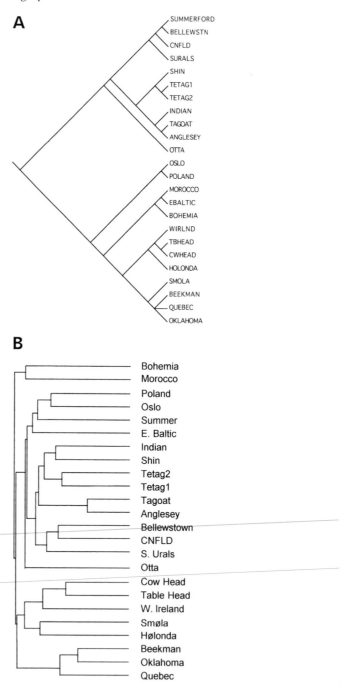

Figure 18.2   *Statistical analysis of sample of 25 assemblages from around and within the Early Ordovician Iapetus Ocean. **A.** Cladistic analysis. **B.** Cluster analysis. **C.** Correspondence analysis. **D.** Seriation on a smaller number of assemblages. Location of the majority of fauna is indicated on Figure 18.3. [Modified from: Harper et al. (1996) and Parkes et al. (1990)]*

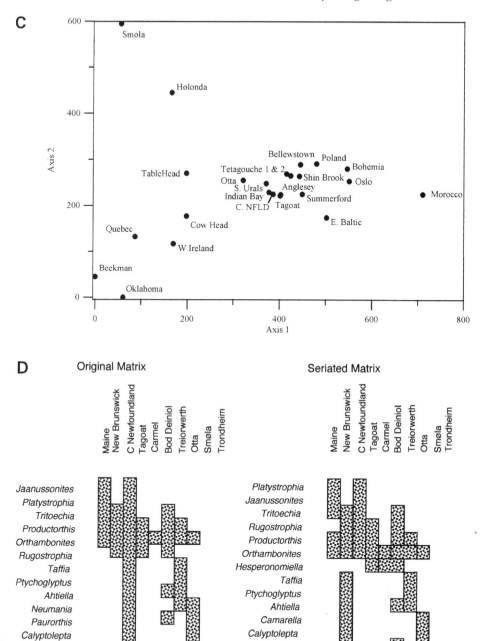

Figure 18.2 *continued*

18.4 STUDIES IN MOUNTAIN BELTS

Biostratigraphical and biogeographical analyses in mountain belts have developed rapidly. Although the use of fossils to establish the age and origin of the rock sequences continues to be fundamental to many investigations in orogenic zones, fossil data are also increasingly used to determine the mode and rates of tectonic processes. Many of the case studies discussed in this section, such as those in the Caledonides (Nowlan & Neuman 1995), highlight the use of palaeontological data in the reconstruction of the palaeogeography together with magmatic and tectonic events in mountain belts, while investigations in the Banda Arcs and the Caribbean Region have targeted biostratigraphic information provided by microfossils in detailed calibrations of the rates of tectonic processes.

Most examples have been drawn from the Phanerozoic mountain belts of Europe. These examples are also linked to tectonic processes; some belts illustrate the products of particular processes better than others. Together the European mountain belts provide more than adequate opportunities to unlock the complexities of orogenic zones and relate stratigraphical data to tectonic processes. The entire structure of Europe is discussed by Derek Ager in a readable article (1975) and book (1980), where emphasis is on the stratigraphical evolution of the continent. The case histories are presented in stratigraphical order but each belt is an individual with its own peculiarities.

18.4.1 Scandinavian Caledonides

The Scandinavian Caledonides display evidence for oceanic opening, closure, terrane development and collision, together with large-scale crustal movements including continental rotations. In addition, events in the mountain belt are reflected in the stratigraphy of the adjacent Baltoscandian platform. The Scandinavian Caledonides stretch some 1800 km from north to south-west Norway (Figure 18.3). The stratigraphy of the Lower Palaeozoic rocks of the orogen has proved of fundamental importance in the reconstruction of events and their location along the belt (Bruton & Harper 1988). Within this mountain belt a pile of thrust sheets, translated from west to east onto the Baltic craton during the Ordovician and Silurian, was constructed as Baltica moved from high to low latitudes (Harper *et al.* 1996). The craton, however, rotated slowly anticlockwise (Torsvik *et al.* 1991), entraining in the lower thrust sheets of the Lower Allochthon the dismembered parts of encratonic marginal basins with Baltic faunas; the higher units of the Middle Allochthon contain parts of the Baltic miogeocline whereas the lower units of the Upper Allochthon contain more exotic volcanogenic units derived from within the Iapetus Ocean. The higher parts of the Upper Allochthon are dominated by Laurentian marginal faunas whereas the Uppermost Allochthon may contain Laurentian basement. North American faunas have long been known from volcanogenic rocks of the higher parts of the Upper Allochthon in the Trondheim Region, in contrast to the Baltic faunas of the Baltoscandian platform and the

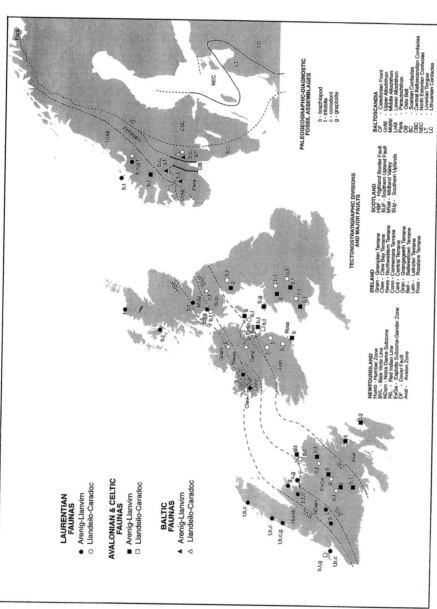

Figure 18.3 *Northern part of the Appalachian–Caledonide Orogen: palaeogeographical affinities of Ordovician faunal assemblages from Newfoundland, Britain, Ireland and Scandinavia. Tectonic sutures and inter-terrane correlations are indicated. [Reprinted from Palaeogeography, Palaeoclimatology, Palaeoecology, 121, Harper et al., The palaeogeography of early Ordovician laepetus terranes: an integration of faunal and palaeomagnetic constraints, 297–312, © 1996, with kind permission of Elsevier Science – NL, Sara BurgerharStaat-25, 1055 KV Amsterdam, The Netherlands]*

Lower Allochthon. Furthermore, serpentine conglomerates in the lower part of the Upper Allochthon contain a remarkable shelly assemblage with confusing provincial signals (Bruton & Harper 1981). The Otta Conglomerate contains serpentine clasts together with an early Llanvirn fauna dominated by brachiopods and trilobites. The brachiopods present are dominated by endemic taxa, such as *Ottadalenites*, together with some Baltic and North American forms. The association probably developed around islands provided by serpentine seamounts within the Iapetus Ocean (Pedersen *et al.* 1992) and is another aspect of the Celtic group of assemblages. In a traverse north-westwards from the Baltoscandian platform, Baltic, Celtic and related (high to mid latitude) faunas together with Toquima-Table Head (low latitude) faunas are sequentially developed during the early Ordovician through the nappe pile which now contains the Iapetus Ocean and its margins. These faunas recur in sequence throughout the Appalachian–Caledonide belt (Neuman 1984) providing a consistent pattern of the palaeogeography of the Iapetus Ocean.

Stratigraphical control of both the allochthons and the adjacent parautochthonous and autochthonous platform successions has permitted an analysis of the relationship between the respective histories of the orogen and the platform (Figure 18.4). Changes in palaeoenvironments and sediment geochemistry together with biotic migrations and the development of bentonites on the Baltic platform have been related to tectonic events in the developing Caledonian Orogeny to the west (Bruton & Harper 1988). Moreover, within the allochthons, individual interactions can be tracked within stratigraphical successions. For example, the development of mid Ordovician turbidite facies in the basins of the Lower Allochthon signalled the eastward advance of the nappes of the Middle Allochthon (Bruton *et al.* 1989).

18.4.2 British and Irish Caledonides

The Caledonides of Britain and Ireland have promoted some of the classic studies on the stratigraphy of mountain belts. A series of Caledonian terranes (Figure 18.3) has now been established in some detail across mainland Britain (Bluck 1990). Recently, considerable attention has been focused on the mutual relationships between adjacent terranes, their origins, history, and time of docking. In particular, the Central Highlands of Scotland have provided a critical test for terrane analysis.

The Highland Border Group was traditionally considered to be the highest part of the Dalradian Supergroup, cropping out along the Highland Boundary Fault. Reliable early Ordovician dates, based mainly on trilobites and brachiopods from the Dounans Limestone (Ingham *et al.* 1985) from the Highland Border Complex, separate this unit from the adjacent Dalradian terrane, along the Highland Boundary Fault, which was simultaneously deformed and uplifted. Moreover, the Ordovician data indicated that the Highland Border basin probably had a quite different history associated with the Laurentian margin, rather than as an exotic Dalradian terrane (Curry *et al.* 1984). Ordovician data from chert blocks within the Clew Bay Supercomplex (Harper *et al.* 1989), south-westwards along strike,

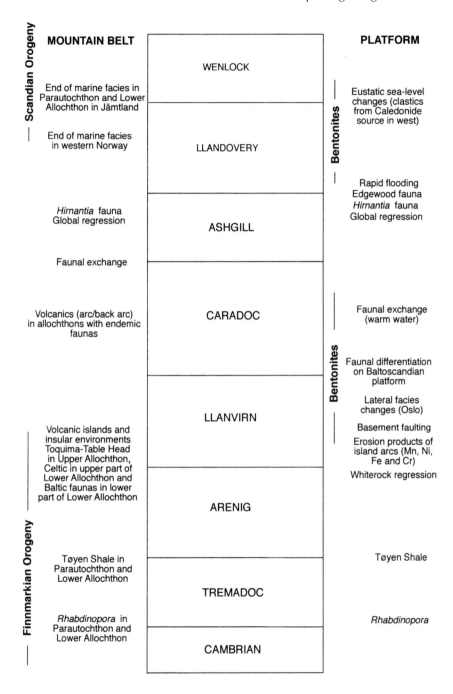

Figure 18.4 *Events on the Baltoscandian platform and in the adjacent Caledonides during the Ordovician–mid Silurian interval based on stratigraphical correlation between the fold belt and platform [Modified from: Bruton & Harper (1988)]*

suggested the extension of the complex into western Ireland. The shaly matrix of the upper part of the supercomplex commonly developed as a melange, however, contains Wenlock palynomorphs implying the zone was a focus for late Caledonian movements (Williams *et al.* 1994). This zone may not have stabilized until the mid Devonian.

Across Ireland, eight terranes have been recognized on the basis of stratigraphical data (Figure 18.3; Harper & Parkes 1989). The changing biogeographical signals in the fossiliferous terranes suggest all were in relatively close proximity, at low latitudes, by the end of the Ordovician (Parkes & Harper 1996). In a transect from north-west to south-east, the Grampian Terrane is dominated by the mainly Proterozoic Dalradian Supergroup. The Northwestern Terrane contains a variety of arc-related successions probably founded on ophiolitic basement and marked by the development of the Toquima-Table Head fauna during the early Ordovician and the Scoto-Appalachian fauna during the late Ordovician. The Central Terrane is the westward extension of the Southern Uplands of Scotland; its Ordovician faunas, particularly in the upper part of the system, are Laurentian marginals (Clarkson *et al.* 1992; Owen *et al.* 1996). The Grangegeeth Terrane is one of two small terranes developed in the Iapetus suture zone; its early Ordovician Gondwanan graptolite assemblages are succeeded by Laurentian marginal brachiopod faunas during the late Ordovician, signalling a transit of the terrane from high to low latitudes during the period (Owen *et al.* 1992; Harper *et al.* 1996). The track of Grangegeeth is therefore parallel to that of the Exploits Subzone of Newfoundland (Williams *et al.* 1996). The Bellewstown Terrane includes early Ordovician Celtic-type faunas whereas Anglo-Welsh assemblages occur in middle Ordovician strata. Abundant and diverse mid Ordovician Anglo-Welsh faunas have been reported from many localities across the Leinster Terrane (Parkes 1992). In the Rosslare Terrane an early Ordovician cover sequence with Celtic province brachiopods rests unconformably on Precambrian Gondwanan basement.

This collage of Iapetus and part of its margins was constructed during the Caledonian orogeny. In the northern parts of the island, disrupted Silurian basins (Williams & Harper 1991) are succeeded by post-orogenic Old Red Sandstone (ORS) molasse systems, whereas in the south-west the Caledonian terranes are overstepped by the marine and marine-marginal deposits associated with the Variscan basins of the Rhenohercynian Zone.

18.4.3 Newfoundland

Newfoundland is the classic laboratory for the study of terranes (Williams 1984), and western Newfoundland displays the key events associated with the collapse and destruction of a continental margin. The island is a mosaic of Caledonian terranes (Figure 18.3), each defined by unique successions and the majority by characteristic faunal assemblages (Williams *et al.* 1996). In some respects the island is the mirror image of the Scandinavian Caledonides. In a traverse from east to west across the island the rocks of the Avalon and Gander zones developed in peri-Gondwana settings. The Avalon Zone exposes high-latitude, Acado-Baltic

Cambrian trilobite faunas together with early Ordovician acritarch, graptolite and trilobite faunas. These faunas show close similarities to other Avalonian-type biotas from north-west Europe. The rocks of the Gander Zone are metamorphic and unfossiliferous. The Exploits Subzone contains, within lower Ordovician volcanogenic rocks, high-latitude, Celtic province fossils, and in the mid Ordovician the terrane was inhabited by Laurentian marginal, Scoto-Appalachian taxa. The early Ordovician brachiopod faunas occur in a range of localities from Baie d'Espoir to New World Island and are variants of the high-latitude Celtic province that occupied peri-Gondwanan and intra-Iapetus terranes (Harper *et al.* 1996). Both the volcaniclastics of the Notre Dame Subzone and the carbonates of the adjacent Humber Zone contain low-latitude, Laurentian faunas. The early Ordovician successions contain typical examples of the Toquima-Table Head shelly fauna. This fauna had an amphicratonic distribution around Laurentia and was the low-latitude equivalent of the Celtic faunas.

Detailed stratigraphical studies in the Humber Zone have permitted the reconstruction of the collapsed Laurentian margin (Pohler *et al.* 1987; Barnes 1991). During the Cambrian and early Ordovician an extensive carbonate platform developed along the passive north-east margin of Laurentia. During the Ordovician the platform foundered as terranes outboard the North American Miogeocline were accreted and loaded onto the Laurentian margin. The platform edge, shelf break and upper slope were dismembered and are now located as allochthonous blocks within deep-water debris flows adjacent to the autochthonous sequences (James *et al.* 1989). The platform collapse was associated with the Taconic Orogeny and the transition, during the Arenig, from passive margin to convergent conditions. Up to nine events have been tracked during the development and collapse of the platform, all related to a biostratigraphical framework (Figure 18.5).

The docking history of the Newfoundland terranes is known in detail and is stratigraphically constrained (Figure 18.6). The earlier docking events (amalgamation of the Notre Dame and Taconic units with the Laurentian Continental Margin and the Exploits Subzone with the Gander Zone) are marked by relatively low-angle contacts associated with melanges and ophiolites, whereas the later events (contact between the Avalon and Gander zones, and the Notre Dame and Exploits subzones) are associated with high-angle structures, commonly with mylonites.

18.4.4 European Variscides

The stratigraphy of the Variscides has revealed mid Palaeozoic episodes of magmatism and microcontinent interactions across Europe through syn- and post-tectonic basinal sequences. The Variscide belt extends roughly E–W across Europe for about 1000 km (Matte 1995). The belt is not continuous, cropping out in a series of isolated massifs from south-east Ireland to Romania. Traditionally the Variscides were considered unusual in comparison with the older Caledonides and the younger Alps. The belt apparently lacked deep-water facies, such as flysch, and ophiolites were rare. Sedimentary sequences developed during the Variscan cycle were dominated by fossiliferous platform carbonates and local, more condensed,

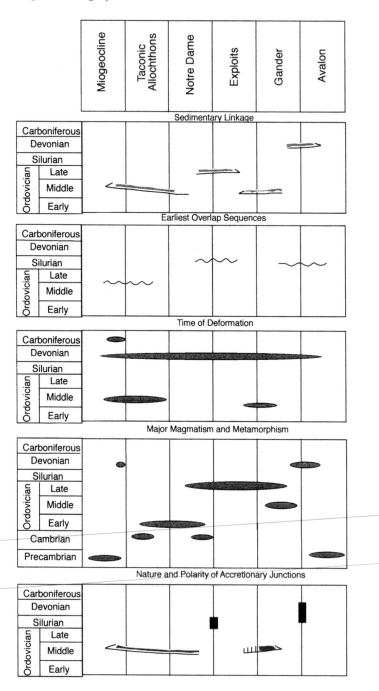

Figure 18.5 *Docking of Newfoundland terranes: accretionary history of the Newfoundland Caledonides based on stratigraphical analysis. Arrows associated with sedimentary linkage indicate the direction of sediment transport. Arrows associated with the nature and polarity of accretionary boundaries indicate the direction of thrust translation; vertical lines on thrust indicate hard contact; vertical bars indicate steep boundaries usually associated with mylonites*

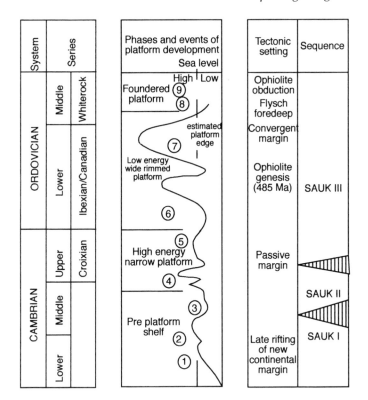

Figure 18.6 *Collapse of the Laurentian miogeocline: four phases and nine events are indicated together with changes in palaeobathymetry. The nine events are (1) clastic shelf sedimentation; (2) carbonate facies established; (3) deeper-water facies developed; (4–5) cyclicity on shelf margin; (6) transgressive cycle with deep-water carbonates; (7) transgressive cycle with deep-water facies covered by flysch; (8) block faulting and subsidence; and (9) encroachment of flysch (see James et al. (1989) for a fuller discussion)*

basinal deposits. Volcanism, where present, is usually acidic and submarine; granites are common along the belt, often associated with mineralization. Faunal provinciality, a particular feature of the Caledonide successions, is not marked in the Variscides. Devonian brachiopods, for example, are part of the Old World Realm and more specifically belong to the Rhenish–Bohemian region (Boucot 1975). Moreover, until recently there was little evidence of large-scale horizontal transport of units. Deformation in the belt was apparently achieved mainly through vertical movements and the passive shearing of cover sequences (Krebs 1976). The absence of characteristic rock sequences associated with sutures, together with a lack of faunal provinciality across most of the belt, suggests that either key areas are not exposed or major sutures were located farther south, and now lie within the Alpine chain.

Much of the pre-Variscan basement to the massifs developed at high latitudes as part of Gondwana and its margins. Widespread carbonates were deposited across Europe during the Devonian prior to syn-orogenic facies sporadically developing during the early Carboniferous and post-orogenic molasse in the late Carboniferous. In contrast to the Caledonides, palaeobiogeography has not been a core subject in Variscan studies. Nevertheless, stratigraphical control based on mainly conodont and goniatite faunas together with palynology has helped unravel a number of aspects of Variscide evolution. Contrasts in stratigraphy have been used to establish a series of tectonostratigraphical zones across the belt (Figure 18.7).

The Rhenohercynian Zone is the most laterally persistent division in the orogen, extending from south-west Ireland through south-west England to eastern Germany and possibly farther eastwards into Moravia and southwards into Portugal. It marks the track of an external basin to the orogen. In most of the zone early Devonian clastics are succeeded by mid Devonian–early Carboniferous platform carbonate facies. Nevertheless flysch sedimentation was initiated in the southern part of the zone in the early Carboniferous, prograding northwards during the mid Carboniferous. Coal-bearing molasse sequences were established by the Namurian.

The Saxothuringian Zone comprises two components: (1) in the north, the Mid-German Crystalline Rise; and (2) in the south, the Saxothuringian basin. The Mid-German Rise is dominated by plutonics and high-grade metamorphics. The ages of these protoliths are uncertain although late Silurian spores have been retrieved from part of the zone. The Saxothuringian basin is underlain by continental basement. Upper Ordovician through lower Carboniferous sequences are mainly hemipelagic shales. Occasional incursions by turbidites and the local development of volcanic seamounts punctuated the background sedimentation. More persistent turbidites such as the Bavarian Flysch were deposited from the early Carboniferous onwards.

Most of the Moldanubian Zone consists of metamorphic and plutonic rocks. However, in the central, depressed part of the zone, the Barrandian Syncline contains the classic fossiliferous Palaeozoic succession of Bohemia. The lower Cambrian to upper Ordovician succession is dominated by a variety of marine clastics with locally abundant Mediterranean province faunas. Silurian through middle Devonian rocks are mainly hemipelagic and shelf carbonates with the development of reefs. Late Devonian turbidites signal an early phase of Variscan deformation. The Mediterranean Zone is developed across southern Europe and is dominated by thinly developed fine-grained siliciclastic rocks including shales and marls together with condensed limestone sequences.

The development of marine platforms and basins across the Rhenohercynian Zone has been charted against a chronostratigraphical framework based on both conodont and goniatite biozones (House 1975a). A broad ecological analysis of the zone recognizes two magnafacies: the shallow-water Rhenish and the more distal, deeper-water Hercynian. This scheme has been enhanced to include an intermediate Eifel magnafacies with calcareous shales and limestones deposited in back-reef environments. All three marine facies contrast with the continental Old Red magnafacies which developed around the Old Red Sandstone continent. Broad biofacies patterns have been established from the disparate outcrops of the Variscan belt (House 1975b) built on the facies analyses of workers on mainland Europe.

511

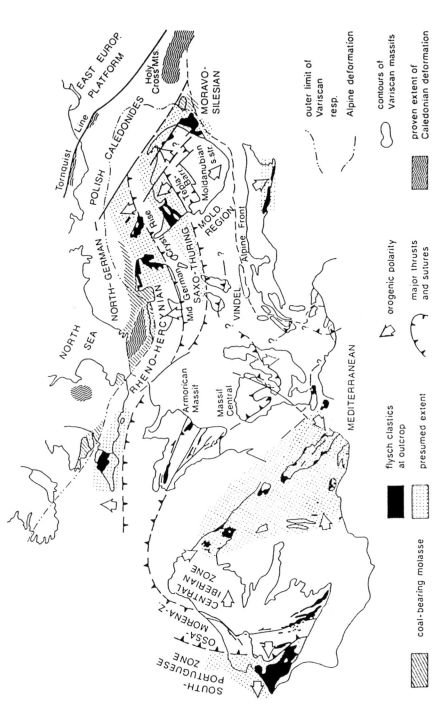

Figure 18.7 *Tectonostratigraphical divisions of the Variscide Orogen in Europe: the divisions are based on stratigraphical contrasts between the zones. [Reproduced with permission from the Geological Society of America; from: Franke (1989)]*

The development of these biofacies, related to topographic depressions (Becken) and rises (Schwellen), has aided the recognition of multi-order basins associated with magmatic and tectonic events in the belt.

Volcanism is well developed across the Rhenohercynian Zone, peaking during the Givetian–Frasnian and Tournaisian intervals. Allochthonous volcanogenic associations occur in the Rhenisch Schiefergebirge and the Harz Mountains. Similar assemblages occur along strike in south-west England and are constrained by fossil data. For example, within the deformed Gramscatho Group of south-west England the timing of volcanism is tightly constrained within a conodont biostratigraphy. Middle and Upper Devonian conodonts from limestones interbedded with lavas have precisely confined two phases of eruption to the Eifelian and the Frasnian (Barnes & Andrew 1986).

Timing of orogenic activity across the belt is well established (Ziegler 1989). Six phases of orogeny of varying intensity and geographic development have been recorded throughout the orogen: the Bretonic (end Devonian), Sudetic (mid Carboniferous), Erzgebirge (mid Silesian), Asturic (late Silesian), Saalic (early Permian) and Pfalzic (end Permian). These phases are recognized in a variety of marine and non-marine facies by various disconformities and unconformities together with the development of flysch, bracketed by a variety of microfossils, and invertebrate and plant macrofossils. The Bretonic phase is fairly widespread throughout the central part of the belt, particularly the Armorican Massif; the Sudetic phase reflects the major deformation of the Saxothuringian basin; while the Asturic phase marks terminal deformation throughout the Rhenohercynian Zone. These could be docking events. Ziegler (1986, 1990), for example, suggested that these events signalled a series of discrete docking phases as Variscan terranes collided and were sequentially added to the southern margin of the Baltica–Laurentia cratonic complex. Alternatively these phases may represent parts of a continuous orogeny occurring throughout the mid Palaeozoic (Franke 1989). Bilateral subduction models involve a pair of southward-dipping subduction zones in the Rhenohercynian and the Saxothuringian zones. Deformation may have been continuous as the orogenic front moved northwards, accreting internal zones within the belt. The development of the European coal and molasse basins reflects the final stages of the orogen and many molasse deposits overlap a number of the Variscan zones.

The main tectonostratigraphical zones of the Variscan Orogen crop out in the Bohemian massif (Figure 18.8) where contrasting stratigraphical sections, intervals of volcanism and tectonism, together with flysch and molasse development, have been documented (Chaloupsky 1989). Moreover, a diachronous overstep sequence of molasse overlaps all the terranes from the Rhenohercynian Zone in the north-west to the Moldanubian Zone in the east.

There is, however, stratigraphical evidence for large-scale megashears along the Variscide belt post-dating the molasse sequences. In southern Europe marine upper Carboniferous and Permian facies are restricted to the Cantabrian Mountains of northern Spain and the Carnic Alps (Martínez-García 1991). If these two areas were adjacent at the end of the Variscan Orogeny then dextral shear in the order of 1500 km has occurred since the Permian. A variety of structures in Iberia

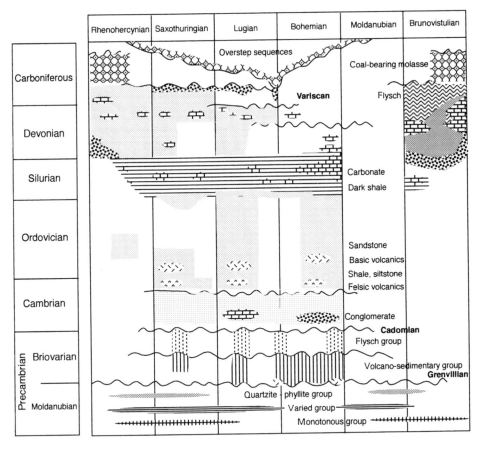

Figure 18.8 *Bohemian terranes: magmatic, tectonic and overstep events are indicated. [Modified from: Chaloupsky (1989)]*

and Montagne Noire have been related to these later strike-slip movements (Martínez-García 1996).

18.4.5 European Alps

The Alps contain classic pre-, syn- and post-orogenic stratigraphical sequences related to ocean opening, closure and continental collision, monitored by the development of flysch and molasse in foreland basins. The Alpide belt is the youngest of Europe's orogenic zones, extending eastwards from the Betics through the Western and Eastern Alps onto the Carpathians. The orogen is sinuous and although commonly less than 200 km wide, it includes the arcs and basins of the ancient Tethys Ocean, now preserved in 10% of its original width. Alpine successions contain ophiolite sequences and oceanic sediments together with syn-orogenic flysch located in far-travelled nappes as well as the post-orogenic foreland sediments in molasse basins. The classic reviews of the dynamics (Trümpy 1960) and the stratigraphy, structure and metamorphism (Ramsay 1963) of the orogen have, of course, been modified by much subsequent research, but nevertheless still provide superb introductions to Alpine geology.

Each part of the belt is characterized by a number of tectonic units defined by distinctive lithofacies and tectonic styles (Trümpy 1982). The pre-orogenic cover sequences are dominated by dolomites, limestones and silty marls, and range in age from Triassic to early Cretaceous. The syn-orogenic sequences range from the late Cretaceous to the early Tertiary and are dominated by flysch. The internal zones, with ophiolites and pelagic sediments, have large recumbent folds and extensive nappe piles constructed during the late Cretaceous to early Tertiary interval. Post-orogenic uplift of the belt helped promote the development of thick molasse successions in a number of foreland basins around the orogen. New and significant age data continue to be generated from both basement (Höll & Reitz 1991; Reitz & Höll 1992) and cover sequences (Kiessling 1992) in the Alpine region.

In simple terms, the Alpine chain, across Switzerland, is divided into foreland successions such as the Jura Mountains and Swiss Molasse Trough, the Helvetic, the Penninic and the Austroalpine nappe units and the Southern Alps (Figure 18.9). These last four units form a tectonostratigraphy that is separated by major tectonic structures, normally large-scale thrusts. In addition, the Prealps is an allochthonous unit derived from the Penninic Zone, and now located on the Helvetic Zone and the Swiss Molasse Trough. The Helvetic Zone, the externalides of the belt, contains the cover, mainly carbonate sequences, to some Variscan massifs in nappes translated a few tens of kilometres towards the European foreland. The Penninic Zone is dominated by oceanic facies, including ophiolites and deep-water sedimentary basins developed on relatively thin crust. The Austroalpine Zone and the Southern Alps have metamorphic complexes covered by low-grade Palaeozoic successions and Mesozoic carbonates developed as a series of giant nappes derived from continental areas south of the Mediterranean Tethys.

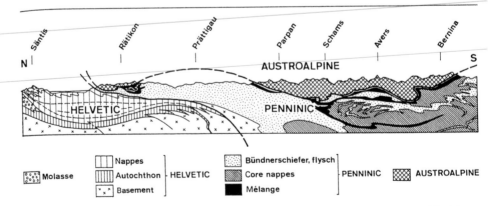

Figure 18.9   *Simplified section through the Swiss part of the Alpine Orogen. [Modified from: Hsü (1994)]*

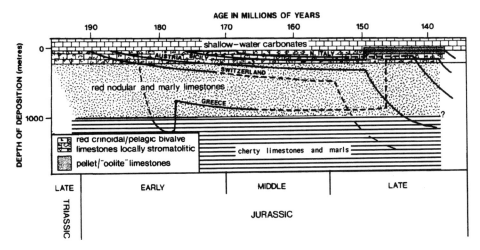

Figure 18.10 *Subsidence curves for parts of the Tethyan continental margins based on relatively shallow-water regions; intense tectonic activity along the belt is evident during the late Jurassic (c. 150 Ma) when margins foundered. [Reproduced with permission from the Geologists' Association, from: Jenkyns (1980)]*

Precise stratigraphical control has helped elucidate at least three important aspects of Alpine geology. First, stratigraphy has provided a matrix within which the changing composition of bio-chemical sediments through time has signalled the development and collapse of platforms along the Alpine belt. Pure chemical sediments such as salt and gypsum are present but rare. Secondly, the timing of orogeny and nappe movement can be monitored by the development of flysch facies. Syn-orogenic basins record the orogeny and uplift in front of deforming belts within the migrating locus of the orogen. Thirdly, sedimentary linkage and eventual overlap in the molasse basins of the belt can track the final emplacement of major nappe terranes.

Stratigraphical constraints have provided an operational framework for investigating the initial collapse of marine carbonate banks along the southern European continental margin during the early Jurassic. Sophisticated studies of changing depth-dependent limestone facies with time chart the transitions from shallow-water carbonates with reefs through pelagic to nodular limestones and into cherty limestones and marls (Jenkyns 1980). These lithostratigraphic changes have been recognized in a number of different segments of the Alpine belt stretching from Austria and Switzerland to Italy and Greece (Figure 18.10). In most sections the track of the margin collapse is similar, with initially a gradual transition from shallow-water environments to slope deposits during an interval of about 40 million years. The subsequent plunge to more bathyal depths was more sudden, achieved during intervals of about 5 million years.

As noted previously, orogeny can be signalled by flysch developed in a number of palaeogeographic belts across the Alps (Figure 18.11). For example, in the Swiss Alps at least seven flysch zones have been established:

1. the North Helvetic Flysch which is about 2 km thick and of late Eocene–early Oligocene age, overlying mid Eocene Nummulitic limestones;
2. the South Helvetic Flysch of late Eocene age, overlying mid–late Eocene Globigerina Marl;
3. the Ultrahelvetic Flysch which is overthrust above the highest Helvetic units, normally between the Helvetic and the Penninic nappes, and which is possibly late Eocene and Oligocene in age;
4. the Wildflysch of central Switzerland, which is really a tectonic melange, contains blocks of pelagic limestone and marl and is of mid to late Eocene age;
5. the Schlieren Flysch of South Penninic origin is Maastrichtian to early Eocene in age with transported foraminifera;
6. the Gurnigel Flysch, which is similar in age and lithology to both the Schlieren Flysch and the Wägital Flysch; and
7. the late Cretaceous Niesen Flysch in the Pre-alps, which is derived from the North-Penninic realm.

The northward translation of the Austroalpine terrane had a profound effect on the sites and timing of both flysch and molasse facies in the evolving orogen. The Austroalpine nappes may have reached the Helvetic margin by late Eocene, moving through the Oligocene and prompting the development of these flysch and molasse basins (Hsü 1994).

Nappe emplacement, however, may also be tracked by sediment linkages obvious in the Molasse Basin (Homewood 1977; Homewood *et al.* 1986). Subsidence of up to 3 km of the Helvetic margin occurred from the latest Triassic to the late Tertiary and was coupled with the onset of foreland basin sedimentation (Pfiffner 1986). Four main phases of molasse sedimentation are documented.

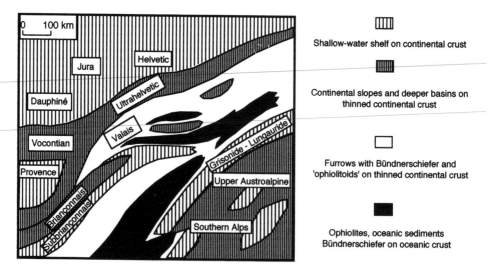

Figure 18.11 *Alpine palaeogeography during the Mesozoic. [Modified from: Hsü (1994)]*

1. Lower Marine Molasse and Northern Helvetic Flysch. These units developed within the Subalpine zone and within Helvetic nappes. Rapid subsidence occurred at the Eocene–Oligocene boundary. The molasse deposits of the Northern Helvetic Flysch contain clasts from the approaching Austroalpine and Penninic thrusts sheets together with clasts from the Prealps; only marls and sandstones are developed in the Lower Marine Molasse. Initial subsidence of the basin was due to emplacement of the Austroalpine nappes.
2. Lower Freshwater Molasse. This reflects a late Oligocene regression with material derived mainly from the Austroalpine nappes within fluviatile fan systems. In the early Miocene, clasts from the Austroalpine basement and Penninic units occur but no clasts from the Helvetic units are present, suggesting that the last units were not yet emergent.
3. Upper Marine Molasse. This unit contains a variety of onshore–offshore marine facies with coarse clastics at the base signalling an orogeny corresponding to the emplacement of the Helvetic nappes onto the Austrian Molasse Basin itself. The younger part of the molasse contains rare Helvetic clasts.
4. Upper Freshwater Molasse. These mid Miocene freshwater deposits contain some pebbles of material from the Helvetic nappes which were then emergent. Nevertheless the material is mainly derived from Austroalpine and Penninic units together with large amounts of sandy flysch of uncertain origin.

Molasse sequences adjacent to the Eastern Alps are regressive; the highest upper Miocene–Pliocene units contain pebbles of the displaced Northern Calcareous Alps (Oxburgh 1968), another exotic and allochthonous unit, like the Prealps, that docked late against the European foreland in the Austrian part of the belt.

18.4.6 Banda Arcs

In more modern orogens, palaeontological data have resolved the rates of tectonic processes. The Banda Arcs are part of a Neogene and Quaternary fold and thrust belt developing along the northern continental margin of Australia. Flat-lying allochthonous thrust sheets are similar in age to the rocks of the underlying parautochthonous sequences (Audley-Charles 1986). A detailed biostratigraphy based on planktonic foraminiferans has determined the start and termination of folding, the emplacement of the thrust sheets themselves, and the timing of imbrication and uplift of the orogen (Carter *et al.* 1976). Although the biostratigraphy has provided a working framework for the analysis of the collision between the Australian continent and the trench system, microfossil data can actually track the rates of movement and eventual uplift of individual thrust sheets. Age constraints are provided by the date of the youngest parautochthonous strata overthrust by the allochthon and the ages of the oldest parts of the autochthon resting on the allochthon and parautochthon. In simple terms, allochthonous thrust sheets must have travelled in the order of 50 km in intervals of between 4 and 5 Ma as these nappes overthrust the Australian plate. The foraminiferan zonal scheme indicated that thrusts were emplaced at rates of between 6.25 and 12.5 cm per year, whereas rates of uplift in the order of 1.5 cm per year were typical.

18.4.7 Cocos Ridge

During the Neogene, in the Caribbean region, eastward subduction of the Cocos Ridge under the Panama microplate promoted the differential uplift of the Burica Peninsula on its southwest margin and the Limón and Bocos del Toro belts to the northeast (Collins *et al.* 1995). The ridge was, in fact, a buoyant, unusually thickened block of oceanic crust and was only partially subducted. The Burica Peninsula was uplifted during the late Pliocene whereas the developing Limón and Bocos del Toro basins sited on the eastward margin of the Isthmus of Panama underwent gradual uplift, later, during the Quaternary (Figure 18.12). Clearly the

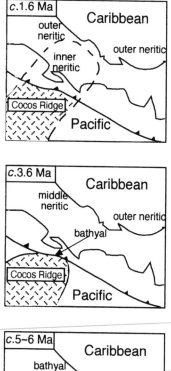

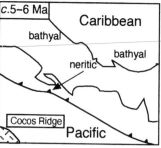

Figure 18.12 *Migration of the Cocos Ridge northeastwards, its partial subduction and influence on basin history of the Neogene of the Isthmus of Panama. The migration of the ridge is shadowed by changes in facies patterns of adjacent sedimentary basins. [Modified from: Collins et al. (1995)]*

north-east-migrating ridge system arrived under the Burica Peninsula on the Pacific margin about 3.6 Ma but did not reach the Caribbean margin until about 1.6 Ma. Tight biostratigraphic control, based on benthic foraminifera, has permitted correlation of the strata with an absolute timescale ranging from approximately 6.5 to 1.5 Ma. In addition, because the majority of the Neogene species are still extant, their current depth ranges together with data from lithofacies can be used to accurately estimate changing palaeobathymetry through the succession. Rates of emergence of about 1000 m Ma^{-1} have been calculated for the rising Burica Peninsula; rates of between 24 and 90 m Ma^{-1} occurred later in the Limón and Bocos del Toro basins.

18.5 SUMMARY

Stratigraphical studies in mountain belts have advanced with increasing precision and sophistication from the tentative realization that substantial vertical and horizontal movements occurred at the earth's surface to the prediction of modes and rates of tectonic processes within orogenic zones. Stratigraphical frameworks have helped correlate and reconstruct magmatic and tectonic events, reassemble collapsed continental margins, predict ancient ocean systems and organize in time and space the perplexing collage of terranes that comprise most Phanerozoic orogenic belts. The establishment of successions by basic stratigraphical principles and their correlation by fossils remain at the core of research in orogens. The most elegant tectonic models may still be demolished by the appearance of a new fossil assemblage with fresh age and biogeographical constraints.

The organization of the history of events in orogenic zones leads directly to an understanding of large-scale tectonic processes. The European mountain belts provide many examples where stratigraphy and tectonics are integrated. The Caledonides in Britain and Ireland, together with those in Newfoundland and Scandinavia, have provided unrivalled opportunities to identify a wide range of fossil terranes, some containing fragments of ancient oceanic crust and parts of collapsed continental margins. Faunal provinciality during the early Palaeozoic, together with identification of tectonic sutures, have helped define the origin, development and closure of the Iapetus Ocean and its island archipelagos. Events in platform basins marginal to the orogen have dated orogenic phases, volcanic eruptions and the approach of allochthonous, deforming and uplifting thrust complexes. A variety of criteria including the age of overstepping sedimentary sequences and stitching plutons has determined the age of docking of various terranes within the Caledonian belt.

The stratigraphical investigation of the many and diverse massifs of the European Variscides has established several tectonostratigraphic units across the continent. Within each sector of the orogen, magmatic and tectonic events during the mid–late Palaeozoic are bracketed by fossil data and local basins and swells have been recognized, although there is scant evidence for oceanic crust and sediments in the part of the orogen exposed north of the Alpine belt. Faunal provinciality in the same order as that in the Caledonides is not apparent.

The Alps contain classic pre-, syn- and post-orogenic sedimentary basins developed during oceanic closure and continental collision. The development of the oceanic crust, during the Mesozoic, along the Tethyan tract, together with the collapse of its margins, are recorded in the stratigraphy of the cover successions to the Variscan massifs. Syn-orogenic flysch and post-orogenic molasse are developed in a variety of foreland basins, of late Mesozoic and Tertiary age, around the orogen. The origin and movement of major thrust sheets can be tracked through the changing environments and sedimentary provenance in these basins.

More recent orogenic activity in the Banda Arcs and on the Panama Peninsula has been monitored by changing faunas and palaeoenvironments in late Tertiary and Pleistocene basins. Detailed stratigraphical data from both sections has allowed the measurement of the rates of tectonic uplift during plate interactions.

ACKNOWLEDGEMENTS

This chapter benefitted from wide-ranging discussion and perceptive comments from Paul Ryan (Galway). Enrique Martínez-García (Oviedo) read the section on the Variscides and offered useful advice, and Hanspeter Funk (Zurich) commented, positively, on the Alpine section.

REFERENCES

Ager, D.V. 1975. The geological evolution of Europe. *Proceedings of the Geologists' Association*, **86**, 127–154.

Ager, D.V. 1980. *Geology of Europe*. Pergamon Press.

Armstrong, H.A., Owen, A.W., Scrutton, C.T., Clarkson, E.N.K. & Taylor, C.M. 1996. Evolution of the Northern Belt, Southern Uplands: implications for the Southern Uplands controversy. *Journal of the Geological Society of London*, **153**, 197–205.

Audley-Charles, M.G. 1986. Rates of Neogene and Quaternary tectonic movements in the southern Banda Arc based on micropalaeontology. *Journal of the Geological Society of London*, **143**, 161–175.

Bailey, E.B. 1935. Switzerland and the Prealps. In Bailey, E.B., *Tectonic essays*. Clarendon Press, Oxford, 30–35, 57–72.

Barnes, C.R. 1991. Unravelling the biogeographical and tectonic puzzles of Lower Paleozoic Orogens. *Terra Abstracts*, **3**, 336–337.

Barnes, R.P. & Andrew, J.R. 1986. Upper Palaeozoic ophiolite generation and obduction in south Cornwall. *Journal of the Geological Society of London*, **143**, 117–124.

Bluck, B.J. 1990. Terrane provenance and amalgamation: examples from the Caledonides. *Philosophical Transactions of the Royal Society of London*, **A331**, 599–609.

Boucot, A.J. 1975. *Evolution and extinction rate controls*. Elsevier, Amsterdam.

Bruton, D.L. & Harper, D.A.T. 1981. Brachiopods and trilobites of the early Ordovician serpentine Otta Conglomerate, south-central Norway. *Norsk Geologisk Tidsskrift*, **61**, 153–181.

Bruton, D.L. & Harper, D.A.T. 1988. Arenig–Llanvirn stratigraphy and faunas across the Scandinavian Caledonides. In Harris, A.L. & Fettes, D.J. (eds) *The Caledonian–Appalachian Orogen*. Geological Society Special Publication No. 38, 247–268.

Bruton, D.L. & Harper, D.A.T. (eds) 1992. Fossils in fold belts. *Terra Nova*, **4**, 178–253.

Bruton, D.L., Harper, D.A.T. & Repetski, J.E. 1989. Stratigraphy and faunas of the Parautochthon and Lower Allochthon of southern Norway. In Gayer, R.A. (ed.) *The Caledonide Geology of Scandinavia*. John Wiley and Sons, Chichester, 231–241.

Carter, D.J., Audley-Charles, M.G. & Barber, A.J. 1976. Stratigraphical analysis of island arc–continental margin collision in eastern Indonesia. *Journal of the Geological Society of London*, **132**, 179–198.

Chaloupsky, J. 1989. Major tectonostratigraphic units of the Bohemian Massif. In Dallmeyer, R.D. (ed.) *Terranes in the circum-Atlantic Paleozoic orogens*. Geological Society of America Special Paper 230, 101–114.

Clarkson, E.N.K., Harper, D.A.T., Owen, A.W. & Taylor, C.M. 1992. Ordovician faunas in mass-flow deposits, southern Scotland. *Terra Nova*, **4**, 245–253.

Collins, L.S., Coates, A.G., Jackson, J.B.C. & Obando, J.A. 1995. Timing and rates of emergence of the Limón and Bocas del Toro basins: Caribbean effects of Cocos Ridge subduction? In Mann, P. (ed.) *Geologic and Tectonic development of the Caribbean plate boundary in southern Central America*. Special Paper of the Geological Society of America 295, 263–289.

Coney, P.J., Jones, D.L. & Monger, J.W.H. 1980. Cordilleran suspect terranes. *Nature*, **288**, 329–333.

Curry, G.B., Bluck, B.J., Burton, C.J., Ingham, J.K., Siveter, D.J. & Williams, A. 1984. Age, evolution and tectonic history of the Highland Border Complex, Scotland. *Transactions of the Royal Society of Edinburgh: Earth Sciences*, **75**, 113–133.

Darwin, C. 1839. *Journal of researches into the geology and natural history of the various countries visited by H.M.S. Beagle, under the command of Captain Fitzroy R.N. from 1832–1836*. Henry Colburn, London.

Dewey, J.F. 1969. Evolution of the Appalachian/Caledonian Orogen. *Nature*, **222**, 124–129.

Fortey, R.A. & Cocks, L.R.M. 1986. Marginal faunal belts and their structural implications, with examples from the Lower Palaeozoic. *Journal of the Geological Society of London*, **143**, 151–160.

Fortey, R.A. & Cocks, L.R.M. 1990. Fossils and tectonics. In Briggs, D.E.G. & Crowther, P.R. (eds) *Palaeobiology: a synthesis*. Blackwells, Oxford, 482–490.

Fortey, R.A. & Mellish, C.J.T. 1992. Are some fossils better than others for inferring palaeogeography? The early Ordovician of the North Atlantic region as an example. *Terra Nova*, **4**, 210–216.

Franke, W. 1989. Tectonostratigraphic units in the Variscan belt of Central Europe. In Dallmeyer R.D. (ed.) *Terranes in the circum-Atlantic Paleozoic orogens*. Geological Society of America Special Paper 230, 67–90.

Hallam, A. 1986. Evidence of displaced terranes from Permian to Jurassic faunas around the Pacific margins. *Journal of the Geological Society of London*, **143**, 209–216.

Harper, D.A.T. 1992. Ordovician provincial signals from Appalachian–Caledonian terranes. *Terra Nova*, **4**, 204–209.

Harper, D.A.T. & Parkes. M. A. 1989. Palaeontological constraints on the definition and development of Irish Caledonide terranes. *Journal of the Geological Society of London*, **146**, 413–415.

Harper, D.A.T., Williams, D.M. & Armstrong, H.A. 1989. Stratigraphical correlations adjacent to the Highland Boundary fault in the west of Ireland. *Journal of the Geological Society of London*, **146**, 381–384.

Harper, D.A.T., Scrutton, C.T. & Williams, D.M., 1995. Mass mortalities on an Irish Silurian seafloor. *Journal of the Geological Society of London*, **152**, 917–922.

Harper, D.A.T., MacNiocaill, C. & Williams, S.H., 1996. The palaeogeography of early Ordovician Iapetus terranes: an integration of faunal and palaeomagnetic constraints. *Palaeogeography, Palaeoclimatology and Palaeoecology*, **121**, 297–312.

Höll, R. & Reitz, E. 1991. Palynological evidence for an Upper Proterozoic age in the older schist cover and for Lower Cretaceous age in the younger schist cover of the Tauern Window, Eastern Alps. *Terra Abstracts*, **3**, 337.

Homewood, P. 1977. Ultrahelvetic and North-Penninic Flysch of the Prealps: a general account. *Eclogae Geologicae Helvetiae*, **70**, 627–641.

Homewood, P., Allen, P.A. & Williams, G.D. 1986. Dynamics of the Molasse Basin of western Switzerland. *Special Publication of the International Association of Sedimentologists,* **8**, 199–217.

House, M.R. 1975a. Facies and time in Devonian tropical areas. *Proceedings of the Yorkshire Geological Society,* **40**, 233–288.

House, M.R. 1975b. Faunas and time in the marine Devonian. *Proceedings of the Yorkshire Geological Society,* **40**, 459–490.

Howell, D.G. 1989. *Tectonics of suspect terranes. Mountain building and continental growth.* Chapman and Hall, London.

Hsü, K.J. 1994. *The geology of Switzerland. An introduction to tectonic facies.* Princeton University Press, Princeton.

Ingham, J.K, Curry, G.B & Williams, A. 1985. Early Ordovician Dounans Limestone fauna, Highland Border Complex, Scotland. *Transactions of the Royal Society of Edinburgh: Earth Sciences,* **76**, 481–513.

James, N.P., Stevens, R.K., Barnes, C.R. & Knight, I. 1989. Evolution of a Lower Paleozoic continental margin carbonate platform, northern Canadian Appalachians. In Crevalla P. *et al.* (eds) *Controls on carbonate platform and basin development,* Special Publication of the Society of Economic Paleontologists and Mineralogists No. 44, 123–146.

Jenkyns, H.C. 1980. Tethys: past and present. *Proceedings of the Geologists' Association,* **91**, 107–118.

Jones, D.L., Silberling, N.J. & Hillhouse, J.W. 1977. Wrangellia: a displaced terrane in north-western North America. *Canadian Journal of Earth Sciences,* **14**, 2565–2577.

Jones, D.L., Howell, D.G., Coney, P.J. & Monger, J.W.H. 1983. Recognition, character and analysis of tectonostratigraphic terranes in western North America. In Hashimoto, S. & Uyeda, S. (eds) *Advances in Earth and Planetary Sciences,* Terra Science Publications, Tokyo, 21–35.

Kiessling, W. 1992. Palaeontological and facial features of the Upper Jurassic Hochstegen Marble (Tauern Window, Eastern Alps). *Terra Nova,* **4**, 184–197.

Krebs, W. 1976. The tectonic evolution of Variscan Meso-Europa. In Ager, D.V. & Brooks, M. (eds) *Europe from crust to core.* John Wiley, London, 119–139.

Lapworth, C. 1878. The Moffat Series. *Quarterly Journal of the Geological Society of London,* **34**, 240–346.

Lyell, C. 1850. Anniversary address of the President. *Quarterly Journal of the Geological Society of London,* **6**, xxvii–lxvi.

Martínez-García, E. 1991. Hercynian syn-orogenic and post-orogenic successions in the Cantabrian and Palentian zones (NW Spain). Comparison with other western Europe occurrences. *Giornale di Geologia Bologna,* **53**, 209–228.

Martínez-García, E. 1996. Correlation of hercynian units of the Iberian massif and southeastern France. *Geogecet,* **20**, 468–471.

Matte, Ph. 1995. Southern Uralides and Variscides: comparisons of their anatomies and evolutions. *Geologie en Mijnbouw,* **74**, 151–166.

Monger, J.W.H. & Ross, C.A. 1971. Distribution of fusilinaceans in the western Canadian Cordillera. *Canadian Journal of Earth Sciences,* **8**, 259–278.

Neuman, R.B. 1984. Geology and paleobiology of islands in the Ordovician Iapetus Ocean. *Bulletin of the Geological Society of America,* **95**, 1188–1201.

Newton, C.R. 1987. Biogeographic complexity in Triassic bivalves of the Wallowa terrane, northwestern United States: oceanic islands, not continents, provide the best analogues. *Geology,* **15**, 1126–1129.

Newton, C.R. 1990. Palaeobiogeography. In Briggs, D.E.G. & Crowther, P.R. (eds) *Palaeobiology: a synthesis.* Blackwells, Oxford, 452–460.

Nowlan, G.S. & Neuman, R.B. 1995. Paleontologic contributions to Paleozoic paleogeographic and tectonic reconstructions. In Williams, H. (ed.) *Geology of the Appalachian, Caledonian Orogen in Canada.* Geological Survey of Canada, Geology of Canada, **6**, 815–842.

Owen, A.W., Harper, D.A.T. & Romano, M. 1992. The Ordovician biogeography of the

Grangegeeth terrane and the position of the Iapetus suture in eastern Ireland. *Journal of the Geological Society of London,* **149**, 3–6.

Owen, A.W., Harper, D.A.T. & Clarkson, E.N.K. 1996. The trilobites and brachiopods of the Wrae Limestone, an Ordovician limestone conglomerate in the Southern Uplands. *Scottish Journal of Geology,* **32**, 133–149.

Oxburgh, E.R. 1968. *The geology of the Eastern Alps.* The Geologists' Association, London.

Parkes, M.A. 1992. Caradoc brachiopods from the Leinster terrane (southeast Ireland) – a lost piece of the Iapetus puzzle? *Terra Nova,* **4**, 223–230.

Parkes, M.A. & Harper, D.A.T. 1996. Ordovician brachiopod biogeography in the Iapetus suture zone of Ireland: provincial dynamics in a changing ocean. In Copper, P. & Jin Jisuo (eds) *Brachiopods.* Balkema, Rotterdam, 197–202.

Parkes, M.A., Ryan P.D. & Harper, D.A.T. 1990. Microcomputer seriation of palaeobiogeographical data. In Bruton, D.L. & Harper, D.A.T. (eds) *Microcomputers in palaeontology.* Palaeontological Contributions from the University of Oslo, 370, 56–66.

Pedersen, R.B., Bruton, D.L. & Furnes, H. 1992. Ordovician faunas, island arcs and ophiolites in the Scandinavian Caledonides. *Terra Nova,* **4**, 217–222.

Pfiffner, O.A. 1986. Evolution of the north Alpine foreland basin in the Central Alps. *Special Publication of the International Association of Sedimentologists,* **8**, 219–228.

Pohler, S.L., Barnes, C.R. & James, N.P. 1987. Reconstructing a lost faunal realm: conodonts from mega-conglomerates of the Ordovician Cow Head Group, western Newfoundland. In Austin, R.L. (ed.) *Conodonts: investigative techniques and applications.* Ellis Horwood, Chichester, 341–362.

Ramsay, J.G. 1963. Stratigraphy, structure and metamorphism in the Western Alps. *Proceedings of the Geologists' Association,* **74**, 357–391.

Ramsay, J.G. 1967. *Folding and fracturing of rocks.* McGraw Hill, New York.

Reitz, E. & Höll, R. 1992. Palynological evidence for Lower Ordovician rocks (Tremadoc and Arenig) in the Northern Greywacke Zone (Eastern Alps). *Terra Nova,* **4**, 198–207.

Ryan, P.D. & Dewey, J.F. 1991. A geological and tectonic cross-section of the Caledonides of western Ireland. *Journal of the Geological Society of London,* **148**, 173–180.

Shi, G.R. 1993. A comparative study of 39 binary similarity coefficients. *Memoir of the Association of Australasian Palaeontologists,* **15**, 329–341.

Smith, P.L. 1990. Paleobiogeography and plate tectonics. *Geoscience Canada,* **15**, 261–279.

Torsvik, T., Ryan, P.D., Trench, A. & Harper, D.A.T. 1991. Cambrian–Ordovician palaeogeography of Baltica. *Geology,* **19**, 7–10.

Trümpy, R. 1960. Paleotectonic evolution of the Central and Western Alps. *Bulletin of the Geological Society of America,* **71**, 843–908.

Trümpy, R. 1971. Stratigraphy in mountain belts. *Quarterly Journal of the Geological Society of London,* **126**, 293–318.

Trümpy, R. 1982. Alpine palaeogeography: a reappraisal. In Hsü, K.J. (ed.) *Mountain Building Processes.* Academic Press, London, 149–156.

Wegener, A. 1929. *The origin of continents and oceans,* 4th edition. Methuen, London.

Williams, A. & Curry, G.B. 1985. Lower Ordovician Brachiopoda from the Tourmakeady Limestone, Co. Mayo, Ireland. *Bulletin of the British Museum (Natural History) Geology,* **38**, 183–269.

Williams, D.M. & Harper, D.A.T. 1991. End Silurian modifications of Caledonian terranes in western Ireland. *Journal of the Geological Society of London,* **148**, 165–171.

Williams, D.M., Harkin, J., Armstrong, H.A. & Higgs, K.T. 1994. A late Caledonian melange in Ireland: implications for tectonic models. *Journal of the Geological Society of London,* **151**, 307–314.

Williams, H. 1984. Miogeoclines and suspect terranes of the Caledonian–Appalachian Orogen: tectonic patterns in the North Atlantic region. *Canadian Journal of Earth Sciences,* **21**, 887–901.

Williams, S.H. & Harper, D.A.T. 1994. Late Tremadoc graptolites from the Lough Nafooey Group, South Mayo, western Ireland. *Irish Journal of Earth Sciences,* **13**, 107–111.

Williams, S.H., Harper, D.A.T., Neuman, R.B., Boyce, W.D. & MacNiocaill, C. 1996. Lower

Paleozoic fossils from Newfoundland and their importance in understanding the history of the Iapetus ocean. *Geological Survey of Canada Special Paper*, **41**, 115–126.

Wilson, J.T. 1966. Did the Atlantic close and then re-open? *Nature*, **211**, 676–681.

Ziegler, P.A. 1986. Geodynamic model for the Palaeozoic crustal consolidation of western and central Europe. *Tectonophysics*, **126**, 303–328.

Ziegler, P.A. 1989. Geodynamic model for Alpine intra-plate compressional deformation in Western and Central Europe. In Cooper, M.A. & Williams, G.D. (eds) *Inversion Tectonics*. Special Publication Geological Society of London No. 44, 63–85.

Ziegler, P.A. 1990. *Geological atlas of Western and Central Europe*, 2nd edition. Shell Internationale Petroleum Maatschappij BV, Netherlands.

Index

Page numbers in italics refer to useful definitions, while those in bold refer to illustrations.